ACTIVITY-DRIVEN CNS CHANGES IN LEARNING AND DEVELOPMENT

ANNALS OF THE NEW YORK ACADEMY OF SCIENCES

Volume 627

ACTIVITY-DRIVEN CNS CHANGES IN LEARNING AND DEVELOPMENT

Edited by Jonathan R. Wolpaw, John T. Schmidt, and Theresa M. Vaughan

The New York Academy of Sciences
New York, New York
1991

Library of Congress Cataloging-in-Publication Data

Activity-driven CNS changes inlearning and development/editors.
Jonathan R. Wolpaw, John T. Schmidt, and Theresa M. Vaughan;
conference organizers, Gregory A. Lnenicka... [et al.].
p. cm. — (Annals of the New York Academy of Sciences. ISSN
0077-8923; v. 627
Based on a conference sponsored by the State University of New
York at Albany's Neurobiological Research Center and Departments of
Biological and Biomedical Sciences and held May 16-20, 1990 in
Rensselaerville, N.Y.
Includes bibliographical references and indexes.
ISBN 0-89766-637-2. — ISBN 0-89766-638-0 (pbk.)
1. Neuroplasticity—Congresses. 2. Neural circuitry—Adaptation—
Congresses. 3. Developmental neurology—Congresses. 4. Learning—
Congresses. I. Wolpaw, Jonathan R. II. Schmidt, John T.
III. Vaughan, Theresa M. IV. State University of New York.
V. Series.
Q11.N5 vol. 627
[QP363.5]
500 s—dc20
[153.1′5] 91-18958
CIP

PCP
Printed in the United States of America
ISBN 0-89766-637-2 (cloth)
ISBN 0-89766-638-0 (paper)
ISSN 0077-8923

ANNALS OF THE NEW YORK ACADEMY OF SCIENCES

Volume 627
August 5, 1991

ACTIVITY-DRIVEN CNS CHANGES IN LEARNING AND DEVELOPMENT[a]

Editors
JONATHAN R. WOLPAW, JOHN T. SCHMIDT, and THERESA M. VAUGHAN

Conference Committee
HAROLD L. ATWOOD, LARRY I. BENOWITZ, GREGORY A. LNENICKA, JOHN T. SCHMIDT, JOHN W. SWANN, SUZANNAH B. TIEMAN, and JONATHAN R. WOLPAW

Conference Organizer
PATRICIA A. HERCHENRODER

CONTENTS

[a]This volume is the result of a conference entitled Activity-Driven CNS Changes in Learning and Development, which was sponsored by the State University of New York at Albany's Neurobiological Research Center and Departments of Biological and Biomedical Sciences, and held on May 16-20, 1990 at the Rensselaerville Institute in Rensselaerville, New York.

Part II. Molecular Mechanisms of Activity-Driven Change

Chair: David L. Martin

Keynote Address

Part III. Silent Synapses/Synaptic Modulation

Chair: Helmut V. B. Hirsh

Part IV. Activity-Driven Anatomical Changes

Chair: Abigail M. Snyder-Keller

Part V. Activity-Dependent Alterations in Hippocampal Networks

Chair: Jonathan S. Carp

Part VI. Activity-Driven Plasticity and Behavioral Change

Chair: Robert J. Brady

Poster Papers

Financial assistance was received from:

- BRISTOL–MYERS SQUIBB PHARMACEUTICAL RESEARCH AND DEVELOPMENT DIVISION
- CIBA–GEIGY CORPORATION
- HOECHST–ROUSSEL PHARMACEUTICALS, INC.
- IBM CORPORATION
- NATIONAL INSTITUTES OF HEALTH
- NORWICH EATON PHARMACEUTICALS, INC.
- PARKE–DAVIS/DIVISION OF WARNER–LAMBERT COMPANY
- RESEARCH FOUNDATION OF THE STATE UNIVERSITY OF NEW YORK
- ROCHE INSTITUTE FOR MOLECULAR BIOLOGY
- STATE UNIVERSITY OF NEW YORK "CONVERSATIONS IN THE DISCIPLINES"
- STERLING DRUG, INC./DIVISION OF EASTMAN KODAK COMPANY
- UNITED STATES AIR FORCE OFFICE OF SCIENTIFIC RESEARCH
- WADSWORTH CENTER FOR LABORATORIES AND RESEARCH

Preface

ACTIVITY-DRIVEN PLASTICITY

Learning and development cause profound changes in the central nervous system (CNS). These changes, which determine the general capacities and specific behaviors of the organism, are in large part a result of neuronal activity. Certain patterns of activity occurring at specific sites, and at crucial times, initiate events that modify synapses so that subsequent CNS function is different.

The sequence of events initiated by activity and leading to altered function comprises phenomena at each of the levels of analysis normally applied to the nervous system. Neurotransmitters released during activity initiate reactions that lead to lasting synaptic modifications. These modifications may be as subtle as an altered ionic conductance or as striking as the formation of a new synapse. Modifications at different synapses combine to modify patterns of activity in networks of neurons within individual structures and throughout the brain. These altered activity patterns find expression as altered behavior, and may also lead to additional activity-driven changes at other sites in the nervous system. In recent years, research at each of these levels has greatly increased understanding of learning and development, and has revealed their close connections to each other. Not only are both driven by neuronal activity, but both depend to a large extent on similar or identical molecular and synaptic processes.

CLOSE RELATIONSHIPS BETWEEN LEARNING AND DEVELOPMENT

Learning results from persistent changes in synaptic connections generated by the pattern of neuronal activity at the time of the learning experience. In the adult organism, these changes are subtle. During development, the nervous system is particularly sensitive to patterns of activity, and striking changes occur. Activity exerts strong control over the development of synaptic connections, by driving their growth, stabilization, and elimination. This control is particularly well documented in the visual system. Thus, development of the synaptic circuitry underlying behavior is not simply controlled by genetic information. Rather it is controlled by genes, by activity, and by the interactions between them. Activity turns specific genes on or off and thereby guides development. Conversely, the genetic endowment defines the developmental response to specific activity patterns.

Learning and development depend on the same set of molecular mechanisms. Studies in the developing visual system and in the mature hippocampus have revealed that strong depolarization of the postsynaptic neuron, which opens the *N*-methyl-D-aspartate (NMDA) receptor channels and thereby admits calcium, is essential for persistent change in synaptic function. Related studies have established the role of protein kinase C in both development and learning. Similarly, the growth-associated protein, GAP-43, which is phosphorylated by protein kinase C, is strongly correlated with both development and adult plasticity. GAP-43 may be involved in the control of growth cones and of transmitter release at the synapse.

Activity-driven synaptic changes during development clearly involve prominent anatomical modifications. The prevalence of such anatomical change in adult learning remains unclear, though studies in *Aplysia* have revealed dramatic anatomical changes. In contrast, studies in other systems support the importance of functional changes in preexisting synapses.

The activity-driven synaptic changes underlying learning and development are manifested as altered behavior. The nature of the behavioral changes depends on the interaction of synaptic changes at multiple sites throughout the CNS. Only recently have these interactions come under careful study, both through theoretical analysis of neural network behavior and through development of anatomically and behaviorally simple experimental models.

THE SUNY CONFERENCE SERIES

In response to the rapid progress of the past 20 years, the State University of New York at Albany began in 1980 to sponsor periodic conferences on activity-dependent plasticity in the nervous system. They emphasized two themes: the importance of activity in initiating and directing plasticity, and the close relationships between learning and development. The first was entitled "From Primate to Cricket: Nature and Nurture in the Development of the Brain." The second meeting, in 1984, was "Activity-Dependent Synaptic Changes: Developmental Changes and Learning—How Similar the Mechanisms?" The proceedings were published in a dedicated issue of *Cellular and Molecular Neurobiology* (**5:** 1-207, 1985).

The conference that formed the basis for the present volume took place in May, 1990 at the Rensselaerville Institute in Rensselaerville, New York near Albany. This last meeting reflected the increased pace and breadth of recent research. Most important, it added a new theme to the two stressed before: its central goal was to discuss, in a connected fashion, the entire sequence of events underlying learning and development. Such a comprehensive and logical format has only become possible in the last few years. Before that, knowledge was too fragmentary to permit meaningful adherence to this framework. At the same time, recent advances have made it imperative to encourage interactions between scientists working at each level in this sequence, if understanding of learning and development is not to remain disjointed and compartmentalized.

To emphasize this theme, the meeting's organization paralleled the progression from neuronal activity to altered behavior. Thus, the first session described receptor-mediated triggers of plasticity, the second discussed accompanying molecular events, the next two evaluated synaptic modifications resulting from these events, and the last two evaluated expression of these synaptic modifications as altered behavior of neural networks and whole animals. Each session occupied a morning or afternoon and consisted of four 30-minute presentations and ample time for discussion. Dr. Byrne's keynote address took place the second evening, and contributed posters were displayed throughout and discussed in a special session on the third day. The meeting finished with an informal breakfast session focused on important recurring issues.

THE PROCEEDINGS

This volume retains the organization of the conference. The six sections correspond to the conference sessions, and progress from neuronal activity to altered behavior. Each begins with a brief introduction that places it in context. Poster summaries are grouped at the end. We hope that this organization will help the reader to appreciate fully the important new information presented in the individual papers and to perceive the comprehensive picture that is just beginning to emerge.

JONATHAN R. WOLPAW
JOHN T. SCHMIDT

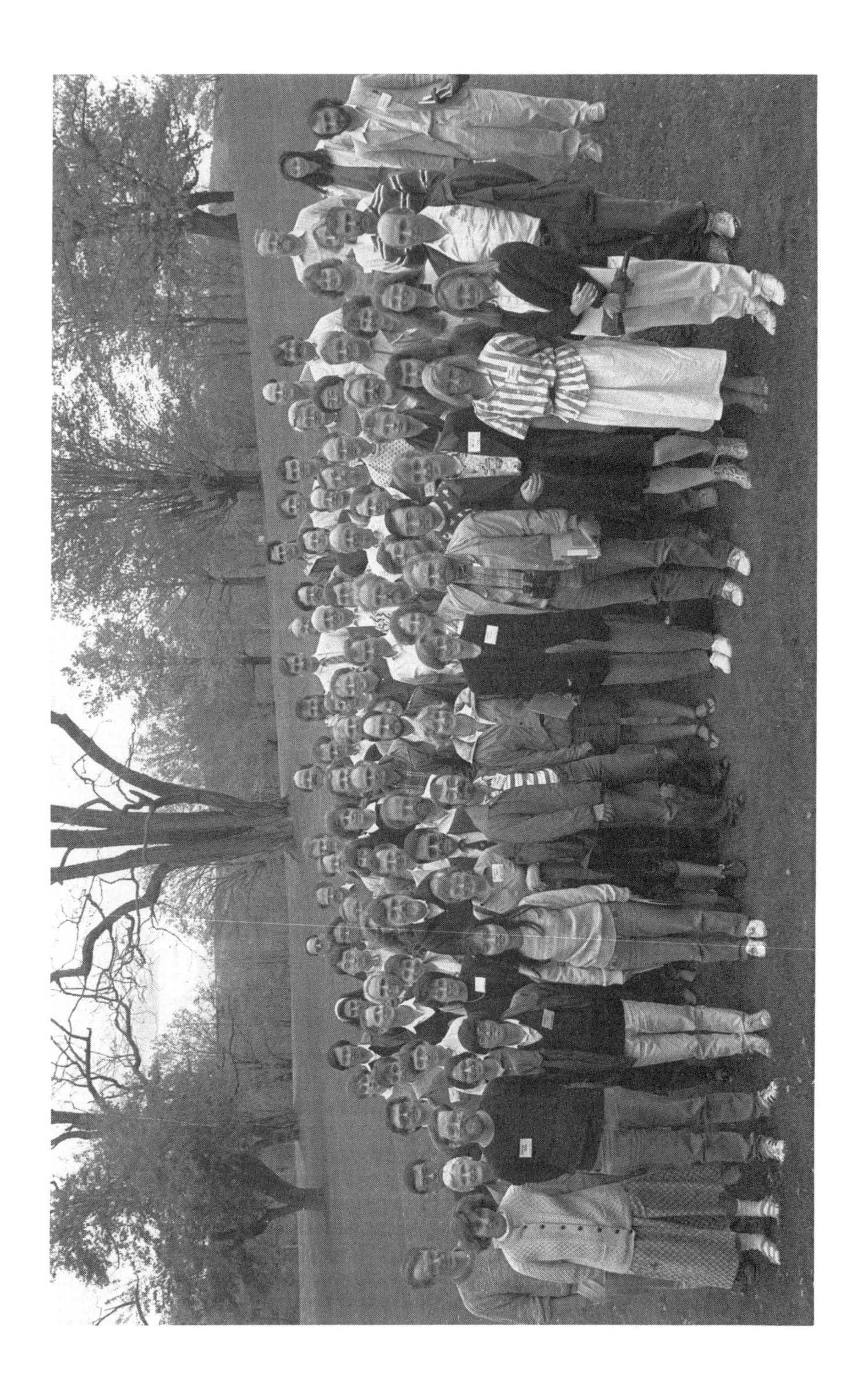

Group photograph from the conference Activity-Driven CNS Changes in Learning and Development (May 16-20, 1990): *1*) Jonathan R. Wolpaw, *2*) Theresa M. Vaughan, *3*) Beverly Bishop, *4*) George Park, *5*) Caroline Gomez, *6*) Wei Chun Lin, *7*) Karen L. Smith, *8*) Chuan Chen, *9*) Suzannah Bliss Tieman, *10*) Helmut V. B. Hirsh, *11*) A. Joffe Mercier, *12*) Sonia Witte, *13*) Donna L. Korol, *14*) W. Michael King, *15*) Luo Sun, *16*) Catharina VanderZee, *17*) Serena M. Dudek, *18*) Karen D. Parfitt, *19*) Harumitsu Hirata, *20*) Donald S. Faber, *21*) Wade Regehr, *22*) Brian Ault, *23*) Harold L. Atwood, *24*) Pedro A. Ferchmin, *25*) Mary Lou Vallano, *26*) Kathleen Egid, *27*) John A. Freeman, *28*) David Tieman, *29*) Larry I. Benowitz, *30*) John H. Byrne, *31*) John T. Schmidt, *32*) Brenda J. Anderson, *33*) Susan B. Udin, *34*) Michael A. Corner, *35*) Michael W. Vogel, *36*) Craig H. Bailey, *37*) Stefan Strak, *38*) Edwin M. Haller, *39*) David Walker, *40*) Abigail M. Snyder-Keller, *41*) James G. McElligott, *42*) Sean Pedini, *43*) Yangu Zhoa, *44*) Linda Shelton, *45*) Vincent Wan, *46*) Sunguon, *47*) Michael M. Merzenich, *48*) Barbara Tolloczko, *49*) Martha Pierson, *50*) John W. Swann, *51*) Deniz Terry, *52*) Michael Fejtl, *53*) Douglas Rasmussen, *54*) Gregory A. Lnenicka, *55*) William T. Greenough, *56*) Victor E. Shashoua, *57*) Robert J. Brady, *58*) Joe Capowski, *59*) Denis Larrivee, *60*) Bradley E. Alger, *61*) Barry W. Peterson, *62*) Robert K. S. Wong, *63*) Roger Davis, *64*) Emily Smits, *65*) Sandy Craner, *66*) Patricia Rosebush, *67*) Roger D. Traub, *68*) Ronald Chase, *69*) Jakob Schmidt, *70*) J. Clancy Leahy, *71*) Andy Dais, *72*) Peter Zarzeiki, *73*) John Brooke, *74*) David Kellerman, *75*) Elwood Henneman, *76*) Jonathan S. Carp, *77*) Staos Aeorganos, *78*) Christopher H. Yeo, *79*) Michael Mazurek, *80*) Adrian Glasser, *81*) James R. Bloedel, *82*) Duncan Byers, *83*) Daniel Madison, *84*) Theodore J. Wiens, *85*) Xiasying Tram, *86*) J. Martin Wojtowicz.

PART I. EXCITATORY AMINO ACID RECEPTORS AS TRIGGERS OF PLASTICITY

Introduction

This section concerns the first steps in activity-driven synaptic changes: the role of neurotransmitter receptors in triggering second messenger production and protein kinase activation. Evidence from several very different systems indicates that the *N*-methyl-D-aspartate (NMDA) type of excitatory amino acid receptor plays a crucial role. Although the NMDA receptor channels carry little current during the activation of single inputs, they have a voltage dependence and open fully only when other synaptic inputs depolarize the postsynaptic neuron. This dependence on simultaneous activation of other inputs is strikingly similar to the essential property of associative conditioning: that an ineffective stimulus becomes effective after it occurs just before or during an effective stimulus. Activated NMDA receptor channels allow entry of calcium ions, which serve as a second messenger. Calcium has numerous intracellular effects, including activation of protein kinase C and calcium/calmodulin-dependent kinase II as well as calcium-dependent proteases. Meffert *et al.* report that activation of both kinases in the postsynaptic cell is necessary for induction of LTP in the hippocampus, and presynaptic kinase C activation may also be necessary for maintenance of long-term potentiation (LTP).

Bear and Dudek focus on a quisqualate receptor that triggers the intracellular messenger inositol trisphosphate (IP_3), and describe its interaction with NMDA receptors in setting the levels of synaptic effectiveness and in producing long-term synaptic changes in developing visual cortex. Specifically, NMDA receptors appear to trigger LTP, while the quisqualate-IP_3 receptors may trigger long-term depression in the absence of significant NMDA receptor activation. These receptor mechanisms appear to be involved in controlling the developmental changes driven by activity in many visual systems. Schmidt explores, in the goldfish visual system, the hypothesis that LTP is involved in the developmental processes of synaptic stabilization and retraction. Udin and Scherer present a striking binocular reorganization in *Xenopus* and the possibility that in the adult the plasticity may be restored by NMDA treatment.

Many questions remain. These concern the interaction between calcium and IP_3 (since IP_3 is thought to mobilize calcium intracellularly), the relevant substrates for the different kinases and the alteration of their functions, the possibility of a retrograde signal to the presynaptic terminal, the pre- and postsynaptic mechanisms producing LTP, and the pre- and postsynaptic contributions to the developmental synaptic stabilization.

Protein Kinases and Long-Term Potentiation

MOLLIE K. MEFFERT, KAREN D. PARFITT, VAN A. DOZE, GAL A. COHEN, AND DANIEL V. MADISON[a]

Department of Molecular and Cellular Physiology
Beckman Center for Molecular and Genetic Medicine
Stanford University School of Medicine
Stanford, California 94305

Excitatory synapses in the hippocampus, and in some other neural tissues, undergo a long-lasting increase in their efficacy when they are heavily used. This potentiating effect is proposed to function as a cellular mechanism for the formation of memory and is very prominent in the hippocampus, a site where the consolidation of experience into long-term memory is thought to occur. When the excitatory synapses are strongly and repetitively activated for brief periods of time, they undergo an increase in strength that has been shown to last several hours or even weeks after induction. This use-dependent increase in synaptic efficacy has been termed long-term potentiation (LTP).[1,2]

LTP is produced by tetanizing the presynaptic fibers with a short-duration train of high-frequency pulses. Typically, the train is on the order of 100 Hz for 1 sec. Immediately following this tetanic stimulation, the strength of the synaptic connection (as tested with single-shock stimuli) increases up to about 5-fold. Most of this increase decays to a level of about 150-200% of baseline within a few minutes after tetanic stimulation. The early short-lasting phase of potentiation is called posttetanic potentiation (PTP) and is believed to be caused by a temporary accumulation of free calcium ions in the presynaptic terminal.[3] The potentiation that persists for hours after the tetanic stimulation is referred to as LTP.

Both the strength and frequency of the tetanic stimulation are critical for the induction of LTP. The strength of the stimulus must reach a critical threshold to elicit LTP. This critical threshold represents the need for a certain number of presynaptic fibers to be activated simultaneously, a property referred to as cooperativity.[4] Frequency and stimulus strength interact such that increasing one decreases the requirement for the other. A weak stimulus that does not produce LTP, for example, can be rendered effective by increasing the stimulus frequency. Conversely, increasing the strength of a stimulus decreases the frequency required to produce LTP.[1,5]

Another striking feature of LTP is its synapse specificity. A given postsynaptic cell receives many afferent synapses, but only a fraction of these afferents receive tetanic stimulation during a typical train of stimuli. While LTP is produced in the tetanized synapses, a test of other nontetanized synapses on the same postsynaptic cell reveals that they remain at pretetanized baseline levels.[6-10]

Despite synapse specificity, different afferent synapses on the same postsynaptic cell do show some ability to interact. By placing two stimulating electrodes in the hippocampal slice, two separate subpopulations of Schaffer collateral afferents terminating on the same postsynaptic cell can be activated. A tetanic stimulus delivered at a stimulus

[a] To whom correspondence should be addressed.

strength too weak to elicit LTP becomes effective if paired with a strong tetanic stimulus delivered through the other electrode (that is, the other population of afferent fibers).[6] This pairing of a weak stimulus with a strong stimulus to generate potentiation of the weak pathway is referred to as associative LTP and demonstrates a form of communication between afferent pathways that occurs, presumably, through the postsynaptic cell. Associative LTP can occur even when the two groups of afferents terminate on portions of the dendrite that are separated by several hundred microns.[9] There is, however, a temporal correlation required for the induction of associative LTP: the activation of weak and strong inputs must not be separated in time by more than about 100 msec for successful potentiation of both pathways to occur.[7,11] This short window of time available for effective pairing rules out many candidates for the associative messenger and makes depolarization of the postsynaptic membrane the most likely associative messenger.

Simply depolarizing cells, whether by injection of current or by causing the cells to fire long-duration calcium action potentials, will not produce LTP. Only when depolarization is paired with activation of the synapse will LTP occur. Depolarization of the postsynaptic membrane, however, reduces the frequency requirement for the induction of LTP such that only single-shock activation is needed to produce small amounts of LTP.[12–14] The factor contributed by synaptic activation appears to be the binding of an excitatory amino acid neurotransmitter to a postsynaptic receptor.[15] At most, if not at all, excitatory synapses in the hippocampus, this transmitter is probably glutamate.[16]

The actions of the excitatory neurotransmitter, glutamate, at the synapse are mediated by at least three receptor subtypes associated with ion channels, the *N*-methyl-D-aspartate (NMDA), the quisqualate, and the kainate receptor (the latter two types will be referred to as Q/K receptors or simply non-NMDA receptors). The Q/K receptor is coupled to a channel that is permeable to sodium and potassium but relatively impermeable to calcium; this channel seems to be responsible for the excitatory postsynaptic potentials (EPSPs) during normal synaptic transmission.[17,18] By contrast, the NMDA-receptor-associated channel is highly permeable to calcium as well as monovalent cations,[19–21] and current flow through this channel exhibits a strong voltage sensitivity. The voltage sensitivity of the NMDA channel is due to a voltage-sensitive block of the channel by physiological levels of magnesium; this block is relieved by depolarization.[22–25] Thus, when a cell is strongly tetanized, depolarizing current flows in through non-NMDA receptor channels, relieves the magnesium block on the NMDA channel, and allows calcium to flow into the postsynaptic cell. Opening of the NMDA ionophore is an absolute requirement for the induction of Schaffer-collateral CA1 LTP; selective antagonists of the NMDA receptor, such as D-2-amino-5-phosphonovalerate (AP5), block the induction of LTP.[15,26] Interestingly, NMDA receptors are not distributed uniformly among hippocampal synapses: the mossy fiber to CA3 synapse has little or none of this receptor type, and appears to show a fundamentally different, AP5-resistant form of LTP.[27]

The morphology of excitatory hippocampal synapses should also be noted in that their structure may be crucial to LTP. The majority of presynaptic terminals contact the postsynaptic cell on small, specialized processes called dendritic spines, which may be both chemically and electrically isolated from the rest of the dendritic shaft.[28–31] The dendritic spine has been suggested to serve several functions; the small intracellular volume of the spine head may allow a relatively small influx of ions to produce a large alteration in local ionic concentrations and membrane potential.[32] The narrow spine neck may serve to diminish the movement of ions between the dendritic shaft and the spine head. This would serve to maximize synaptic depolarizations in the spine head while protecting the spine from calcium generated extrasynaptically by voltage-dependent calcium channels (VDCCs), as well as enabling calcium influx through the NMDA channel to cause significant elevations of intracellular calcium in the dendritic spine head.[32]

A role in LTP for calcium influx through the NMDA channel is supported by the observation that postsynaptic injection of the calcium chelators EGTA or nitr-5 will block the induction of LTP.[33] In addition, Malenka *et al.*[33] found that release of calcium into postsynaptic cells by the caged calcium chelator nitr-5 produces a potentiation of synaptic transmission similar to LTP.[33] These experiments not only show the importance of calcium influx, but also point out the critical involvement of the postsynaptic cell in the induction of LTP. Additional evidence supporting a role for calcium is the report by some investigators that application of high concentrations of extracellular calcium can produce an AP5-sensitive LTP essentially indistinguishable from that produced by high-frequency stimulation.[34,35] In addition to the NMDA-associated channels, CA1 dendrites also possess VDCCs; several observations, however, argue against the possibility that calcium influx through VDCCs might be sufficient to produce LTP. First, as mentioned previously, large depolarizations alone are insufficient to induce LTP in the absence of concomitant synaptic stimulation.[7,12,26] Second, selective activation of VDCCs by strong postsynaptic depolarization in the presence of AP5 does not induce LTP.[26] Third, repeated activation of calcium action potentials, delivering large calcium loads to the postsynaptic neuron, does not produce potentiation.[36] It appears, then, that it is the calcium influx through NMDA channels and not through VDCCs that triggers LTP induction in area CA1 of the hippocampus. To explain this, many investigators have proposed a model in which NMDA receptors, Q/K receptors, and the calcium-dependent triggering mechanism are all localized to the dendritic spine, while voltage-gated calcium channels are located on the dendritic shaft. If the flow of ions between spine and shaft is restricted, then, only calcium influx through NMDA-receptor-associated channels could induce LTP.

Considering that a rise in calcium within the dendritic spine may be an essential step in the induction of LTP, many investigators have begun to look at biochemical processes activated by calcium that might be responsible for LTP. Accumulating evidence suggests that calcium-activated protein kinases are required for the induction and/or the maintenance of LTP. A number of nonselective kinase inhibitors including H-7, sphingosine, polymyxin B, and k-252b have been shown to block LTP.[37,38] In addition, intracellular injection of H-7 into the postsynaptic cell inhibits LTP,[39,40] suggesting that postsynaptic kinase activity, presumably activated by the rise in calcium in the dendritic spine, is required for LTP. Two favored candidates for the persistent signal underlying LTP are calcium/calmodulin-dependent protein kinase II (CaMKII) and protein kinase C (PKC), the calcium- and phospholipid-dependent protein kinase. Although these kinases have some substrate phosphorylation sites in common, other sites are unique to each of the kinases. The factors that delineate substrate specificity remain unclear but seem to be produced, at least in part, by the amino acids surrounding the target proteins' acceptor hydroxy amino acids.[41]

CaMKII constitutes 20-40% of the protein found in isolated postsynaptic densities and is found in great abundance in dendritic spines of the hippocampus[42,43]; thus, it is in a strategic location to be exposed to the rise in calcium following tetanic stimulation. Bath application of calmodulin antagonists has been shown to block LTP,[44-47] but many of these compounds lack selectivity. More direct and convincing evidence of a role for CaMKII in LTP comes from experiments in which LTP was blocked by peptides that bind to and inhibit calmodulin[39] or by a peptide that directly blocks CaMKII activity.[40] These peptides are effective when they are injected into the postsynaptic cell.

Because CaMKII is capable of autophosphorylation, which makes it constitutively active, this kinase is a particularly attractive candidate to underlie persistent potentiation of synaptic efficacy. Following initial activation of calmodulin by a rise in intracellular calcium, calmodulin-dependent autophosphorylation occurs such that CaMKII no longer requires calmodulin to maintain its kinase activity.[48,49] Several years ago it was proposed that a kinase capable of such constitutive activity could serve as a "molecular switch" that could store information for long periods of time.[48,50]

Another kinase capable of constitutive activity is the calcium-dependent, phospholipid-stimulated PKC. PKC is enriched in neural tissue[51] and is capable of phosphorylating a wide variety of neuronal proteins including receptors, metabolic enzymes, ion pumps, and cytoskeletal proteins.[52] When PKC is activated, the molecule is cleaved and the active subunit is translocated from the cytosol to the membrane. LTP has been shown to be associated with a translocation of PKC from cytosol to membrane as well as with increased phosphorylation of a PKC substrate protein, GAP-43 (also known as B-50, F-1, or P-57).[51,53] It has also been reported in studies *in vivo* that classical conditioning in rabbits produces a translocation of PKC activity from the cytosolic to the membrane compartments of hippocampal CA1 neurons.[54] Interestingly, brain autoradiographs taken in rabbits three days after learning suggest a further redistribution of PKC within CA1 from the cell soma to the dendrites.[54]

In addition to calcium, PKC can also be activated by 1,2-diacylglycerol (DAG), a second messenger that dramatically reduces the concentration of calcium required to activate PKC. DAG is transiently produced *in vivo* by receptor-mediated hydrolysis (by the enzyme phospholipase C) of phosphatidylinositol-4,5-bisphosphate (PIP_2) into DAG and inositol-1,4,5-trisphosphate (IP_3). IP_3 might also contribute to the activation of PKC by promoting the release of calcium from internal stores. An increase in phosphoinositide (PI) turnover, perhaps involving the activation of a G protein, is associated with LTP.[55] Whether a pertussis toxin-sensitive G protein is required for LTP generated in the CA1 region is controversial.[56,57] If receptor-mediated PI turnover is required for LTP, a receptor that may mediate this turnover is the metabotropic quisqualate receptor.[58]

Experimentally, PKC may be activated by compounds that mimic DAG, such as synthetic DAGs or phorbol esters. Activation of PKC by phorbol esters enhances synaptic transmission,[59] as does injection of PKC directly into postsynaptic cells.[60] In addition, pretreatment with phorbol ester occludes tetanus-induced LTP.[59] Although phorbol-induced potentiation resembles LTP in many ways, it is not long term, lasting only as long as the agonist is present (approximately 60 min).[38] This difference does not, however, exclude a role for PKC in LTP. On the contrary, the finding that injection of a specific pseudosubstrate peptide inhibitor of PKC (PKC 19-31) into postsynaptic cells blocks LTP[40] provides direct evidence supporting a critical role for PKC in LTP.

Another pathway leading to the activation of PKC, which might play a role in LTP, involves the production of *cis* fatty acids such as oleate and arachidonate.[61] Recent work by Routtenberg and co-workers[62] has confirmed that oleate and arachidonate can activate purified PKC in the absence of calcium and phospholipid. This same signaling pathway may also exist *in vivo.* Free *cis* fatty acids could be released from the 2-acyl position of membrane phospholipids by the enzyme phospholipase A_2; these *cis* fatty acids would then activate PKC. With the recent separation of PKC into a family of subtypes has come the finding that several subtypes can be selectively activated by *cis* fatty acids.[63]

Several discrete subspecies of PKC have already been well defined.[64] In the brain tissues, for example, at least seven subspecies can be distinguished, one of which is expressed only in tissues of the central nervous system. Biochemical and immunocytochemical studies with subspecies-specific antibodies indicate that the PKC subspecies may be differentially located in particular tissues and cell types, although many cell types express multiple subspecies that may be differentially distributed within the same cell. The subspecies PKC_γ appears to be expressed solely in the brain and spinal cord and is particularly localized in the pyramidal cells of the hippocampus.[64] Interestingly, the γ subspecies develops postnatally and reaches maximum activity, in the rat, at around three weeks after birth, the time when the expression of LTP also appears to become maximal. Although this correlation is suggestive, it seems premature to discuss the relationship of the γ subspecies with any specific neuronal function. PKC_γ shows less activation by DAG than other subspecies, but is significantly activated by micromolar concentrations of free arachidonic acid.[65] Activation of PKC_γ by arachidonic acid does not depend upon DAG and phospholipid, nor does it require calcium.

The γ subspecies of PKC is also very susceptible to proteolysis by the physiologically occurring enzyme, calpain I. Calpain I and calpain II are calcium-dependent, thiol proteases that are proposed to become activated following an increase in intracellular calcium concentration and can cause proteolysis of protein kinases, cytoskeletal proteins, and other substrates. Cleavage of PKC by calpain results in the release of a constitutively active subunit, known as PKM, as well as a regulatory subunit.[66,67] The active subunit is subsequently removed from the cell quite rapidly. The physiological significance of PKC_γ proteolysis by calpain is not yet fully understood, and it is unclear whether calpain is actually found in hippocampal synapses made onto dendritic spines.[68] Nevertheless, calpain has been suggested to play an important role in LTP.[69] Cleavage by calpain is a mechanism for the activation of PKC, and the resulting active subunit has been proposed to play a role in the control of cellular function. The regulatory fragment resulting from PKC_γ cleavage may also have some role in cellular function as it does have a DNA-binding motif.[64] It is also possible, however, that the proteolysis of PKC by calpain serves primarily as a first step in the degradation process of the kinase. Calpain, like prolonged application of phorbol esters, may elicit the translocation of the active subunit from the soluble fraction to the membrane where the amount of degradation by proteases is greater than in the cytosol,[70] thus disrupting the balance of synthesis to degradation and resulting in the depletion of the kinase from the cell.

Whatever the mode of activation, there are still many unresolved questions concerning the exact function that PKC may fulfill in the production of LTP. Specifically, the question of whether PKC is involved in the induction and/or maintenance phase of LTP is not yet fully resolved. Working with the perforant path synapses onto dentate gyrus granule cells, Lovinger *et al.*[71] first used PKC inhibitors to show that application immediately prior to tetanus produced a potentiated response that decayed to the baseline value within 60 min. A critical window for PKC activation in LTP maintenance was suggested by the observation that the PKC inhibitors, mellitin and polymyxin B, applied 10 min after LTP induction produced a decay to baseline in 20-60 min.[47,71] On the other hand, application of these PKC inhibitors 240 min after LTP induction was ineffective.[71] In the Schaffer collateral to the CA1 synapse, Malinow *et al.*[38] found that pretreatment with 10 μM sphingosine or 300 μM H-7 caused potentiated responses to decay to baseline within about 45 min. Application of 300 μM H-7 within 15 min after the induction of LTP had the same effect as application of the drug before LTP induction.[38,71] Application of sphingosine after the tetanic stimulation, however, did not reduce LTP,[38] strongly suggesting that a sphingosine-sensitive kinase is required for the induction of LTP. Sphingosine acts on the regulatory sites of PKC and CaMKII, whereas H-7 acts on the kinases' catalytic sites. This raises the possibility that, following induction, LTP is maintained by a constitutively active kinase such as the catalytic subunit of PKC or autophosphorylated CaMKII, both of which are spingosine insensitive.

Although both postsynaptic CaMKII and PKC may be required for the generation of LTP,[40] the exact mechanisms of their actions remain to be determined. As discussed above, both CaMKII and PKC have properties that allow them to exhibit constitutive activity, which may contribute to the persistent signal underlying LTP. Additionally, synergistic interactions between these kinases have been observed in purified systems[72] and may occur in hippocampal pyramidal neurons. An important step toward the understanding of CaMKII and PKC involvement in LTP will undoubtedly be the identification of their relevant substrate proteins. Substrate proteins suggested thus far include GAP-43 (B-50, F-1, P-57)[72] and synapsin.[74]

The requirement of postsynaptic CaMKII and PKC activity for the generation of LTP does not rule out the involvement of presynaptic kinase activity. Indeed, accumulating evidence points to a presynaptic component of LTP. If LTP were maintained by a postsynaptic constitutively active kinase, then previously established LTP would be blocked by postsynaptic delivery of protein kinase inhibitor. Malinow *et al.*,[40] however, found that although bath application of H-7 blocked LTP, postsynaptic injection of H-7 failed to attenuate established LTP. A possible explanation for this is that a presynaptic kinase ultimately causes a persistent increase in neurotransmitter release that underlies LTP. Supporting this hypothesis is the observation that phorbol esters mimic LTP and are known to increase neurotransmitter release in some systems, including the hippocampus.[75,76] Consistent with this is the observation that phorbol esters enhance the phosphorylation of hippocampal synapsin, a process that is thought to result in enhanced transmitter release.[77] In addition, activation of PKC by phorbol ester causes an increase in the activity of VDCCs in cultured hippocampal neurons. Single-channel recordings have shown that this increase probably occurs in both L- and N-type channels and appears to involve a shift in the voltage dependence of the channel activation to more negative voltages.[78] If such a kinase-induced change occurs in the calcium channels of hippocampal excitatory presynaptic terminals after tetanic stimulation, then this could result in an increase in transmitter release. Such an increase in transmitter release could conceivably underlie the expression of LTP.

Because induction has a critical postsynaptic component, postulating a presynaptic mechanism for the expression of LTP leads necessarily to the conclusion that the postsynaptic cell can somehow communicate with the presynaptic terminal via the release of a messenger factor that acts selectively on recently activated terminals, thus influencing further transmitter release.[79] Arachidonic acid or some of its metabolites are attractive candidates for such a factor because these agents readily cross cell membranes,[80] because inhibitors of their synthesis have been reported to block LTP,[79,81,82] because increases in extracellular arachidonic acid have been observed during LTP,[81] and because application of arachidonic acid paired with synaptic stimulation can enhance synaptic transmission.[79] In any case, the debate regarding a pre- versus postsynaptic locus for LTP maintenance and expression remains highly controversial. Evidence is growing, however, to indicate that both sides of the synapse may contribute to this complex phenomenon.

In summary, excitatory synapses in the hippocampus and other neural tissues undergo a use-dependent, long-lasting increase in efficacy known as long-term potentiation. Although the mechanics of this phenomenon are not yet completely understood, recent experiments have shed a great deal of light on the potential physiological and biochemical mechanisms that underlie this form of synaptic plasticity. Current models hold that the induction of LTP depends on the conjunction of ligand binding to the postsynaptic NMDA subtype of glutamate receptor and the depolarization of the postsynaptic membrane potential. Following induction, a complex and poorly understood series of events, probably involving both protein kinase C and calcium/calmodulin-dependent protein kinase II leads to the formation of a persistent signal that maintains the potentiated transmission. The final expression of LTP may reside in the presynaptic terminal. Future studies on LTP will likely focus on the nature of the mechanisms that maintain and express LTP.

REFERENCES

1. BLISS, T. V. P. & T. LOMO. 1973. J. Physiol. **232:** 311-356.
2. LOMO, T. 1966. Acta Physiol. Scand. **68**(Suppl. 277): 128.
3. DELANEY, K. R., R. S. ZUCKER & D. W. TANK. 1989. J. Neurosci. **9:** 3558-3567.
4. MCNAUGHTON, B. L., R. M. DOUGLAS & G. V. GODDARD. 1978. Brain Res. **157:** 277-293.
5. DUNWIDDIE, T. V. & G. LYNCH. 1978. J. Physiol. **276:** 353-367.
6. BARRIONUEVO, G. & T. H. BROWN. 1983. Proc. Natl. Acad. Sci. USA **80:** 7347-7351.
7. GUSTAFSSON, B. & H. WIGSTROM. 1986. J. Neurosci. **6:** 1575-1582.
8. KELSO, S. R., A. H. GANONG & T. H. BROWN. 1986. Proc. Natl. Acad. Sci. USA **83:** 5326-5330.
9. LARSON, J. & G. LYNCH. 1986. Science **232:** 985-988.
10. SASTRY, B. R., J. W. GOH & A. AUYEUNG. 1986. Science **232:** 988-990.
11. LEVY, W. B. & O. STEWARD. 1983. Neuroscience **8:** 791-797.
12. WIGSTROM, H., B. GUSTAFSSON & Y.-Y. HUANG. 1985. Acta Physiol. Scand. **124:** 475-478.
13. WIGSTROM, H., B. GUSTAFSSON, Y.-Y. HUANG & W. C. ABRAHAM. 1986. Acta Physiol. Scand. **126:** 317-319.
14. GUSTAFSSON, B. & H. WIGSTROM. 1988. Trends Neurosci. **11:** 156-162.
15. COLLINRIDGE, G. L., S. J. KEHL & H. MCLENNAN. 1983. J. Physiol. **334:** 19-31.
16. COTMAN, C. W. & J. V. NADLER. 1981. *In* Glutamate: Transmitter in the Central Nervous System. P. J. Roberts, J. Storm-Mathisen & G. A. R. Johnston, Eds.: 117-154. Wiley. Chichester.
17. COLLINGRIDGE, G. L. 1985. Trends Pharmacol. Sci. **6:** 407-411.
18. HERRON, C. E., R. A. J. LESTER, E. J. COAN & G. L. COLLINGRIDGE. 1985. Neurosci. Lett. **60:** 19-23.
19. JAHR, C. E. & C. F. STEVENS. 1987. Nature **325:** 522-525.
20. MACDERMOTT, A. B., M. L. MAYER, G. L. WESTBROOK. S. J. SMITH & J. L. BARKER. 1986. Nature **321:** 519-522.
21. MAYER, M. L., A. B. MACDERMOTT, G. L. WESTBROOK, S. J. SMITH & J. L. BARKER. 1987. J. Neurosci. **7:** 3230-3244.
22. MAYER, M. L. & G. L. WESTBROOK. 1987. J. Physiol. **394:** 501-528.
23. MAYER, M. L., G. L. WESTBROOK & P. B. GUTHRIE. 1984. Nature **309:** 261-263.
24. NOWAK, L., P. BREGESTOVSKI, P. ASCHER, A. HERBET & A. PROCHIANTZ. 1984. Nature **307:** 462-465.
25. COLLINGRIDGE, G. L. & T. V. P. BLISS. 1987. Trends Neurosci. **10:** 288-293.
26. GUSTAFSSON, B., H. WIGSTROM, W. C. ABRAHAM & Y.-Y. HUANG. 1987. J. Neurosci. **7:** 774-780.
27. HARRIS, E. W. & C. W. COTMAN. 1986. Neurosci. Lett. **70:** 132-137.
28. HARRIS, K. M. & D. M. LANDIS. 1986. Neuroscience **19:** 857-872.
29. HARRIS, K. M. & J. K. STEVENS. 1989. J. Neurosci. **9:** 2982-2997.
30. LANDIS, D. M. & T. S. REESE. 1983. J. Cell Biol. **97:** 1169-1178.
31. LANDIS, D. M. 1988. J. Electron Microsc. Technol. **10:** 129-151.
32. GAMBLE, E. & C. KOCH. 1987. Science **236:** 1311-1315.
33. MALENKA, R. C., J. A. KAUER, R. S. ZUCKER & R. A. NICOLL. 1988. Science **242:** 81-84.
34. REYMANN, K. G., H. K. MATTHIES, U. FREY, V. S. VOROBYEV & H. MATTHIES. 1986. Brain Res. Bull. **17:** 291-296.
35. BLISS, T. V. P., M. L. ERRINGTON & M. A. LYNCH. 1987. *In* Excitatory Amino Acid Transmission. T. P. Hicks & D. Lodge, Eds.: 337-340. Alan R. Liss. New York, NY.
36. MALENKA, R. C., J. A. KAUER, D. J. PERKEL & R. A. NICOLL. 1989. Trends Neurosci. **12:** 444-450.
37. LOVINGER, D. M., K. L. WONG, K. MURAKAMI & A. ROUTTENBERG. 1987. Brain Res. **436:** 177-183.
38. MALINOW, R., D. V. MADISON & R. W. TSIEN. 1988. Nature **335:** 820-824.
39. MALENKA, R. C., J. A. KAUER, D. J. PERKEL, M. D. MAUK, P. T. KELLY, R. A. NICOLL & M. N. WAXHAM. 1989. Nature **340:** 554-557.
40. MALINOW, R., H. SCHULMAN & R. W. TSIEN. 1989. Science **245:** 862-866.
41. SCHWARTZ, J. H. & S. M. GREENBERG. 1987. Annu. Rev. Neurosci. **10:** 459-476.

42. KELLY, P. T., T. L. MCGUINNESS & P. GREENGARD. 1984. Proc. Natl. Acad. Sci. USA **81:** 945-949.
43. KENNEDY, M. B., M. K. BENNETT & N. E. ERONDU. 1983. Proc. Natl. Acad. Sci. USA **80:** 7357-7361.
44. DUNWIDDIE, T. V., N. L. ROBERTSON & T. WORTH. 1982. Pharmacol. Biochem. Behav. **17:** 1257-1264.
45. FINN, R. C., M. BROWNING & G. LYNCH. 1980. Neurosci. Lett. **19:** 103-108.
46. TURNER, R. W., K. G. BAIMBRIDGE & J. J. MILLER. 1982. Neuroscience **7:** 1411-1416.
47. REYMANN, K. G., U. FREY, R. JORK & H. MATTIES. 1988. Brain Res. **440:** 305-314.
48. SAITOH, T. & J. H. SCHWARTZ. 1985. J. Cell Biol. **100:** 835-839.
49. MILLER, S. G. & M. B. KENNEDY. 1986. Cell **44:** 861-870.
50. LISMAN, J. E. 1985. Proc. Natl. Acad. Sci. USA **82:** 3055-3057.
51. AKERS, R. E., P. A. COLLEY, D. J. LINDEN & A. ROUTTENBERG. 1986. Science **231:** 587-589.
52. NISHIZUKA, Y. 1986. Nature **233:** 305-312.
53. NELSON, R. B. & A. ROUTTENBERG. 1985. Exp. Neurol. **89:** 213-224.
54. OLDS, J. L., M. L. ANDERSON, D. L. MCPHIE, L. D. STATEN & D. L. ALKON. 1989. Science **245:** 866-868.
55. LYNCH, M. A., M. P. CLEMENTS, M. L. ERRINGTON & T. V. P. BLISS. 1988. Neurosci. Lett. **84:** 291-296.
56. ITO, I., D. OKADA & H. SUGIYAMA. 1988. Neurosci. Lett. **90:** 181-185.
57. GOH, J. W. & P. S. PENNEFATHER. 1989. Science **244:** 980-982.
58. MILLER, R. J. & S. N. MURPHY. 1988. Proc. Natl. Acad. Sci. USA **85:** 8737-8741.
59. MALENKA, R. C., D. V. MADISON & R. A. NICOLL. 1986. Nature **321:** 175-177.
60. HU, G.-Y., O. HVALBY, S. I. WALAAS, K. A. ALBERT, P. SKJEFLO, P. ANDERSON & P. GREENGARD. 1987. Nature **328:** 426-429.
61. MURAKAMI, K., S. Y. CHAN & A. ROUTTENBERG. 1986. J. Biol. Chem. **261:** 15424-15429.
62. LINDEN, D. J. & A. ROUTTENBERG. 1989. Brain Res. Rev. **14:** 279-296.
63. NISHIZUKA, Y. 1988. Nature (London) **334:** 661-665.
64. KIKKAWA, U., A. KISHIMOTO & Y. NISHIZUKA. 1989. Annu. Rev. Biochem. **58:** 31-44.
65. NAOR, Z., M. S. SHEARMAN, A. KISHIMOTO & Y. NISHIZUKA. 1988. Mol. Endocrinol. **2:** 1043-1048.
66. TAKAI, Y., A. KISHIMOTO, M. INOUE & Y. NISHIZUKA. 1977. J. Biol. Chem. **252:** 7603-7609.
67. INOUE, M., A. KISHIMOTO, Y. TAKAI & Y. NISHIZUKA. 1977. J. Biol. Chem. **252:** 7610-7616.
68. HAMAKUBO, T., R. KANNAGI, T. MURACHI & A. MATUS. 1986. J. Neurosci. **6:** 3103-3111.
69. STAUBLI, J., J. LARSON, O. THIBAULT, M. BAUDRY & G. LYNCH. 1988. Brain Res. **444:** 153-158.
70. BALLESTER, R. & O. M. ROSEN. 1985. J. Biol. Chem. **260:** 15194-15199.
71. LOVINGER, D. M., K. WONG, K. MURAKAMI & A. ROUTTENBERG. 1987. Brain Res. **436:** 177-183.
72. ALEXANDER, K. A., B. M. CIMLER, K. E. MEIER & D. R. STORM. 1987. J. Biol. Chem. **262:** 6108.
73. LOVINGER, D. M., P. A. COLLEY, R. F. AKERS, R. B. NELSON & A. ROUTTENBERG. 1986. Brain Res. **399:** 205-211.
74. PARFITT, K. D., B. J. HOFFER & M. D BROWNING. 1990. Proc. Nat. Acad. Sci. USA: in press.
75. ZURGIL, N. & N. ZISAPEL. 1985. FEBS Lett. **185:** 257.
76. MALENKA, R. C., G. S. AYOUB & R. A. NICOLL. 1987. Brain Res. **403:** 198-203.
77. BARNES, E. & M. D. BROWNING. 1989. Soc. Neurosci. Abstr. **15:** 84.
78. MADISON, D. V. 1989. Soc. Neurosci. Abstr. **15:** 16.
79. WILLIAMS, J. H., M. L. ERRINGTON, M. A. LYNCH & T. V. P. BLISS. 1989. Nature **341:** 739-742.
80. PIOMELLI, D., A. VOLTERRA, N. DALE, S. A. SIEGELBAUM, E. R. KANDEL, J. H. SCHWARTZ & F. BELARDETTI. 1987. Nature **328:** 38-43.
81. LYNCH, M. A., M. L. ERRINGTON & T. V. P. BLISS. 1989. Neuroscience **30:** 693-701.
82. OKADA, D., S. YAMAGISHI & H. SUGIYAMA. 1989. Neurosci. Lett. **100:** 141-146.

Long-Term Potentiation during the Activity-Dependent Sharpening of the Retinotopic Map in Goldfish[a]

JOHN T. SCHMIDT

Department of Biological Science
and
Neurobiology Research Center
State University of New York at Albany
Albany, New York 12222

INTRODUCTION

In visual development, many of the fine details of synaptic circuitry are established by the selective stabilization of appropriate synapses from a more diffuse set of initial connections, a process that requires normal patterns of activity during the sensitive period. Increasingly, this selective stabilization of connections, like long-term potentiation (LTP), has been found to be triggered by *N*-methyl-D-aspartate (NMDA) receptors that detect synchronous activation of inputs onto common postsynaptic cells.[1–4] In addition, there is intriguingly often a greater capacity for LTP during these sensitive periods of development.[3,5,6] These findings put experience-dependent development and learning on a continuum. In this paper, I will review the evidence from the development and regeneration of the retinotectal projection of fish and frog, and discuss it in relation to similar findings in the developing cat visual system.

Several features of the development of visual projections appear to be activity driven; these include the segregation of visual afferents into eye-specific patches or stripes in visual cortex[7] or tectum[8–10] and into lamina in lateral geniculate nucleus (LGN),[11] the segregation of receptive field types in LGN,[12] and the refinement of retinotopic maps in LGN[13] and tectum.[14–16] Two of these, the retinotopic sharpening and the formation of eye-specific patches, are easily investigated in the retinotectal projection, which is direct and accessible. In addition, the projection regenerates following optic nerve crush, allowing the use of larger, more easily manipulated, adult animals. The projection is crossed at the chiasm so that each retina innervates its contralateral tectum. Crush of the optic nerve reproducibly results in axonal outgrowth from the crush site, synaptogenesis after arrival back at the tectum, and a period of synapse elimination and stabilization during maturation. Thus, the regenerating projection recapitulates the major features seen during development. In addition, recent studies have shown that activity refines retinotopic order during development as well.[17,18]

RETINOTOPIC SHARPENING OF THE REGENERATED MAP

Part of the mechanism of the map formation involves chemoaffinity between optic fibers and different regions of tectum, as envisaged by Sperry,[19] but this is sufficient to

[a] This work was supported by the National Institutes of Health (EY03736).

organize only a very rough map. Thus, in the horizontal dimension, axons from temporal retina grow into rostral tectum, whereas those from nasal retina grow around the edges to caudal tectum. In the second dimension, dorsal retinal fibers grow into ventrolateral tectum, whereas those from ventral retina grow into dorsomedial tectum. Chemoaffinity information, however, is not precise enough to tell the axons exactly where to grow their arbors, and a second mechanism, an activity-driven retinotopic sharpening, returns the map to the mature level of organization.[14–16,20] This process has been reviewed in detail,[21] and only the main points are considered here.

Anatomical Studies

Staining of individual fibers with horseradish peroxidase shows that axons arriving back at the tectum make many exploratory branches within the plane of the map.[22,23] The average normal arbor is about 220 μm across, but these early arbors average about 1200 μm, a substantial fraction of the tectal length of 3 to 4 mm.[22] Many branches are subsequently eliminated, as the arbors return toward normal appearance, both in extent and in number of branches. One of the cues used to determine which of the branches get eliminated is impulse activity, for regenerating arbors that were silenced by intraocular tetrodotoxin (TTX) or synchronized in their firing by strobe illumination were roughly two times larger than normal in extent and also failed to coalesce into a single cluster of branches.[24] This lack of sharpening is also reflected in the map that is recorded electrophysiologically from the tectum.

Electrophysiological Recordings

In superficial tectal neuropil, extracellular unit recordings are mostly from the arbors of the retinal ganglion cells.[15,20] Several arbors are usually recorded simultaneously, and the area of the visual field from which they can be driven is called the multiunit receptive field. In normal goldfish tectum, this area is approximately 11°, almost exactly the same average as for single ganglion cell receptive fields, and this shows the normal degree of order of the projection.[15,20] Following nerve crush, the fibers first arrive at the tectum at around 14 days, but orderly maps can first be recorded only at 35 to 40 days, about the time that small arbors are first seen anatomically. Before that point, the multiunit activity seems to be driven from wide areas of the retina, but the receptive fields are difficult to define because the units are very low amplitude and fatigue extremely rapidly.[15,20] In frog, however, the entire process of sharpening is easily followed.[20,25] Although these recordings are largely presynaptic, several studies have also shown that the diffuse projections make effective synaptic connections.[20,26]

In goldfish, if the projection is silenced with intraocular TTX or alternatively synchronized by strobe illumination in a white, featureless environment[27] from 14 to 35 days after nerve crush, then the diffuse map is retained (FIG. 1, compare to the normal map in FIG. 7, left). The units recorded from each point are driven from retinal areas about 30° or more across. The centers of these areas are approximately in the right place; it is their extent that reflects the uncorrected errors in targeting of the arbor branches. Both the separation of single units by spike height and the recording of single ganglion cells from retina showed that single units retained normal receptive

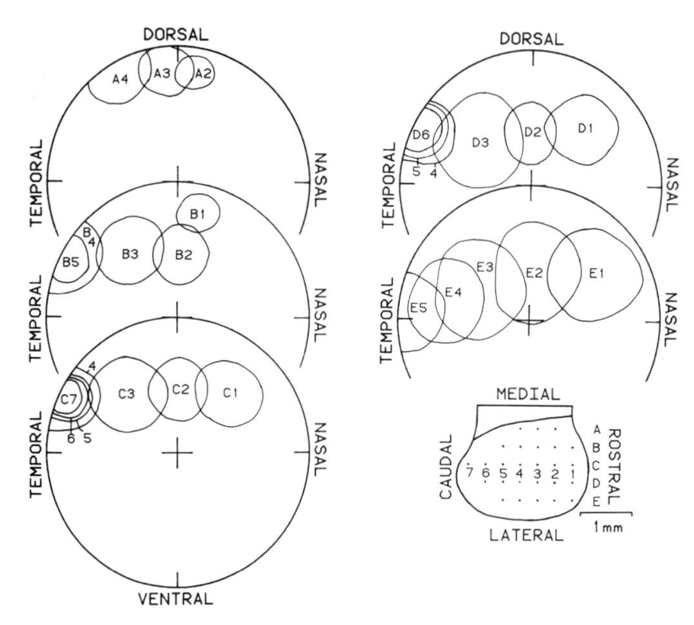

FIGURE 1. Map of a retinotectal projection regenerated under stroboscopic illumination and recorded 55 days after nerve crush. At the lower right is a drawing of the tectal surface viewed from above. Each point in the tectal array is an electrode penetration. The receptive fields recorded at these points, although enlarged, fall into an orderly array in the visual field similar to the array seen on the tectal surface. The receptive fields are numbered accordingly. For clarity, five separate representations of the hemispherical visual field, one for each row of points A through E on the tectum, are presented. Circles within the visual field define the boundaries of the multiunit receptive fields. For convenience, the drawing of the tectum has been inverted about one axis so that the arrays are oriented in the same direction: rostral on the tectum corresponds to nasal in the visual field, medial to dorsal, lateral to ventral, and caudal to temporal. Reprinted with permission from reference 27.

fields after TTX or strobe treatment; the enlarged multiunit receptive fields were therefore due to abnormal convergence onto each tectal point of ganglion cell arbors from wide retinal areas.[15,24,28] The effects of strobe, which causes the firing of all ganglion cells,[27] point to the importance of the normal pattern of activity that must contain information that is instructive for the process of sharpening. Normally, activity is correlated between neighboring, but not distant, ganglion cells of the same type (ON, OFF, or ON-OFF); this is especially true in the light, but also occurs in the dark due to retinal circuitry.[29–31]

A Hebbian Model Based on NMDA Receptors

The model for retinotopic sharpening is based upon 1) the correlated firing of neighboring ganglion cells, 2) the initially large and highly overlapping arbors, 3) the resultant summation of their excitatory postsynaptic potentials (EPSPs) in postsynaptic tectal cells, and 4) a Hebbian strengthening of the most effective synapses.[32] FIGURE 2 shows how this would work for two sets of neighboring ganglion cells. On each postsynaptic cell, the inputs from neighboring cells firing with the highest degree of correlation would experience the greatest summation of their EPSPs and be stabilized. Of course arbors from some more distant ganglion cells would also overlap on cells in this region, but would not have correlated firing and would not be stabilized. On other postsynaptic cells, a different set of inputs would have the best summation and capture exclusive control. The mechanism whereby that state is reached could be either 1) a stimulated growth of new contacts in the immediate region of stabilized synapses, combined with the withdrawal of unstabilized synapses, or 2) an iterative process of random reinsertion followed by further stabilization.

From this model, it is easy to see that if no cells had activity, no synapses would be differentially stabilized. Conversely, if all ganglion cells were to fire in synchrony because of the strobe, the cue for finding neighboring cells would be removed and all synapses would be stabilized equally regardless of their origin.

Eye-Specific Stripes

A second activity-driven phenomenon that may operate through the same mechanism is the segregation of visual afferents by eye of origin.[7–10] Although each eye normally innervates its contralateral tectum exclusively, co-innervation can be produced when one tectal lobe is removed and the stump of the tract deflected toward the opposite tectum as shown in FIGURE 3. Alternatively, Cline *et al.*[2] transplanted a third eye to the forehead of the embryonic frog, and it grew optic fibers out to innervate one of the two tecta. The projection of the contralateral eye is normally continuous over the surface. Initially, in the goldfish,[9] the invading ipsilateral eye also forms a continuous projection, but shortly thereafter both projections break up into exclusively occupied patches and stripes. These can be seen when the fibers from both eyes are labeled—but with different markers (in FIG. 3, one cross-section through the tectum is labeled via

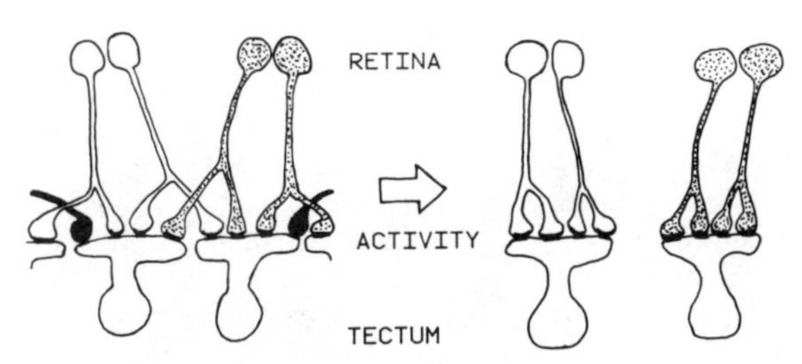

FIGURE 2. A model for the activity-dependent stabilization of retinotectal synapses in the retinotopically correct order. On the left, ganglion cells from the retina regenerate scattered arbors contacting many tectal cells. At some postsynaptic cells, their arbors overlap with those of their neighbors. Because of locally correlated activity in retinal ganglion cells, these inputs may sum postsynaptically and therefore be more effective. Others from distant synapses would not sum. If the most effective synapses were preferentially retained and the others removed, the mechanism would result in a stable, retinotopically ordered arrangement as seen on the right. Reprinted with permission from reference 27.

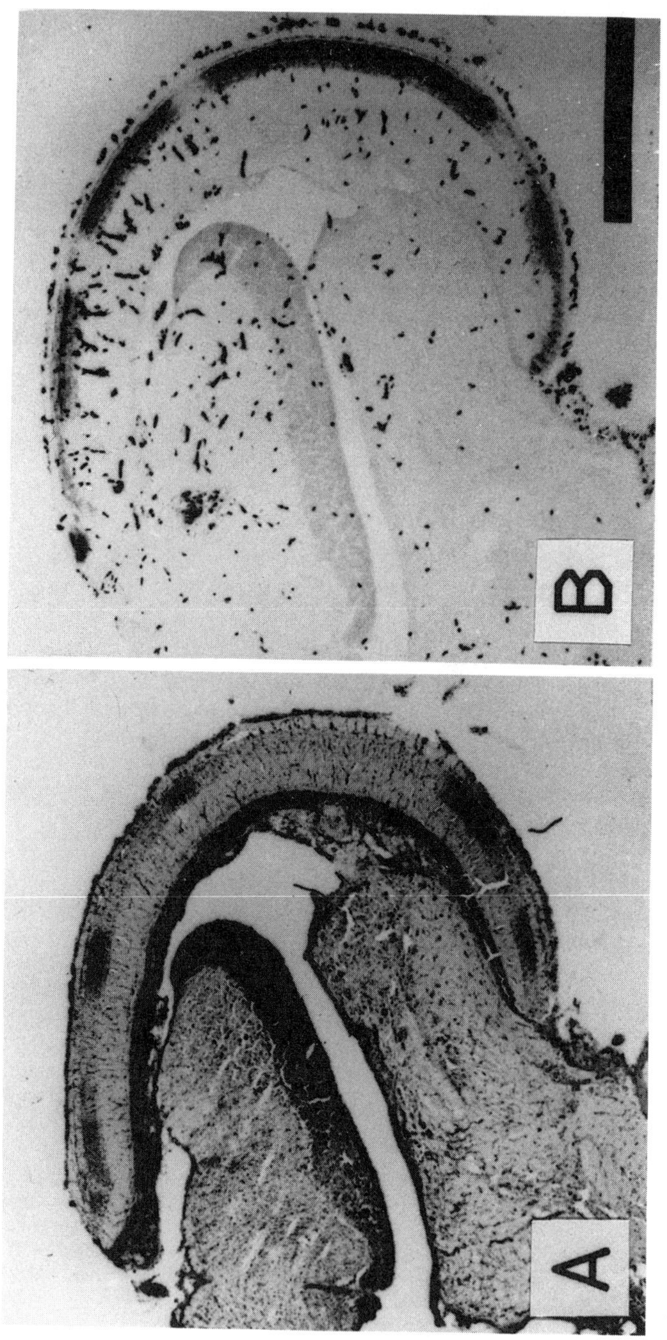

FIGURE 3. Sharp left eye and right eye ocular dominance patches in adjacent cross-sections from one tectum of a fish 8.5 months after removal of the other tectum. **(A)** A section showing the ipsilateral projection labeled by radioautography (thionin counterstain). **(B)** Adjacent section showing the contralateral projection labeled by horseradish peroxidase (no counterstain). Note how the patches of one eye fit into the holes of the other. Calibration bar: 1 mm. Reprinted with permission from reference 9.

a horseradish peroxidase procedure, and the adjacent cross-section is labeled via a radioautography procedure).

When both eyes are silenced by intraocular injection of TTX, the two projections remain mingled and no patches form, thus establishing that segregation is activity driven.[9] It is easy to envisage the same model giving rise to ocular dominance patches. The fibers within each eye, especially after retinotopic sharpening, should have highly correlated activity, whereas those from different eyes should not. Thus, in FIGURE 2, the two neighboring ganglion cells would be from the right eye and the other two from the left eye. The fibers from each eye should cooperatively stabilize each other's synapses, driving out the other eye's synapses and consolidating their hold on their areas. Indeed, the segregation generally occurs after the period of retinotopic sharpening, and this may offer some clue as to why the tectum is divided up into stripes that are usually several hundred micrometers wide. The projections probably form this way because each map is constrained to cover the whole tectum by chemoaffinity (and also initially by activity-driven stabilization) and because activity later drives them to separate completely. In this case, it is obvious that synapse stabilization is also associated with synapse elimination because large gaps are produced in a once continuous projection. Thus, the mechanism that normally sharpens the map here produces great discontinuities.

NMDA RECEPTORS AS TRIGGERS OF HEBBIAN STABILIZATION

Retinotectal transmission appears to be mediated primarily by glutamate (or some similar excitatory amino acid) acting on several types of postsynaptic receptor. Most of the EPSP amplitude seems to be mediated by non-NMDA receptors, often referred to as quisqualate and kainate receptors, that operate in a conventional fashion. The NMDA receptors, although fewer in number, appear to play a tremendously important role in controlling the development and maintenance of synaptic connections.

NMDA receptors, which have been implicated in triggering LTP in the CA1 region of the hippocampus,[33] are well suited to the task of signaling that the synapse has participated in a substantial postsynaptic depolarization. When activated by a neurotransmitter, the channel is fully conducting only when the cell is sufficiently depolarized to remove a magnesium block,[34] when it then allows substantial calcium entry.[35] Thus, NMDA receptors both have the appropriate trigger properties and generate a second messenger that could signal Hebbian stabilization.

If NMDA receptors play such a role in the stabilization of retinotectal synapses, then 1) the transmission should be carried largely by an excitatory amino acid with NMDA receptors as one of the receptor types, 2) the retinotopic sharpening should be blocked by NMDA receptor blockers, and 3) there might also be an enhanced capacity for LTP during the period of sharpening, and this too should be blocked by NMDA receptor blockers.

NMDA and Non-NMDA Receptors in the Retinotectal Transmission

NMDA receptors have been detected in goldfish tectum by binding[36] and localized by radioautography to the optic terminal layers in frog tectum.[37] In the retinotectal projection of goldfish, most transmission appears to be mediated by an excitatory amino acid.[3,38] 6-Cyano-2,3-dihydroxy-7-nitroquinoxaline (CNQX), a selective blocker of non-NMDA excitatory amino acid receptors, topically applied at 20 μM effectively blocks more than 90% of the amplitude of field potentials elicited by optic nerve shock (FIG. 4). The sources and sinks of synaptic current were decremented even more strongly. A 50% decrement of field potential amplitude occurs at 5 μM and an 80% decrement at 10 μM. All concentrations above 25 μM produced a virtually complete block. In contrast, 2-carboxypiperazine-4-yl-propyl-1-phosphonic acid (CPP), a selective NMDA receptor blocker, at 1 μM blocks only about 5-10% of the field potential amplitude (FIG. 5), and has a similarly small effect on the sources and sinks of synaptic

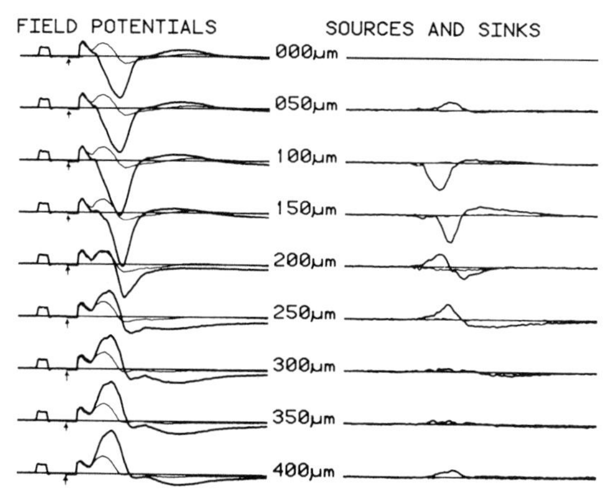

FIGURE 4. Effect of topically applied CNQX on field potentials and the sources and sinks of synaptic current elicited by optic nerve shock in a normal goldfish tectum. Arrows mark the time of stimulation (artifact electronically suppressed). Depths of recording are shown next to each trace, and each trace is an average of five responses. The larger responses (heavy traces) are the controls, and the smaller ones (lighter traces) are the responses after 20 μM CNQX. Negative is downward. The calibration pulse at the beginning of each trace is 1 mV in amplitude and 2 msec in duration. Total trace length is 50 msec. On the right are the sources and sinks of synaptic current calculated by a second differencing technique. Sinks are downward and reflect the entry of current into the postsynaptic cells at excitatory synapses. Sources (upward deflections) represent exit of current from cells, usually passive return for excitatory synapses but also possibly occurring at inhibitory feed-forward synapses.

current. Increasing the concentration to 10 μM produces no further decrement, indicating very little contribution from NMDA receptors. In confirmation of this, both 2-amino-5-phosphonovalerate (AP5) and 2-amino-7-phosphonoheptanoate (AP7) block the same small percentage of the response amplitude at 25-50 μM (FIG. 6), concentrations at which they should be selective for NMDA receptors. At higher millimolar concentrations, however, they block all of the field potentials (FIG. 6). Thus, as in the hippocampus,[33] the NMDA receptors are present, but they appear to carry very little of the synaptic current as assayed by single stimuli.

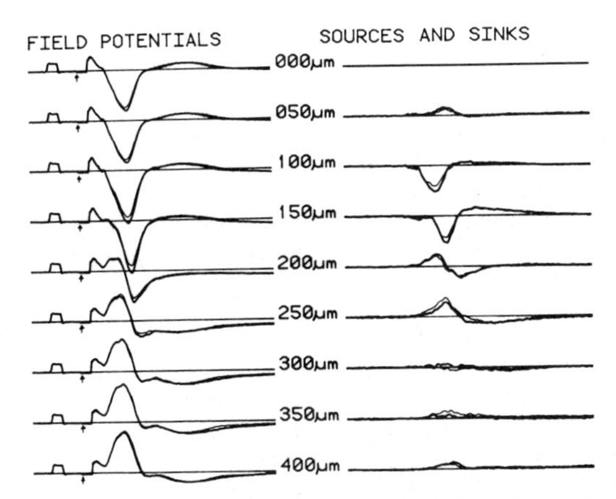

FIGURE 5. Effect of CPP on field potentials and sources and sinks of synaptic current in normal goldfish tectum: control (heavy traces) and after 1 µM CPP (light traces). Conventions are as in FIGURE 4. Note the small 5-10% decrement of both field potential amplitude and the underlying sources and sinks.

NMDA Receptor Blockers Prevent Sharpening

In order to block NMDA receptors over a long period, I infused AP5 or AP7 into the tectal ventricles using an osmotic minipump attached to the fish's head.[3] The pumps were attached at 21 days postcrush when field potentials were first detectable, and delivered an average of 3.5 µl/day over the 12 to 24 days that they were attached. The concentration in the pumps (usually 500 µM) was substantially diluted within the 200 µl of fluid spaces within and around the brain, and I estimate the effective concentration was probably less than 10 µM (and certainly no more than 50 µM) within tectum. These concentrations should not have reduced retinotectal transmission to any substan-

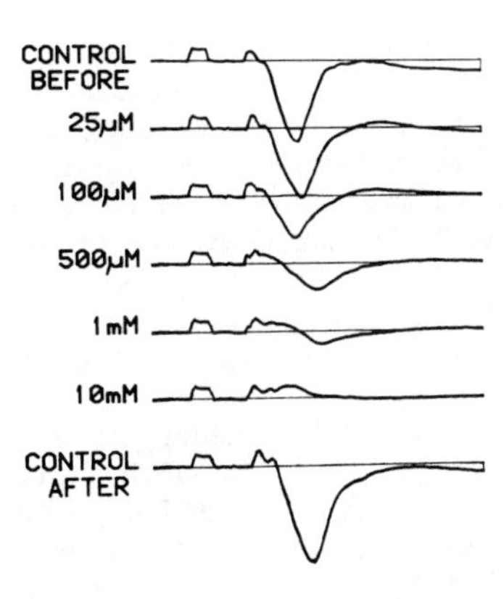

FIGURE 6. Effects of different doses of AP5 on the amplitude of the field potentials in a normal tectum. All recordings were made at a depth of 125 µm in one fish. Increasingly higher concentrations were applied topically to the tectal surface, followed by a washout with repeated changes of Ringer's solution. The equilibration time between increasing doses was 15 min, but was 2 hr for the washout. Reprinted with permission from reference 3.

tial degree, and indeed fish injected intracranially with the same or larger amounts of AP5 or AP7 showed no performance deficit in a tectally mediated behavioral assay, the optomotor response to a rotating drum.[3]

In contrast with the eight control infusion fish whose projections sharpened normally, the projections of seven fish infused with NMDA receptor blockers all failed to sharpen (FIG. 7). The results were completely comparable to those of strobe exposure for the same period.[27] The gross organization of the map was correct, but multiunit receptive fields were much enlarged, averaging 28°. These enlarged values in the blocked projections must reflect uncorrected errors in targeting of optic arbors, because single units in these projections had normal receptive fields of about 11°. In contrast, all of the projections receiving control infusions (either Ringer's solution alone, or with the chemically similar but inactive 2-amino-6-phosphonohexanoate (AP6), or with AP5 at too low a concentration) had multiunit receptive fields averaging 11.6°, indistinguishable from uninfused regenerated projections that sharpened normally.

FIGURE 7 also shows the lack of effect on the intact projection. Because the canula delivered the AP7 at the midline, the intact projection must have received the same concentration of the blocker as the regenerating one, but it was not rendered unsharpened as a result. Thus, the NMDA receptor blockers did not produce general disorder but interfered more selectively with the active processes of synapse stabilization and elimination going on in the regenerating projection. This parallels the results from strobe and TTX experiments.[27]

Enhanced Capacity for LTP during Sharpening

In studying the return of field potentials during regeneration, I noticed that, although the amplitude after maximal optic nerve stimulation was very small during the early phase, it tended to increase markedly after a series of maximal stimuli were administered to it, suggesting that the projection was undergoing LTP.[3] An example of the LTP is shown in FIGURE 8. Initially, the amplitude elicited by maximal stimuli was only 1 mV, but after the application of a standard train of stimuli, 20 stimuli delivered at 0.1 Hz, the response grew to more than 3 mV. Further trains of stimuli caused further increases in amplitude, and the potentiation was stable for hours and also overnight in the three cases tested.

The responses initially fatigued rapidly even at 1 Hz, necessitating the lower 0.1 Hz frequency for potentiating trains. After LTP, the responses became more resistant to fatigue although the latency (reflecting the slow conduction velocity) remained constant. The rapid fatigue as well as the long latency of the response was due to the immature state of the regenerated fibers. In general, it was necessary for the postsynaptic response during the trains to maintain more than half their initial amplitude or the potentiation would not occur. If the responses decremented substantially, a depression lasting about 30 min resulted. In addition, submaximal stimulating trains also failed to trigger LTP. The test stimuli were kept to a very low rate (one per 15 min) in order to avoid potentiation. In seven control cases, the responses increased an average of only 21% after 5-8 hr.

The frequencies used here are much lower than generally used in the hippocampus and preclude temporal summation of postsynaptic responses. However, LTP in cortex can be elicited by low-frequency trains such as 2 Hz for 60 min.[5,39,40] In addition, LTP in hippocampus can also be elicited by single shocks if sufficient depolarization is assured either by blocking inhibition with picrotoxin,[41] by pairing with intracellular injection of depolarizing current,[42] or by pairing with trains to another input.[43] The exact reason why the tectal cells might be sufficiently depolarized or disinhibited during regeneration has not been studied, but it is known that tectal cells become more excitable and fire in bursts until reinnervated.[44] The development of potentiation is also much slower than in hippocampus (30 min versus 5 min) under similar conditions,[42] but much of this can be attributed to a 20° difference in temperature. In addition, potentiation in the cortex is also much slower than in the hippocampus.[5,39]

LTP was greatest early in regeneration when the fibers were forming synapses at the fastest rate and sharpening was taking place. Fish between 20 and 30 days and those between 30 and 40 days both averaged 380% potentiation, while those between 40 and 70 days averaged 80-90%. Because a sharp map can first be recorded between

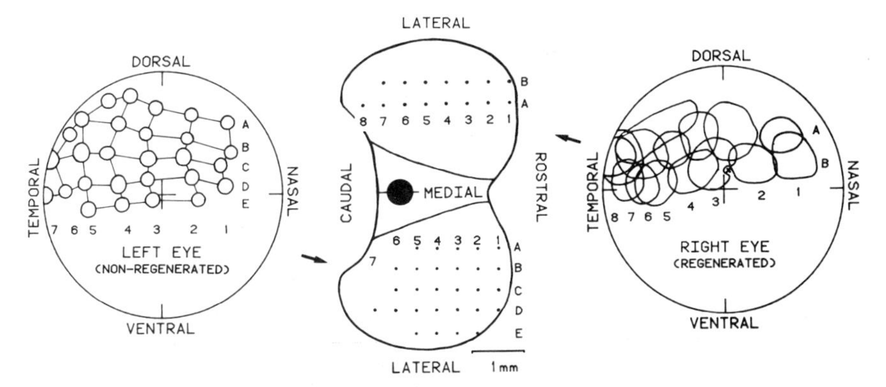

FIGURE 7. Maps of two retinotectal projections (one regenerating and one not regenerating) from a fish infused with AP7 from 21 to 35 days postcrush. The pump infused an average of 3.1 μl/day of 500 μM AP7, and the effective concentration after dilution in the fluids in and around the brain was estimated at 10-20 μM. On the *left* is the orderly map of the nonregenerating projection from the left eye to the right tectum, which was not affected by the block of NMDA receptors. On the opposite tectum of this fish, the regenerating projection from the right eye (*right*) was prevented from sharpening by the block of NMDA receptors (average multiunit receptive field size of 27.4° versus 10.8°). Conventions are as in FIGURE 1, except that the visual field of the right eye was reversed so that the array would directly correspond to that on the tectum. The large blackened circle at the midline shows the point of insertion of the cannula through the tectal commissure to deliver the AP7. Reprinted with permission from reference 3.

35 and 40 days,[15,28] the match to the period of sharpening is particularly good. This time period also corresponds to the period of greatest expression in the retinal ganglion cells of GAP-43,[45] a protein that is phosphorylated by the C kinase during LTP in the hippocampus.[46] The level of LTP after 40 days is still much higher than that found in the mature projection (5-10%) and remains high for at least a year for unknown reasons. In general, it is not clear exactly why there is an enhanced capacity for LTP during the period of sharpening. The increased excitability or possible disinhibition of the tectal cells are likely possibilities along with the reexpression of GAP-43. The field potentials showed little evidence of any greater expression of NMDA receptors in the projection (see below), as was suggested by Tsumoto *et al.*[47] in visual cortex.

These results directly parallel those from developing kitten visual cortex. Komatsu *et al.*[5] found that during the sensitive period for development of ocular dominance

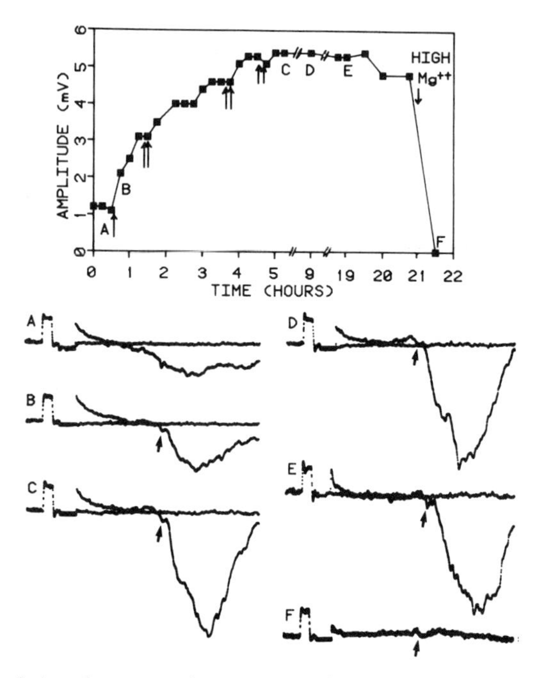

FIGURE 8. Potentiation of postsynaptic responses as demonstrated by field potential recordings in a fish at 39 days postcrush. *Top:* Amplitude of the response is plotted versus time. Arrows indicate the time of the potentiating volleys (20 stimuli at 0.1 Hz). *Bottom:* Sample responses recorded from the optic fiber terminal layer at times marked by letters in the plot above. The traces are 50 msec long and begin with a calibration pulse (1 mV × 2 msec). The stimulus artifact, which would occur just after the calibration pulse, was electronically suppressed. The arrows mark the presynaptic fiber volley preceding the negative going postsynaptic field potential. Note the breaks in the time axis of the upper plot as the potentiation was tracked into the second day. The arrow labeled High Mg^{++} denotes the time of topical application of Ringer's with 0 mM Ca^{++} and 40 mM Mg^{++} (isoosmotic replacement of Na^{+}) briefly to the tectum to block postsynaptic responses and leave the presynaptic fiber volleys (arrow). Two traces were superimposed here to show the reproducibility of the presynaptic fiber volley. Note the long latency indicating the slow conduction velocity in these regenerating fibers (< 0.5 m/sec). Reprinted with permission from reference 3.

there was a significant capacity for LTP that was not present in the adult. Some capacity for LTP can be uncovered in the adult, however, by blocking inhibition with bicuculine.[48] In addition, a capacity for LTP is also present in developing rat visual cortex, and it can be blocked by NMDA receptor blockers.[6]

LTP Is Blocked by NMDA Receptor Blockers

In these experiments,[3] LTP was attempted twice in each projection, first with an NMDA receptor blocker present, and later after the blocker was washed out. Thus each projection served as its own control. A typical result with topically applied AP7 is shown in FIGURE 9. The initial amplitude was about 0.7 mV and was not affected by the application of the AP7, indicating that NMDA receptors still constituted a very small fraction of the total. In general, the amplitude did not decrease noticeably after either AP5 or AP7 was applied. Two trains of 20 stimuli were given after the AP7 had equilibrated, but these did not result in any significant increase in amplitude. After the AP7 was washed out for 2 hr, the identical two trains of stimuli potentiated the response to nearly 3 mV. Overall, in the six cases, the increase elicited after the trains with blockers present was 2.7% (4.3% SEM) whereas the increase elicited after the blocker was washed out was 103.8% (35.9% SEM) in the same fish. The chemically similar but inactive compound AP6 did not prevent LTP at the same concentration.

Two factors therefore link the LTP and the activity-driven retinotopic sharpening. First, the greatest capacity for LTP occurs precisely during the time when the sharpening occurs, and second, both are blocked by low concentrations of NMDA receptor blockers. The exact relationship between the two phenomena is not completely defined,

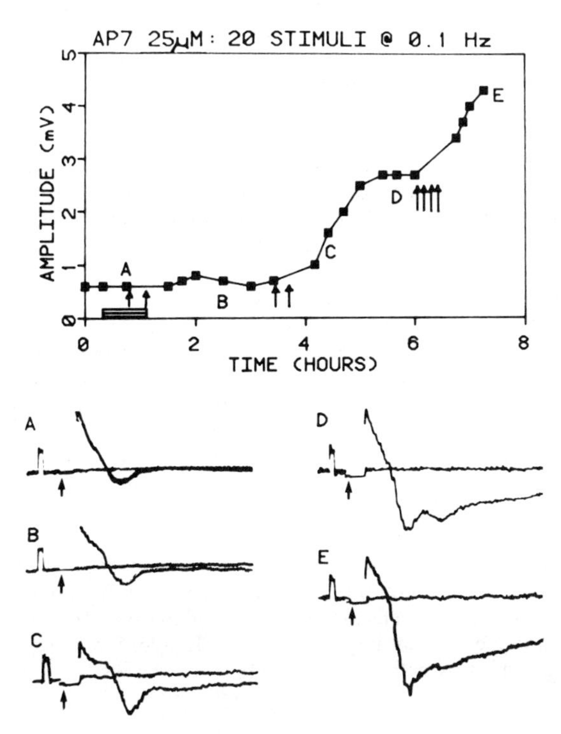

FIGURE 9. Experiment showing that AP7 blocks potentiation in a regenerating fish. Conventions are as in FIGURE 8. *Top:* Amplitude of the response is plotted versus time. *Bottom:* Sample responses from times marked by the letters in the plot above. Arrows in the plot show the times of the potentiating trains of stimuli (20 stimuli at 0.1 Hz). The first time that these trains were applied, the tectal NMDA receptors were blocked by topical application of AP7 (striped bar), and these two trains elicited no consistent increase in the synaptic gain (compare A and B). Two hours after washout of AP7, the same trains of stimuli resulted in very large increases in gain (C & D). Further trains of stimuli elicited further gains (E). Reprinted with permission from reference 3.

but the intriguing possibility now exists that LTP or an increase in the gain of individual synapses may constitute the first step in a stabilization process, one that leads to the later anatomical rearrangements.

NMDA Receptors in the Formation of Eye-Specific Patches

Using the three-eyed tadpole model, Cline and Constantine-Paton[2,49] have explored the role of NMDA receptors in the segregation of visual afferents into eye-specific stripes. They placed plastic Elvax pellets impregnated with blockers onto the tectal surface to produce a slow release into the tectal tissue. The stripes were assayed by labeling the fibers from the ectopic third eye with horseradish peroxidase. The stripes were expected to be present at the time of treatment, but after several weeks of treatment with D-AP5 there were no stripes in the area around the plastic implant but rather a continuous distribution of labeled fibers. Control projections in which inactive L-AP5 was used still had stripes. More recently, they have shown that the NMDA channel blocker MK-801 could also desegregate the stripes, but only if it were applied with NMDA.[49] The NMDA was presumably necessary to activate the receptors, because MK-801 binds only within the open receptor channel. In contrast to the antagonists, NMDA itself caused the stripes to become sharper than in controls with 75% fewer fibers crossing the open areas between the stripes.[49] The NMDA and AP5 also had effects on individual arbor morphology, causing the arbors to be less highly branched. Thus, there is also very strong evidence that segregation into eye-specific stripes depends upon NMDA receptor activation, as would be predicted from the Hebbian model above.

In cat visual cortex, infusion of AP5 prevents the shift in ocular dominance caused by monocular lid suture.[1] In control cases, lid suture causes the open eye to dominate responses of cortical cells at the expense of the closed eye, which loses effectiveness. In the area infused with AP5, the closed eye remained effective in driving cells, although the orientation selectivity was decreased. In addition, Gu *et al.*[50] showed that AP5 infusion in cases of reverse suture prevents the shift back to the reopened eye. These experiments have been criticized because the AP5 can lower visual responsiveness of cortical neurons, but no loss of responsiveness was found in at least one case where the effect was present.[1] In addition, no detectible decrease in responsiveness was noted in goldfish tectum or frog tectum.[51]

Although visual afferents from the two eyes often segregate because they carry differing patterns of information, in some cases they carry the same information and end up connected to the same cells to mediate binocular vision. Most neurons in normal cat visual cortex in fact remain largely binocular. In addition, binocular convergence is set up in frog tectum by activity-driven cues.[51]

Beyond the Calcium Entry Signal

NMDA receptor activation causes the entry of calcium into the postsynaptic cell as a signal that the synapse participated in a successful depolarization of the cell. This postsynaptic signal must somehow mediate the presynaptic changes seen in the branch retraction and eye-specific segregation. Therefore, one would have to postulate some kind of retrograde signaling mechanism to account for the presynaptic terminal's reacting appropriately. This retrograde signal might take one of several forms[18]: 1) a soluble signal released by the postsynaptic element, 2) an addition or modification of adhesion molecules mediating a stronger attachment, or 3) the polymerization of ependymin in the synaptic cleft due to calcium depletion as calcium enters the postsynaptic cell.[52,53] LTP is also thought to require a retrograde signal, because part of the mechanism may be increased presynaptic release of glutamate.[54] One candidate for this signal is arachidonic acid,[55,56] because LTP induction is prevented by nordihydroguaiaretic acid, an inhibitor of the lipoxygenase pathway of arachidonic acid metabolism.

Coming back to the immediate effects of the calcium entry into the postsynaptic element, we find that kinases are very important in LTP; both the C kinase and the calcium/calmodulin kinase II are required.[57] Blocking either one via intracellular injection prevents induction of LTP. Expression, however, is not dependent on either kinase in the postsynaptic cell, but may be dependent on the C kinase in the presynaptic terminal, where it is known to phosphorylate GAP-43 and where it increases transmitter release by increasing Ca^{2+} entry.[57] If LTP is indeed an important step in the stabilization of synaptic connections, then such stabilization might be expected to depend on the proper functioning of these kinases. Infusion of agents that block these kinases (such as sphingosine and H-7) or that indiscriminately activate and downregulate the C kinase (phorbol esters such as tetradecanoylphorbol acetate (TPA)) should block the activity-driven stabilization. Cline and Constantine-Paton[58] recently reported that chronic release of such agents from Elvax pellets did not cause the eye-specific stripes to disappear from the tectum of three-eyed frogs, although the arbors became much smaller and had fewer branches. They concluded that protein kinase activity is not required for selective stabilization of coactive inputs in this case. If this were true, LTP would be shown to be a very different phenomenon from synapse stabilization, sharing only the common starting point of NMDA receptor activation. In fact, Cline and Constantine-Paton have argued that the first step in stabilization is the slowly decaying potentiation (SDP) that in hippocampus is independent of kinase activation.[57]

In recent experiments, I have tested the effects of the phorbol ester TPA in goldfish during the period of sharpening with results that suggest a different conclusion. Whether the TPA was placed into the eye (six injections of 2 μl of 100 μM over 2 weeks) or infused into the tectum with minipumps (3 μl/day for 2 weeks), the projection failed to undergo activity-driven retinotopic sharpening. The resulting projections were much like those seen with NMDA blockers with enlarged multiunit receptive fields (unpublished results). In the eye, the TPA could have stimulated phosphorylation of substrate proteins such as GAP-43 that are transported to tectum, and probably also greatly downregulated C kinase expression in ganglion cells. In tectum, it might have triggered phosphorylation of these or other proteins involved in many different processes. In controls, TPA was injected into the opposite eye without effect, supporting the conclusion that we have two separate sites of action. In addition, intracranial H-7 and sphingosine both blocked sharpening. Although these results are preliminary, it is striking that the results appear to be quite different from those reported for maintenance of the eye-specific stripes. Indeed, retinotopic sharpening, because it is such an active process, may be much more sensitive to the block of kinases than the stripe assay, which in fact only requires the maintenance of preexisting stripes. The fact that the kinase blockers caused the arbors to be smaller suggests that the kinase blockers might have interfered with growth, preventing branches from entering the interstripe zones. This possibility, although it would be consistent with my findings, presents a complication of the interpretation, and further experiments are underway to examine this question.

ACKNOWLEDGMENTS

I thank Drs. Suzannah Tieman and Greg Lnenicka for comments on the manuscript.

REFERENCES

1. BEAR, M., A. KLEINSCHMIDT & W. SINGER. 1990. J. Neurosci. **10:** 909-925.
2. CLINE, H. T., E. A. DEBSKI & M. CONSTANTINE-PATON. 1987. Proc. Natl. Acad. Sci. USA **84:** 4342-4345.
3. SCHMIDT, J. 1990. J. Neurosci. **10:** 233-246.
4. SCHERER, W. J. & S. B. UDIN. 1989. J. Neurosci. **9:** 3837-3843.
5. KOMATSU, Y., K. FUJII, J. MAEDA, H. SAKAGUCHI & K. TOYOMA. 1988. J. Neurophysiol. **59:** 124-141.
6. KIMURA, F., A. NISHIGORI, T. SHIROKAWA & T. TSUMOTO. 1989. J. Physiol. **414:** 125-144.
7. STRYKER, M. P. & W. A. HARRIS. 1986. J. Neurosci. **6:** 2117-2133.
8. MEYER, R. L. 1982. Science **218:** 589-591.
9. BOSS, V. C. & J. T. SCHMIDT. 1984. J. Neurosci. **4:** 2891-2905.
10. REH, T. A. & M. CONSTANTINE-PATON. 1985. J. Neurosci. **5:** 1132-1143.
11. SHATZ, C. & M. P. STRYKER. 1988. Science **242:** 87-89.
12. DUBIN, M. W., L. A. STARK & S. M. ARCHER. 1986. J. Neurosci. **6:** 1021-1036.
13. ARCHER, S. M., M. DUBIN & L. A. STARK. 1982. Science **217:** 743-745.
14. MEYER, R. L. 1983. Dev. Brain Res. **6:** 292-298.
15. SCHMIDT, J. T. & D. L. EDWARDS. 1983. Brain Res. **269:** 29-39.
16. COOK, J. E. & E. C. C. RANKIN. 1986. Exp. Brain Res. **63:** 421-430.
17. CLINE, H. T. & M. CONSTANTINE-PATON. 1989. Neuron **3:** 413-426.
18. SCHMIDT, J. T. & M. BUZZARD. 1991. Manuscript in preparation.
19. SPERRY, R. W. 1963. Proc. Natl. Acad. Sci. USA **50:** 703-709.
20. ADAMSON, J., J. BURKE & P. GROBSTEIN. 1984. J. Neurosci. **4:** 2635-2649.
21. SCHMIDT, J. T. 1991. *In* Advances in Neural and Behavioral Development. V. A. Casagrande, Ed. Vol. 4: in press. Ablex. Norwood, NJ.
22. SCHMIDT, J. T., J. C. TURCOTTE, M. BUZZARD & D. G. TIEMAN. 1988. J. Comp. Neurol. **269:** 565-591.
23. STUERMER, C. A. O. 1988. J. Comp. Neurol. **267:** 69-91.
24. SCHMIDT, J. T. & M. BUZZARD. 1990. J. Neurobiol. **21:** 900-917.
25. HUMPHREY, M. F. & L. D. BEAZLEY. 1982. Brain Res. **239:** 595-602.
26. MATSUMOTO, N., M. KOMETANI & K. NAGANO. 1987. Neuroscience **22:** 1103-1110.
27. SCHMIDT, J. T. & L. E. EISELE. 1985. Neuroscience **14:** 535-546.
28. EISELE, L. E. & J. T. SCHMIDT. 1988. J. Neurobiol. **19:** 395-411.
29. ARNETT, D. W. 1978. Exp. Brain Res. **32:** 49-53.
30. GINSBURG, K. S., J. A. JOHNSEN & M. W. LEVINE. 1984. J. Physiol. **351:** 433-451.
31. MASTRONARDE, D. N. 1989. Trends Neurosci. **12:** 75-80.
32. WILLSHAW, D. J. & C. VON DER MALSBURG. 1976. Proc. R. Soc. London Ser. B **194:** 431-445.
33. HARRIS, E. W., A. H. GANONG & C. W. COTMAN. 1984. Brain Res. **323:** 132-137.
34. NOWAK, L., P. BREGESTOVSKI, P. ASCHER, A. HERBERT & A. PROCHIANZ. 1984. Nature **307:** 462-465.
35. MACDERMOTT, A. B., M. L. MAYER, G. L. WESTBROOK. S. J. SMITH & J. L. BARKER. 1986. Nature **321:** 519-522.
36. HENLEY, J. M. & R. E. OSWALD. 1988. J. Neurosci. **8:** 2101-2107.
37. MCDONALD, J. W., H. T. CLINE, M. CONSTANTINE-PATON, W. F. MARAGOS, M. V. JOHNSON & A. B. YOUNG. 1989. Brain Res. **482:** 155-158.
38. LANGDON, R. & J. A. FREEMAN. 1986. Brain Res. **398:** 169-174.

39. IRIKI, A., C. PAVLIDES, A. KELLER & H. ASANUMA. 1989. Science **245:** 1385-1387.
40. TEYLER, T. J. 1989. J. Neurosci. Methods **28:** 101-128.
41. ABRAHAM, W. C., B. GUSTAFSSON & H. WIGSTROM. 1986. Neurosci. Lett. **70:** 217-222.
42. GUSTAFSSON, B., H. WIGSTROM, W. C. ABRAHAM & Y.-Y. HUANG. 1987. J. Neurosci. **7:** 774-780.
43. SASTRY, B. R., J. W. GOH & A. AUYEUNG. 1986. Science **232:** 988-990.
44. BRINK, D. L. & R. L. MEYER. 1986. Neurosci. Abstr. **12:** 438.
45. BENOWITZ, L. I. & J. T. SCHMIDT. 1987. Brain Res. **417:** 112-118.
46. LOVINGER, D. M., R. F. AKERS, R. B. NELSON, C. A. BARNES, B. L. MCNAUGHTON & A. ROUTTENBERG. 1985. Brain Res. **343:** 137-143.
47. TSUMOTO, T., K. HAGIHARA, H. SATO & Y. HATA. 1987. Nature **327:** 513-514.
48. ARTOLA, A. & W. SINGER. 1987. Nature **330:** 649-652.
49. CLINE, H. T. & M. CONSTANTINE-PATON. 1990. J. Neurosci. **10:** 1197-1216.
50. GU, Q., M. BEAR & W. SINGER. 1989. Dev. Brain Res. **47:** 281-288.
51. UDIN, S. B. & W. J. SCHERER. 1990. Ann. N.Y. Acad. Sci. This volume.
52. SHASHOUA, V. E. 1990. Ann. N.Y. Acad. Sci. This volume.
53. SCHMIDT, J. T. & V. E. SHASHOUA. 1988. Brain Res. **446:** 269-284.
54. ERRINGTON, M. L., M. A. LYNCH & T. V. P. BLISS. 1987. Neuroscience **20:** 279-284.
55. WILLIAMS, J. H. & T. V. P. BLISS. 1988. Neurosci. Lett. **88:** 81-85.
56. OKADA, D., S. YAMAGISHI & H. SUGIYAMA. 1989. Neurosci. Lett. **100:** 141-146.
57. MEFFERT, M. K., K. D. PARFITT, V. A. DOZE, G. A. COHEN & D. V. MADISON. 1990. Ann. N.Y. Acad. Sci. This volume.
58. CLINE, H. T. & M. CONSTANTINE-PATON. 1990. Neuron **4:** 899-908.

Experience-Dependent Formation of Binocular Maps in Frogs

Possible Involvement of *N*-Methyl-D-Aspartate Receptors

SUSAN B. UDIN AND WARREN J. SCHERER

Department of Physiology
State University of New York at Buffalo
Buffalo, New York 14214

INTRODUCTION

During early life, the developing central nervous system can be profoundly affected by sensory input. Abnormal experience, such as visual deprivation, can alter the way that axons make their connections. In contrast to the dramatic susceptibility of many growing systems to altered activity, plasticity in the mature central nervous system is far more restricted, and many sets of axons seem to have no capacity for readjusting their connections in response to altered input. In this paper, we will describe one such system, the ipsilateral visual projection to the optic tectum of the frog, *Xenopus laevis,* in which alterations in eye position can induce rearrangements of the ipsilateral map during a critical period of development but not, under normal circumstances, at later ages. We will present evidence that the *N*-methyl-D-aspartate (NMDA)-type glutamate receptor plays a key role in mediating the influence of vision on the connectivity of tectal maps, and we shall offer some data indicating that the usual loss of plasticity with age can be reversed by chronic application of NMDA.

THE IPSILATERAL PATHWAY

Each lobe of the tectum receives a visuotopic binocular input. Input from the ipsilateral eye is relayed to the tectum via the nucleus isthmi (NI), and this ipsilateral map is normally in register with the map that is relayed from the contralateral eye via the optic nerve (FIG. 1).[1–3] This pattern of connections was determined first with electrophysiological recording methods and later confirmed using horseradish peroxidase tracing. An example of the sort of maps that are used for assessing binocular connections to the tectum is illustrated in FIGURE 2. In order to detect action potentials of retinotectal (contralateral eye) or isthmotectal (ipsilateral eye) terminals, a metal microelectrode is placed in the superficial neuropil of the tectum, where the retinotectal and isthmotectal axons arborize. Activity is then elicited by moving an appropriate visual stimulus along the inner surface of a calibrated hemisphere. The location at which the object activates the axons is noted. By making an orderly sequence of penetrations in a grid pattern on the tectum, one can assemble a picture of the way in which different retinal positions are represented on the tectum.

THE CHALLENGE OF MATCHING MAPS DURING DEVELOPMENT

How do the ipsilateral and contralateral maps come into register during development? This is no simple issue for *Xenopus,* because that species exhibits an unusual pattern of eye migration during metamorphosis and juvenile stages. The eyes start off facing laterally, with very little binocular overlap (FIG. 3). Near the end of the tadpole stage of life, the eyes begin to move dorsally and frontally; this movement continues until the binocular field is about 170° in extent. During this prolonged period of change, the relationship between the visual fields of the two eyes changes and the locations of isthmotectal arbors change correspondingly. The caudal shift of isthmotectal terminals keeps the receptive fields of the two eyes' projections to the tectum in register,[4] but only if normal visual input is available during this critical period.[5] Thus, axons bringing ipsilateral input to a given tectal location will respond to a visual stimulus at the same

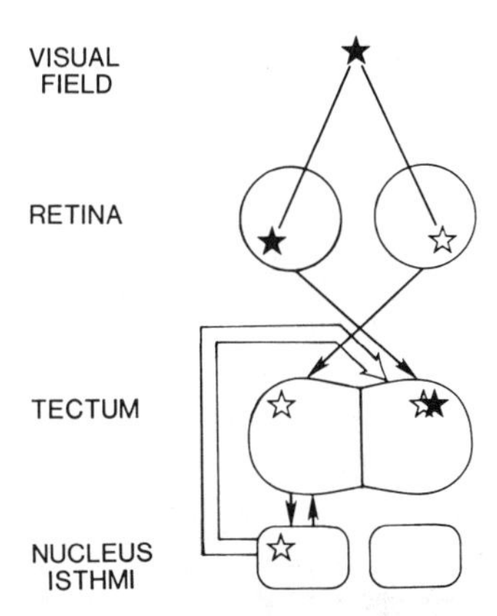

FIGURE 1. Pathway for binocular input to right lobe of tectum.

position that activates the axons bringing contralateral input to that location. The receptive field sizes tend to be large, with the average ipsilateral receptive field being 13.7° in diameter in adults, and with the average contralateral receptive field being 10.3°.[6]

MECHANISMS UNDERLYING TOPOGRAPHY OF THE IPSILATERAL PROJECTION

Axons from the nucleus isthmi begin to enter the tectum quite early, weeks before the eyes begin to change position.[7] When they arrive, the axons select mediolateral positions along which to enter the tectum.[8] The mediolateral location chosen by a particular axon is appropriate for its later growth, and each axon normally will not deviate more than a few hundred microns medial or lateral from this point during

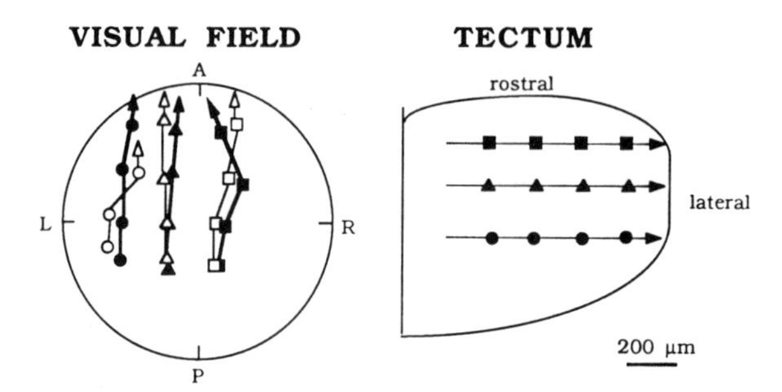

FIGURE 2. Normal binocular maps. Each symbol on tectum (right) shows an electrode penetration site. The frog was positioned beneath a calibrated hemisphere for presentation of visual stimuli. The receptive field centers recorded from the terminals of retinotectal and isthmotectal axons progress in an orderly fashion from posterior to anterior visual field as one moves from medial to lateral tectum and from right to left as one moves from rostral to caudal. Filled symbols: contralateral eye fields; open symbols: ipsilateral eye fields.

development (FIG. 4). This choice of mediolateral position may be governed by chemoaffinity cues, much the same as those that assist the developing retinotectal projection to establish initial order.[9] Timing may also be important, because the first-born isthmotectal axons enter the first-born part of the tectum, and successively newer axons enter progressively newer regions of the tectal border.[7] Once the axons reach the tectum, they normally grow straight toward the caudal border; each axon shows remarkably little wandering outside of its "corridor."[8] Prior to about stage 60 (when the tadpoles

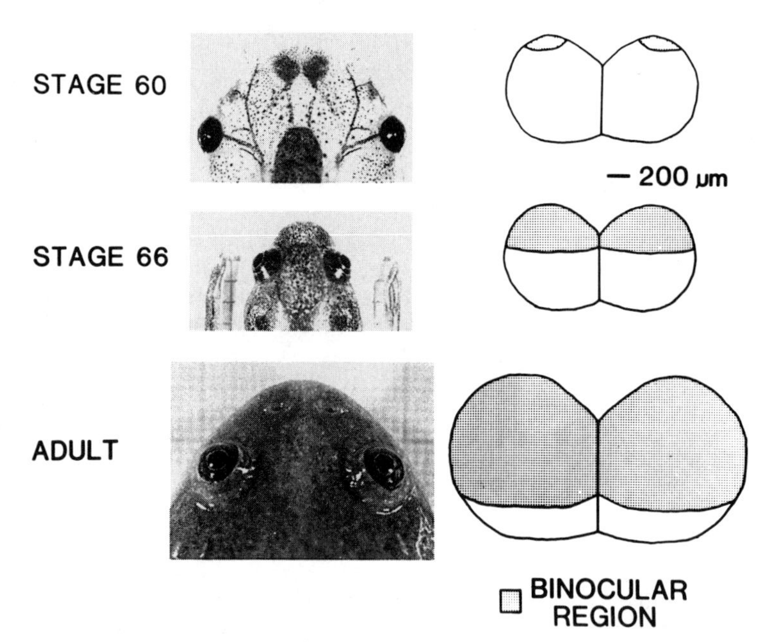

FIGURE 3. Changes in eye position (*left*) increase the size of the binocular portion of the maps on the tectum (*right*).

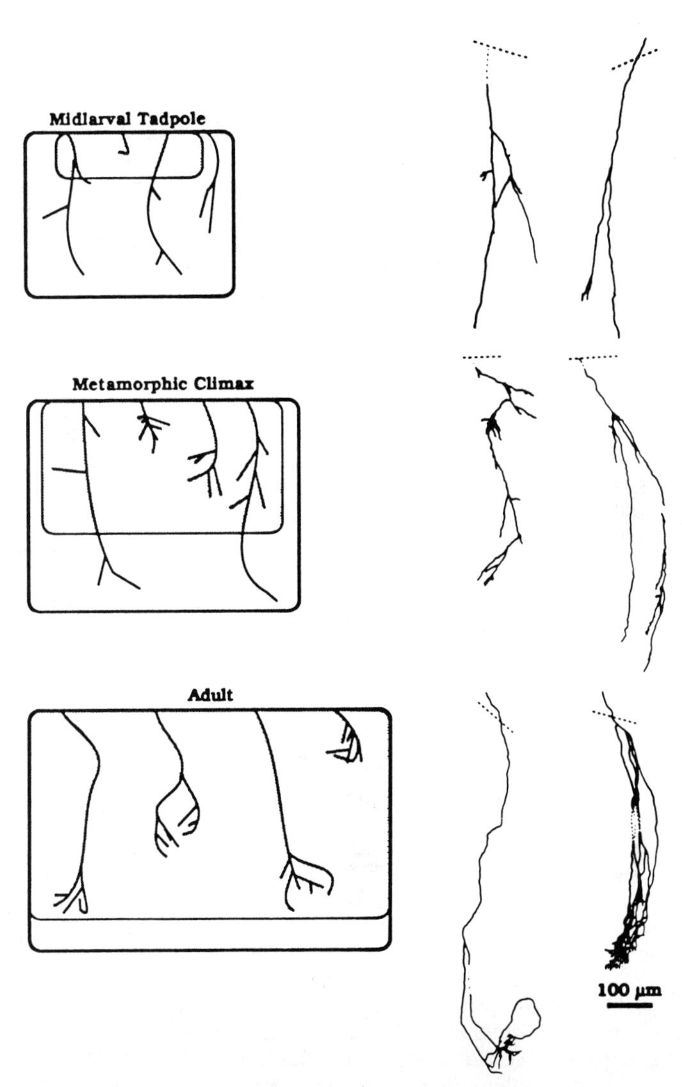

FIGURE 4. The sketches on the left illustrate a range of typical isthmotectal axons found at various stages of normal development. Note that the axons extend over a large proportion of the rostrocaudal extent of the tectum but rarely span more than 20% of the mediolateral extent. The small rectangles indicate the binocular zones. The pairs of axons shown at right are camera lucida tracings of horseradish peroxidase-filled axons at the corresponding stages.

are beginning the stages of metamorphic climax), the eyes are still facing laterally. Because of this lack of binocularity, the receptive fields of the isthmotectal axons do not match any of the receptive fields of the retinotectal axons that occupy the tectum, and the majority of isthmotectal axons really do not have a proper destination. At this point, the isthmotectal axons have few branches. This situation changes when the remarkable shift in eye position brings about an extensive overlap of the two monocular visual fields. The previously tiny binocular zone rapidly increases in extent. This expansion of binocular overlap means that more and more isthmotectal axons' receptive fields match the fields of the retinotectal projection from the opposite eye. The axons begin to produce terminal arbors, and electrophysiological recording indicates that the isthmotectal axons are forming terminals at locations where their receptive fields overlap the retinotectal receptive fields.[4] The zone of binocular overlap continues to expand through the first few years of postmetamorphic life (most rapidly in the first month or so), and the binocular zone of the tectum correspondingly increases, all the while maintaining matching maps from the two eyes. These changes take place primarily after the completion of both cell proliferation and cell death in the nucleus isthmi and tectum.[7]

The model that has been invoked to explain how the growing axons can arborize appropriately during development and shift in the proper direction at the proper time

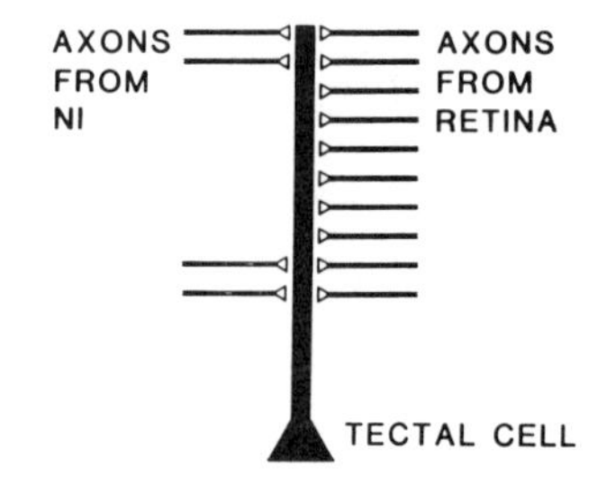

FIGURE 5. Isthmotectal and retinotectal axons probably converge on common postsynaptic dendrites, although individual cells do not necessarily receive input from axons in all laminae represented here.

is a variant of the "Hebb synapse."[10] The key aspect of this model is that correlated pre- and postsynaptic activity stabilizes the connections between the coactive elements. In the binocular system of *Xenopus,* the crucial interactions are between the retinotectal and isthmotectal axons, by way of common tectal cell dendrites (FIG. 5). According to this model, isthmotectal axons initially make branches at random, even in inappropriate regions of the tectum, and those branches form temporary synapses (Udin, Fisher, and Norden, unpublished observations). When an isthmotectal axon happens to make connections onto a dendrite that receives input from retinotectal axons with the same receptive field location, then the isthmotectal axon will be stabilized. When an isthmotectal axon connects to a dendrite receiving noncorresponding retinotectal input, then the isthmic connection is weakened and is likely to be retracted.

A major piece of evidence in favor of this hypothesis is that the maintenance of map congruence is critically dependent on visual input during the critical period.[5,6,11] Dark-reared frogs have normally organized contralateral (retinotectal) maps, but their ipsilateral maps are poorly organized and are misaligned with respect to the contralateral maps. This observation is one of the bases for hypothesizing that visual activity provides the cues that signal to isthmotectal axons if they have found correct terminal zones, with "correctness" being defined by temporal correlations of the visually elicited firing of the retinotectal and isthmotectal axons. To coin a phrase, "Axons that fire together, stay together."

This idea gets further support from the observation that abnormal alignment of the eyes can actually cause a redirection of isthmotectal axonal growth. For example, by rotating one eye during mid-tadpole life, one can induce the ipsilateral projection from the normal eye to rotate by a corresponding amount, thus bringing the ipsilateral map back into alignment with the contralateral map.[12,13] Similarly, the ipsilateral map corresponding to the rotated eye will shift to come into register with the contralateral map from the normal eye. These adjustments restore congruence in both tecta, even though one eye is abnormally positioned. The isthmotectal axons thus can be induced to terminate in regions of the tectum that they normally would never even enter.

In order to understand this phenomenon, consider first what happens to the retinotectal projection of the rotated eye. This manipulation does not change the connectivity of the retinotectal axons. The rotated position of the eye, however, causes the ganglion cells to relay a rotated map (FIG. 6). For example, a cell in the original nasal part of the right retina that used to relay information about the arrowhead will instead relay information about the tail of the arrow after the eye has been rotated by 180°. What happens to the input from the normally oriented left eye? The retinotectal map from the left eye to the right tectum is normal, as is the tectoisthmic map. The axons from the right nucleus isthmi, however, do not have a normal arrangement: they are rerouted so that an isthmic axon that has a receptive field corresponding to the arrowhead now will project to caudal tectum rather than to rostral tectum. This new topography brings the isthmotectal map into register with the retinotectal map. Similar rearrangements bring the ipsilateral map from the rotated right eye into register with the normal retinotectal map on the right tectal lobe.

What is the evidence for such rearrangements of connections? First, electrophysiological recordings show altered ipsilateral topography. No matter what the degree of misalignment of the two eyes, each ipsilateral map conforms to the contralateral map on each tectum (FIGS. 7A & 7B). Second, matched focal injections of horseradish peroxidase in normal frogs and in frogs with unilateral eye rotation back-label cells in different locations in the NI.[13] In addition, anterograde labeling of isthmotectal cells after injections of horseradish peroxidase demonstrate that eye rotation induces the axons to take very abnormal routes in the tectum.[14] Instead of the usual orderly rostrocaudal trajectories of normal isthmotectal axons, the axons in experimental frogs exhibit considerable criss-crossing and turning en route to their final sites of termination.

THE ROLE OF THE NMDA RECEPTOR

What is the underlying mechanism that allows the developing ipsilateral map to stay in register with the contralateral map during the dramatic changes of eye position characterizing normal development and to change so radically in response to alterations in relative eye position? As mentioned above, correlations of activity could provide the information necessary to communicate to a branch of an isthmotectal axon with a given receptive field location as to whether it is synapsing in the same tectal region as retinotectal terminals with the same receptive field locations. Increasing evidence points to the NMDA-type glutamate receptor as the mediator of this communication. These correlations probably involve convergence of glutamatergic retinotectal terminals[15–17]

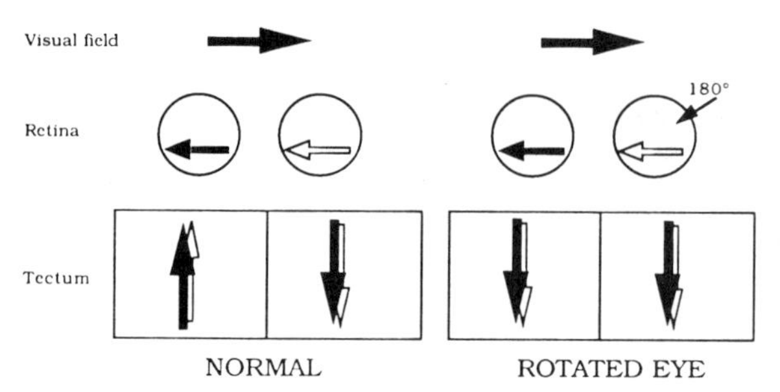

FIGURE 6. Schematic showing the changes in representation of the visual field on the tectum resulting from rotation of the right eye in a tadpole.

and cholinergic isthmotectal terminals[18,19] onto tectal cell dendrites. The role of glutamate is of particular interest in this regard because the NMDA-type glutamate receptor is a pivotal component of activity-dependent synaptic modification in the hippocampus,[20,21] visual cortex,[22,23] and tectum[24–26] (see reference 27 for a review).

The NMDA receptor is one of at least four receptors that bind glutamate in the nervous system. It is named for the agonist, *N*-methyl-D-aspartate, which binds to it with high selectivity. The NMDA receptor has two notable properties: first, that it opens only when the appropriate ligand binds to it *and* it is in a depolarized environment,[28] and second, that it admits calcium ions.[29] If a stimulus causes release of glutamate onto a dendrite whose membrane is near resting potential, then the NMDA channels are blocked by magnesium ions.[28] However, if the same input occurs when other excitatory inputs are active as well, then the local depolarization ejects the magnesium ion and unblocks the NMDA channel. Single synaptic events are unlikely to generate sufficient depolarization to allow NMDA channels to open. However, if several synapses are active within a short period of time, the temporal and spatial summation of their effects depolarizes the dendrite enough to allow the NMDA receptor channel to open. The resulting calcium flux is no doubt of great significance because of the many intracellular processes that are triggered by or dependent upon the free calcium concentration. Some of these processes, such as the activation of calcium/calmodulin kinase or protein kinase C, could result in synaptic stabilization.

In most systems where the NMDA receptor has been implicated in activity-dependent modification of connections, it is the glutamatergic synapses themselves that are affected. In the binocular system of *Xenopus,* however, activation of NMDA receptors by glutamatergic retinotectal axons may trigger processes that stabilize not only those synapses but also other synapses onto the same dendrites. If, as seems likely, isthmotectal axons terminate on the same tectal cells that receive direct retinotectal input, then isthmotectal synapses could be influenced by changes occurring within the dendrite. For example, NMDA could trigger the dendrite to release some trophic substance that would be available to all inputs to the dendrite. In order for such a mechanism to selectively stabilize "appropriate" isthmotectal axons, a further postulate is necessary, namely that the recent history of activity of the isthmotectal axon itself determines whether it is susceptible to the trophic influence of the dendrite. This mechanism would predict that some event, such as calcium influx into the isthmotectal synapse, would prime it to be stabilized by the trophic substance. If the isthmotectal input to the dendrite failed to be active during the proper time window (as would tend to occur if the isthmotectal axon were in the wrong location relative to the contralateral map), then it would fail to be stabilized; perhaps it even would be actively destabilized. Over time, such a mechanism would promote the formation of matching maps.

In order to test the possibility that the NMDA receptor is required for matching of the ipsilateral and contralateral maps in the tectum of *Xenopus,* we have investigated the effects of blocking the NMDA receptor during the critical period.[30] We chose to perform this study using animals with a rotated eye rather than otherwise normal frogs because we wanted to be able to distinguish the roles of activity mechanisms from chemoaffinity cues. Because the blockage of NMDA receptors seemed unlikely to interfere with the latter, we suspected that such treatment might produce a result similar to that observed in dark-reared frogs, in which the isthmotectal projection forms a map that retains a recognizable degree of order, particularly along the mediolateral tectal axis, probably reflecting the initial order the axons display when they first enter the tectum.[5] After eye rotation, it is a straightforward matter to determine whether the isthmotectal axons have responded to the altered visual input due to the eye rotation, because such axons must ignore chemoaffinity cues and enter regions of the tectum where they normally would never be found.

We compared the maps in frogs that had had 90° clockwise rotations of the left eye as late tadpoles, followed by several months of treatment with either NMDA itself or the NMDA antagonists DL-2-amino-5-phosphonovaleric acid (APV) or 3-(($\pm$)-2-carboxypiperazin-4-yl)-propyl-1-phosphonic acid (CPP).[30] The drugs were administered by implanting slabs of the slow-release polymer Elvax, which had been prepared with 0.1 mM of the appropriate drug or, as a control, with distilled water. As FIGURE 7B shows, the untreated frogs showed the expected rearrangements of their ipsilateral maps. In contrast, the APV produced a different outcome (FIGS. 7C & 7D). The contralateral map was rotated by 90°, in accord with the rotation of the left eye, but

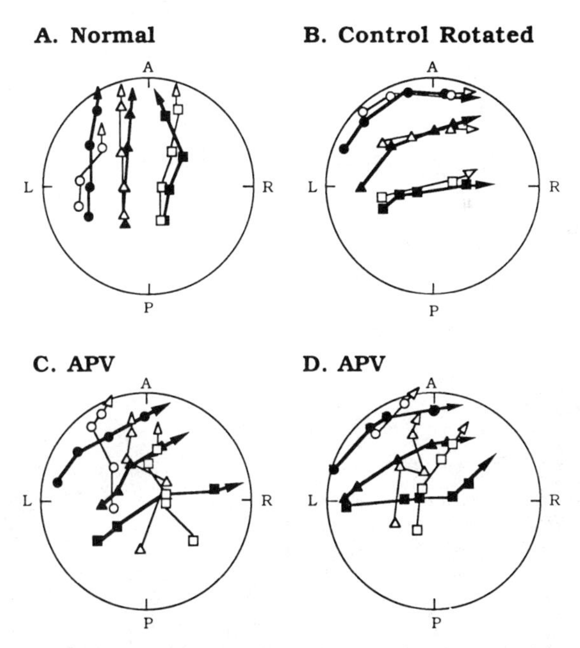

FIGURE 7. (A & B) Rotation of the left eye by 90° in an otherwise normal tadpole induces the ipsilateral map by rotate by 90° to match the contralateral map. **(C & D)** Chronic treatment with the NMDA antagonist APV prevents the ipsilateral map from changing its orientation to match the contralateral map.

the ipsilateral map had not rotated to match. Instead, the ipsilateral map showed a recognizable, albeit degraded, topography, and that topography was essentially normal in orientation. (These ipsilateral maps look very much like the ipsilateral maps seen in dark-reared frogs.) CPP also prevented congruence of the binocular maps but had the as yet unexplained additional effect of sometimes disrupting the contralateral map, which itself is still slowly shifting (see below). The maps from the NMDA-treated frogs were indistinguishable from untreated ones. This result thus indicates that blocking the NMDA receptor does not prevent isthmotectal axons from forming a crudely organized map, but it does prevent them from responding to altered visual input by shifting their connections in such a way as to bring the ipsilateral map into register with the contralateral map.

The use of slow-release vehicles is complicated by the fact that it is difficult to assess the effective concentration of the administered drug in the tissue. In particular, one must worry whether the dosage is so high as to produce unwanted effects such as total suppression of activity. In order to assess whether the dosage of APV in our implants could be so high as to block retinotectal transmission, we performed some experiments to determine whether retinotectal axons still activate tectal cells when APV is being administered. We implanted slices of APV-impregnated Elvax into the right tecta of young juveniles, waited for 2 weeks, and then recorded the ipsilateral visual activity in the left tectum (FIG. 8). Because the only pathway for ipsilateral input to the left tectum depends upon transmission through the right tectum, successful recording of ipsilateral input to the left lobe would prove that at least some components of the retinotectal projection are still able to activate tectal cells in the presence of APV on the right lobe. We did in fact find that the ipsilateral units could be recorded with no difficulty. In order to have a quantitative measure of the responsiveness of the units, we stimulated the units with flashes of light and collected the resulting spike data. The results indicated that there is no statistically significant difference in the response of the tectum after chronic treatment with APV when compared with age-matched controls (FIG. 9).[31]

THE CRITICAL PERIOD

Xenopus frogs normally lose plasticity during the first few months after metamorphosis.[32] This phenomenon is demonstrated by comparing the results of rotating one

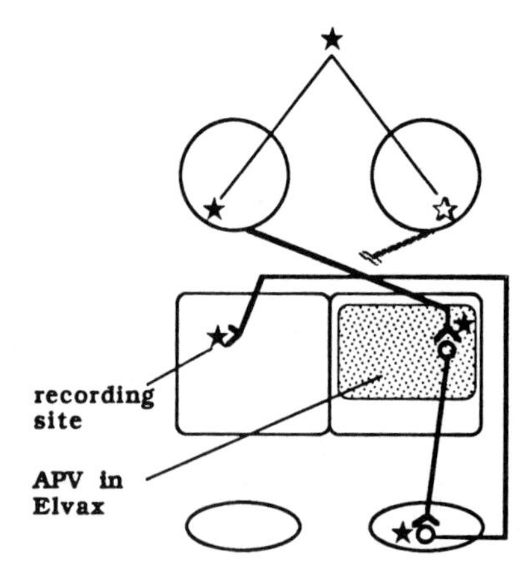

FIGURE 8. Method for assessing effects of chronic drug treatment of the right lobe of tectum by recording tecto-isthmo-tectal output to the left lobe.

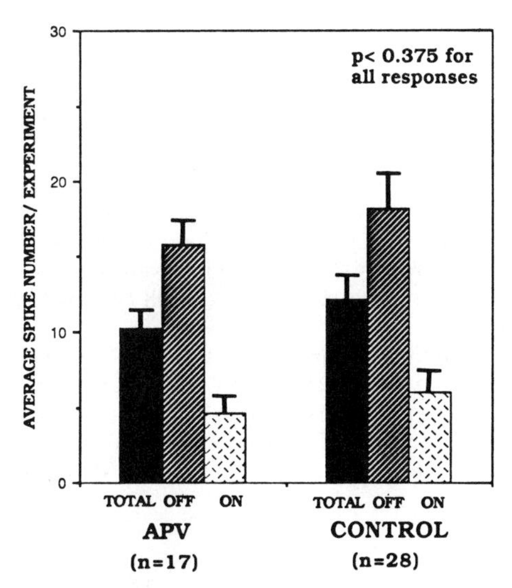

FIGURE 9. Flashes of light elicit normal amounts of firing from ipsilateral units in young juvenile frogs chronically treated with APV. The TOTAL columns indicate the average of OFF and ON.

eye by 90° or 180° at different ages. Such rotations performed before eye migration begins lead to complete compensation by the isthmotectal projection. However, eye rotation performed just a few days later, at the end of metamorphosis, will elicit compensation only for the 90° rotations. Some low-level response to eye rotation does persist, however. For example, we find that 90° eye rotation at 8 months, followed by 3 months of survival, sometimes leads to partial disorganization of the ipsilateral map. Some remnants of plasticity might be expected, given that the *Xenopus* eye continues to grow throughout postmetamorphic life, leading to a continued, gradual "sliding" of the retinal projection across the tectum, for which the ipsilateral projection must compensate.

What could account for the loss of the magnificent degree of plasticity displayed by tadpoles as the animals mature? Perhaps there is a reduction in the number or activity of NMDA receptors. Our result with NMDA receptor blockers showed that we could, in a sense, mimic the normal loss of plasticity by diminishing NMDA receptor function, so we reasoned that normal loss of plasticity could result from the loss of NMDA receptor function. There are several ways in which this could come about, such as a decrease in NMDA receptor number, a change in NMDA channel kinetics or voltage dependence,[33] or an increase in the relative strength of inhibitory circuits. Because there are already data indicating that NMDA receptor function decreases with age in other systems,[34–37] we embarked on a study to determine whether NMDA treatment could restore plasticity in frogs that had passed the normal end of the critical period.[38] We implanted slabs of slow-release Elvax polymers containing 0.1 or 0.4 mM NMDA under the tectal pia in frogs with 90° rotation of the left eye. Other frogs with eye rotations received drug-free implants as controls. As FIGURE 10A shows, the maps recorded 3 months later in the drug-free tecta remained out of alignment, but the maps from the NMDA-treated tecta (FIG. 10B) showed that the NMDA had indeed restored plasticity. Excellent registration was found between the ipsilateral and contralateral maps, whereas the maps from the control frogs showed essentially no congruence. This result supports our hypothesis that NMDA receptor function may play a role in normal developmental loss of plasticity.

We used the same sort of quantitative analysis described above to test whether chronic treatment with NMDA altered the responsiveness of the tectum. We found that it did do so: the chronic treatment for 3 weeks yielded about 35% more responses per stimulus than normal. We do not yet know whether the increase in activity and the restoration of plasticity are causally related, but we suspect not, because there is no difference between the responsiveness of the untreated tectum of critical period juvenile frogs and the untreated tectum of older frogs that have lost their plasticity. Thus, for normal frogs, we see no changes in overall activity that are correlated with changes in plasticity.

There are several plausible interpretations of the possible mechanism by which the NMDA could alter the tectum in such a way as to promote plasticity. As stated above, it is possible that there is a normal decrease in NMDA receptors during development. Such a decrease could bring the NMDA-receptor-mediated calcium fluxes below a threshold level necessary to trigger changes in synaptic stabilization. In that case, the

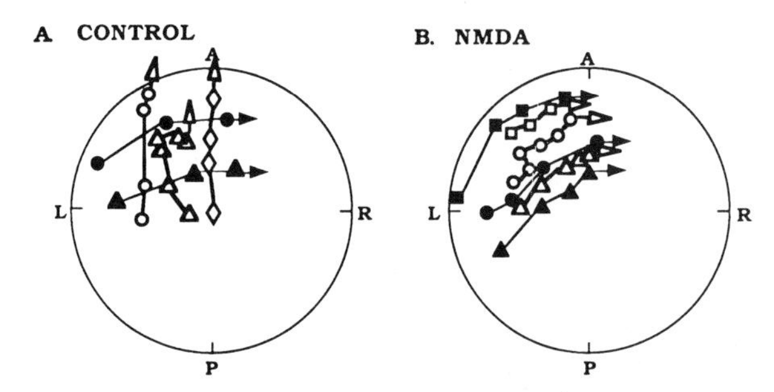

FIGURE 10. (**A**) Rotation of one eye in a frog at 8 months past metamorphosis usually does not lead to realignment of the ipsilateral map, as assessed by recording 3 months later. (**B**) The same manipulation leads to realignment of the ipsilateral map if the tectum is chronically treated with NMDA during the period after the eye rotation. Filled symbols: contralateral units; open symbols: ipsilateral units.

addition of exogenous NMDA might help to keep open any NMDA receptors that are opened by visual input for a bit longer than they would be otherwise and thus allow extra influx of calcium, bringing the calcium level to threshold.

A quite different sort of effect of chronic NMDA has been suggested by Cline and Constantine-Paton,[39] based on the work of Debski *et al.*[40] They find that chronic NMDA treatment of *Rana pipiens* reduces the responsiveness of the tectum to additional application of NMDA. In order to explain how chronic NMDA treatment can promote activity-dependent sharpening of ocular dominance bands in surgically produced three-eyed frogs, they suggest that this desensitization makes the criteria for activating the receptors more stringent in that the correlation of activity of the retinotectal terminals on a given dendrite must be quite strong in order to open the receptors. Thus, only the best matches are retained, and the less well correlated connections are not. This model then implies that normal levels of NMDA receptor function are sufficient for moderate sharpening of maps, whereas reductions in receptor sensitivity, perhaps coupled with exogenous NMDA, lead to increased precision. In interpreting our results on restoration of plasticity with NMDA, we could apply this model if we assumed that NMDA receptors are actually more easily triggered in postcritical-period frogs than in younger ones or that the subsequent events triggered by NMDA receptor activation occur more readily in older animals. Thus, in older animals, eye rotation would not lead to successful reorientation of the ipsilateral map because many inappropriate connections made by isthmotectal axons would be capable of being stabilized. Chronic NMDA and resulting desensitization of the NMDA receptors would permit reorganization of the map by decreasing the likelihood that inappropriate branches would persist. Only branches that had successfully located sites with appropriate retinotectal activity would be stabilized.

Another way of manipulating the duration of plasticity in *Xenopus* is to rear the animals in the dark. Dawes *et al.* have shown that this manipulation remarkably prolongs plasticity,[41] even two years beyond the normal end of the critical period (S. Grant, personal communication). Although dark-rearing in mammals has been found to prolong plasticity,[42] such plasticity seems to be confined primarily to physiological changes in synaptic efficacy rather than to structural rearrangements of the locations of connections.[43] In contrast, the prolonged plasticity in *Xenopus* involves the capacity for alteration of axonal arbor locations, because rotation of an eye of a dark-reared frog leads to rearranged maps, which come about only when axons are able to grow to regions of the tectum where they normally never are found. It will be of great interest to learn whether dark-rearing keeps the NMDA receptors as they normally are during the critical period.

THE PROBLEM OF SIMULTANEITY VERSUS CORRELATION OF FIRING

The Hebbian mechanism has been invoked in several other systems, such as the retinotectal projection of frogs and fishes and the geniculocortical projection of mammals.[44–46] In these systems, the matching mechanism is generally assumed to depend upon essentially simultaneous firing of the afferent axons. In contrast, in the binocular system of *Xenopus,* there is a time delay of as much as 13 msec between the arrival of the first retinotectal impulses and the onset of isthmotectal firing.[47] Is this a major obstacle for the model we have presented? Probably not, for several reasons, which we will present next.

First, we have examined the possibility that the correlations that take place in the tectum are not between the isthmotectal axons and the retinotectal axons, but instead involve another set of axons that we have not yet described, namely the uncrossed isthmotectal projection. The NI projects not just to the opposite tectum (via what we will refer to as the crossed isthmotectal pathway), but also to the tectum on the same side of the brain (via the uncrossed isthmotectal pathway) (FIG. 1).[1] The uncrossed projection could serve as a delay line, because visual input to the tectum via the retinotectal projection is relayed to the NI, which then sends it back in a reciprocal topographic pattern. Thus, the contralateral eye's activity arrives at the same place in the tectum twice, first via the optic nerve and second via the uncrossed isthmotectal pathway. This isthmic projection terminates in the superficial layers of the tectum and thus could mediate interactions with the crossed isthmotectal axons. The relative delay between the firing of the two isthmic inputs is probably only a few milliseconds, because both are subjected to comparable synaptic delays at the retinotectal and tectoisthmic synapses. In order to test whether the crossed isthmotectal axons (ipsilateral eye input) are dependent upon interactions with the uncrossed isthmotectal axons (contralateral eye input), one NI was ablated in tadpoles that had had one eye rotated.[48] Thus, the crossed axons from the spared NI would encounter intact retinotectal input but would not be able to interact with uncrossed isthmotectal axons. Electrophysiological mapping several months later revealed that the crossed isthmotectal fibers had been able to shift in response to the eye rotation, despite the absence of the reciprocal set of fibers connecting the tectum and NI. This result strengthens the hypothesis that the crucial interactions are between retinotectal and crossed isthmotectal axons that converge on common postsynaptic targets, but the possibility still remains that thalamotectal axons or intratectal collaterals could play a role.

There are other reasons for suspecting that the 13-msec delay is not an insurmountable obstacle to our working hypothesis. First, the kinetics of the NMDA receptor are relatively slow: it is slow to open and it can remain open for tens or even hundreds of milliseconds. For example, in hippocampal CA1 cells, associative long-term potentiation may be induced if paired stimuli are separated by as much as 40 msec.[49] Thus, the burst of spikes coming from a retinotectal input could open NMDA channels and keep them open for at least as long as it takes for isthmotectal firing to begin. Second, the calcium influx resulting from NMDA receptor opening could persist beyond the retinotectal firing duration. Moreover, the processes resulting from calcium influx, such as phosphorylation of proteins, could be quite prolonged. Moreover, the ultimate effect of stabilizing or destabilizing the isthmotectal terminal may be mediated by a retrograde messenger released by the dendrite, and such a retrograde signal would also have the potential for persisting for tens of milliseconds.

SYNAPTIC RELATIONSHIPS IN THE TECTUM

In order to understand the possible means by which isthmotectal axons are influenced by retinotectal activity, we have investigated the connections of the isthmic axons that bring input to the tectum from the ipsilateral eye. By filling those axons with horseradish peroxidase and using electron microscopy, we have determined that isthmotectal axons make very ordinary asymmetric synapses onto dendrites.[50] We see no signs of more exotic connections, such as axo-axonal synapses between retinotectal and isthmotectal terminals (although the lack of morphologically identifiable structures certainly does not prove that there are no nonclassical interactions). Thus, our model is in accord with what we currently know about the ultrastructure of the tectum.

CONCLUSION

We have presented evidence that the NMDA receptor is a crucial component of the process by which visual activity promotes the orderly matching of binocular maps in the visual system of developing *Xenopus* frogs, and we suggest that changes in NMDA receptor function may underlie the normal loss of plasticity that occurs with maturation.

REFERENCES

1. GRUBERG, E. R. & S. B. UDIN. 1978. Topographic projections between the nucleus isthmi and the tectum of the frog, *Rana pipiens.* J. Comp. Neurol. **179:** 487-500.
2. GROBSTEIN, P. & C. COMER. 1983. The nucleus isthmi as an intertectal relay for the ipsilateral oculotectal projection in the frog, *Rana pipiens.* J. Comp. Neurol. **217:** 54-74.
3. GLASSER, S. & D. INGLE. 1978. The nucleus isthmus as a relay station in the ipsilateral visual projection to the frog's optic tectum. Brain Res. **159:** 214-218.
4. GRANT, S. & M. J. KEATING. 1989. Changing patterns of binocular visual connections in the intertectal system during development of the frog, *Xenopus laevis.* I. Normal maturational changes in response to changing binocular geometry. Exp. Brain Res. **75**(1): 99-116.
5. GRANT, S. & M. J. KEATING. 1989. Changing patterns of binocular visual connections in the intertectal system during development of the frog, *Xenopus laevis.* II. Abnormalities following early visual deprivation. Exp. Brain Res. **75**(1): 117-132.
6. KEATING, M. J. & C. KENNARD. 1987. Visual experience and the maturation of the ipsilateral visuotectal projection in *Xenopus laevis.* Neuroscience **21:** 519-527.
7. UDIN, S. B. & M. D. FISHER. 1985. The development of the nucleus isthmi in *Xenopus.* I. Cell genesis and formation of connections with the tecta. J. Comp. Neurol. **232:** 25-35.
8. UDIN, S. B. 1989. The development of the nucleus isthmi in *Xenopus.* II. Branching patterns of contralaterally projecting isthmotectal axons during maturation of binocular maps. Vis. Neurosci. **2:** 153-163.
9. SPERRY, R. W. 1944. Optic nerve regeneration with return of vision in anurans. J. Neurophysiol. **7:** 57-69.
10. HEBB, D. O. 1949. The Organization of Behavior. Wiley. New York, NY.
11. KEATING, M. J. & J. FELDMAN. 1975. Visual deprivation and intertectal neuronal connections in *Xenopus laevis.* Proc. R. Soc. London Ser. B **191:** 467-474.
12. GAZE, R. M., M. J. KEATING, G. SZÉKELY & L. BEAZLEY. 1970. Binocular interaction in the formation of specific intertectal neuronal connexions. Proc. R. Soc. London Ser. B **175:** 107-147.
13. UDIN, S. B. & M. J. KEATING. 1981. Plasticity in a central nervous pathway in *Xenopus:* Anatomical changes in the isthmotectal projection after larval eye rotation. J. Comp. Neurol. **203:** 575-594.
14. UDIN, S. B. 1983. Abnormal visual input leads to development of abnormal axon trajectories in frogs. Nature **301:** 336-338.
15. DEBSKI, E. A. & M. CONSTANTINE-PATON. 1988. The effects of glutamate receptor agonists and antagonists on the evoked tectal potential in *Rana pipiens.* Soc. Neurosci. Abstr. **14:** 674.
16. LANGDON, R. B. & J. A. FREEMAN. 1987. Pharmacology of retinotectal transmission in the goldfish: Effects of nicotinic ligands, strychnine, and kynurenic acid. J. Neurosci. **7:** 760-773.
17. ROBERTS, P. J. & R. A. YATES. 1976. Tectal deafferentation in the frog: Selective loss of L-glutamate and γ-aminobutyrate. Neuroscience **1:** 371-374.
18. DESAN, P. H., E. R. GRUBERG, K. M. GREWELL & F. ECKENSTEIN. 1987. Cholinergic innervation of the optic tectum in the frog *Rana pipiens.* Brain Res. **413:** 344-349.
19. RICCIUTI, A. J. & E. R. GRUBERG. 1985. Nucleus isthmi provides most tectal choline acetyltransferase in the frog *Rana pipiens.* Brain Res. **341:** 399-402.
20. COLLINGRIDGE, G. L., S. J. KEHL & H. MCLENNAN. 1983. Excitatory amino acids in synaptic transmission in the Schaffer collateral-commissural pathway of the rat hippocampus. J. Physiol. (London) **334:** 33-46.
21. HARRIS, E. W., A. H. GANONG & C. W. COTMAN. 1984. Long-term potentiation in the hippocampus involves activation of *N*-methyl-D-aspartate receptors. Brain Res. **323:** 132-137.
22. KLEINSCHMIDT, A., M. F. BEAR & W. SINGER. 1987. Blockade of "NMDA" receptors disrupts experience-dependent plasticity of kitten striate cortex. Science **238:** 355-358.
23. BEAR, M. F., A. KLEINSCHMIDT, Q. GU & W. SINGER. 1990. Disruption of experience-

dependent synaptic modifications in striate cortex by infusion of an NMDA receptor antagonist. J. Neurosci. **10**(3): 909-925.
24. CLINE, H. T. & M. CONSTANTINE-PATON. 1989. NMDA receptor antagonists disrupt the retinotectal topographic map. Neuron **3:** 413-426.
25. SCHMIDT, J. T. 1990. Long-term potentiation and activity-dependent retinotopic sharpening in the regenerating retinotectal projection of goldfish: Common sensitive period and sensitivity to NMDA blockers. J. Neurosci. **10**(1): 233-246.
26. CLINE, H. T., E. A. DEBSKI & M. CONSTANTINE-PATON. 1987. *N*-Methyl-D-aspartate receptor antagonist desegregates eye-specific stripes. Proc. Natl. Acad. Sci. USA **84:** 4342-4345.
27. CONSTANTINE-PATON, M., H. T. CLINE & E. DEBSKI. 1990. Patterned activity, synaptic convergence, and the NMDA receptor in developing visual pathways. Annu. Rev. Neurosci. **13:** 129-154.
28. NOWAK, L., P. BREGESTOVSKI, A. H. P. ASCHER & A. PROCHIANTZ. 1984. Magnesium gates glutamate-activated channels in mouse central neurones. Nature **307:** 462-465.
29. MACDERMOTT, A. B., M. L. MAYER, G. L. WESTBROOK, S. J. SMITH, J. L. BARKER. 1986. NMDA receptor activation increases cytoplasmic calcium concentration in cultured spinal cord neurones. Nature **321:** 519-522.
30. SCHERER, W. J. & S. B. UDIN. 1989. *N*-Methyl-D-aspartate antagonists prevent interaction of binocular maps in *Xenopus* tectum. J. Neurosci. **9**(11): 3837-3843.
31. SCHERER, W. J. & S. B. UDIN. 1991. Chronic effects of NMDA and APV on retinotectal transmission in *Xenopus laevis.* Vis. Neurosci.: in press.
32. KEATING, M. J., L. BEAZLEY, J. D. FELDMAN & R. M. GAZE. 1975. Binocular interaction and intertectal neuronal connexions: Dependence upon development stage. Proc. R. Soc. London Ser. B **191:** 445-466.
33. BEN-ARI, Y., E. CHERUBINI & K. KRNJEVIC. 1988. Changes in voltage dependence of NMDA currents during development. Neurosci. Lett. **94**(1-2): 88-92.
34. FOX, K., H. SATO & N. DAW. 1989. The location and function of NMDA receptors in cat and kitten visual cortex. J. Neurosci. **9**(7): 2443-2454.
35. TREMBLAY, E., M. P. ROISIN, A. REPRESA, C. CHARRIAUT-MARLANGUE & Y. BEN-ARI. 1988. Transient increased density of NMDA binding sites in the developing rat hippocampus. Brain Res. **461:** 393-396.
36. TSUMOTO, T., K. HAGIHARA, H. SATO & Y. HATA. 1987. NMDA receptors in the visual cortex of young kittens are more effective than those of adult cats. Nature **327:** 513-514.
37. BODE-GREUEL, K. M. & W. SINGER. 1989. The development of *N*-methyl-D-aspartate receptors in cat visual cortex. Dev. Brain Res. **46**(2): 197-204.
38. UDIN, S. B. & W. J. SCHERER. 1990. Restoration of the plasticity of binocular maps by NMDA after the critical period in *Xenopus.* Science **249:** 669-672.
39. CLINE, H. T. & M. CONSTANTINE-PATON. 1990. NMDA receptor agonist and antagonists alter retinal ganglion cell arbor structure in the developing frog retinotectal projection. J. Neurosci. **10**(4): 1197-1216.
40. DEBSKI, E. A., H. T. CLINE & M. CONSTANTINE-PATON. 1989. Chronic application of NMDA or APV affects the NMDA sensitivity of the evoked tectal response in *Rana pipiens.* Soc. Neurosci. Abstr. **15:** 495.
41. DAWES, E. A., S. GRANT & M. J. KEATING. 1984. Visual deprivation extends the 'critical period' for intertectal plasticity in the frog, *Xenopus laevis.* J. Physiol. (London) **350:** 18P.
42. CYNADER, M. & D. E. MITCHELL. 1980. Prolonged sensitivity to monocular deprivation in dark-reared cats. J. Neurophysiol. **43**(4): 1026-1040.
43. MOWER, G. D., C. J. CAPLAN, W. G. CHRISTEN & F. H. DUFFY. 1985. Dark rearing prolongs physiological but not anatomical plasticity of the cat visual cortex. J. Comp. Neurol. **235:** 448-466.
44. MEYER, R. L. 1983. Tetrodotoxin inhibits the formation of refined retinotopography in goldfish. Dev. Brain Res. **6:** 293-298.
45. SCHMIDT, J. T. & D. L. EDWARDS. 1983. Activity sharpens the map during the regeneration of the retinotectal projection in goldfish. Brain Res. **269:** 29-39.
46. COOK, J. E. 1988. Topographic refinement of the goldfish retinotectal projection: Sensitivity to stroboscopic light at different periods during optic nerve regeneration. Exp. Brain Res. **70:** 109-116.

47. SCHERER, W. J. & S. B. UDIN. 1991. Latency and temporal overlap of visually elicited contralateral and ipsilateral firing in *Xenopus* tectum during and after the critical period. Dev. Brain. Res. **58:** 129-132.
48. UDIN, S. B. 1990. Plasticity in the ipsilateral visuotectal projection persists after lesions of one nucleus isthmi in *Xenopus.* Exp. Brain Res. **79**(2): 338-344.
49. GUSTAFSSON, B. & H. WIGSTRÖM. 1986. Hippocampal long-lasting potentiation produced by pairing single volleys and brief conditioning tetani evoked in separate afferents. J. Neurosci. **6:** 1575-1582.
50. UDIN, S. B., M. D. FISHER & J. J. NORDEN. 1990. Ultrastructure of the crossed isthmotectal projection in *Xenopus* frogs. J. Comp. Neurol. **292**(2): 246-254.

Stimulation of Phosphoinositide Turnover by Excitatory Amino Acids

Pharmacology, Development, and Role in Visual Cortical Plasticity[a]

MARK F. BEAR AND SERENA M. DUDEK

Center for Neural Science
Brown University
Providence, Rhode Island 02912

INTRODUCTION

It is a widely held belief that the modification of brain, especially neocortex, by sensory experience is a likely basis for memory in mammals. One model system in which such experience-dependent changes can be studied is the ascending visual pathway in the cat. In this system, the frontally placed retinae give rise to two independent streams of sensory information that are combined at the level of the striate cortex (area 17) to form a single percept of visual space. Wiesel and Hubel demonstrated 25 years ago that the binocular connections in cat visual cortex can be dramatically altered by simple forms of sensory deprivation, such as the temporary closure of one eyelid (monocular deprivation; MD) or the misalignment of the two eyes (strabismus).[1] The modification of binocular connections by sensory experience does not occur throughout life, however; it is restricted to a "critical period" of early postnatal development.[2] The sensitivity of cat visual cortex to brief (7-14 day) MD develops at about 3 weeks of age and disappears by 3 months of age.[3] This is a period of rapid head and body growth in which plasticity of binocular connections evidently is normally necessary to maintain proper retinal correspondence.

We divide the problem of visual cortical plasticity into three parts. *First,* what controls the onset and duration of the critical period? The answer to this question is unknown at present, but some interesting possibilities include specific patterns of gene expression[4,5] that may be under hormonal control.[6] *Second,* within the critical period, what factors enable synaptic modification to proceed? This question is prompted by the observation that many experience-dependent modifications of visual cortex seem to require that animals attend to visual stimuli and use vision to guide behavior.[7] The best candidates for "enabling factors" are the neuromodulators acetylcholine and norepinephrine that are released in visual cortex by fibers arising from neurons in the basal forebrain and brain stem.[8] *Third,* when modifications are allowed to occur during the critical period, what controls their direction and magnitude? This is the part of the problem where we believe that excitatory amino acid (EAA) receptors play a central role.[9-13]

Consider a single cortical neuron receiving input (via the lateral geniculate nucleus) from homotypic points on the two retinae. When patterns of retinal activity in the

[a] This work was supported by the National Institutes of Health and the Office of Naval Research.

two eyes occur in register, then the binocular connections in cortex consolidate and strengthen during the critical period. However, when the patterns of retinal activity are brought out of register, either by MD or strabismus, then the binocular connections are disrupted: effective inputs are retained and ineffective inputs are lost, where "effectiveness" is determined by whether an input drives the cortical neuron beyond some threshold. These theoretical considerations have led naturally to the hypothesis that excitatory synapses in visual cortex are consolidated during development when their activity consistently correlates with target depolarization beyond a critical level.[14-16] What mechanism can support this form of "Hebbian" modification?

One strong candidate is the *N*-methyl-D-aspartate (NMDA) receptor mechanism. NMDA receptors are thought to coexist postsynaptically with other types of EAA receptors.[17,18] Together, both NMDA- and non-NMDA-type EAA receptors mediate excitatory synaptic transmission in the kitten striate cortex.[19] The ionic conductances activated by non-NMDA receptors at any instant depend only on input activity, and are independent of the postsynaptic membrane potential. However, the ionic channels linked to NMDA receptors are blocked with Mg^{2+} at the resting potential, and become effective only upon membrane depolarization.[20,21] Another distinctive feature of the NMDA receptor channel is that it will conduct calcium ions.[22,23] Hence, the passage of Ca^{2+} through the NMDA channel could specifically signal when pre- and postsynaptic elements are concurrently active. This property led naturally to the hypothesis that synaptic consolidation in the visual cortex occurs when the NMDA receptor-mediated Ca^{2+} flux exceeds some critical level.[9]

As the other papers in this volume attest, this hypothesis is supported by work on a wide range of model systems. Besides the pioneering studies on long-term potentiation (LTP) in adult hippocampus,[24,25] work on the retinotectal projection in goldfish,[26] and on both the retino- and isthmotectal projections in frogs,[27-30] indicates that NMDA receptor activation is an essential step in the strengthening of coactive synapses. This hypothesis is also supported by several lines of research in the visual cortex. For example, it is now well documented that a form of LTP can be elicited in visual cortex and that this depends on NMDA receptor activation.[31-34] Furthermore, blockade of visual cortical NMDA receptors *in vivo* disrupts the physiological[13,35,36] and anatomical[37] consequences of visual deprivation during the critical period. Thus, there appears to be strong support for the idea that NMDA receptor mechanisms subserve a Hebbian form of synaptic modification in many locations in the vertebrate brain, including the cat visual cortex.

A second form of modification, however, appears to be required to explain visual cortical plasticity. Recall that, according to our reasoning above, a theoretical consequence of input activity that *fails* to correlate with postsynaptic depolarization is a weakening of synaptic efficacy.[16] Indeed, there is direct evidence that retinal activity promotes synaptic weakening in striate cortex under conditions where cortical neurons are unable to respond normally.[13,38] What mechanism can support this form of use-dependent synaptic depression?

Clearly, activation of non-NMDA receptors can signal input activity regardless of the level of postsynaptic depolarization. This line of reasoning has led us to investigate further the consequences of activating non-NMDA receptors in the visual cortex, with the specific aim of identifying a possible mechanism of synaptic weakening. We have become interested in a type of non-NMDA receptor, which we call the Q_2 receptor, that is linked, probably via a G protein, to the enzyme phospholipase C (PLC). PLC catalyzes the hydrolysis of membrane phosphoinositides to form diacylglycerol (DAG) and inositol trisphosphate (IP_3), both of which serve as intracellular second messengers (FIG. 1). In this paper, we discuss our current understanding of the pharmacology and development of the Q_2 receptor in the neocortex. In addition, we present preliminary evidence suggesting that activation of this receptor may be required for some forms of experience-dependent modification in kitten visual cortex.

PHARMACOLOGICAL CHARACTERIZATION OF THE Q_2 RECEPTOR

The ability of certain EAAs to stimulate the hydrolysis of phosphoinositides was first demonstrated in cultures of striatal neurons.[39] Subsequently, EAA-stimulated phosphoinositide (PI) turnover was shown to occur in a number of tissues, including hippocampal slices, cerebellar granule cells in culture, cortical synaptoneurosomes, and frog oocytes injected with rat brain mRNA.[40–49] Although differences between systems exist, there is general agreement that PI turnover in most locations is potently stimulated by quisqualate, ibotenate, glutamate, and *trans*-1-amino-1,3-cyclopentane dicarboxylic acid (*trans*-ACPD). Sensitivities to other EAAs have been reported, but this evidently depends on the source of tissue. For example, NMDA stimulates PI turnover in striatal and cerebellar neurons in culture.[39,42]

To investigate the relative effectiveness of different EAAs in neocortex, we assayed PI turnover in synaptoneurosomes prepared from rat and cat.[11,48] In both tissues we found ibotenate, quisqualate, and glutamate to be effective agonists. We also found that NMDA was ineffective in stimulating PI turnover (FIG. 2), and that the NMDA receptor antagonist 2-amino-5-phosphonovaleric acid (AP5) did not reduce glutamate-stimulated PI turnover. Thus, in neocortex, PI turnover is stimulated by EAAs specifically at a *non*-NMDA receptor site. Work using hippocampus[40] and frog oocytes injected with rat brain mRNA[43] has led to a similar conclusion.

It was originally proposed that two different EAA receptors stimulate PI turnover in neurons: a receptor that prefers ibotenate, and one that prefers quisqualate.[50] Evidence leading to this hypothesis is that ibotenate is most effective in stimulating PI turnover in hippocampus, whereas quisqualate is the more potent agonist in striatum, cerebellum, and rat whole forebrain. Schoepp and Johnson,[45] however, have demonstrated in hippocampal slices that maximal stimulation of PI hydrolysis by ibotenate and quisqualate

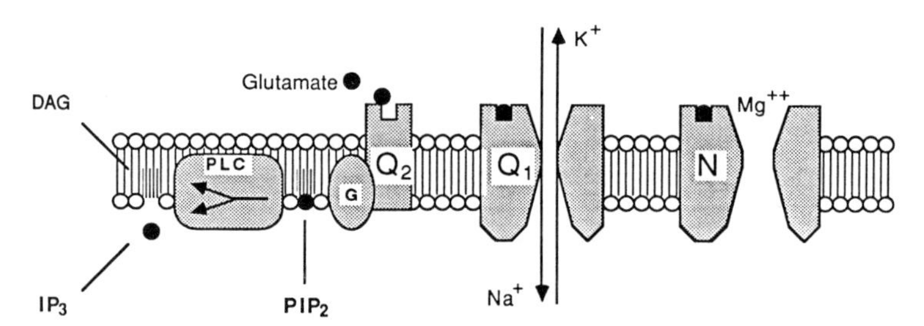

FIGURE 1. A hypothetical postsynaptic response to glutamate. Upon release as a neurotransmitter, glutamate binds to the *N*-methyl-D-aspartate receptor (N), which is blocked by magnesium at resting membrane potentials, and to a quisqualate receptor associated with an ion channel (Q_1). Additionally, glutamate may bind to a second type of quisqualate receptor linked to phosphoinositide metabolism (Q_2). Upon the binding of glutamate to the Q_2 receptor, phospholipase C (PLC) is activated via a G protein (G); PLC breaks down phosphatidyl inositol-4,5-bisphosphate (PIP_2) into second messengers diacylglycerol (DAG) and inositol-1,4,5-trisphosphate (IP_3).

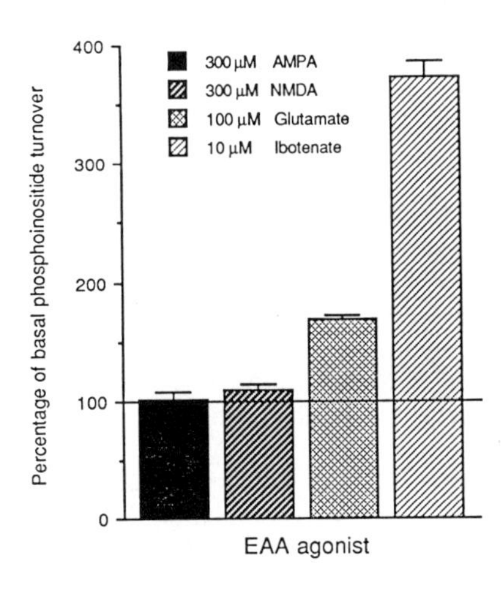

FIGURE 2. The effects of different excitatory amino acids on [^{3}H]inositol phosphate accumulation in synaptoneurosomes prepared from 5-week-old kittens. Glutamate and ibotenate stimulate PI turnover while *N*-methyl-D-aspartate (NMDA) and α-amino-3-hydroxy-5-methylisoxazole-4-propionic acid (AMPA) are without effect. Results are means ± SEMs from at least three experiments. Methods in reference 11.

is not additive, suggesting that they act at a single site (in contrast, stimulation by quisqualate and acetylcholine is additive, confirming that these act at separate sites). Similarly, Desai and Conn[51] have presented evidence that quisqualate and ibotenate both act as partial agonists at the same site stimulated by *trans*-ACPD. This conclusion is also supported by work by Sugiyama *et al.*[43] in the oocyte preparation. It thus seems reasonable to conclude that in cortex PI turnover is stimulated by a single type of non-NMDA EAA receptor.

The pharmacological profile of EAA-stimulated PI turnover in hippocampus and neocortex suggests that the quisqualate (Q) receptor is specifically involved. However, the selective Q agonist α-amino-3-hydroxy-5-methylisoxazole-4-propionic acid (AMPA) does not stimulate PI turnover (FIG. 2), and the Q antagonists 6-cyano-7-nitroquinoxaline (CNQX) and kynurenate do not block quisqualate-stimulated PI turnover. For this reason, we divide Q receptors into two classes, Q_1 and Q_2 (FIG. 1). Q_1 receptors are those sensitive to AMPA, CNQX, and kynurenate; activation of these receptors leads to an ionic conductance. Q_2 receptors are those sensitive to *trans*-ACPD; activation of these receptors leads to PI hydrolysis. It should be noted that other nomenclatures have been proposed: Q_p or IBO_p receptor,[50] metabotropic glutamate receptor (mGluR or Glu_BR),[52] and the ACPD receptor.[53] We favor Q_2 because of its simplicity (and because we thought of it).

It has been shown that quisqualate and glutamate potently stimulate PI turnover in cultured astrocytes.[54,55] How do we know what fraction of the EAA-stimulated PI turnover we measure in cortex is neuronal in origin? Palmer *et al.*[56] have shown in rat hippocampal slices that NMDA inhibits quisqualate-stimulated PI turnover in a calcium-dependent fashion. These data suggest that the bulk of the EAA-stimulated PI turnover measured in hippocampus occurs in neurons because astrocytes do not express NMDA receptors.[57] It should be noted that Baudry *et al.*[58] found that NMDA also inhibits carbachol-, histamine-, and depolarization-induced stimulation of PI turnover in hippocampal slices. We have performed similar experiments in visual cortical slices, and also find that NMDA can completely block quisqualate stimulation of PI turnover (FIG. 3). The inhibition of PI turnover by NMDA in slices may be solely due to excitotoxicity, but the interesting possibility remains that this may be a physiologically relevant regulatory mechanism.[56] In any case, these data support the idea that Q_2 receptors are expressed on visual cortical neurons.

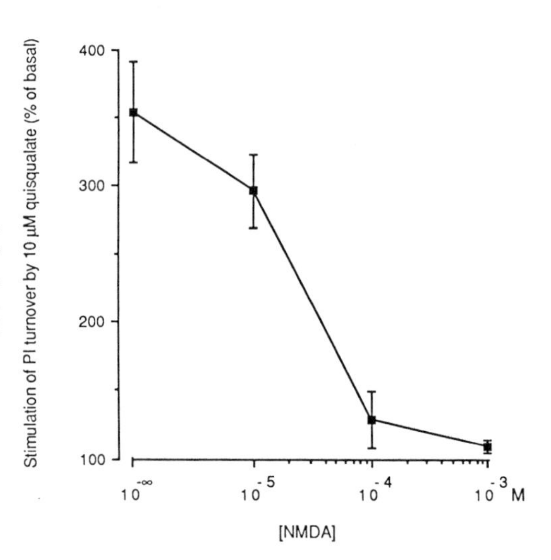

FIGURE 3. Inhibition of quisqualate-stimulated PI hydrolysis by NMDA in slices of kitten visual cortex. Methods are described in reference 56. Results are from four experiments and represent means ± SEMs.

Certainly, the elucidation of the function of EAA-stimulated PI turnover depends on the development of potent and selective antagonists of the Q_2 receptor. Several compounds with antagonist-like properties have been described, including 2-amino-4-phosphonobutyrate (AP4), 2-amino-3-phosphonopropionate (AP3), and L-serine-*O*-phosphate (LSOP). Nicoletti *et al.*[41] have shown that both AP4 and LSOP inhibit stimulation of PI turnover by ibotenate in adult rat hippocampal slices. These compounds, however, are not selective for the Q_2 receptor.[46] Moreover, AP4 and LSOP actually *stimulate* PI turnover in hippocampal slices from neonatal rats. The use of AP3 as a Q_2 inhibitor will be discussed in detail below, when we consider the involvement of these receptors in visual cortical plasticity.

DEVELOPMENT OF THE Q_2 RECEPTOR

One of the most striking features of EAA-mediated PI turnover is its dependence on the age of the animal. First to describe the developmental profile of the Q_2 receptor was Nicoletti *et al.*[41] Using rat hippocampal slices, they showed that stimulation of PI hydrolysis by ibotenate and glutamate is at least 3 times greater in neonates than in adults. A similar developmental profile was found using rat neocortical synaptoneurosomes.[48] We found that glutamate-stimulated PI turnover is maximal at 1 week of age, when it is approximately 5 times the adult value (FIGS. 4A & 5A). By 5 weeks of age, however, EAA-stimulated PI turnover declines to the low adult level. It is interesting to note that Q_2 receptors appear to develop postnatally, as glutamate was found to be ineffective in stimulating PI turnover in newborn rat neocortex. Thus, the increase in EAA-stimulated PI turnover occurs coincident with the period of most rapid synaptogenesis in rat neocortex.[59] These observations have led to speculation that EAA-stimulated PI turnover plays a special role in cortical development and plasticity.[41,48]

As was discussed earlier, modification of binocular connections in visual cortex by sensory experience is also restricted to a critical period of early postnatal development.

In kittens, sensitivity to lid closure begins at about 3 weeks of age, peaks at 5 weeks, and disappears between 12 and 16 weeks.[3] We were, of course, keenly interested in how well the developmental profile for EAA-stimulated PI turnover maps onto this profile of the critical period. A synaptoneurosome preparation of kitten area 17 was used to investigate this possibility.[11]

As illustrated in FIGURE 4B, the developmental differences in glutamate-stimulated PI turnover in kitten striate cortex were similar to those observed in rat neocortex; stimulation in the neonate was greater than 3 times the adult value. The more potent agonist ibotenate was used to stimulate PI turnover in the striate cortex at different postnatal ages, and the results are shown in FIGURE 5B. The postnatal changes in ibotenate-stimulated PI turnover show a remarkable correlation with the development of ocular dominance plasticity in kitten striate cortex. Thus, when the striate cortex is most sensitive to sensory deprivation, it is also most sensitive to stimulation of PI hydrolysis by certain excitatory amino acids.

It is interesting to note that other receptors linked to PI turnover, such as the M_1 acetylcholine receptor and the α_1 norepinephrine receptor, do not share this same developmental profile. For example, in rat hippocampus, Nicoletti *et al.*[41] found that norepinephrine does not potently stimulate PI turnover until *after* 3 weeks of age. In kitten striate cortex, the muscarinic agonist carbachol stimulates PI hydrolysis at all ages, indicating that the transient increase in stimulated PI turnover during the critical period is relatively specific to a mechanism linked to the Q_2 receptor.

The normal development of Q_2 receptors in striate cortex evidently requires visual experience. We found that the transient rise in ibotenate-stimulated PI turnover at 5 weeks of age did not occur in the striate cortex of dark-reared kittens. In contrast, stimulation by carbachol was unaffected by visual deprivation. These findings further support the idea that EAA-stimulated PI turnover may play a specific role in develop-

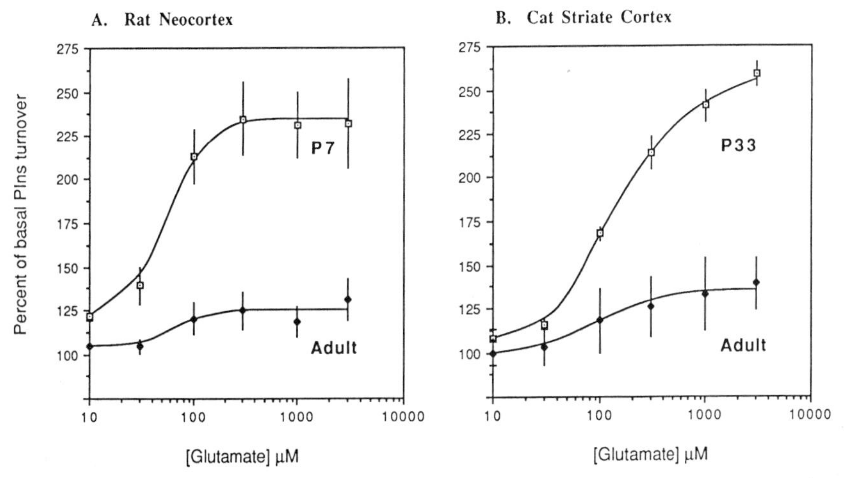

FIGURE 4. Stimulation of PI turnover by glutamate. Accumulation of [^{3}H]inositol phosphate was measured in synaptoneurosomes prepared from **(A)** rat neocortex and **(B)** cat striate cortex. In both cases, glutamate was more effective in young animals (postnatal day 7 in the rat, and 33 in the cat) than in adults. Results are means ± SEMs from at least three experiments. Data in **A** redrawn from reference 48; data in **B,** from reference 11, Table 1.

A. Rat Neocortex

Stimulation of PI turnover by 300 μM glutamate (% basal)

AGE (days)

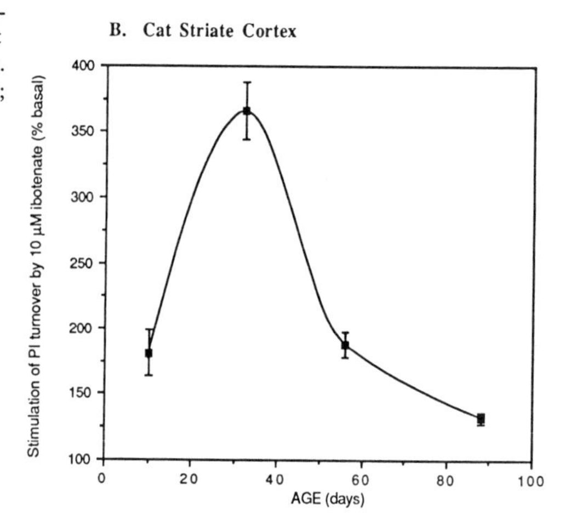

FIGURE 5. Age dependence of EAA-stimulated PI turnover. [^{3}H]Inositol phosphate was measured in synaptoneurosomes prepared from (**A**) rat neocortex and (**B**) cat striate cortex. PI turnover was stimulated with 300 μM glutamate in the rat and 10 μM ibotenate in the cat. Results are means ± SEMs from at least three experiments. Data in **A** redrawn from reference 48; data in **B,** from reference 11.

mental plasticity because dark rearing is known to postpone the onset of the critical period.[60]

Q_2 RECEPTORS AND OCULAR DOMINANCE PLASTICITY

The evidence discussed in the preceding section suggests that at the same ages that retinal activity drives synaptic modification in kitten striate cortex, excitatory synaptic activity also stimulates a specific second-messenger cascade in visual cortical neurons. These data are therefore consistent with our hypothesis that Q_2 receptors play a central role in the activity-dependent modification of striate cortex during the critical period. Direct tests of this hypothesis have been hampered, however, by the lack of a potent and selective antagonist.

One compound that has shown some promise as an inhibitor of EAA-stimulated PI turnover is 2-amino-3-phosphonopropionate (AP3). Using hippocampal minces prepared from adult rats, Schoepp and Johnson[46] found that D,L-AP3 inhibits PI turnover stimulated by 100 μM ibotenate with an IC_{50} of approximately 120 μM. It is of interest to note that AP3 had long been considered inert at EAA receptors, which is not surprising, as AP3 has no effect on the specific binding of ^{3}H-labeled ligands to the other EAA receptor types (kainate,[61] AMPA,[62] *N*-acetylaspartylglutamate,[46] various NMDA receptor ligands[63]). Thus, it appears that the action of AP3 is selective for the Q_2 receptor.

There are several indications, however, that the mechanism of AP3 action is complex. First, the effectiveness of AP3 as an inhibitor is age dependent. Using hippocampus from 1-week-old rats, Schoepp and Johnson[47] found that AP3 inhibited ibotenate-stimulated PI turnover with an IC_{50} of ~370 μM, more than 3 times the adult value. Second, AP3 itself has been shown to stimulate PI turnover, weakly in comparison with quisqualate and ibotenate, but nonetheless significantly (~200% of basal with 1 mM AP3; Schoepp and Johnson[47]). Finally, the inhibitory effects of AP3 seem to vary with the potency and concentration of the EAA agonist used to stimulate PI turnover.[48,51] These considerations have led to the hypothesis that AP3 inhibits EAA-stimulated PI turnover by acting as a "partial agonist" at the Q_2 receptor.[46,47,51]

Our own experiments using AP3 to antagonize EAA-stimulated PI turnover in visual cortex have yielded mixed results. In synaptoneurosomes, we find that AP3 inhibits PI turnover stimulated by 10 μM ibotenate with an apparent IC_{50} of about 80 μM (FIG. 6). Using slices of visual cortex, however, we find that the stimulation of PI turnover by 10 μM quisqualate is not inhibited by AP3 in concentrations up to 1 mM[64]; in fact, the stimulation of PI turnover by quisqualate and AP3 appears to be additive. There are a number of possible explanations for our results (differences in the preparation of the tissue, for example). They are consistent, however, with the hypothesis that AP3 is a partial agonist where, by definition, the inhibitory effect depends on the concentration and efficacy of the "pure" agonist used.[65]

These caveats notwithstanding, at the present time AP3 remains the most selective pharmacological tool available to test the hypothesis that EAA-stimulated PI turnover plays a specific role in visual cortical plasticity. Therefore, we performed a series of

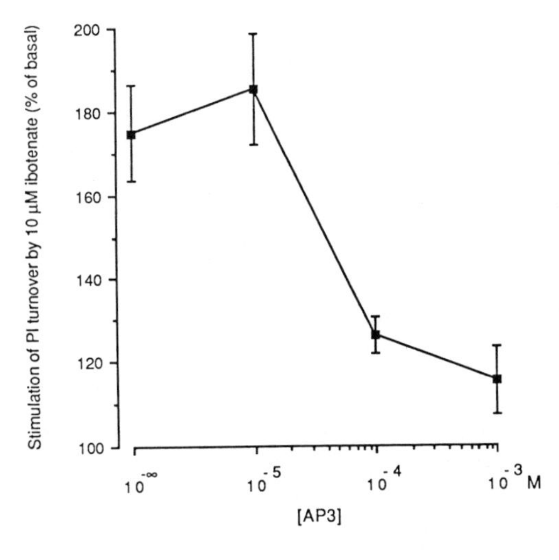

FIGURE 6. 2-Amino-3-phosphonopropionate (AP3) inhibition of ibotenate-stimulated PI turnover. AP3 inhibits stimulation of PI turnover by 10 μM ibotenate in synaptoneurosomes from kitten striate cortex (kittens aged 5-7 weeks). Results are from five experiments and are expressed as means ± SEMs.

experiments to see if the chronic infusion of AP3 into the visual cortex would interfere with the synaptic modifications that are elicited by brief MD during the critical period. In our initial experiments, kittens were reared normally to approximately 8 weeks of age, at which time one eyelid was sutured closed (TABLE 1). At the same time, minipumps were implanted that could deliver solutions directly to the visual cortex at a rate of 1 μl/hr. AP3 (50 mM) dissolved in sterile Ringer was infused into the hemispheres contralateral to the deprived eye; Ringer alone was infused into hemispheres ipsilateral to the deprived eye. Then, 7-8 days later, the animals were prepared for the physiological assessment of cortical binocularity according to routine procedures.[13]

Electrophysiological recordings from the Ringer-treated (control) hemispheres confirmed that the period of MD was sufficient to cause an ocular dominance shift; the large majority of visual cortical neurons responded preferentially to stimulation of the eye that had been open. In striking contrast, most neurons responded vigorously through both eyes in the contralateral, AP3-treated hemispheres (FIG. 7). These experiments suggest that AP3 treatment can block the effects of MD in the visual cortex contralateral to the deprived eye in 2-month-old kittens. Preliminary work, however, suggests that AP3 treatment is less effective under conditions where the effects of MD are more severe, for example, in the hemispheres ipsilateral to the deprived eye in younger kittens. Although more work needs to be done, our impression is that the effectiveness of AP3 in blocking the ocular dominance shift after MD varies as a function of age. This perhaps is not surprising considering the age dependency of AP3's inhibition of EAA-stimulated PI turnover.[47,48]

It should be noted that in cases where AP3 treatment blocks the ocular dominance shift, it does not appear to disrupt other receptive field properties such as orientation and direction selectivity. This result contrasts with the effects of chronic NMDA receptor blockade, where most affected neurons lack normal selectivity and responsiveness.[13] This observation leads us to believe that AP3 is probably not exerting its effects on visual cortical plasticity by blocking NMDA receptors.[b] Of course, the interpretation of our results using AP3 is still complicated by all the uncertainties concerning the action of AP3 on the Q_2 receptor. There is clearly a need for a pure antagonist of EAA-stimulated PI turnover. Nonetheless, our results are consistent with the hypothesis that the Q_2 receptor is centrally involved in the experience-dependent modification of visual cortex during the critical period.

Assuming that this hypothesis is in fact correct, then what is the specific contribution of the Q_2 receptor to the mechanisms of visual cortical plasticity? Theoretical analysis led us to propose that Q_2 activation might play a specific role in the process of activity-dependent synaptic weakening.[10-12] Evidence in support of this hypothesis has been obtained recently by Stanton and colleagues in slices of hippocampus. They find that activation of excitatory afferents coincident with strong hyperpolarization of the postsynaptic target neuron leads to a lasting depression of synaptic effectiveness.[66] And, induction of this form of long-term depression, or "LTD," evidently is blocked by application of AP3.[67]

Of course it is possible that Q_2-mediated PI turnover plays an entirely different role in the mechanisms of synaptic plasticity than the one we propose here. For example, Q_2-stimulated PI turnover might lead to the release of intracellular Ca^{2+} and thus

[b] This also raises a question of theoretical interest. Is the absence of AP3 effects on selectivity to be expected if the Q_2 receptor specifically mediates use-dependent decreases in synaptic strength? Although this question requires further study, post hoc reasoning suggests that the blockade of a synaptic weakening signal would not necessarily have an effect on receptive field properties that had been established previously.

TABLE 1. Effects of AP3 on Ocular Dominance Plasticity

Animal	Age at Implant/MD (days)	Age at Recording (days)	Hemisphere	Treatment	Open Eye Dominance[a]	Binocularity[b]	N^c
SE1	53	60	L	Ringer	0.74	0.36	27
			R	AP3	0.44	0.65	33
SE4	58	66	L	Ringer	0.89	0.19	38
			R	AP3	0.28	0.76	37
UO4	57	64	L	Ringer	0.96	0.08	27
			R	AP3	0.19	0.78	34

[a] Open eye dominance is the number of cells in ocular dominance group 5 plus 0.5 times the number of cells in group 4 divided by the total number of classifiable cells (group 5, by convention, is the monocular group dominated by the open eye). A value of 1.0 would represent a complete ocular dominance shift to the open eye.

[b] Binocularity is defined as the total number of cells in OD groups 2 to 4 divided by the total number of classifiable cells and represents the proportion of cells that are binocular.

[c] *N* refers to the number of units recorded in each hemisphere.

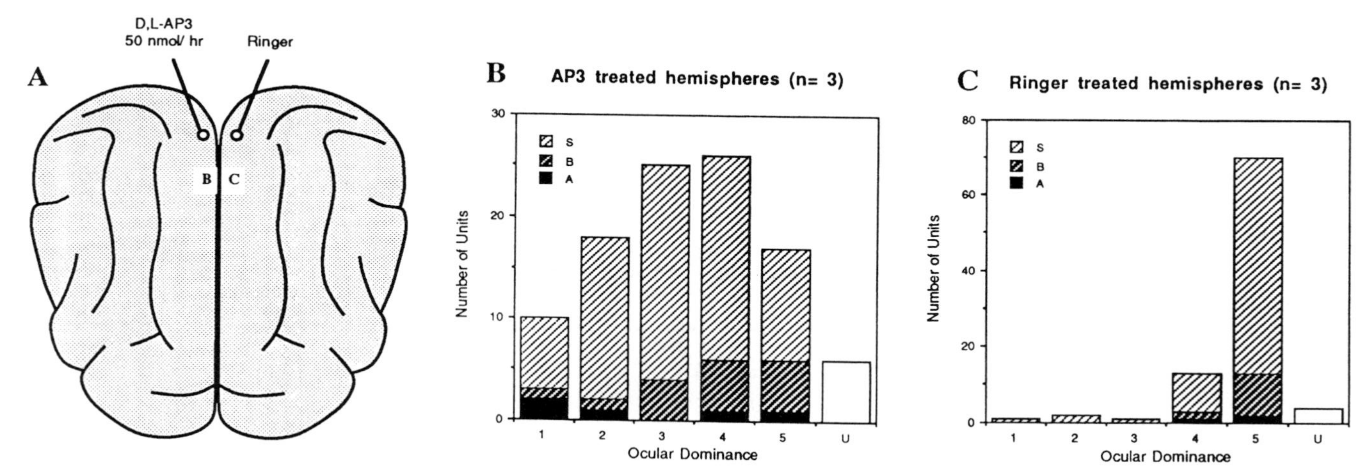

FIGURE 7. Effect of AP3 on ocular dominance plasticity. (**A**) Dorsal view of a kitten brain to illustrate the experimental design. Miniature osmotic pumps (ALZA 2001) were used to deliver either 50 mM AP3 dissolved in Ringer, or Ringer alone, into striate cortex at a rate of 1 μl/hr via chronically implanted stainless steel canulae fasioned from 27-gauge hypodermic needles. At the same time, the left eyelid was sutured closed. Then, seven to eight days later, recordings were made 3-6 mm anterior to the infusion canulae. Based on previous measurements,[13] we estimate that extracellular [AP3] is $\sim$200 μM in the sample zone (marked "B" in **A**). Control recordings indicated that neurons in the sample zone retained visual responses during AP3 infusion. (**B** & **C**) Ocular dominance (OD) and selectivity histograms compiled from recordings in the AP3-or Ringer-treated cortices in three experiments (see TABLE 1). Multi-unit and single-unit recordings were sampled every 100 μ along multiple tangential electrode tracks, and were classified for OD and orientation selectivity. Neurons assigned to OD categories 1 and 5 were activated by either the left or the right eyes, respectively, but not both. Cells assigned to OD category 3 were activated equally strongly by stimulation of either eye. Neurons in OD categories 2 and 4 were binocular, but their responses were clearly dominated by either the left or right eyes, respectively. Neurons for which no receptive field could be identified were recorded as unclassifiable (U). Filled bars represent the number of cells within each OD group that were aspecific (A) for stimulus orientation; darkly hatched bars, cells that showed a weak bias (B) for stimulus orientation; and lightly hatched bars, neurons that were selective (S) for orientation.

substitute for NMDA-receptor-mediated Ca^{2+} signals. Alternatively, Q_2 receptor activation may simply mimic the effects of other modulatory inputs that are known to stimulate PI turnover. For example, acetylcholine stimulates PI turnover in cortical neurons, and both ACh and the Q_2 agonist *trans*-ACPD decrease a Ca^{2+}-activated potassium conductance in pyramidal cell dendrites.[68] Thus, it is possible that Q_2 receptors affect synaptic plasticity by modulating postsynaptic excitability. On the other hand, because the neuron is so highly compartmentalized, it may not be correct to assume that a consequence of PI turnover at one location is equivalent to that at another. If Q_2 receptors are in fact located beneath modifiable excitatory synapses, it is the PI turnover occurring in dendritic spines that is of special interest. Here, either the diacylglycerol or inositol trisphosphate arms of this biochemical pathway could selectively trigger decreases in synaptic strength.

SUMMARY AND CONCLUSIONS

Theoretical analysis suggests that in the visual cortex during early postnatal development, afferent activity can yield either an increase or a decrease in synaptic strength depending on the pattern of EAA receptor activation in cortical neurons. This motivated us to study the mechanism of EAA-stimulated phosphoinositide turnover in visual cortex. Available evidence suggests that PI hydrolysis is stimulated by EAAs primarily at a single receptor site (Q_2 receptor), and that this site is distinct from both the traditional quisqualate (Q_1) receptor and the NMDA receptor. NMDA does, however, inhibit EAA-stimulated PI turnover in visual cortex, confirming that the Q_2 receptor is on visual cortical neurons (as opposed to glia). We find that Q_2 receptors in the neocortex are expressed transiently during postnatal development. The developmental time-course of EAA-stimulated PI turnover correlates precisely with the critical period when synaptic modifications are most readily elicited in visual cortex by changes in sensory experience. The compound AP3 can inhibit EAA-stimulated PI turnover, probably by acting as a partial Q_2 agonist, and under some circumstances AP3 evidently can interfere with experience-dependent synaptic modifications.

Increases in synaptic strength in visual cortex, as elsewhere, have been linked specifically to activation of NMDA receptors. We propose that decreases in synaptic strength may be specifically related to activation of the Q_2 receptor. Further tests of this hypothesis will require the development of selective and potent antagonists.

ACKNOWLEDGMENT

The authors wish to thank Amy Bohner for her valuable contributions to these studies.

REFERENCES

1. WIESEL, T. N. & D. H. HUBEL. 1965. Comparison of the effects of unilateral and bilateral eye closure on cortical unit responses in kittens. J. Neurophysiol. **28:** 1029-1040.

2. HUBEL, D. H. & T. N. WIESEL. 1970. The period of susceptibility to the physiological effects of unilateral eye closure in kittens. J. Physiol. (London) **206:** 419-436.
3. OLSON, C. R. & R. D. FREEMAN. 1980. Profile of the sensitive period for monocular deprivation in kittens. Exp. Brain Res. **39:** 17-21.
4. NEVE, R. L. & M. F. BEAR. 1989. Visual experience regulates gene expression in the developing striate cortex. Proc. Natl. Acad. Sci. USA **86:** 4781-4784.
5. SUR, M., D. O. FROST & S. HOCKFIELD. 1988. Expression of a surface-associated antigen on Y-cells in the cat lateral geniculate nucleus is regulated by visual experience. J. Neurosci. **8:** 874-882.
6. DAW, N. W., H. SATO & K. FOX. 1988. Effect of cortisol on plasticity in the cat visual cortex. Neurosci. Abstr. **14:** 81.11.
7. SINGER, W. 1979. Central core control of visual cortex functions. *In* The Neurosciences Fourth Study Program. F. Schmitt & F. Worden, Eds.: 1093-1109. MIT. Cambridge, MA.
8. BEAR, M. F. & W. SINGER. 1986. Modulation of visual cortical plasticity by acetylcholine and noradrenaline. Nature **320:** 172-176.
9. BEAR, M. F., L. N. COOPER & F. F. EBNER. 1987. A physiological basis for a theory of synapse modification. Science **237:** 42-48.
10. BEAR, M. F. 1988. Involvement of excitatory amino acid receptors in the experience-dependent development of visual cortex. *In* Frontiers in Excitatory Amino Acid Research. E. Cavalheiro, J. Lehmann & L. Turski, Eds. Vol. 46: 393-401. Alan R. Liss. New York, NY.
11. DUDEK, S. M. & M. F. BEAR. 1989. A biochemical correlate of the critical period for synaptic modification in the visual cortex. Science **246:** 673-675.
12. BEAR, M. F. & L. N. COOPER. 1990. Molecular mechanism for synaptic modification in the visual cortex: Interaction between theory and experiment. *In* Neuroscience and Connectionist Theory. M. Gluck & D. Rumelhart, Eds.: 65-93. Lawrence Erlbaum Associates. Livermore, NJ.
13. BEAR, M. F., A. KLEINSCHMIDT, Q. GU & W. SINGER. 1990. Disruption of experience-dependent synaptic modifications in striate cortex by infusion of an NMDA receptor antagonist. J. Neurosci. **10**(3): 909-925.
14. HEBB, D. O. 1949. The Organization of Behavior. John Wiley & Sons. New York, NY.
15. STENT, G. S. 1973. A physiological mechanism for Hebb's postulate of learning. Proc. Natl. Acad. Sci. USA **70:** 997-1001.
16. BIENENSTOCK, E. L., L. N. COOPER & P. W. MUNRO. 1982. Theory for the development of neuron selectivity: Orientation specificity and binocular interaction in visual cortex. J. Neurosci. **2:** 32-48.
17. FOSTER, A. C. & G. E. FAGG. 1985. Amino acid binding sites in mammalian neuronal membranes: Their characteristics and relationship to synaptic receptors. Brain Res. Rev. **7:** 103-164.
18. STEVENS, C. F. & J. M. BEKKER. 1989. NMDA and non-NMDA receptors are co-localized at individual excitatory synapses in cultured rat hippocampus. Nature **341:** 230-233.
19. TSUMOTO, T., H. MASUI & H. SATO. 1986. Excitatory amino acid neurotransmitters in neuronal circuits of the cat visual cortex. J. Neurophysiol. **55:** 469-483.
20. NOWAK, L., P. BREGOSTOVSKI, P. ASCHER, A. HERBERT & A. PROCHIANTZ. 1984. Magnesium gates glutamate-activated channels in mouse central neurones. Nature **307:** 462-465.
21. MAYER, M. L. & G. L. WESTBROOK. 1987. The physiology of excitatory amino acids in the vertebrate central nervous system. Prog. Neurobiol. **28:** 197-276.
22. DINGLEDINE, R. 1983. *N*-Methylaspartate activates voltage-dependent calcium conductance in rat hippocampal pyramidal cells. J. Physiol. **343:** 385-405.
23. MCDERMONT, A. B., M. L. MAYER, G. L. WESTBROOK. S. J. SMITH & J. L. BARKER. 1986. NMDA receptor activation increases cytoplasmic calcium concentration in cultured spinal cord neurones. Nature **321:** 519-522.
24. COLLINGRIDGE, G. L., S. L. KEHL & H. MCLENNAN. 1983. Excitatory amino acids in synaptic transmission in the Schaffer collateral-commisural pathway of the rat hippocampus. J. Physiol. **334:** 33-46.
25. MEFFERT, M. K., K. D. PARFITT, V. A. DOZE, G. A. COHEN & D. V. MADISON. 1991. Protein kinases and long-term potentiation. Ann. N.Y. Acad. Sci. This volume.
26. SCHMIDT, J. T. 1990. Long-term potentiation and activity-dependent retinotopic sharpening

in the regenerating retinotectal projection of goldfish: Common sensitive period and sensitivity to NMDA blockers. J. Neurosci. **10**(1): 233-246.
27. CLINE, H. T., E. DEBSKI & M. CONSTANTINE-PATON. 1987. NMDA receptor antagonist desegregates eye-specific stripes. Proc. Natl. Acad. Sci. USA **84:** 4342-4345.
28. CLINE, H. T. & M. CONSTANTINE-PATON. 1990. NMDA receptor agonist and antagonists alter retinal ganglion cell arbor structure in the developing frog retinotectal projection. J. Neurosci. **10**(4): 1197-1216.
29. SCHERER, W. J. & S. B. UDIN. 1989. *N*-Methyl-D-aspartate antagonists prevent interaction of binocular maps in *Xenopus* tectum. J. Neurosci. **9:** 3837-3843.
30. UDIN, S. B. & W. J. SCHERER. 1991. Experience-dependent formation of binocular maps in frogs: Possible involvement of *N*-methyl-D-aspartate receptors. Ann. N.Y. Acad. Sci. This volume.
31. ARTOLA, A. & W. SINGER. 1987. Long-term potentiation and NMDA receptors in rat visual cortex. Nature **330:** 649-652.
32. CONNORS, B. W. & M. F. BEAR. 1988. Pharmacological modulation of long-term potentiation in slices of visual cortex. Neurosci. Abstr. **14:** 298.8.
33. KIMURA, F., T. TSUMOTO, A. NISHIGORI & T. SHIROKAWA. 1988. Long-term potentiation and NMDA receptors in the rat pup visual cortex. Neurosci. Abstr. **14:** 81.10.
34. PRESS, W. A. & M. F. BEAR. 1990. Effects of disinhibition on LTP induction in slices of visual cortex. Neurosci. Abstr. **16:** 348.9.
35. KLEINSCHMIDT, A., M. F. BEAR & W. SINGER. 1987. Blockade of "NMDA" receptors disrupts experience-dependent modifications of kitten striate cortex. Science **238:** 355-358.
36. GU, Q., M. F. BEAR & W. SINGER. 1989. Blockade of NMDA receptors prevents ocularity changes in kitten visual cortex after reversed monocular deprivation. Dev. Brain Res. **47:** 281-288.
37. COLMAN, H. & M. F. BEAR. 1989. Blockade of visual cortical NMDA receptors prevents the shrinkage of lateral geniculate neurons following monocular deprivation. Neurosci. Abst. **15:** 4.8.
38. REITER, H. O. & M. P. STRYKER. 1988. Neural plasticity without action potentials: Less active inputs become dominant when kitten visual cortical cells are pharmacologically inhibited. Proc. Natl. Acad. Sci. USA **85:** 3623-3627.
39. SLADECZEK, F., J.-P. PIN, M. RECASENS, J. BOCKAERT & S. WEISS. 1985. Glutamate stimulates inositol phosphate formation in striatal neurones. Nature **317:** 717-719.
40. NICOLETTI, F., J. L. MEEK, M. J. IADAROLA, D. M. CHUANG, B. L. ROTH & E. COSTA. 1986. Coupling of inositol phospholipid metabolism with excitatory amino acid recognition sites in rat hippocampus. J. Neurochem. **46**(1): 40-46.
41. NICOLETTI, F., M. J. IADAROLA, J. T. WROBLEWSKI & E. COSTA. 1986. Excitatory amino acid recognition sites coupled with inositol phospholipid metabolism: Developmental changes and interaction with α_1-adrenoceptors. Proc. Natl. Acad. Sci. USA **83:** 1931-1935.
42. NICOLETTI, F., J. T. WROBLEWSKI, A. NOVELLI, H. ALHO, A. GUIDOTTI & E. COSTA. 1986. The activation of inositol phospholipid metabolism as a signal-transducing system for excitatory amino acids in primary cultures of cerebellar granule cells. J. Neurosci. **6**(7): 1905-1911.
43. SUGIYAMA, H., I. ITO & C. HIRONO. 1987. A new type of glutamate receptor linked to inositol phospholipid metabolism. Nature **325:** 531-533.
44. RECASENS, M., I. SASSETTI, A. NOURIGAT, F. SLADECZEK & J. BOCKAERT. 1987. Characterization of subtypes of excitatory amino acid receptors involved in the stimulation of inositol phosphate synthesis in rat brain synaptoneurosomes. Eur. J. Pharmacol. **141:** 87-93.
45. SCHOEPP, D. D. & B. G. JOHNSON. 1988. Excitatory amino acid agonist-antagonist interactions at 2-amino-4-phosphonobutyric acid-sensitive quisqualate receptors coupled to phosphoinositide hydrolysis in slices of rat hippocampus. J. Neurochem. **50**(5): 1605-1613.
46. SCHOEPP, D. D. & B. G. JOHNSON. 1989. Comparison of excitatory amino acid-stimulated phosphoinositide hydrolysis and *N*-[^{3}H]acytylaspartylglutamate binding in rat brain: Selective inhibition of phosphoinositide hydrolysis by 2-amino-3-phosphonopropionate. J. Neurochem. **53**(1): 273-278.
47. SCHOEPP, D. D. & B. G. JOHNSON. 1989. Inhibition of excitatory amino acid-stimulated phosphoinositide hydrolysis in the neonatal rat hippocampus by 2-amino-3-phosphonopropionate. J. Neurochem. **53**(6): 1865-1870.

48. DUDEK, S. M., W. D. BOWEN & M. F. BEAR. 1989. Postnatal changes in glutamate stimulated phosphoinositide turnover in rat neocortical synaptoneurosomes. Dev. Brain Res. **47:** 123-128.
49. PALMER, E., D. T. MONAGHAN & C. W. COTMAN. 1989. *trans*-ACPD, a selective agonist of the phosphoinositide-coupled excitatory amino acid receptor. Eur. J. Pharmacol. **166:** 585-587.
50. SLADECZEK, F., M. RECASENS & J. BOCKAERT. 1988. A new mechanism for glutamate receptor action: Phosphoinositide hydrolysis. TINS **11**(12): 545-549.
51. DESAI, M. A. & P. J. CONN. 1990. Selective activation of phosphoinositide hydrolysis by a rigid analogue of glutamate. Neurosci. Lett. **109:** 157-162.
52. SUGIYAMA, H., I. ITO & M. WATANABE. 1989. Glutamate receptor subtypes may be classified into two major categories: A study on *Xenopus* oocytes injected with rat brain mRNA. Neuron **3:** 129-132.
53. MONAGHAN, D. T., R. J. BRIDGES & C. W. COTMAN. 1989. The excitatory amino acid receptors: Their classes, pharmacology, and distinct properties in the function of the central nervous system. Annu. Rev. Pharmacol. Toxicol. **29:** 365-402.
54. PEARCE, B., J. ALBRECHT, C. MORROW & S. MURPHY. 1986. Astrocyte glutamate receptor activation promotes inositol phospholipid turnover and calcium flux. Neurosci. Lett. **72:** 335-340.
55. JENSEN, A. M. & S. Y. CHIU. 1990. Fluorescence measurement of changes in intracellular calcium induced by excitatory amino acids in cultured cortical astrocytes. J. Neurosci. **10**(4): 1165-1175.
56. PALMER, E., D. T. MONAGHAN & C. W. COTMAN. 1988. Glutamate receptors and phosphoinositide metabolism: Stimulation via quisqualate receptors is inhibited by *N*-methyl-D-aspartate receptor activation. Mol. Brain Res. **4:** 161-165.
57. USOWICZ, M. M., V. GALLO & S. G. CULL-CANDY. 1989. Multiple conductance channels in type 2 cerebellar astrocytes activated by excitatory amino acids. Nature **339:** 380-383.
58. BAUDRY, M., J. EVANS & G. LYNCH. 1986. Excitatory amino acids inhibit stimulation of phosphatidylinositol metabolism by aminergic agonists in hippocampus. Nature **319:** 329-331.
59. PARNAVELAS, J. G. & M. E. BLUE. 1982. The role of the noradrenergic system on the formation of synapses in the visual cortex of the rat. Dev. Brain Res. **3:** 140-144.
60. MOWER, G. D., C. J. CAPLAN, W. G. CHRISTEN & F. H. DUFFY. 1985. Dark rearing prolongs physiological but not anatomical plasticity of the cat visual cortex. J. Comp. Neurol. **235:** 448-466.
61. NIETO-SAMPEDRO, M., D. SHELTON & C. W. COTMAN. 1980. Specific binding of kainic acid to purified subcellular fractions from rat brain. Neurochem. Res. **5:** 591-604.
62. MURPHY, D. E., E. W. SNOWHILL & M. WILLIAMS. 1987. Characterization of quisqualate recognition sites in rat brain tissue using DL-[^{3}H]α-amino-3-hydroxy-5-methylisoxazole-4-propionic acid (AMPA) and a filtration assay. Neurochem. Res. **12:** 775-783.
63. FOSTER, A. C. & G. E. FAGG. 1987. Comparison of L-[^{3}H]glutamate, D-[^{3}H]aspartate, DL-[^{3}H]AP5 and [^{3}H]NMDA as ligands for NMDA receptors in crude postsynaptic densities in from rat brain. Eur. J. Pharmacol. **133:** 291-300.
64. DUDEK, S. M., A. P. BOHNER & M. F. BEAR. 1990. Effects of AP3 on EAA-stimulated PI turnover and ocular dominance plasticity in the kitten visual cortex. Neurosci. Abstr. **16:** 331.6.
65. BOWMAN, W. C., M. J. RAND & G. B. WEST. 1968. Textbook of Pharmacology. Blackwell. Oxford.
66. STANTON, P. K. & T. J. SEJNOWSKI. 1989. Associative long-term depression in the hippocampus induced by Hebbian covariance. Nature **339:** 215-218.
67. CHATTARJI, S., P. K. STANTON & T. J. SEJNOWSKI. 1990. 2-Amino-3-phosphonopropionate (AP3) blocks induction of associative long-term depression (LTD) in hippocampal field CA1. Neurosci. Abstr. **16:** 276.20.
68. STRATTON, K. R., P. F. WORLEY & J. M. BARABAN. 1990. Excitation of hippocampal neurons by stimulation of glutamate Q_p receptors. Eur. J. Pharmacol. **173:** 235-237.

PART II. MOLECULAR MECHANISMS OF ACTIVITY-DRIVEN CHANGE

Introduction

This section focuses on cellular and molecular mechanisms that may be involved in altering synaptic relationships. Neurons that are growing or regenerating their axons express the neuron-specific phosphoprotein GAP-43 at high levels and transport it via the rapid transport system to their growth cones and nerve terminals. After establishing stable synaptic connections, most cells express very little GAP-43, but some, particularly those in associative cortex and hippocampus, express it throughout life. Benowitz and Perrone-Bizzozero discuss the correlation of its expression with synaptic development and areas where learning may occur, and also address the mechanisms controlling its expression. Norden *et al.* describe studies of the physiological role of GAP-43 in nerve endings, the possible role in calcium regulation and in the control of membrane fusion events that underlie both neurotransmitter release and insertion of new membrane into growth cones.

Shashoua discusses another group of brain proteins, the ependymins, that are known to be released from nonneuronal cells (because of a drop in extracellular calcium) and that may aggregate at synapses participating in learning. He hypothesizes that these depositions of matrix material are linked to the elaboration of synaptic contacts during development and learning. This concept is supported by findings that ependymin antibodies interfere with both and also stain the synaptic cleft in hippocampus.

Finally, Schilling *et al.* discuss the link between physiological activity and changes in gene transcription, which could produce enduring changes. Second messengers (for example, calcium) that are produced by activity alter the binding of proteins to regulatory elements in genes and thereby enhance or repress their rates of transcription. Physiological stimulation in hippocampus may trigger a cascade of changes in which early onset genes (for example, c-*fos*) are activated within minutes, their products influencing the expression of other genes. Such cascades may change the functional state and perhaps also the structure of the neurons involved.

Together, these lines of investigation are providing insights into one of the most intriguing and daunting issues in neuroscience, how transient patterns of physiological activity, such as those that accompany a fleeting experience, can bring about the long-lasting changes in synaptic properties that become evident whenever we have learned something new.

The Relationship of GAP-43 to the Development and Plasticity of Synaptic Connections[a]

LARRY I. BENOWITZ AND
NORA I. PERRONE-BIZZOZERO

Department of Psychiatry and Program in Neuroscience
Harvard Medical School
and
Mailman Research Center
McLean Hospital
Belmont, Massachusetts 02178

In *The Histology of the Nervous System of Man and Vertebrates,* Ramón y Cajal[1] proposed that learning and memory utilize cellular mechanisms similar to those involved in the initial development of synaptic relationships:

> The extension, growth, and multiplication of neural processes do not stop at the time of birth; they continue beyond that; and nothing is more striking than the difference that exists between the newborn and the adult from the point of view of the length and number of second and third order neuronal processes. Usage is without doubt a basic feature of these modifications...
>
> ...The expansion of newly formed cellular processes do not advance haphazardly; they are determined by the dominant patterns of neural activity, or again by the intercellular associations that result from voluntary (mental) associations. It is thought that the expansion of these new associations are accompanied by an active growth.

One molecule that may be common to the initial development of synaptic relationships and to their modification later on in life is a phosphoprotein of the nerve terminal membrane that was discovered independently in several labs and given a variety of names, including GAP-43 (which will be used here), B-50, F1, pp46, 48k 4.8, P-57 (neuromodulin), and GAP-48 (or protein 4). The present article will attempt to emphasize aspects of this protein's structure, functions, and regulation not covered in other recent reviews.[2,3]

EXPRESSION OF GAP-43 IN THE DEVELOPING NERVOUS SYSTEM

Prior to undergoing final cell division and differentiation, neurons express very little GAP-43.[4,5] When differentiation is stimulated in culture, or when this process begins

[a]This work was supported by the National Eye Institute (EY05690), the National Institute of Neurological Disease and Stroke (NS25830), and the Scottish Rite Foundation.

in vivo, levels of the mRNA and protein rise rapidly coincident with the onset of process outgrowth.[4,6–8] In hippocampal pyramidal cells in culture, the protein is initially abundant throughout the soma and its nascent processes when growth begins, but at the point at which one process begins to extend more rapidly and becomes an axon, GAP-43 withdraws from processes destined to become dendrites and gets redirected down the axon.[9] The close temporal relationship between GAP-43 redistribution and axonal differentiation indicates that the protein may play a causal role in this process. *In vivo,* GAP-43 is densely concentrated along axons as they are elongating.[5,10,11] As development proceeds, the protein becomes less concentrated in the proximal portions of the axon and moves out distally,[12] becoming one of the most abundant proteins of the growth cone membrane.[12–15] *In vivo,* GAP-43 disappears from axons rather abruptly, though for a brief period of time it remains concentrated in the immature synapses of the neuropil. In much of the brain, this stage also lasts for only a short while, before levels of the protein in the neuropil begin to drop sharply.[5,10,11] It may be significant that the time of this transition coincides with the known time course of the critical period, in which the pattern of synaptic relationships is amenable to being altered by physiological activity (FIG. 1). In the rat cerebral cortex, for example, GAP-43 is concentrated in the neuropil for only a few days in the first postnatal week,[16] the time at which the thalamocortical projection of the facial vibrissae in the parietal cortex is amenable to being altered by manipulating ascending aspects of the trigeminal pathway. Likewise, in the hamster visual pathway, the time at which GAP-43 immunostaining remains high in the superior colliculus coincides with the time in which synapses are forming and are susceptible to being altered by experimental manipulations.[10] In the cat striate cortex, the period of intense neuropil staining similarly coincides with the critical period for the formation of feature detectors.[17] Although levels of the protein decline markedly in most regions of the brain, there are areas where high levels persist throughout life, as reviewed below.

GAP-43 IN NERVE REGENERATION

As in the initial development of the brain's circuitry, GAP-43 also appears to be involved in the regeneration of neural connections. In the optic pathway of lower vertebrates, retinal ganglion cells increase their synthesis of the mRNA about 20-fold within a few days of transecting the nerve.[18,19] Levels of the protein increase approximately 100-fold during the first few weeks of regeneration and begin to decline when the axons reach the brain, though they remain above the levels in the intact pathway throughout the period in which synaptic organization takes place.[20–23] Regeneration of peripheral nerves in mammals is likewise accompanied by large increases in the synthesis of GAP-43 mRNA and protein.[24–26] During the course of this regeneration, the protein is synthesized in the ganglionic neurons and conveyed not only down the regenerating peripheral nerve, but down the unlesioned dorsal root as well.[27,28] It is possible that this indiscriminate transport could have undesirable consequences if the presence of GAP-43 (and perhaps other growth-associated molecules) makes the central projections susceptible to being altered in their functional and structural relationships. This could be a possible basis for the hyperalgesia that results when peripheral nerves regenerate after injury in man.

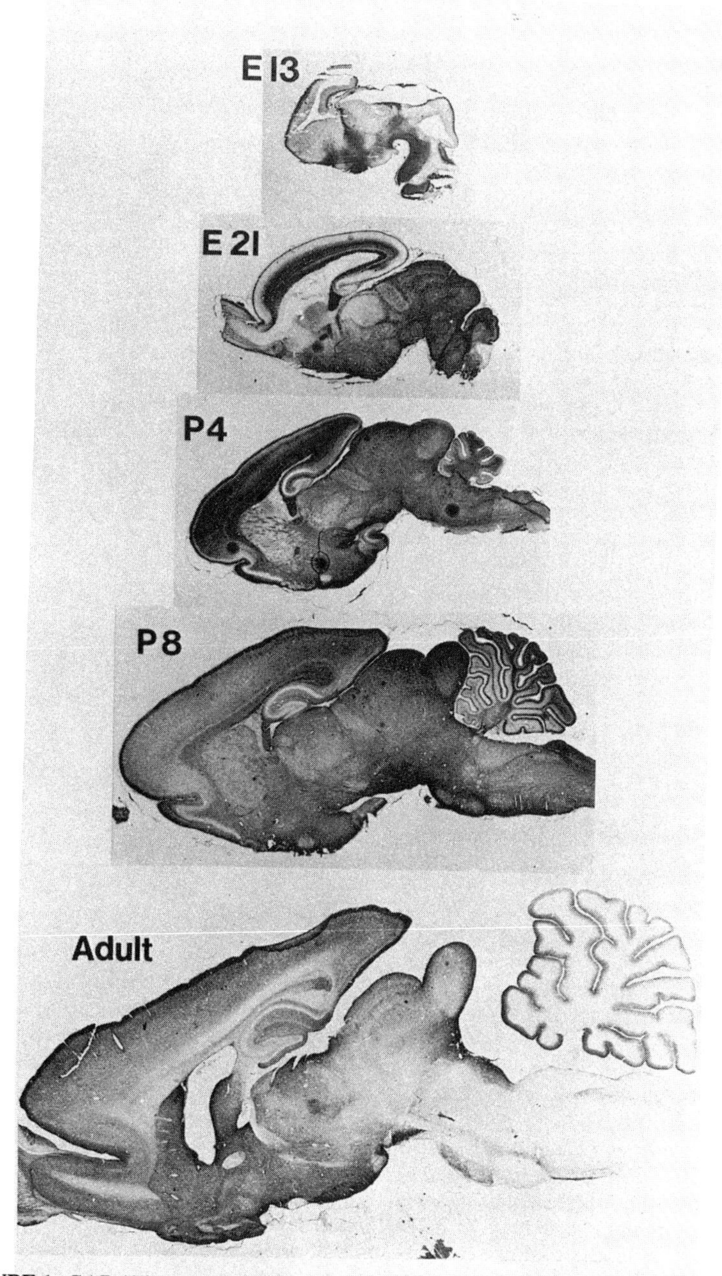

FIGURE 1. GAP-43 is concentrated in growing axons and nerve endings during brain development, then declines in most brain regions after synapses mature. Levels peak in the neocortex at postnatal day 4, at the height of synaptogenesis (and the critical period), then drop off rapidly. Moderately high levels, however, remain in the molecular layer of the neocortex, basal forebrain, and some other regions throughout life.

GAP-43 IN THE ADULT NERVOUS SYSTEM

Although overall levels of GAP-43 decline sharply in most parts of the brain after synaptogenesis has occurred,[29] there are regions where appreciable levels persist throughout life.[30] In the adult rat, there is a sharp concentration gradient along the neuraxis, with very little GAP-43 in the spinal cord or brainstem below the diencephalic level, but with high levels of GAP-43 in the cerebral cortex, particularly in limbic system structures, the striatum, the medial substantia nigra, and in a continuum of subcortical structures from the olfactory bulb through the olfactory tubercle, bed nucleus of the stria terminalis, amygdala, preoptic area, and hypothalamus.[31] In the primate brain, including that of man, the regional differences are even more striking, perhaps reflecting a greater degree of functional specialization for synaptic plasticity. Nelson *et al.*[32] reported a greater *in vitro* incorporation of radioactive phosphate into GAP-43 (that is, protein F1) in membrane preparations from higher integrative areas of the visual system than in membrane preparations from primary or secondary visual cortex. In the human brain, we found that these differences reflected a general pattern of regional variations in the levels of the mRNA and protein. Whereas GAP-43 expression is low in primary sensory and motor cortex, it is extremely high in much of the associative neocortex, such as the inferior temporal cortex, perisylvian cortex, prefrontal cortex, and cingulate (FIG. 2). In the hippocampal formation, high levels of the mRNA are found in CA3 pyramidal cells, and the protein is enriched in the molecular layer of the dentate gyrus and in other subfields.[33,34] Observations from clinical neuropsychology and physiological psychology indicate that higher integrative functions, and perhaps the storage of memories, may involve just these structures, and it is therefore intriguing to speculate that the presence of GAP-43 may be part of the synaptic machinery that endows certain neurons with the potential for functional and structural remodeling.[2,3,11,29,32,35]

SYNAPTIC REMODELING

Although the initial visualization of GAP-43 was in relation to axonal growth, its abundance in all nerve endings during development, and its persistence in certain synapses of the adult nervous system, suggest that it may be involved in synaptic growth and change. One well-known instance of synaptic remodeling occurs in the rat hippocampal formation (FIG. 3), where lesioning the perforant pathway induces projections from hippocampal pyramidal cells (in the CA3 and CA4 fields) and other areas to sprout collaterals that form new synapses in place of the ones lost by the surgery.[36] During the course of this, GAP-43 levels increase greatly in the newly sprouting terminals.[37,38] Similarly, lesions of the fimbria-fornix have been reported to induce changes in the levels of GAP-43 mRNA in neurons undergoing collateral sprouting.[39] Thus, even in the absence of extensive axonal outgrowth, the formation of new synapses involves increased synthesis of GAP-43. This in turn implies that neurons which constitutively express high levels of the protein under normal circumstances may continuously be undergoing synaptic remodeling.

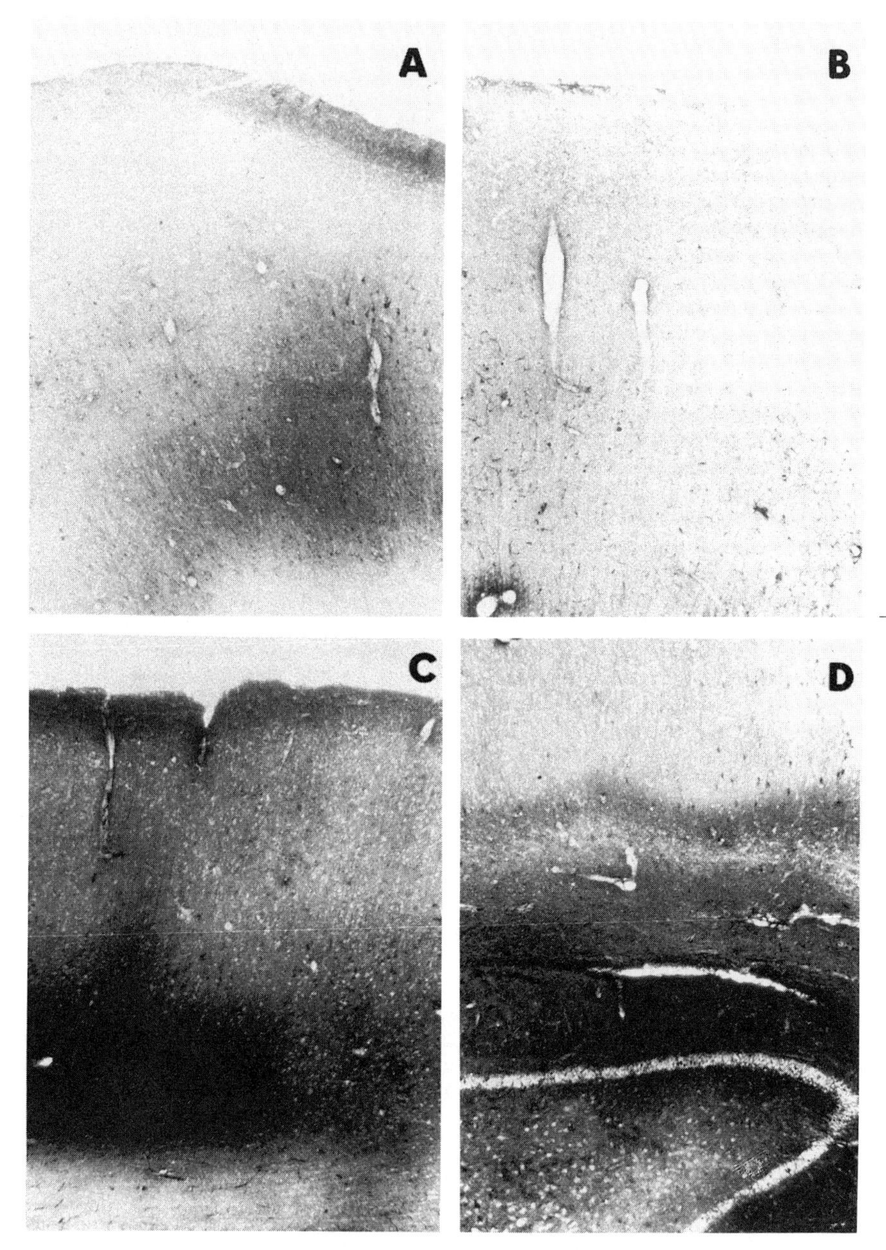

FIGURE 2. GAP-43 remains abundant in associative and limbic portions of the human brain throughout life. Levels are very low in primary sensory (**A:** visual cortex, Brodmann area 17) and motor cortex (**B:** motor cortex, area 4), but are high in associative areas (**C:** inferior temporal cortex) and parts of the hippocampal formation (**D**). These findings are consistent with a role for the protein in associative mechanisms, as is also suggested by its implication in long-term potentiation.

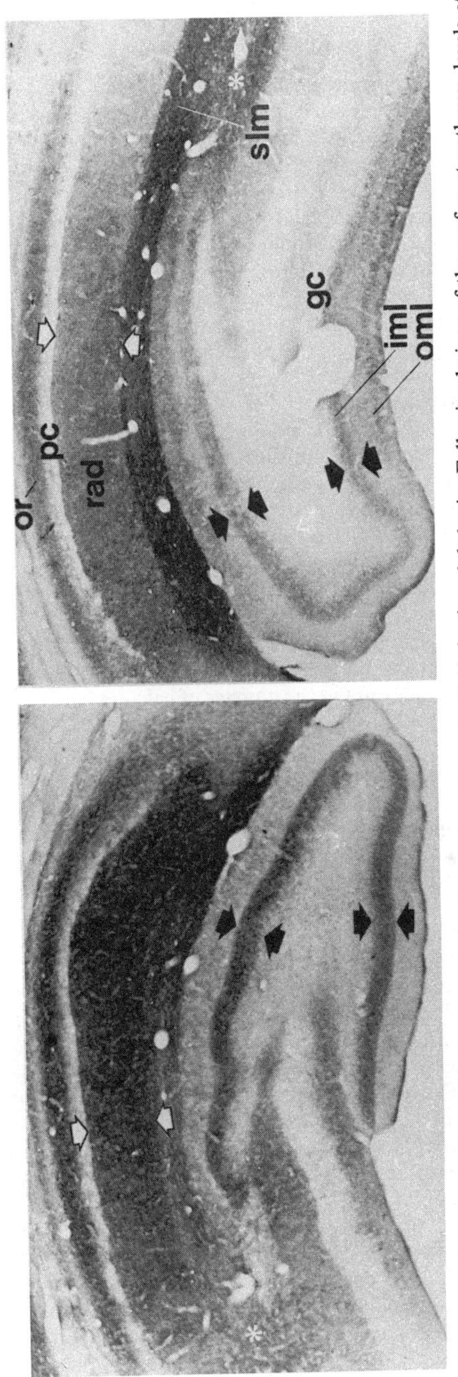

FIGURE 3. GAP-43 is involved in the reorganization of synaptic connections in the adult brain. Following lesions of the perforant pathway, levels of GAP-43 increase markedly in the inner molecular layer of the dentate gyrus (black arrows) ipsilateral to the lesion (left), coincident with the collateral sprouting of commissural-associational axons originating from pyramidal cells in the hilus and CA3 region.

MOLECULAR INTERACTIONS OF GAP-43

If GAP-43 is indeed related to the formation of neuronal relationships, it is important to understand how this occurs in terms of the protein's interactions with other cellular constituents. Toward this end, a number of studies have examined the relationship of GAP-43 to major signal transduction systems of the nerve ending and the effects of manipulating the protein on synaptic properties. The earliest mention of this protein was in the context of it being a synaptic phosphoprotein, sensitive to changes in calcium concentration but not cAMP.[40,41] Subsequent work showed that the phosphorylation of GAP-43 is regulated by the calcium/phospholipid-dependent kinase, protein kinase C (PKC),[42,43] and that the protein is one of the principal substrates of this kinase.[13,14] In procedures used to isolate the protein, both its kinase (that is, PKC) and the kinase responsible for the phosphorylation of phosphatidylinositol phosphate (PIP) to phosphatidylinositol 4,5-bisphosphate (PIP_2) co-purify with GAP-43 (that is, B-50) for several steps.[44] Moreover, whereas antibodies to GAP-43, or dephosphorylating the protein, both *increase* the activity of PIP kinase (thus producing more PIP_2), enhancing the phosphorylation of the protein diminishes the activity of the enzyme.[44,45] Because PIP_2 is subsequently hydrolyzed by phospholipase C to form diacylglycerol and inositol trisphosphate, GAP-43 would be in a strategic position to regulate the activity of protein kinase C and other Ca^{2+}-dependent events,[46] thereby controlling any number of physiological properties of the nerve terminal.

Another line of investigation has suggested that GAP-43 (that is, P-57 or neuromodulin) interacts with the calcium-binding protein, calmodulin.[47] The interaction of the two proteins was first visualized by the photoaffinity labeling of GAP-43 with a calmodulin derivative after the protein was extracted from the membrane.[48] This binding was found to take place only in the *absence* of calcium, and to occur only when GAP-43 was dephosphorylated.[47] The portion of GAP-43 that seems to be responsible for the calmodulin binding[49] is adjacent to the phosphorylation site (serine 41: see below) and represents the most highly conserved portion of the molecule.[18] Based upon these observations, it was proposed that GAP-43 binds calmodulin and sequesters it locally at certain sites along the nerve terminal membrane, releasing it to activate other effector molecules (for example, calcium/calmodulin kinase) when calcium levels build up (or when GAP-43 is phosphorylated). Conversely, calmodulin might serve to regulate the functions of GAP-43 in response to changes in calcium levels. It is difficult to reconcile these ideas, however, with the observations that 1) the binding has not been demonstrated when GAP-43 is still attached to the membrane,[48] and 2) whereas the binding requires the presence of an anionic detergent (Zwiers and Coggins, personal communication), GAP-43, as a peripheral protein on the cytoplasmic face of the membrane, presumably exists in an aqueous environment.

More recently, an interaction of GAP-43 with the signal transduction molecule, G_o, has been described.[50] The α and β subunits of G_o are major constituents of neuronal growth cone membranes, and in *in vitro* assays, purified GAP-43, or an NH_2-terminal peptide representing its first 24 amino acids, stimulated the binding of GTP to G_o. Several approaches failed to reveal a strong association of GAP-43 to G protein subunits; however, the possibility that this interaction occurs *in vivo* is strengthened by the observation that one portion of GAP-43 that is involved in the interaction shares a consensus sequence with other proteins known to interact with G proteins.[50] Through its interactions with G_o, GAP-43 could be involved in transducing intracellular information, either to influence the generation of second messengers or to directly influence an ion channel. Along these lines, the introduction of antibodies to GAP-43 into pheochromocytoma cells has been reported to alter calcium uptake.[51]

The sequence of GAP-43 reveals a region at the NH_2-terminal that is relatively more hydrophobic than the rest of this very hydrophilic protein.[7] Removal of the NH_2-terminal region, through genetic engineering, renders the protein unable to bind to the membrane, while conversely, adding the first 10 amino acids of GAP-43 to a soluble bacterial protein causes the latter to associate with the nerve terminal membrane. The two cysteines in positions 3 and 4 of the amino acid sequence are particularly important for the membrane attachment.[52] What has remained unclear, however, is the basis of this linkage. Reducing agents that disrupt disulfide bonds do not affect the binding of GAP-43 to the membrane, nor do chelating agents or high salt concentrations; on the other hand, alkaline conditions and modest concentrations of anionic detergents will dissociate the protein.[45,53,54] Two studies examined whether the linkage of GAP-43 to the membrane involves fatty acid acylation: whereas one of these reported palmitic acid incorporation into the protein,[55] another did not find this.[56] One possible basis for this discrepancy lies in the fact that the two studies allowed the embryonic neurons to develop under somewhat different culture conditions, one of which may have been more conducive to acylation than the other. In any event, the observations that 1) the acylation is dynamic and reversible, and 2) that GAP-43 seems to attach to growth cone membranes regardless of acylation, suggest that this reaction is not obligatory for the membrane attachment of the protein, but may have another modulatory role.

PHYSIOLOGICAL ROLE OF GAP-43

Several studies point to a possible role for GAP-43 (that is, B-50) in regulating neurotransmitter release. In the hippocampal slice preparation, membrane depolarization causes a rapid and transient change in GAP-43 phosphorylation that parallels the time course of transmitter release.[57] Moreover, the introduction of antibodies to GAP-43 into isolated synaptosomes interferes with the calcium-dependent release of norepinephrine.[58] It is possible that this regulation could be a consequence of the reported effect of GAP-43 on calcium influx.[51] Because some synapses contain extremely low levels of GAP-43, whereas others have high concentrations,[31,59] it seems likely that whatever role the protein plays in transmitter release is not a prerequisite for this process to occur, though it may modulate release from a limited subset of synapses. In this regard, it is interesting to note that the macroscopic distributional pattern of GAP-43 in the rat brain is rather similar to the distribution of *N*-methyl-D-aspartate (NMDA) receptors.[60] This would suggest that in the adult nervous system, the protein may be linked particularly to the modulation of glutaminergic transmission.

Routtenberg and his colleagues have provided evidence that long-term potentiation (LTP), induced in the hippocampal formation following high-frequency stimulation of the perforant pathway, may involve an enduring increase in the phosphorylation of GAP-43 (that is, protein F1).[61–63] In this work, the perforant pathway was stimulated in the living animal, after which phosphorylation was examined by labeling membrane fractions of the hippocampal formation *in vitro.* This has raised the question of whether the observed changes in GAP-43 phosphorylation actually occur *in vivo.* It has been shown repeatedly that manipulations of protein kinase C activity do alter LTP,[64,65] although it is not yet certain whether it is the phosphorylation of GAP-43 per se, rather than other substrates of the kinase, that is most relevant.

Whereas GAP-43 has been linked to presynaptic aspects of synaptic regulation, NMDA receptors are involved in the postsynaptic induction of LTP (but not its persistence).[66] When LTP is blocked using an NMDA receptor antagonist, changes in GAP-43 phosphorylation that normally accompany high-frequency stimulation of the perforant pathway fail to appear.[67] One suggested route by which postsynaptic changes (involving NMDA receptor activation and calcium influx) could influence presynaptic physiology would be through the release of arachidonic acid or its metabolites; such release has been reported to occur after the induction of LTP in hippocampal slices,[68] and in turn, arachidonic acid can influence the activation of protein kinase C.[69]

STRUCTURE AND REGULATION OF THE GENE

Through the isolation and sequencing of GAP-43 cDNAs and genomic DNA, several interesting properties of the protein, mRNA, and gene have emerged. The GAP-43 proteins of the goldfish, chick, mouse, rat, cow, and man, are highly hydrophilic, charged molecules with a striking conservation of amino acid sequences in the NH_2-terminal region (FIG. 4).[6,7,18,19,70–75] This region includes the membrane-attachment site (and site of interaction with G_o: amino acids 1-10: references 50 and 52), the sequence surrounding the protein kinase C phosphorylation site (serine 41: reference 76), and the putative calmodulin-binding site.[49] Past this point, whereas the mammalian proteins are highly similar, goldfish GAP-43 diverges and shows only occasional sequence similarity to the mammalian and avian proteins.[18,19] In all cases, the sequence predicts a protein of only 210-230 amino acids with a molecular size of about 24 kDa. This is far smaller than indicated by its migration position on SDS-polyacrylamide gels, but predictable from its anomalous electrophoretic migration properties and its sedimentation in density gradients.[77,78]

The 1.5-kilobase messenger RNA that encodes GAP-43 contains a very long untranslated 3′ end, portions of which are as highly conserved throughout evolution as the translated portion of the mRNA.[19,74] Certain sequences in the 3′ end resemble regions known to influence the rate of mRNA degradation (for example, AUUUA regions and stem-loop structures, which are thought to bind proteins that modulate ribonuclease activity). Consistent with this, GAP-43 levels appear to be controlled not by transcriptional mechanisms but by the rate of mRNA degradation. In pheochromocytoma cells, levels of GAP-43 mRNA increase substantially when cells are induced to differentiate with nerve growth factor.[6,8,79] These cells, however, show active transcription of the GAP-43 gene even in the control, undifferentiated state, and the rate of transcription does not change with nerve growth factor treatment.[79,80] Similarly, in goldfish retinal ganglion cells undergoing axonal regeneration, although levels of GAP-43 mRNA increase approximately 20-fold within the first few days of crushing the optic nerve,[18] the gene is transcribed at a moderate rate even in the intact state, and this does not change during regeneration.[80] In studies on the rate of mRNA degradation, the half-life of GAP-43 mRNA was found to increase markedly when cells were stimulated to differentiate. This suggests that the key control point is the regulation of the stability

FIGURE 4. Much of the structure of GAP-43 has remained unaltered for a half a billion years. Analysis of GAP-43 cDNAs from several species shows a high degree of conservation of the first 58 amino acids, which are thought to include the site that is phosphorylated by protein kinase C (S-41, denoted with an asterisk), a region that binds the protein calmodulin (box with dashed lines), and the membrane-binding domain (box with solid lines). GAP-43 is a highly charged, acidic protein with no hydrophobic domains.

```
              5    10   15   20   25   30   35   40
Human     MLCCMRRTKQVEKN-DDDQKIEQDGIKPEDKAHKAATKIQ
Bovine    MLCCMRRTKQVEKN-DEDQKIEQDGIKPEDKAHKAATKIQ
Rat       MLCCMRRTKQVEKN-DEDQKIEQDGVKPEDKAHKAATKIQ
Mouse     MLCCMRRTKQVEKN-DEDQKIEQDGVKPEDKAHKAATKIQ
Chicken   MLCCMRRTKQVEKNEDGDQKIEQDGIKPEDKAHKAATKIQ
Goldfish  MLCCIRRTKPVEKNEEADQEIKQDGTKPGENAHKAATKIQ

          41*  45   50   55   60   65   70   75   80
Human     ASFRGHITRKKLKGEKKDDVQAAEAEANKKDEAPVADGV-
Bovine    ASFRGHITRKKLKGEKKGDAPAAEAEANEKDEAAVAEGT-
Rat       ASFRGHITRKKLKDEKKGDAPAAEAEAKEKDDAPVADGV-
Mouse     ASFRGHITRKKLKGEKKGDAPAAEAEAKEKDDAPVADGV-
Chicken   ASFRGHITRKKLKGEKKADAPASESEAADKKDEGPAGGRA
Goldfish  ASFRGHITRKKMKDEDKD----------GENDTAP-----

          81   85   90   95   100  105  110  115  120
Human     EKK-GEGTTTAEAAPATGSKPDE-PGKAGETPSEEKKGEG
Bovine    EKKEGEGSTPAEAAPGAGPKPEEKTGKAGETP---KKGEG
Rat       EKKEGDGSATTDAAPATSPKAEE-PSKAGDAPSEEKKGEG
Mouse     EKKEGDGSATTDAAPATSPKAEE-PSKAGDAPSEEKKGEG
Chicken   ENKESEASTATEA-SAADSAQLDEGSKDSSVPTEEKKGNG
Goldfish  -----DESAET----------EEK--KERVSPSEEKPVEV

          121  125  130  135  140  145  150  155  160
Human     --DAA-TEQAAPQA--PA-SSEEK-AGSAETESATKASTD
Bovine    APDAA-TEQAAPQA--PA-PSEEK-AGSAETESATKASTD
Rat       --DAA--------------PSEEK-AGSAETESAAKATTD
Mouse     --DAA--------------PSEEK-AGSAETESAAKATTD
Chicken   AADTG-SEQPAPQAATPAASSEEKPAAAAETESATKASTD
Goldfish  STETAEESKPAEQPNSPAA---EAP-PTATTDSA--PS--

          161  165  170  175  180  185  190  195  200
Human     NSPSSKAEDAPAK--------------EEPKQADVPAAVT
Bovine    NSPSSKAEDAPAK--------------EEPKQADVPAAVT
Rat       NSPSSKAEDGPAK--------------EEPKQADVPAAVT
Mouse     NSPSSKAEDGPAK--------------EEPKQADVPAAVT
Chicken   NSPSLKADE--AKTKRSLNKPTCLLLT--------PLATT
Goldfish  DTPT-K-EEAQEQLQDA----------EEPKEAENTAAAD

          201  205  210  215  220  225  230  235  240
Human     ------------------------AAAATTPAAEDAAAKA
Bovine    ------------------------AAAATAPAAEDAAAMA
Rat       ------------------------DAAATTPAAEDAA-KA
Mouse     ------------------------DAAATTPAAEDAATKA
Chicken   -----------------------------TPAAEDATAKA
Goldfish  DITTQKEEEKEEEEEEEEEEAKRADVPDDTPAA-------

          241
Human     TAQPPTETGESSQAEENIEAVDETKPKESARQDEGKE-EP
Bovine    TAQPPTETAESSQAEEKIEAVDETKPKDSARQDEGKG-ER
Rat       -AQPPTETAESSQAEEEKEAVDEAKPKESARQDEGKE-DP
Mouse     -AQPPTETAESSQAEEEKDAVDEAKPKESARQDEGKE-DP
Chicken   RLQPQMETVESSQTEEKTDAVEETKPTESAQQEEVKEEES
Goldfish  -----TESQETDQT-DKKEALDDSKPAEEA----GKDQNV

          281
Human     EADQEHA
Bovine    EADQEHA
Rat       EADQEHA
Mouse     EADQEHA
Chicken   KADQENA
```

of the mRNA.[80] In view of the high degree of conservation for sequences in the 3′ untranslated portion of the mRNA, it is likely that stabilization is mediated by the binding of specific proteins, in a growth factor dependent manner, to certain portions of the 3′ end, thereby influencing ribonuclease activity.

Induction of GAP-43 expression appears to be synonymous with the onset of process outgrowth in neurons.[3] Therefore, it might be concluded that any of a number of signals that induce differentiation can stimulate the expression of GAP-43 mRNA and protein. As is true for most aspects of neuronal differentiation, the transduction mechanisms that underlie this are unknown. However, one source of signals controlling GAP-43 expression appears to consist of factors that are retrogradely transported up the axon. Many studies have shown that axotomy in peripheral nerves induces the expression of GAP-43 mRNA and protein.[24–26] If peripheral nerves are ligated, so that growing axons are prevented from reaching their target cells, levels of GAP-43 mRNA do not return to baseline levels, but remain high for a long period of time.[81] This suggests that the production of GAP-43 mRNA levels normally becomes suppressed by a factor that travels up the axon when synapses are made with target cells.[81] Support for this hypothesis comes from studies that interfere with axonal transport using agents that disrupt microtubules. When vinblastine is applied to the rat sciatic nerve, levels of GAP-43 and its mRNA increase to the same degree as after crushing the nerve (FIG. 5).[27,82]

Studies on the structure of the GAP-43 gene[83,84] show no recognizable enhancer or promoter regions within 10 kilobases of the transcription start site. The gene, however, does have several prominent GA-rich stretches, which are believed to serve as potential sites for the formation of H-DNA.[83] This is a triple-stranded conformation that leaves one strand free and potentially open for continuous transcription.[85] This would be consistent with the finding that GAP-43 mRNA levels are not controlled transcriptionally, but involve a different mechanism.

OUTSTANDING ISSUES

Despite the increasing number of studies on the structure, biochemistry, and functions of GAP-43, we still do not have a clear picture as to how this protein contributes to the regulation of the function and structure of the nerve terminal. The observations relating the protein to calmodulin binding, phosphatidylinositol metabolism, and G protein stimulation, have all been made in isolated systems: whether these interactions occur *in vivo,* and whether they are a subset of a larger set of molecular interactions of the protein, remain to be determined. With regard to synaptic physiology, the finding that antibodies to GAP-43 interfere with calcium influx and transmitter release are certainly intriguing, but again, the molecular bases of these effects and their relationship to other events in the synapse remain to be elucidated. Of particular interest is the issue of whether long-term potentiation, a model for the synaptic changes that underlie learning and memory, involves GAP-43 phosphorylation. Again, the fact that these observations utilized post hoc phosphorylation in isolated membrane fractions, where the extent of prior protein phosphorylation *in vivo* is unknown, makes interpretation difficult. More direct observations through *in vivo* phosphate labeling would be desirable.

With regard to the mechanisms that control GAP-43 expression, this too is an area just beginning to be explored. As indicated above, GAP-43 expression appears to depend more on posttranscriptional mechanisms than on transcriptional changes. The signal transduction mechanisms that control this regulation must be complex, since in some neurons GAP-43 continues to be expressed at high levels even into maturity (for example, in pyramidal cells in layer 2 of associative neocortex); in others it can be induced by axotomy alone (as in the peripheral nervous system); in some neurons it can be induced by making new target areas available (for example, reactive synaptogenesis in the hippocampal formation); in some cases treatment with growth factors is effective (for example, nerve growth factor in the case of PC12 cells or ganglionic neurons); and in some cases none of these suffices (for example, in the case of axotomized retinal ganglion cells). However, the fact that the latter can be reversed by presenting the cut nerve ending to the environment of the peripheral nervous system[86] argues for a strong contribution by glial-derived factors.

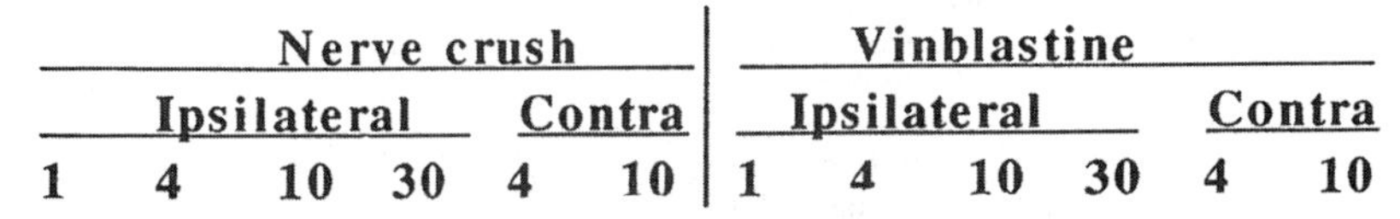

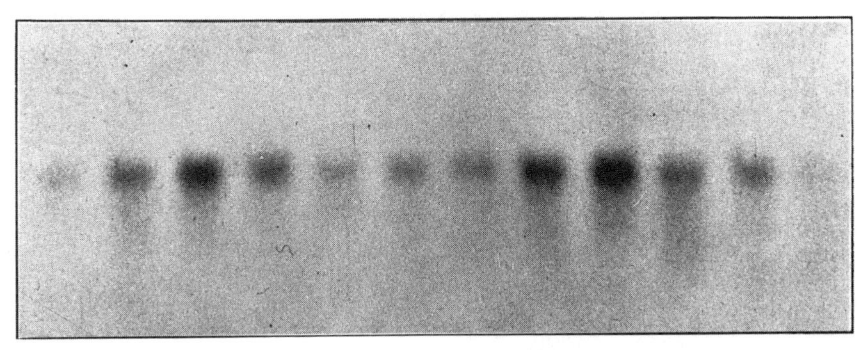

FIGURE 5. GAP-43 mRNA levels are regulated by a factor that is retrogradely transported from the periphery to the cell body. Rat sciatic nerves were either crushed or treated with vinblastine. At the times indicated, dorsal root ganglia (ipsilateral and contralateral to the treated nerve) were dissected and the mRNA was isolated, run on gels, transferred to a nylon membrane, and probed using ^{32}P-labeled GAP-43 cDNA. Both treatments caused equally large changes in levels of GAP-43 mRNA in ganglia ipsilateral to the treated nerves, with similar time courses. These results indicate that the effect of nerve crush may be to interrupt a retrogradely transported inhibitory factor that is normally conveyed from the periphery.

In the light of current observations, a more complete understanding of GAP-43 regulation would require the identification of 1) specific factors that derive from such sources as target interactions and the glial environment, whose presence or absence stimulates 2) a transduction mechanism that regulates the binding of 3) specific proteins to 4) particular segments of the mRNA, either enhancing or retarding its degradation. In view of the observations that transfection of the GAP-43 gene into nonneuronal cells stimulates process outgrowth,[87] and that oligonucleotides complementary to portions of GAP-43 mRNA interfere with process outgrowth (L. I. Benowitz, L. Dawes, N. I. Perrone-Bizzozero, N. Irwin, C. O'Brien, and R. L. Neve, unpublished observations), it would appear that GAP-43 may indeed be a central player in neuronal differentiation and synaptic remodeling. In this case, an understanding of its physiological role and the mechanisms controlling its expression are essential for understanding neuronal development and plasticity.

REFERENCES

1. RAMÓN Y CAJAL, S. 1972. Histologie du système nerveux de l'homme & des vertébrés, V. II. Consejo Superior de Investigaciones Cientificas. Instituto Ramón y Cajal. Madrid.
2. BENOWITZ, L. I. & A. ROUTTENBERG. 1987. A membrane phosphoprotein associated with neuronal development, axonal regeneration, phospholipid metabolism, and synaptic plasticity. Trends Neurosci. **10:** 527-532.
3. SKENE, J. H. P. 1989. Axonal growth-associated proteins. Annu. Rev. Neurosci. **12:** 127-156.
4. PERRONE-BIZZOZERO, N. I., S. P. FINKLESTEIN & L. I. BENOWITZ. 1986. Synthesis of a growth-associated protein by embryonic cortical neurons *in vitro.* J. Neurosci. **6:** 3721-3730.
5. DANI, J. W., D. M. ARMSTRONG & L. I. BENOWITZ. 1991. Mapping the development of the rat brain by GAP-43 immunocytochemistry. Neuroscience **40:** 277-287.
6. KARNS, L. R., S. H. NG, J. A. FREEMAN & M. C. FISHMAN. 1987. Cloning of complementary DNA for GAP-43, a neuronal growth-related protein. Science **236:** 597-600.
7. BASI, G. S., R. D. JACOBSON, I. VIRAG, J. SCHILLING & J. H. P. SKENE. 1987. Primary structure and transcriptional regulation of GAP-43, a protein associated with nerve growth. Cell **49:** 785-791.
8. COSTELLO, B., A. MEYMANDI & J. A. FREEMAN. 1990. Factors influencing GAP-43 gene expression in PC12 pheochromocytoma cells. J. Neurosci. **10:** 1398-1406.
9. GOSLIN, K., D. J. SCHREYER, J. H. P. SKENE & G. BANKER. 1990. Changes in the distribution of GAP-43 during the development of neuronal polarity. J. Neurosci. **10:** 588-602.
10. MOYA, K. L., S. JHAVERI, G. E. SCHNEIDER & L. I. BENOWITZ. 1989. Immunohistochemical localization of GAP-43 in the developing hamster retinofugal pathway. J. Comp. Neurol. **288:** 51-58.
11. MCGUIRE, C. B., G. J. SNIPES & J. J. NORDEN. 1988. Light-microscopic immunolocalization of the growth-associated protein GAP-43 in the developing brain. Dev. Brain Res. **41:** 277-291.
12. MEIRI, K. F., M. WILLARD & M. I. JOHNSON. 1988. Distribution and phosphorylation of the growth-associated protein GAP-43 in regenerating sympathetic neurons in culture. J. Neurosci. **8:** 2571-2581.
13. DE GRAAN, P. N. E., C. O. M. VAN HOOFF, B. C. TILLY, A. B. OESTREICHER, P. SCHOTMAN & W. H. GISPEN. 1985. Phosphoprotein B-50 in nerve growth cones from fetal rat brain. Neurosci. Lett. **61:** 235-241.
14. KATZ, F., L. ELLIS & K. H. PFENNINGER. 1985. Nerve growth cones isolated from fetal rat brain. III. Calcium-dependent protein phosphorylation. J. Neurosci. **5:** 1402-1411.
15. MEIRI, K., K. H. PFENNINGER & M. WILLARD. 1986. Growth-associated protein, GAP-43, a polypeptide that is induced when neurons extend axons, is a component of growth cones and corresponds to pp46, a major polypeptide of a subcellular fraction enriched in growth cones. Proc. Natl. Acad. Sci. USA **83:** 3537-3541.
16. ERZURUMLU, R. S., S. JHAVERI & L. I. BENOWITZ. 1990. Transient patterns of GAP-43 expression during the formation of barrels in the rat somatosensory cortex. J. Comp. Neurol. **292:** 443-456.
17. BENOWITZ, L. I., W. R. RODRIGUEZ, G. T. PRUSKY & M. S. CYNADER. 1989. GAP-43 levels in cat striate cortex peak at the height of the critical period. Neuroscience Abstr. **15:** 796.
18. LABATE, M. & J. H. P. SKENE. 1989. Selective conservation of GAP-43 structure in vertebrate evolution. Neuron **3:** 299-310.
19. PERRONE-BIZZOZERO, N. I., R. L. NEVE, E. R. FRANCK, J. M. GOSSELS, N. IRWIN & L. I. BENOWITZ. 1990. GAP-43 in the regenerating goldfish optic pathway: Sequence and regulation of mRNA levels. Submitted for publication.

20. BENOWITZ, L. I. & E. R. LEWIS. 1983. Increased transport of 44,000- to 49,000-Dalton acidic proteins during regeneration of the goldfish optic nerve: A two-dimensional gel analysis. J. Neurosci. **3:** 2153-2163.
21. SKENE, J. H. P. & M. WILLARD. 1981. Changes in axonally transported proteins during axon regeneration in toad ganglion cells. J. Cell Biol. **89:** 86-95.
22. PERRY, G. W., D. W. BURMEISTER & B. GRAFSTEIN. 1987. Fast axonally transported proteins in regenerating goldfish optic axons. J. Neurosci. **7:** 792-806.
23. BENOWITZ, L. I. & J. T. SCHMIDT. 1987. Activity-dependent sharpening of the regenerating retinotectal projection in goldfish: Relationship to the expression of growth-associated proteins. Brain Res. **417:** 118-126.
24. SKENE, J. H. P. & M. WILLARD. 1981. Axonally transported proteins associated with axon growth in rabbit central and peripheral nervous system. J. Cell Biol. **89:** 96-103.
25. VAN DER ZEE, C. E. E. M., H. B. NIELANDER, J. P. VOS, S. LOPES DA SILVA, J. VERHAAGEN, A. B. OESTREICHER, L. H. SCHRAMA, P. SCHOTMAN & W. H. GISPEN. 1989. Expression of growth-associated protein B-50/GAP-43 in dorsal root ganglia and sciatic nerve during regenerative sprouting. J. Neurosci. **9:** 3505-3512.
26. HOFFMAN, P. N. 1989. Expression of GAP-43, a rapidly transported growth-associated protein, and class II β-tubulin, a slowly transported cytoskeletal protein, are coordinated in regenerating neurons. J. Neurosci. **9:** 893-897.
27. WOOLF, C. J., M. L. REYNOLDS, C. MOLANDER, C. O'BRIEN, R. M. LINDSAY & L. I. BENOWITZ. 1990. GAP-43, a growth-associated protein, appears in rat dorsal root ganglion cells and in the dorsal horn of the spinal cord following peripheral nerve injury. Neuroscience **34:** 465-478.
28. ERZURUMLU, R. S., S. JHAVERI, K. L. MOYA & L. I. BENOWITZ. 1989. Peripheral nerve regeneration induces elevated expression of GAP-43 in the brainstem trigeminal complex of adult hamster. Brain Res. **498:** 135-139.
29. JACOBSON, R. D., I. VIRAG & J. H. P. SKENE. 1986. A protein associated with axon growth, GAP-43, is widely distributed and developmentally regulated in rat CNS. J. Neurosci. **6:** 1843-1855.
30. KRISTJANSSON, G. I., H. ZWIERS, A. B. OESTREICHER & W. H. GISPEN. 1982. Evidence that the synaptic phosphoprotein is localized exclusively in nerve tissue. J. Neurochem. **39:** 371-378.
31. BENOWITZ, L. I., P. J. APOSTOLIDES, N. I. PERRONE-BIZZOZERO, S. P. FINKLESTEIN & H. ZWIERS. 1988. Anatomical localization of the growth-associated protein GAP-43 (B-50) in the adult rat brain. J. Neurosci. **8:** 339-352.
32. NELSON, R. B., D. P. FRIEDMAN, J. B. O'NEILL, M. MISHKIN & A. ROUTTENBERG. 1987. Gradients of protein kinase C substrate phosphorylation in primate visual system peak in visual memory storage areas. Brain Res. **416:** 387-392.
33. NEVE, R. L., E. A. FINCH, E. D. BIRD & L. I. BENOWITZ. 1988. The growth-associated protein GAP-43 (B-50, F1) is expressed selectively in associative regions of the adult human brain. Proc. Natl. Acad. Sci. USA **85:** 3638-3642.
34. BENOWITZ, L. I., N. I. PERRONE-BIZZOZERO, S. P. FINKLESTEIN & E. D. BIRD. 1989. Localization of the growth-associated phosphoprotein GAP-43 in the human cerebral cortex. J. Neurosci. **9:** 990-995.
35. SNIPES, G. J., C. B. MCGUIRE, S. CHAN, B. R. COSTELLO, J. J. NORDEN, J. A. FREEMAN & A. ROUTTENBERG. 1987. Evidence for the co-identification of GAP-43, a growth-associated protein, and F1, a plasticity-associated protein. J. Neurosci. **7:** 4066-4075.
36. COTMAN, C. W., D. A. MATTHEWS, D. TAYLOR & G. LYNCH. 1973. Synaptic rearrangement in the dentate gyrus: Histochemical evidence of adjustments after lesions in immature and adult rats. Proc. Natl. Acad. Sci. USA **70:** 3473-3477.
37. BENOWITZ, L. I., W. R. RODRIGUEZ & R. L. NEVE. 1990. The pattern of GAP-43 immunostaining changes in the rat hippocampal formation during reactive synaptogenesis. Mol. Brain Res. **8:** 17-23.
38. NORDEN, J. J., R. WOLTJER & O. STEWARD. 1988. Changes in the immunolocalization of the growth- and plasticity-associated protein GAP-43 during lesion-induced sprouting in the rat dentate gyrus. Neurosci. Abstr. **14:** 116.
39. SCHRAMA, L. H., B. J. C. EGGEN, H. B. NIELANDER, P. SCHOTMAN, A. B. OESTREICHER, B. M. SPRUIJT & W. H. GISPEN. 1989. Increased expression of the growth-associated protein B-50 (GAP-43) in unilateral fimbria-fornix-lesioned rats. Neurosci. Abstr. **15:** 319.

40. ROUTTENBERG, A. & Y. H. EHRLICH. 1975. Endogenous phosphorylation of four cerebral cortical membrane proteins: Role of cyclic nucleotides, ATP and divalent cations. Brain Res. **92:** 415-430.
41. ZWIERS, H., V. M. WIEGANT, P. SCHOTMAN & W. H. GISPEN. 1978. ACTH-induced inhibition of endogenous rat brain phosphorylation *in vitro:* Structure activity. Neurochem. Res. **3:** 455-463.
42. ALOYO, V. J., H. ZWIERS & W. H. GISPEN. 1983. Phosphorylation of B-50 protein by calcium-activated, phospholipid-dependent protein kinase and B-50 protein kinase. J. Neurochem. **41:** 649-653.
43. CHAN, S. Y., K. MURAKAMI & A. ROUTTENBERG. 1986. Phosphoprotein F1: Purification and characterization of a brain kinase C substrate related to plasticity. J. Neurosci. **6:** 3618-3627.
44. JOLLES, J., H. ZWIERS, C. J. VAN DONGEN, P. SCHOTMAN, K. W. A. WIRTZ & W. H. GISPEN. 1980. Modulation of brain phosphoinositide metabolism by ACTH-sensitive protein phosphorylation. Nature **286:** 623-625.
45. OESTREICHER, A. B., C. J. VAN DONGEN, H. ZWIERS & W. H. GISPEN. 1983. Affinity-purified anti-B-50 protein antibody: Interference with the function of the phosphoprotein in synaptic plasma membranes. J. Neurochem. **41:** 331-340.
46. GISPEN, W. H., L. M. LEUNISSEN, A. B. OESTREICHER, A. J. VERKLEIJ & H. ZWIERS. 1985. Presynaptic localization of B-50 phosphoprotein: The ACTH-sensitive protein kinase substrate involved in rat brain phosphoinositide metabolism. Brain Res. **328:** 381-385.
47. ALEXANDER, K. A., B. M. CIMLER, K. E. MEIER & D. R. STORM. 1987. Regulation of calmodulin binding to P-57. J. Biol. Chem. **263:** 7544-7549.
48. ANDREASEN, T. J., C. H. KELLER, D. C. LA PORTE, A. M. EDELMAN & D. R. STORM. 1981. Preparation of azidocalmodulin: A photoaffinity label for calmodulin-binding proteins. Proc. Natl. Acad. Sci. USA **78:** 2782-2785.
49. WAKIM, B. T., K. A. ALEXANDER, H. R. MASURE, B. M. CIMLER, D. R. STORM & K. A. WALSH. 1987. Amino acid sequence of P-57, a neurospecific calmodulin-binding protein. Biochemistry **26:** 7466-7470.
50. STRITTMATTER, S. M., D. VALENZUELA, T. E. KENNEDY, E. J. NEER & M. C. FISHMAN. 1990. G_o is a major growth cone protein subject to regulation by GAP-43. Nature **344:** 836-841.
51. FREEMAN, J. J., A. A. LETTES & B. COSTELLO. 1988. Possible role of GAP-43 in calcium regulation/transmitter release. Neurosci. Abstr. **14:** 1126.
52. ZUBER, M. X., S. M. STRITTMATTER & M. C. FISHMAN. 1989. A membrane targeting signal in the amino terminus of the neuronal protein GAP-43. Nature **341:** 345-348.
53. SKENE, J. H. P. & M. WILLARD. 1981. Characteristics of growth-associated polypeptides in toad retinal ganglion cell axons. J. Neurosci. **1:** 419-426.
54. PERRONE-BIZZOZERO, N. I., D. WEINER, G. HAUSER & L. I. BENOWITZ. 1988. Extraction properties of major Ca^{++}-dependent synaptic phosphoproteins. J. Neurosci. Res. **20:** 346-350.
55. SKENE, J. H. P. & I. VIRAG. 1989. Posttranslational membrane attachment and dynamic fatty acid acylation of a neuronal growth cone protein. GAP-43. J. Cell Biol. **108:** 613-624.
56. PERRONE-BIZZOZERO, N. I., L. I. BENOWITZ, S. P. FINKLESTEIN, P. J. APOSTOLIDES & O. A. BIZZOZERO. 1989. Major acylated proteins in developing cortical neurons. J. Neurochem. **52:** 1149-1152.
57. DEKKER, L. V., P. N. E. DE GRAAN, D. H. G. VERSTEEG, A. B. OESTREICHER & W. H. GISPEN. 1989. Phosphorylation of B-50 (GAP-43) is correlated with neurotransmitter release in rat hippocampal slices. J. Neurochem. **52:** 24-30.
58. DEKKER, L. V., P. N. E. DE GRAAN, A. B. OESTREICHER, D. H. G. VERSTEEG & W. H. GISPEN. 1989. Inhibition of noradrenaline release by antibodies to B-50 (GAP-43). Nature **342:** 74-76.
59. DIFIGLIA, M., R. ROBERTS & L. I. BENOWITZ. 1990. Immunoreactive GAP-43 in adult rat caudate neuropil: Localization in unmyelinated fibers, axon terminals, and dendritic spines. J. Comp. Neurol. **302:** 992-1001.
60. MONAGHAN, D. T. & C. W. COTMAN. 1985. Distribution of NMDA-sensitive L-[^{3}H]glutamate binding sites in the rat brain. J. Neurosci. **6:** 2909-2919.
61. AKERS, R. F. & A. ROUTTENBERG. 1985. Protein kinase C phosphorylates a 47 M_r protein directly related to synaptic plasticity. Brain Res. **334:** 147-151.

62. NELSON, R. B. & A. ROUTTENBERG. 1985. Characterization of protein F1 (47 kDa, 4.5 pI): A kinase C substrate directly related to neural plasticity. Exp. Neurol. **89:** 213-224.
63. LOVINGER, D. M., R. F. AKERS, R. B. NELSON, C. A. BARNES, B. L. MCNAUGHTON & A. ROUTTENBERG. 1985. A selective increase in phosphorylation of protein F1, a protein kinase C substrate, directly related to three-day growth of long-term synaptic enhancement. Brain Res. **343:** 137-143.
64. MALENKA, R. C., D. V. MADISON & R. A. NICOLL. 1986. Potentiation of synaptic transmission in the hippocampus by phorbol esters. Nature **321:** 175-177.
65. LOVINGER, D. M., K. L. WONG, K. MURAKAMI & A. ROUTTENBERG. 1987. Protein kinase C inhibitors eliminate hippocampal long-term potentiation. Brain. Res. **436:** 177-183.
66. COTMAN, C. W., D. T. MONAGHAN & A. H. GANONG. 1988. Excitatory amino acid neurotransmission: NMDA receptors and Hebb-type synaptic plasticity. Annu. Rev. Neurosci. **11:** 61-80.
67. LINDEN, D. J., K. L. WONG, F.-S. SHEU & A. ROUTTENBERG. 1988. NMDA receptor blockade prevents the increase in protein kinase C substrate (protein F1) phosphorylation produced by long-term potentiation. Brain Res. **458:** 142-146.
68. BLISS, T. V. P., M. P. CLEMENTS, M. L. ERRINGTON, M. A. LYNCH & J. H. WILLIAMS. 1988. Long-term potentiation is accompanied by an increase in extracellular release of arachidonic acid. Neurosci. Abstr. **14:** 564.
69. MURAKAMI, K. & A. ROUTTENBERG. 1985. Direct activation of purified protein kinase C by unsaturated fatty acids (oleate and arachadonate) in the absence of phospholipids and calcium. FEBS Lett. **192:** 189-193.
70. ROSENTHAL, A., S. Y. CHAN, W. HENZEL, C. HASKELL, W.-J. KUANG, E. CHEN, J. N. WILCOX, A. ULLRICH, D. V. GOEDDEL & A. ROUTTENBERG. 1987. Primary structure and mRNA localization of protein F1, a growth-related kinase C substrate associated with synaptic plasticity. EMBO J. **6:** 3641-3646.
71. NIELANDER, H. B., L. H. SCHRAMA, A. J. VAN ROZEN, M. KASPERAITIIS, A. B. OESTREICHER, P. N. E. DE GRAAN, W. H. GISPEN & P. SCHOTMAN. 1987. Primary structure of the neuron-specific phosphoprotein B-50 is identical to growth-associated protein GAP-43. Neurosci. Res. Commun. **1:** 163-172.
72. CIMLER, B. M., D. H. GIEBELHAUS, D. R. WAKIM, D. R. STORM & R. T. MOON. 1987. Characterization of murine cDNAs encoding P-57, a neuronal-specific calmodulin-binding protein. J. Biol. Chem. **262:** 12158-12163.
73. KOSIK, K. S., L. D. ORECCHIO, G. A. P. BRUNS, L. I. BENOWITZ, G. P. MACDONALD, D. R. COX & R. L. NEVE. 1988. Human GAP-43: Its deduced amino acid sequence and chromosomal localization in mouse and human. Neuron **1:** 127-132.
74. NG, S.-C., S. M. DE LA MONTE, G. L. CONBOY, L. R. KARNS & M. C. FISHMAN. 1988. Cloning of human GAP-43: Growth association and ischemic resurgence. Neuron **1:** 133-139.
75. BAIZER, L., S. ALKAN, K. STOCKER & G. CIMENT. 1990. Chicken growth-associated protein (GAP-43): Primary structure and regulated expression of the mRNA during embryogenesis. Mol. Brain Res. **7:** 61-68.
76. COGGINS, P. J. & H. ZWIERS. 1989. Evidence for a single protein kinase C-mediated phosphorylation site in rat brain protein B-50. J. Neurochem. **53:** 1895-1901.
77. BENOWITZ, L. I., N. I. PERRONE-BIZZOZERO & S. P. FINKLESTEIN. 1987. Molecular properties of the growth-associated protein GAP-43 (B-50). J. Neurochem. **48:** 1640-1647.
78. MASURE, H. R., K. A. ALEXANDER, B. T. WAKIM & D. T. STORM. 1986. Physicochemical and hydrodynamic characterization of P-57, a neurospecific calmodulin-binding protein. Biochemistry **25:** 7553-7560.
79. FEDEROFF, H. J., E. GRABCZYK & M. C. FISHMAN. 1988. Dual regulation of GAP-43 gene expression by nerve growth factor and glucocorticoids. J. Biol. Chem. **263:** 19290-19295.
80. PERRONE-BIZZOZERO, N. I., N. IRWIN, S. E. LEWIS, I. FISCHER, R. L. NEVE & L. I. BENOWITZ. 1990. Posttranscriptional regulation of GAP-43 mRNA levels during process outgrowth. Neurosci. Abstr. **16:** 814.
81. BISBY, M. A. 1988. Dependence of GAP-43 (B-50, F1) transport on axonal regeneration in rat dorsal root ganglion neurons. Brain Res. **458:** 157-161.
82. BENOWITZ, L. I., C. O'BRIEN, N. I. PERRONE-BIZZOZERO, N. IRWIN & C. J. WOOLF. 1990. The expression of GAP-43 mRNA is regulated by an axonally transported factor. Neurosci. Abstr. **16:** 338.

83. Grabczyk, E., M. X. Zuber, H. Federoff, S.-C. Ng, A. Pack & M. C. Fishman. 1989. GAP-43 gene structure. Neurosci. Abstr. **15:** 957.
84. Nedivi, E., G. S. Basi & J. H. P. Skene. 1989. Structural and potential regulatory domains of a rat GAP-43 gene. Neurosci. Abstr. **15:** 1269.
85. Wells, R. D., D. A. Collier, J. C. Hanvey, M. Shimizu & F. Wohlrab. 1988. The chemistry and biology of unusual DNA structures adopted by oligopurine-oligopyrimidine sequences. FASEB J. **2:** 2939-2949.
86. Aguayo, A., S. David, P. Richardson & G. Bray. 1982. Axonal elongation in peripheral and central nervous system transplants. *In* Advances in Cellular Neurobiology. S. Federoff & L. Hertz, Eds. Vol 6: 215-234. Academic Press. New York, NY.
87. Zuber, M. X., D. W. Goodman, L. R. Karns & M. C. Fishman. 1989. The neuronal growth-associated protein GAP-43 induces filopodia in nonneuronal cells. Science **244:** 1193-1195.

Possible Role of GAP-43 in Calcium Regulation/Neurotransmitter Release[a]

JEANETTE J. NORDEN,[b] ANDREW LETTES, BRIAN COSTELLO, LI-HSIEN LIN, BEN WOUTERS, SUSAN BOCK, AND JOHN A. FREEMAN

Department of Cell Biology
Vanderbilt University School of Medicine
Nashville, Tennessee 37232

INTRODUCTION

The molecular mechanisms that regulate the growth of axons and the formation of synapses are of central importance to our understanding of the development of the nervous system and its response to injury. Great interest has recently centered around the identification of proteins involved in these processes. Prominent among these identified proteins which are likely candidates for mediating neurite growth is the growth-associated protein GAP-43, a nervous system-specific and fast-axonally transported phosphoprotein whose expression and phosphorylation state are correlated with axon growth.[1–6] GAP-43 is identical to the plasticity-associated protein F1,[7] the B-50 phosphoprotein implicated in phosphatidylinositol turnover,[8] the calmodulin-binding protein P-57 (neuromodulin),[9] and the growth cone-associated protein pp46.[10,11] GAP-43 is of special interest inasmuch as it appears to be involved in neurite growth during development and regeneration as well as in synaptic plasticity in the mature nervous system. In the adult mammalian brain, GAP-43 is concentrated within presynaptic terminals,[12] and is particularly abundant in areas such as the molecular layers of the hippocampus, neocortex, and cerebellum,[13] all areas known to be highly plastic in function. Some insight into the significance of this finding for the role GAP-43 might play in the adult central nervous system was provided by our demonstration that GAP-43 was identical to protein F1,[7] a protein kinase C (PKC) substrate that had been shown to undergo a major change in phosphorylation state during long-term potentiation (LTP) in the hippocampus.[14–16] In LTP, the delivery of high-frequency pulse trains to presynaptic cells induces long-term changes in the efficacy of individual synapses, which is reflected in changes in both the latency and spike amplitude of evoked responses in postsynaptic cells. The correlation of a change in GAP-43 phosphorylation and LTP, and the enrichment of this specific protein within so-called associative areas of the brain,[13,17–20] suggest that GAP-43 may play a direct role in synaptic plasticity in the adult central nervous system.

A number of studies suggest that a PKC substrate or substrates could play a role in synaptic plasticity by modulating neurotransmitter release in the nervous system.[21]

[a] This work was supported by the National Institutes of Health (NS18103 and EY01117 to J. A. F. and NS25150 to J. J. N.).

[b] To whom correspondence should be addressed.

For example, phorbol esters, which directly stimulate PKC activity, increase the spontaneous release of transmitter, and augment evoked transmitter release at nerve-muscle synapses.[22] In *Hermissenda,* intracellular injection of PKC or activation of PKC by phorbol esters induces an enhancement of Ca^{2+} conductance mechanisms and augmentation of neurotransmitter release similar to changes seen during associative learning.[23] Similarly, phorbol esters and PKC have been shown to enhance Ca^{2+} currents in *Aplysia* neurons.[24] In mammalian hippocampus, the induction[25] or maintenance[26] of LTP can be induced by phorbol esters, and PKC inhibitors block the induction of LTP.[27] That this is related to neurotransmitter release is suggested by the findings that LTP is correlated with an augmented release of glutamate,[28-31] which has been shown recently to occur as the result of an increase in the probability of transmitter release.[32]

These findings prompted us to investigate whether GAP-43 might be a PKC substrate involved in neurotransmitter release, and if so, to determine how it might influence this process. We have used electron microscopic immunolocalization in tissue sections and subcellular fractionation to determine the subcellular localization of the protein in presynaptic terminals, and fura-2 measurements to investigate changes in Ca^{2+} fluxes during neurotransmitter release in PC12 cells. Our results suggest that GAP-43 is localized to plasma membrane and possibly to synaptic vesicles (discussed below), and that GAP-43 may play a role in neurotransmitter release by regulating Ca^{2+} flux in presynaptic terminals.

Some of this work has been reported previously.[33-35]

METHODS

Electron Microscopic Immunolocalization

Vibratome-sectioned tissue slices (50 μm) of adult rat cerebellar cortex fixed with 0.2-0.5% glutaraldehyde and 4% paraformaldehyde were incubated in normal goat serum (1% in phosphate buffer) for 30 min and left overnight at 4 °C in affinity-purified GAP-43 antibody (1:10) or in antibody preabsorbed with purified GAP-43. The next day, the tissue was rinsed thoroughly and reacted using an ABC kit (Vector Labs) and 3,3′-diaminobenzidine (DAB) as the chromagen. The tissue was processed for electron microscopy according to standard protocols, sectioned using an LKB IV ultramicrotome, and examined unstained in a Hitachi H600 electron microscope.

Subcellular Fractionation

Two different methods were used for subcellular fractionation. In the first method, a purified fraction of synaptic vesicles was obtained from adult rat brain essentially as described by Huttner *et al.*[36] This method selects for an ultrapure sample of clear, ~60 nm synaptic vesicles. In the second method, subcellular fractionation was performed on adult rat neocortex, cerebellar cortex, and hippocampus. In this latter method, crude synaptosomes are purified using a Ficoll gradient, lysed, and fractionated on a discontinuous sucrose gradient.[37,38] For both fractionation procedures, aliquots at various stages of fractionation were run on standard one-dimensional gels. The proteins from these gels were then transferred electrophoretically to nitrocellulose or to nylon, and the blots were probed with GAP- 43 antibody or with antibody preabsorbed with purified GAP-43. The blots were developed using ABC kits and 4-chloronapthol as the chromagen. The purity of the synaptic vesicle fractions was verified by electron microscopy, by probing duplicate Western blots with antibodies to synaptophysin, or by measuring Na^+/K^+- ATPase activity in various fractions.

Measurement of Intracellular Free Ca^{2+} and Neurotransmitter Release

For these experiments, we used cell cultures of the rat pheochromocytoma or PC12 cell line. PC12 cells were maintained in 85% RPMI 1640 medium (Sigma) supplemented with 10% horse serum and 5% fetal calf serum. Some dishes were pretreated with nerve growth factor (NGF) at 100 ng/ml for up to 10 days to induce neurite outgrowth.

A high-resolution, thermoelectrically cooled CCD camera system (described below) was used to record Ca^{2+} fluxes in single cells utilizing the fluorescent Ca^{2+}-binding dye fura-2 (fura-2 AM loading method). PC12 cultures were placed on the stage of a Zeiss IM-35 inverted microscope equipped with a Peltier-effect constant temperature module (maintained at 35 °C) and epi-illuminator. Cells were bathed in the loading solution for 30 min (allowing the permeant form of the indicator to enter the cells), and were then rinsed and postincubated for 2 hr in culture medium to allow the indicator to be deesterified. We employ a high-resolution CCD camera (Photometrics IC200, Tucson, AZ) equipped with a Thompson TH7882 CCD array (which was thermoelectrically cooled to −45 °C to minimize noise) and with 340- and 380-nm two-cavity interference filters (Melles-Girot, 10-nm ½ B.W.) to measure fluorescence using a ratio method.[39] An electronically controlled shutter (Uniblitz) was interposed in the light path to limit exposure to bright ultraviolet light. Exposure times of 0.2 sec were typically used to integrate fluorescence intensity at each of the two wavelengths. The time required to change excitation filters was approximately 1.5 sec. Measurements were typically collected at 30-sec intervals following application of the pharmacological agents described below. Each digitized fluorescent image was first corrected for dark current, background fluorescence, and optical nonlinearities by subtracting a reference image obtained at the same wavelength from a culture dish filled to the appropriate depth with culture medium. The approximate thickness of PC12 cell somas was estimated to be between 4 and 8 μm by differential focusing with Nomarski optics. The intracellular Ca^{2+} concentration was computed from[40]

$$[Ca^{2+}]_i = K_D T F_o/F_s$$

$$\text{where } T = (R - R_{min})/(R_{max} - R).$$

The quantity K_D is the equilibrium dissociation constant for the indicator (225 μM). The R terms are ratios at 340- and 380-nm excitation. R_{max} (9.0-10.0) is the ratio obtained for saturating levels of Ca^{2+}, and R_{min} (0.39-0.42) is the minimum fluorescence ratio for test solutions with no added Ca^{2+} and with 5 mM EGTA. F_o/F_s (7.5) is the fluorescence ratio obtained at low and at saturating Ca^{2+} concentrations using only the longer wavelength excitation. Fluctuations of the ratio image due to the low light levels of noncellular areas were eliminated by setting background areas to a uniform black level. A pseudocoloring scheme was used to represent different Ca^{2+} concentrations, in which increasing values of the image ratio increase from blue to red.

PC12 cells were incubated with affinity-purified GAP-43 antibody (1:10 in 1% DMSO) for 12 hr at 37 °C, and loaded with [^{3}H]norepinephrine ([^{3}H]NE; specific activity: 14.2 Ci/mmole). In order to determine if the antibody had become internally localized, some cells were incubated with fluorescein-tagged antibody. Control cells were incubated in 1% DMSO without antibody. The effect of GAP-43 antibody on fura-2 fluorescence and on the release of [^{3}H]NE was assessed after stimulation with carbachol (0.5 mM), or after stimulation with high potassium (50 mM) or high potassium and the phorbol ester 12-*O*-tetradecanoylphorbol 13-acetate (TPA; 1 μM). The degree to which GAP-43 antibody blocked phosphorylation was determined by using an *in vitro* phosphorylation assay. Homogenates of rat hippocampus were incubated with different dilutions of affinity-purified GAP-43 antibody for 2 hr at 4 °C, and phosphorylated by the addition of γ-[^{32}P]ATP (5 μM; specific activity: 10 μCi/nmole) for 2 min. GAP-43 was immunoprecipitated and subjected to one-dimensional gel electrophoresis and autoradiography. Autoradiograms were scanned using a Microscan 1000 (Technology Resources, Nashville, TN).

RESULTS

Electron Microscopic Immunolocalization

Numerous GAP-43-positive synapses could be identified in the molecular layer of the cerebellar cortex. FIGURE 1 shows the typical pattern of labeling within two presynaptic profiles. GAP-43 immunoreactivity was associated with the external face of synaptic vesicles and, more prominently, with the cytoplasmic face of presynaptic plasma membrane. Near the active zone of synapses positive for GAP-43, immunoreactivity associated with synaptic vesicles appeared to be more intense (FIG. 1 and inset). In labeled presynaptic terminals, mitochondrial membrane also appeared immunoreactive. Postsynaptic densities were inherently electron dense,[41] but appeared slightly more dense when apposed by labeled presynaptic terminals, although this was not seen in all cases (FIG. 1). Other GAP-43-positive profiles resembling presynaptic boutons (but not in synaptic contact) could also be identified scattered throughout the neuropil. No significant staining of any positively identified postsynaptic profiles was seen, and no labeling of pre- or postsynaptic profiles was observed in sections where the antibody had been preabsorbed with purified GAP-43.

Subcellular Fractionation

Using the fractionation procedure of Huttner *et al.*,[36] a highly purified and homogeneous fraction of clear, round synaptic vesicles was obtained as shown in FIGURE 2A. The Western blot of various subcellular fractions is shown in FIGURE 2B. GAP-43 immunoreactivity is present within the crude synaptosomal fraction and fractions enriched in plasma membrane (SGP, CM). In the ultrapurified synaptic vesicle fraction (CV), a weakly immunoreactive GAP-43 band is present, which increases in intensity with increasing concentration of protein loaded on the gels (CV; 10, 20, 40 μg).

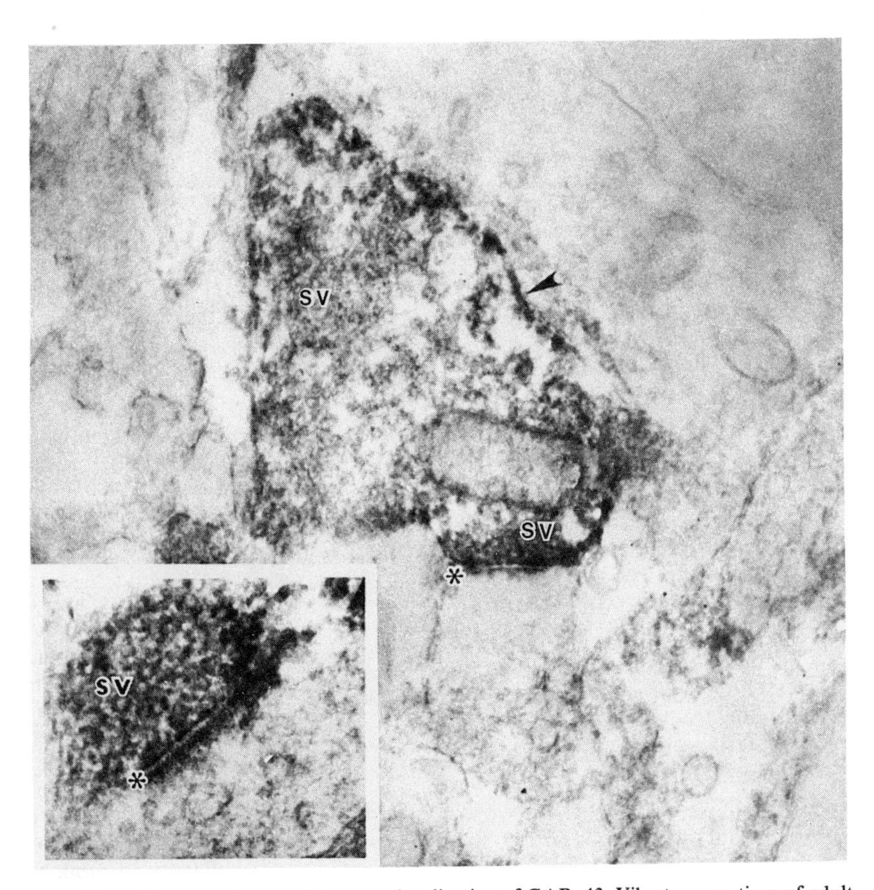

FIGURE 1. Electron microscopic immunolocalization of GAP- 43. Vibratome sections of adult rat cerebellum were incubated in affinity-purified GAP-43 antibody and processed as described in the text. The tissue was osmicated but not stained following sectioning. GAP-43-positive presynaptic profiles could be identified throughout the cerebellar molecular layer. The pattern of staining indicated that GAP-43 is associated primarily with the cytoplasmic face of the plasma membrane (arrow), where it appears to outline presynaptic terminals, and possibly with the external face of synaptic vesicles (SVs). The immunoreactivity associated with mitochondria in labeled presynaptic terminals is likely to be an artifact (see text for discussion). The inset shows the active zone of another GAP-43-positive presynaptic terminal. Near the active zones of synapses (*), there was intense GAP-43 immunoreactivity surrounding the synaptic vesicles. Postsynaptic densities were inherently electron dense, and in most synapses in which the presynaptic terminal was labeled, they showed no immunoreactivity (main figure). At some synapses, however, such as that shown in the inset, the postsynaptic densities appeared to be immunoreactive.

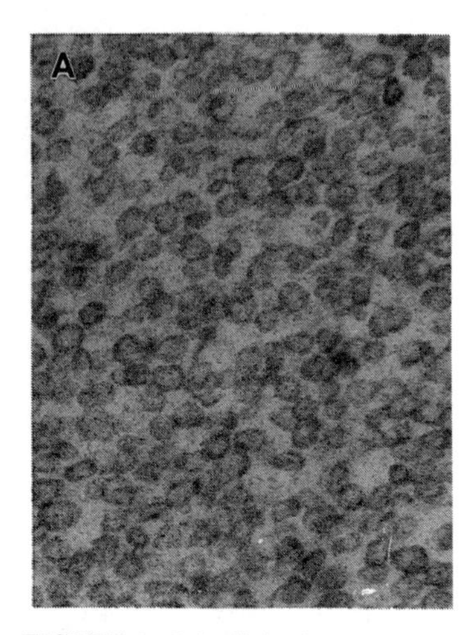

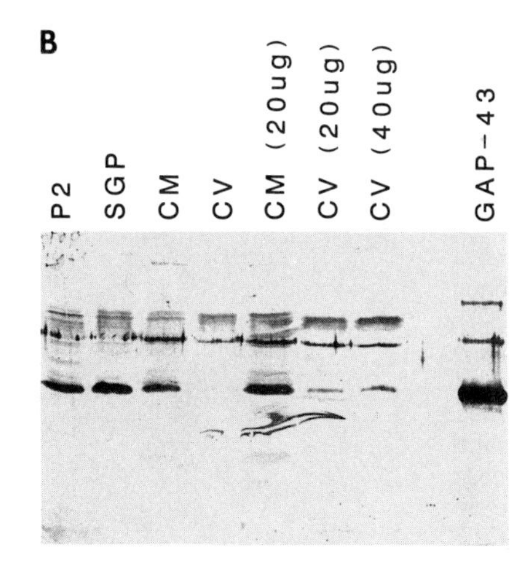

FIGURE 2. Subcellular fractionation of synaptic vesicles. A highly purified fraction of synaptic vesicles was isolated from adult rat brain by the procedure of Huttner *et al.*[36] **(A)** The purity and homogeneity of the synaptic vesicle fraction was verified by electron microscopy. Every tenth section through a pelleted synaptic vesicle fraction was examined. Only small, clear vesicular structures with an approximate diameter of 60 nm were present. No contamination by larger membrane fragments was seen in any of the sections through the pellet. Magnification: 108,000 ×. **(B)** Western blot of various subcellular fractions probed with GAP-43 antiserum. Ten micrograms of protein were loaded per lane, except where indicated otherwise. From left to right, the lanes represent crude synaptosomes (P2); small to medium-sized plasma membranes (SGP); the first membranes off the exclusion column (CM); the last fraction off the exclusion column (CV); 20 μg of protein loaded (CM 20 μg); 20 μg of protein loaded (CV 20 μg); 40 μg of protein loaded (CV 40 μg); a blank; and 1 μg of purified GAP-43. The CV lanes represent purified fractions of synaptic vesicles. With increasing amounts of protein loaded, an increasingly dense GAP-43 band can be identified in the Western blot.

A second fractionation procedure was used for three reasons. First, we wanted to enrich the fractions for synaptic terminals containing GAP-43. We did this by using only neocortex, cerebellar cortex, and hippocampus, all areas known to contain relatively high amounts of GAP-43 in the adult brain.[13] Second, we wanted to use a more gentle fractionation procedure to try to insure that GAP- 43 would not be dissociated from the synaptic vesicles, or from other molecules that might be associated with synaptic vesicles. Finally, we wanted to obtain a synaptic vesicle fraction that did not select for only a subpopulation of small, clear, round vesicles. It is known, for example, that synapsin, a synaptic vesicle associated protein distributed within nerve terminals in widespread areas of the brain, is not associated with all types of synaptic vesicles,[42,43] and GAP-43 could be associated with an even smaller fraction of vesicles because it is enriched within only a few neuropil areas.[13] For this fractionation, we used the method outlined by Morgan *et al.*[37,38] To insure that no subpopulation of vesicles was excluded, we included the standard synaptic vesicle fraction D and the D-E interface in the samples that were loaded onto the gels. According to Morgan,[44] both fractions D and E are enriched for synaptic vesicles, but also contain a significant plasma membrane component. The Western blots from this fractionation, probed with antibodies to the integral synaptic membrane protein synaptophysin[45] or to GAP-43, are shown in FIGURE 3. The pattern of staining on the blot probed with monoclonal antibodies to synaptophysin indicates that the synaptic vesicle fraction was highly enriched for synaptic vesicles but that synaptic vesicles were also present in the G_L and G_H fractions, which contain predominantly small and medium-sized plasma membranes. The duplicate blot probed with GAP- 43 antiserum shows that GAP-43 is present in synaptosome and plasma membrane fractions, and in the synaptic vesicle (SV) fraction. To determine the extent to which the various fractions were contaminated with plasma membrane, we used an enzyme assay for oubain-sensitive Na^+/K^+-ATPase.[46] Results of this assay indicated that the synaptic vesicle, light membrane, and heavy membrane fractions contained approximately equal amounts of Na^+/K^+-ATPase activity (12.3 (SV), 11.6 (G_L), and 10.2 (G_H), expressed as μmoles/hr/mg protein). Although GAP-43 is associated with plasma membrane, GAP-43 staining in the Western blot was most positively correlated with the presence of synaptic vesicles as indicated by the duplicate blot probed with synaptophysin, and not with Na^+/K^+-ATPase activity. These data suggest that the presence of a GAP-43 band in the synaptic vesicle fraction is not likely to be due solely to contamination of the fraction by plasma membrane. Thus, GAP-43 may be associated with a subset of synaptic vesicles. Caveats regarding the interpretation of these data are discussed below.

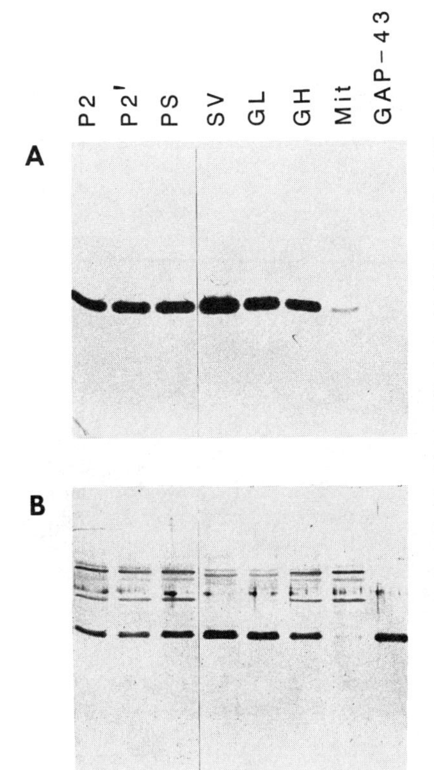

FIGURE 3. Subcellular fractionation of synaptic vesicles. Subcellular fractionation of adult rat neocortex, cerebellar cortex, and hippocampus was also prepared as described by Morgan *et al.*[37,38] **(A)** Western blot of various subcellular fractions probed with monoclonal antibodies to synaptophysin, an integral membrane protein of synaptic vesicles.[45] **(B)** A duplicate blot probed with polyclonal antibodies to GAP-43. Twenty micrograms of protein were loaded in each lane, except for purified GAP-43 (1 μg loaded). From left to right, the lanes represent crude synaptosomes (P2); purified synaptosomes (P2′); a presucrose or lysate fraction loaded onto the discontinuous sucrose gradient (PS); the synaptic vesicle fraction D and the D-E interface (SV)[44]; plasma membranes, light (G_L); plasma membranes, heavy (G_H); a mitochondrial fraction (Mit); purified GAP-43. These blots show that the SV fraction was enriched for synaptic vesicles and that GAP-43 immunoreactivity is associated with both plasma membrane and synaptic vesicle fractions.

Effect of GAP-43 Antibody on Neurotransmitter Release and Ca^{2+} Levels

To investigate the possible role of GAP-43 in neurotransmitter release and Ca^{2+} level regulation, we developed a sensitive assay for [^{3}H]NE release and employed a quantitative fura-2 imaging system to measure intracellular free Ca^{2+} levels. In preliminary experiments, we have examined the effects of various agents on [^{3}H]NE release and Ca^{2+} levels in the presence and absence of GAP-43 antibody. Access of GAP-43 antibody to the intracellular compartment was achieved by partial membrane permeabilization using DMSO. Control experiments were performed to determine a minimum concentration of DMSO that resulted in GAP-43 antibody internalization and no measurable decrement in the ability of PC12 cells to generate action potentials following intracellular depolarization via micropipette. As shown in FIGURE 4, incubation of PC12 cells in medium containing fluorescein-labeled antibody and 1% DMSO results in antibody being internalized primarily to the cell body and to neuritic growth cones. Thus, under the conditions used for the experimental studies, GAP-43 antibody was internalized by the cells. The results of control experiments performed to determine the extent to which GAP-43 antibody blocks the *in vitro* phosphorylation of GAP-43 are shown in TABLE 1.

In the first set of experiments, PC12 cells were pretreated for 7 days with NGF, and then incubated overnight in medium, supplemented with 1% DMSO and 100 ng/ml NGF, with or without affinity-purified GAP-43 antibody (1:10). After priming with

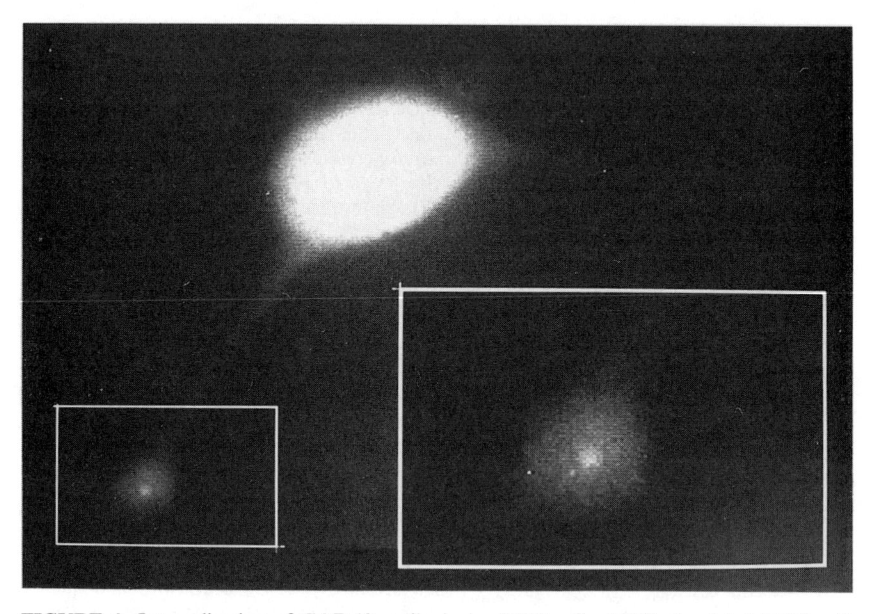

FIGURE 4. Internalization of GAP-43 antibody by PC12 cells. NGF-stimulated PC12 cells were incubated for 1 hr in medium containing fluorescein-labeled antibody and 1% DMSO. Fluorescence was most intense in cell bodies and in growth cones. The growth cone delimited by the small rectangle is shown at higher magnification in the inset in the lower right hand corner. These images demonstrate that GAP-43 antibody is internalized by PC12 cell growth cones under the conditions used for the experimental studies.

TABLE 1. Effect of GAP-43 Antibody on Phosphorylation[a]

Anti-GAP-43 Dilution (affinity purified)	Anti-GAP-43 Dilution (serum)	Phosphorylation (percentage of control)
1:2.5	1:38	73.7
1:10	1:150	68.2
1:40	1:600	80.0
1:160	1:2400	87.3

[a] Using an *in vitro* phosphorylation assay, we determined the degree to which different concentrations of affinity-purified antibody blocked the phosphorylation of GAP-43. The approximate concentration of GAP-43 antiserum required to block phosphorylation to the same extent as affinity-purified antibodies is shown in the middle column.

NGF, PC12 cells express GAP-43 and acetylcholine receptors, and release norepinephrine in a Ca^{2+}-dependent manner. We first tested the effect of carbachol, an agonist for norepinephrine release. PC12 cells were loaded with [^{3}H]NE, and the release of tritium was measured in the presence of increasing concentrations of carbachol. Carbachol stimulated [^{3}H]NE release with an ED_{50} of approximately 0.1 mM. Carbachol (0.5 mM) produced an approximate 2-fold increase in both [^{3}H]NE release and in the integrated Ca^{2+} fluorescence ratio value of growth cones ($N = 14$) over control levels. If cells were incubated in the presence of GAP-43 antibody, the [^{3}H]NE release was decreased by approximately 37%, and the Ca^{2+} levels decreased by 35% ($N = 17$).

GAP-43 Phosphorylation and Ca^{2+} Channel Conductance in PC12 Cells

In the preceding experiments we found that transmitter release was correlated with free Ca^{2+} levels in PC12 cell growth cones, and that there was a direct correlation between GAP-43 phosphorylation, intracellular free Ca^{2+}, and neurotransmitter release. The decrease in both the free Ca^{2+} levels and the release obtained in the presence of GAP-43 antibody, which decreases the phosphorylation of GAP-43, suggests that the conductance of transmembrane Ca^{2+} channels might be regulated by the phosphorylation of GAP-43.

In another series of experiments, undifferentiated PC12 cells were incubated in 1% DSMO and depolarized by potassium in the presence or absence of the phorbol ester TPA, which stimulates the phosphorylation of GAP-43 by PKC. Duplicate cultures were preincubated overnight with 1% DMSO and affinity-purified GAP-43 antibody (1:10). Depolarization with 50 mM KCl caused a 3.9-fold increase in [^{3}H]NE release (three experiments, FIG. 5) and a 2.5-fold increase in integrated intracellular free Ca^{2+} measured over PC12 cell bodies (from a mean level of 54 nM to 135 nM, 20 cells). The phorbol ester TPA alone had little effect on neurotransmitter release, but augmented K^+-stimulated release and intracellular Ca^{2+} levels by approximately 40%. When PC12 cells were preincubated with GAP-43 antibody, the potentiated tritium release was decreased by approximately 30% as shown in FIGURE 5, whereas the TPA-potentiated Ca^{2+} level increase was completely abolished, presumably as a result of

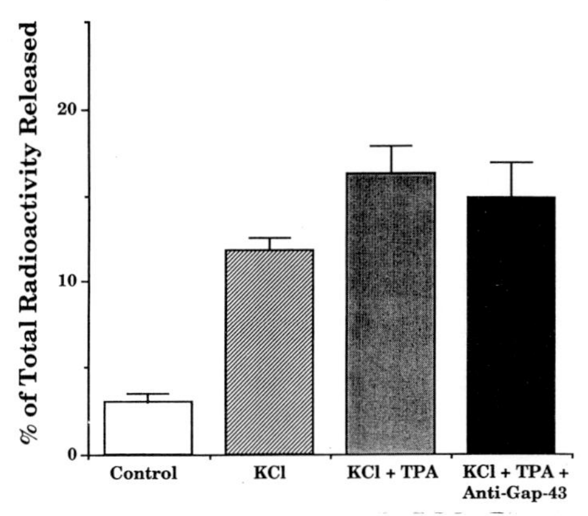

FIGURE 5. Inhibition of [^{3}H]NE release by GAP-43 antibody in undifferentiated PC12 cells. Cells were loaded with [^{3}H]NE (specific activity: 14.2 Ci/mmole) for 2 hr in medium supplemented with 12 μM nialamide, a monoamine oxidase inhibitor. Cells were then washed three times with Krebs-Ringer-HEPES buffer (KRHB), and incubated for an additional 2 hr in medium. The medium was replaced with 1.5 ml KRHB, and incubated for 15 min. The buffer was replaced with fresh KRHB, and the cells incubated for 1 min. This solution was removed and filtered (basal release). Cells were then incubated with test solutions for 1 min. The solution was then removed and filtered (stimulated release). The cells were solubilized in 1 M NaOH overnight at 50 °C. Samples were dissolved in scintillation cocktail, and the radioactivity was measured by scintillation counting. Release is expressed as the percentage of the initial total radioactivity present prior to the measurement of the basal release. For analysis of the effect of GAP-43 antibody on release, cultures were treated similarly except that two days after they were plated, PC12 cells were incubated overnight with GAP-43 affinity-purified antibody (1:10), then depolarized by 50 mM K^+ + KRHB in the presence or absence of 1 μM TPA. TPA markedly potentiated the release of ^{3}H, and this effect was attenuated by pretreatment with GAP-43 antibody.

decreased Ca^{2+} influx. At the dilution of affinity-purified antibody (1:10) used for these experiments, the phosphorylation of GAP-43 was blocked by approximately 40% (TABLE 1), suggesting that the changes in release and in Ca^{2+} flux were due to the effect of the antibody on phosphorylation.

In differentiated PC12 cells, Ca^{2+} levels were highest within actively extending growth cone filopodia, ranging from 100 nM to as high as 500 nM. A ratio image from a representative PC12 cell is shown in FIGURE 6. To test the possibility that GAP-43 phosphorylation might regulate Ca^{2+} channel conductance and thereby intracellular Ca^{2+} levels in growth cones, we depolarized 1% DMSO-treated cells with potassium (50 mM KCl), which caused a 35% increase in intracellular free Ca^{2+} levels in actively extending growth cones, and then stimulated GAP-43 phosphorylation by the addition of 1 μM TPA, which caused an additional 20% increase in intracellular free Ca^{2+} in the same growth cones. In contrast, this incremental increase in Ca^{2+} levels was completely blocked in growth cones that had been preincubated with GAP-43 antibody, as shown in FIGURE 7. The results of all of these studies on [^{3}H]NE release and Ca^{2+} flux are summarized in TABLE 2. Although these experiments are preliminary, taken together they suggest that the conductance of voltage-sensitive Ca^{2+} channels in the PC12 cell membrane may be regulated either directly or indirectly by GAP-43 phosphorylation.

DISCUSSION

Immunolocalization of GAP-43/Subcellular Fractionation

The results of our immunolocalization suggest that GAP-43 is preferentially localized within presynaptic terminals to areas where it could play a role in neurotransmitter release. Both the localization in tissue sections and the subcellular fractionation data indicate that GAP-43 is associated with the plasma membrane of presynaptic terminals. It is somewhat more difficult to interpret the tissue and subcellular fractionation data with regard to an association of GAP-43 with synaptic vesicles. With respect to the tissue immunolocalization, it is well recognized that there can be an artifactual translo-

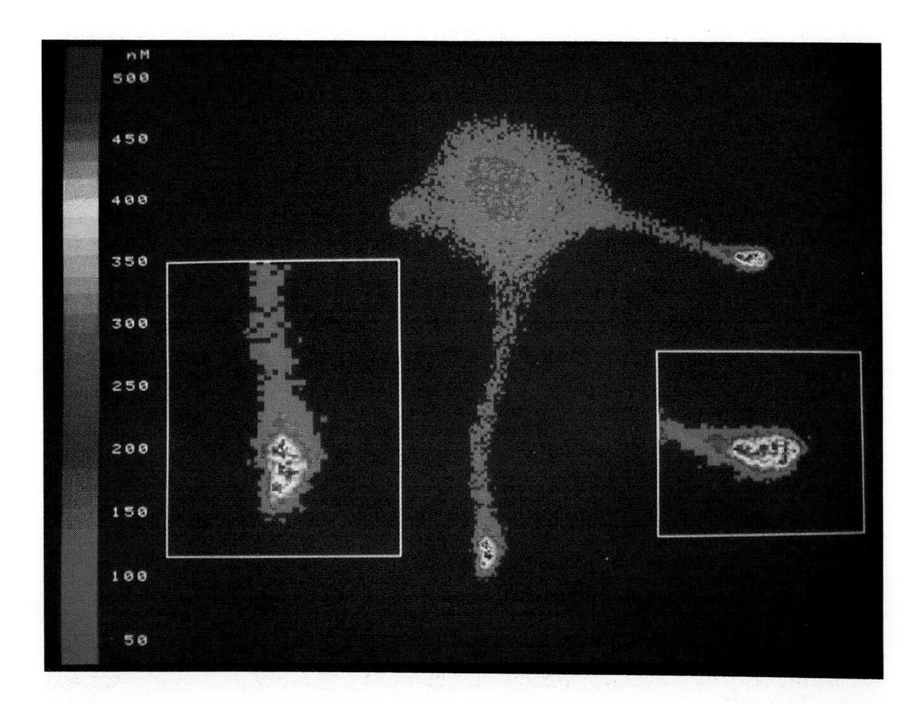

FIGURE 6. Measurement of Ca^{2+} levels in PC12 cells. A high-resolution, thermoelectrically cooled CCD camera was used to record Ca^{2+} fluxes in single cells utilizing the Ca^{2+}-binding dye fura-2. PC12 cultures were placed on the stage of a Zeiss IM-35 inverted microscope equipped with a Peltier-effect constant temperature module and epi-illuminator. Cells were bathed in the loading solution for 30 min, allowing the permeant form of the indicator to enter the cells, and then rinsed and postincubated for 2 hr in culture medium to allow the indicator to be deesterified. The Ca^{2+} levels computed from the fluorescence ratio measured at 340 nM and 380 nM have been pseudocolored (colorscale: blue = 50 nM, red = 500 nM). This PC12 cell had two actively extending neurites. The growth cones have high levels of free Ca^{2+}, shown in more detail in the insets.

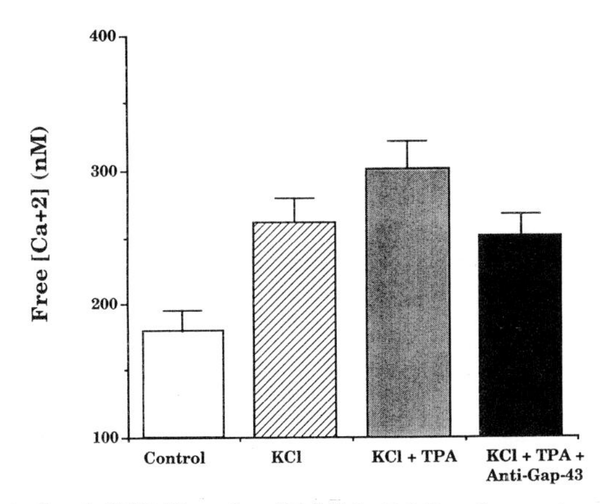

FIGURE 7. Effect of anti-GAP-43 on free [Ca^{2+}] in PC12 cell growth cones. In this set of experiments, the mean intracellular free Ca^{2+} level was computed for a number of actively extending growth cones by dividing the integrated fura-2 fluorescence ratio by the growth cone area, which in turn was measured from the digitized frame-buffer image with a quantitative imaging system (ImageMaster 2000, Technology Resources). PC12 cells in four separate cultures were used: two control cultures incubated with 1% DMSO only, and two cultures treated with both 1% DMSO and anti-GAP-43 (affinity purified 1:10). Calcium level measurements were repeated for the same growth cones following application of KCl or TPA. Depolarization by 50 mM KCl caused an average 35% increase in Ca^{2+} levels over resting levels in both sets of cultures (N = 22, 15 sec postapplication). Addition of 0.1 μM of the phorbol ester TPA caused an additional 20% increase in Ca^{2+} levels in the control cultures (N = 12), but in the cultures treated with anti-GAP-43, this additional increase was completely abolished (N = 10).

cation of DAB reaction product from the site of oxidation to adjacent membranous structures.[47] This is particularly a problem when the tissue is not well fixed before processing, when the antigen of interest is a soluble protein, or when development times in DAB are long.[47,48] This diffusion artifact is widely believed to result in the frequently seen "labeling" of structures such as mitochondria.[47] No artifactual translocation is seen distant to true antigenic sites, however, and peroxidase methods using DAB have been used at the electron microscopic level to localize specific antigens to intracellular organelles, including synaptic vesicles.[41,43] With reference to GAP-43, although it is predominantly a membrane-associated protein, a small amount of GAP-43 is cytosolic,[49] and thus it could potentially become associated artifactually with nearby membranes. To minimize translocation of reaction product in the present study, we used very short reaction times in DAB,[41] but some translocation may have occurred.

In support of an association of GAP-43 with at least some subset of synaptic vesicles in presynaptic terminals is the association of calmodulin with synaptic vesicles.[50] Synaptic vesicles are known to have multiple calmodulin binding sites.[50,51] Because GAP-43 binds calmodulin with low affinity under physiological conditions,[52] GAP-43 may not be associated with the synaptic vesicle membrane per se but might be bound to calmodulin. During subcellular fractionation, these molecules may become dissociated, which would account for the loss of GAP-43 immunoreactivity in highly purified synaptic vesicle fractions.

It is important to acknowledge that electron microscopic immunolocalizations using gold/cryo-ultramicrotomy of growth cones and synaptosomes have not found a significant association of GAP- 43 with synaptic vesicles.[53] In these studies, however, there is always the possibility that GAP-43 could become largely dissociated from the synaptic vesicles or from calmodulin during the subcellular fractionation procedure (as discussed above). It has been proposed that GAP-43 is tightly bound to plasma membrane,[54] and, thus, less would be lost under these conditions. Although electron microscopic immunolocalization of GAP- 43 (B-50) using gold/ultramicrotomy methods on tissue sections appears to show a similar pattern,[55] there are problems associated with the interpretation of the actual location of antigenic sites with this method. For example, the gold particles often appear to be associated with the external, rather than the internal, face of the plasma membrane. In this regard, the peroxidase methods appear to yield a picture more consistent with the existing data showing that GAP-43, which has no potential membrane-spanning domains[56] and is an intracellular protein, is associated with the cytoplasmic face of the plasma membrane at presynaptic terminals. In addition, fractionation/cryo-ultramicrotomy results in poor preservation of membranous structures. For these reasons, we are now attempting to use other gold methods directly on tissue sections in order to more adequately address the question as to the subcellular localization of this protein.

One of the problems in interpreting data on the electron microscopic immunolocalization of GAP-43 obtained in different labs is the observation that the various polyclonal antibodies against GAP-43 result in slightly different patterns of immunostaining, even at the light microscopic level. For example, whereas we reported previously that the molecular layer of the cerebellar cortex was intensely immunoreactive using our polyclonal antibody[13] (which is why cerebellar cortex was chosen for the present electron microscopic study), a similar light microscopy study by Benowitz *et al.*[17] using polyclonal antisera revealed very little immunoreactivity in the cerebellum. Thus in addition to the problems in interpreting the results of different electron microscopic methods, antiserum raised against GAP-43 isolated by various procedures may also

TABLE 2. Summary of Experimental Data on the Effect of GAP-43 Antibodies on [^{3}H]NE Release and Ca^{2+} Levels in PC12 Cells[a]

	[^{3}H]NE Release	Ca^{2+} Influx
Differentiated Cells		
Carbachol	2-fold ↑	2-fold ↑
Carbachol + Anti-GAP-43	37% ↓	35% ↓
KCl		35% ↑
KCl + TPA		20% ↑↑
KCl + TPA + Anti-GAP-43		100% ↓↓
Undifferentiated Cells		
KCl	3.9-fold ↑	2.5-fold ↑
KCl + TPA	40% ↑↑	40% ↑↑
KCl + TPA + Anti-GAP-43	30% ↓↓	100% ↓↓

[a] [^{3}H]NE release and Ca^{2+} influx were measured in PC12 cells under various conditions as described in the text. Changes in [^{3}H]NE release were closely paralleled by changes in Ca^{2+} influx as measured by quantitative fluorescence. In cells stimulated with carbachol, GAP-43 antibodies decreased both the release of [^{3}H]NE and Ca^{2+} influx to the same extent. For PC12 cells stimulated with KC1 and the phorbol ester TPA, GAP-43 antibodies blocked a portion of the augmented [^{3}H]NE release and, to a greater extent, the augmented Ca^{2+} influx induced by TPA. Note that GAP-43 antibody treatment caused a 30% reduction in the TPA-augmented [^{3}H]NE release, and a 100% reduction in the TPA-augmented Ca^{2+} influx (↑↑: increase in TPA-augmented response over control; ↓↓: decrease in TPA-augmented response).

result in different patterns of staining and confound interpretation of the subcellular localization of this protein.

Finally, interpretation of the subcellular fractionation data is made difficult by the recognition that synaptic vesicle fractions that include all classes of synaptic vesicles are likely to be contaminated with small pieces of resealed plasma membrane. Here we found that although contamination by plasma membrane was present in some of the synaptic vesicle fractions, the GAP-43 immunoreactivity of the synaptic vesicle fraction in Western blots could not be accounted for solely in terms of this contamination. We interpret this to mean that GAP-43 may be associated with some subset of synaptic vesicles, either directly, or by association with calmodulin.

GAP-43, Neurotransmitter Release, and Ca^{2+} Fluxes

Before discussing the results from our fura-2 measurements, several possible caveats deserve mention. Whereas we observed high levels of Ca^{2+} in growth cones similar to those reported by Connor[40] and Cohan *et al.*,[57] who also used the fura-2-AM loading method, Silver *et al.*[58] reported substantially lower levels obtained in growth cones of a neuroblastoma cell line in which fura-2 had been applied by pressure microinjection at the soma. In those experiments it is not clear what the actual fura-2 concentration was in the growth cones. Some of the differences between those results and our present results might thus be due to differences in 1) cell loading technique, 2) fura-2 concentration, and 3) clonal cell lines. In the present studies, we were less interested in the absolute values of Ca^{2+} in growth cones, or in the correlation of Ca^{2+} levels and different growth states, than we were in the relative changes caused by different pharmacological manipulations. If, as Silver *et al.*[58] suggest, the AM loading method results in artifactually high Ca^{2+} levels (due to binding to intracytoplasmic organelles), this would imply that the relative differences in Ca^{2+} levels that we observed following different pharmacological treatments are an underestimate of the actual changes.

The results from our fura-2 measurements indicate that in PC12 cells pretreated with GAP-43 antibody, the influx of Ca^{2+} into presynaptic terminals or growth cones is markedly reduced compared to controls treated with DMSO alone. Furthermore, treatment with GAP-43 antibodies also decreases the carbachol- or K^+-induced release of [^{3}H]NE under the same conditions. Using an *in vitro* phosphorylation assay, we demonstrated that GAP-43 affinity-purified antibody blocks the phosphorylation of GAP-43. Thus, the phosphorylation of GAP-43 could be involved in the regulation of Ca^{2+} fluxes and in regulating the change in neurotransmitter release that is observed. It is likely that Ca^{2+} flux and neurotransmitter release in these experiments could have been decreased even more if the effective internal concentration of GAP-43 antibody had been higher.

Our results are consistent with the growing body of data suggesting that a PKC substrate is involved in the modulation of neurotransmitter release (reviewed in the Introduction). Since our initial interest in investigating whether GAP-43 might be a specific PKC substrate involved in neurotransmitter release, a number of reports have appeared that support a role for GAP-43 in this process. For example, Dekker *et al.*[59] have reported that the incorporation of phosphate into GAP-43 (B-50) was enhanced by depolarization by K^+ and phorbol esters in hippocampal slices. This depolarization-induced GAP-43 phosphorylation was inhibited by polymyxin B, an inhibitor of PKC. In another paper, Dekker *et al.*[60] reported complete inhibition by GAP-43 antibodies of Ca^{2+}- dependent [^{3}H]NE release from rat cortical synaptosomes permeabilized with streptolysin-O. The results from our fura-2 measurements would indicate that GAP-43 may play a role in neurotransmitter release, at least in part, by regulating Ca^{2+} influx into presynaptic terminals.

Calcium has long been recognized to play a critical role in neurotransmitter release in the nervous system.[61,62] Data has also accumulated suggesting that PKC and PKC substrate(s) are involved in the modulation of neurotransmitter release, probably by regulating voltage-sensitive Ca^{2+} channels.[21] Such studies also point to the involvement of GTP-binding proteins in this process. In dorsal root ganglion cells, application of NE or GABA results in a decrease in the recorded Ca^{2+} current through L-type, voltage-sensitive Ca^{2+} channels and a decrease in the amount of neurotransmitter released.[63] This neurotransmitter modulation of the Ca^{2+} channel is indirect in that it requires the activation of a GTP-binding protein[64] and PKC.[65] Both phorbol esters and diacylglycerol analogues mimic the modulatory effects of neurotransmitters on Ca^{2+} currents,[66] and PKC inhibitors block the neurotransmitter modulation of the Ca^{2+} current.[65]

In other cells, including PC12 cells, stimulation with phorbol esters or PKC results in a striking enhancement of Ca^{2+} action potentials evoked by depolarizing stimuli and potentiation of neurotransmitter release.[21] Our results indicate that the augmented neurotransmitter release induced by TPA is associated with an increase in Ca^{2+} influx into PC12 cells. Both the augmented neurotransmitter release and Ca^{2+} influx can be blocked specifically by GAP- 43 antibodies, suggesting that GAP-43 plays a role in the modulation of neurotransmitter release by regulating Ca^{2+} channels. There is also reason to believe that a GTP-binding protein may be involved in this process. Recently, Strittmatter *et al.*[67] have reported that GAP- 43 increases the binding of GTP to G_o, a G protein linked to the regulation of Ca^{2+} channels in neuronal cells.[68,69] Thus, in both dorsal root ganglion cells and in PC12 cells, the modulation of neurotransmitter release appears to involve protein phosphorylation by PKC and a G protein.

Our results would indicate that in PC12 cells, GAP-43 is likely to be a PKC substrate involved in modulating neurotransmitter release, possibly by regulating Ca^{2+} flux through Ca^{2+} channels. It is not known whether the changes observed are mediated by both voltage-sensitive and receptor-operated Ca^{2+} channels in PC12 cells, or by multiple channel types. In dorsal root ganglion cells, a dihydropyridine-sensitive L-type Ca^{2+} channel is implicated in the modulation of neurotransmitter release.[70] In differentiated PC12 cells (which resemble sympathetic neurons), and in sympathetic neurons, neurotransmitter release appears to involve primarily dihydropyridine-insensitive, voltage-sensitive Ca^{2+} channels.[71–73] It is interesting that in dorsal root ganglion cells activation of PKC inhibits Ca^{2+} currents and decreases the evoked release of neurotransmitter, whereas in PC12 cells and sympathetic neurons activation of PKC increases Ca^{2+} currents and augments neurotransmitter release. Although much remains to be investigated about the differences between these two systems, it is clear that the process of exocytosis and neurotransmitter release is modulated largely through the regulation of Ca^{2+} channels in presynaptic terminals. Modulation of neurotransmitter release in different cell types may also involve multiple types of Ca^{2+} channels. There is some evidence, for example, that part of the effect of GAP-43 on Ca^{2+} flux may be mediated through its effect on dihydropyridine-sensitive Ca^{2+} channels. It has been proposed recently that calcineurin, a Ca^{2+}-activated phosphatase, limits Ca^{2+} influx through dihydropyridine-sensitive Ca^{2+} channels in plasma membrane by dephosphorylating the channel or a closely associated protein.[74] GAP-43 could be such a protein because it is a specific substrate protein for calcineurin,[75] and our results implicate the phosphorylation of GAP-43 in the regulation of Ca^{2+} flux.

GAP-43 may also influence neurotransmitter release through its interaction with calmodulin or by a role in phospholipid metabolism. In nerve terminals, calmodulin is known to mediate the effects of Ca^{2+} on several important enzyme systems involved in protein phosphorylation and dephosphorylation.[76] Calmodulin also plays a role in the regulation of the intracellular Ca^{2+} concentration by activating Ca^{2+} pumps in the plasma membrane.[77] The phosphorylation of synaptic vesicle/plasma membrane proteins by calmodulin is believed to play a direct role in neurotransmitter release.[78] Finally, GAP-43 could influence Ca^{2+}-mediated events in nerve terminals by influencing phosphatidylinositol turnover.[79]

GAP-43 was first identified as one of a class of fast-axonally transported proteins whose expression was significantly increased during regeneration of the toad optic nerve.[80] It is now well established that axon growth during development and regeneration in both mammalian and nonmammalian vertebrates is accompanied by a significant increase in the expression of this protein.[6] A number of correlations suggest a role for GAP-43 in a common mechanism underlying both axon growth and the modulation of neurotransmitter release in the adult nervous system. For example, 1) GAP-43 is concentrated within growth cones and in mature synaptic terminals, particularly within so-called plastic areas of the brain,[13] and 2) stimulation of PKC activity by phorbol esters augments neurotransmitter release (reviewed above) and can stimulate neuritic outgrowth.[81]

Our present results prompt us to speculate that the processes of axonal growth and of facilitated neurotransmitter release associated with long-term potentiation might be linked to a common mechanism involving in part the regulation of Ca^{2+} channel conductance by GAP-43 phosphorylation. Several additional lines of evidence support the likely importance of Ca^{2+} channels and their regulation in these two processes. For example, N-type, voltage-sensitive Ca^{2+} channels are known to be concentrated within synaptic terminals, where they are believed to play a role in neurotransmitter release.[72] L-type, voltage-gated Ca^{2+} channels are also present in nerve terminals, and they may also play a role in neurotransmitter release, or augmentation of release, by increasing Ca^{2+} concentration under some conditions.[72] In previous studies using a vibrating probe, we obtained evidence that actively extending goldfish retinal ganglion cell growth cones also generate steady inward Ca^{2+} currents at their tips.[82] These currents are presumably produced by leaky Ca^{2+} channels on the growth cone filopodia, and would be expected to set up both a Ca^{2+} gradient and an electric field that might play an important role in the growth process. Fura-2 imaging and patch-clamp recordings in sympathetic neurons indicate that both L-type and N-type voltage-gated Ca^{2+} channels are present on growth cones,[83] and could play a role in spontaneous or evoked neurotransmitter release from growth cones.[84–86] Garber *et al.*[87] recently found that during differentiation PC12 cells express an increased fraction of low-threshold T-type Ca^{2+} channels, and Streit and Lux[88] found that during initial sprouting, PC12 cell growth cones generate current densities through Ca^{2+} channels that are 5.4 times higher than those in the somata. They conclude that growing neurites of PC12 cells express high densities of Ca^{2+} channels in the growth cones, whereas in the neuritic shaft the Ca^{2+} channel density is initially low, and increases later during neurite consolidation. These findings are in substantial agreement with our present results, and serve to confirm our earlier work on growth cone physiology obtained with a two-dimensional vibrating probe[82] in which we postulated the existence of a relatively high density of Ca^{2+} channels on growth cone filopodia.

Growth cones elongate by the addition of membrane at their tips, and neurotransmitter release in growth cones or mature synaptic terminals must necessarily involve the fusion of plasma and synaptic vesicle membrane. During periods of axon growth, there is an accelerated fusion of membrane at the extending tip, and during augmented neurotransmitter release, an increase in exocytosis within the presynaptic terminal. Long-term potentiation in the dentate gyrus, for example, is associated with an increase in the number of front-line synaptic vesicles at potentiated synapses[89,90] and with an increase in the probability of neurotransmitter release.[32] As an intracellular protein concentrated within growth cones and some presynaptic terminals, GAP-43 could influence Ca^{2+}-mediated events by its interaction with calmodulin and/or by playing a role in phospholipid metabolism. The present results suggest that GAP-43 is also likely to influence these processes by regulating Ca^{2+} flux through Ca^{2+} channels.

REFERENCES

1. Freeman, J., S. Bock, M. Deaton, C. B. McGuire, J. J. Norden & G. J. Snipes. 1986. Exp. Brain Res. **13**(Suppl.): 34-47.
2. Benowitz, L. I. & A. Routtenberg. 1987. Trends Neurosci. **10:** 527-532.
3. Snipes, G. J., B. Costello, C. B. McGuire, B. N. Mayes, S. S. Bock, J. J. Norden & J. A. Freeman. 1987. Prog. Brain Res. **71:** 155-175.
4. Skene, J. H. P. 1989. Annu. Rev. Neurosci. **12:** 127-156.
5. Liu, Y. & D. R. Storm. 1990. Trends Physiol. **11:** 107-111.
6. Norden, J. J., B. Wouters, J. Knapp, S. Bock & J. A. Freeman. 1991. *In* Vision and Visual Dysfunction. J. Cronly-Dillon, Ed. Vol. 11: in press. Macmillan. London.
7. Snipes, G. J., S. Y. Chan, C. B. McGuire, B. R. Costello, J. J. Norden, J. A. Freeman & A. Routtenberg. 1987. J. Neurosci. **7:** 4066-4075.
8. Jacobson, R. D., I. Virag & J. H. P. Skene. 1986. J. Neurosci. **6:** 1843-1855.
9. Wakim, B. T., K. A. Alexander, H. R. Masure, B. M. Cimler, D. R. Storm & K. A. Walsh. 1987. Biochemistry **26:** 7466-7470.
10. Meiri, K. F., K. H. Pfenninger & M. B. Willard. 1986. Proc. Natl. Acad. Sci. USA **83:** 3537-3541.
11. Nelson, R. B., D. J. Linden, C. Hyman, K. H. Pfenninger & A. Routtenberg. 1989. J. Neurosci. **9:** 381-389.
12. Skene, J. H. P., R. D. Jacobson, G. J. Snipes, C. B. McGuire, J. J. Norden & J. A. Freeman. 1986. Science **233:** 783-786.
13. McGuire, C. B., G. J. Snipes & J. J. Norden. 1988. Dev. Brain Res. **41:** 277-291.
14. Lovinger, D. M., R. F. Akers, R. B. Nelson, C. A. Barnes, B. L. McNaughton & A. Routtenberg. 1985. Brain Res. **343:** 137-143.
15. Routtenberg, A., D. M. Lovinger & O. Steward. 1985. Behav. Neural Biol. **43:** 3-11.
16. Lovinger, D. M., P. A. Colley, R. F. Akers, R. B. Nelson & A. Routtenberg. 1986. Brain Res. **399:** 205-211.
17. Benowitz, L. I., P. J. Apostolides, N. Perrone-Bizzozero, S. P. Finklestein & H. Zwiers. 1988. J. Neurosci. **8:** 339-352.
18. Benowitz, L. I., N. I, Perrone-Bizzozero, S. P. Finklestein & E. D. Bird. 1989. J. Neurosci. **9:** 990-995.
19. Neve, R. L., E. A. Finch, E. D. Bird & L. I. Benowitz. 1988. Proc. Natl. Acad. Sci. USA **85:** 3638-3642.
20. De la Monte, S. M., H. J. Federoff, S. C. Ng, E. Grabczyk & M. Fishman. 1989. Dev. Brain Res. **46:** 161-168.
21. Kaczmarek, L. K. 1987. Trends Neurosci. **10:** 30-34.
22. Shapira, R., S. D. Silberberg, S. Ginsburg & R. Rahamimoff. 1987. Nature **325:** 58-60.
23. Farley, J. & S. Auerbach. 1986. Nature **319:** 220-223.
24. DeRiemer, S. A., J. A. Strong, K. A. Albert, P. Greengard & L. K. Kaczmarek. 1985. Nature **313:** 313-316.
25. Malenka, R. C., D. V. Madison & R. A. Nicoll. 1986. Nature **321:** 175-177.
26. Routtenberg, A., P. Colley, D. Linden, D. Lovinger, K. Murakami & F.-S. Sheu. 1986. Brain Res. **378:** 374-378.
27. Lovinger, D. M., K. L. Wong, K. Murakami & A. Routtenberg. 1987. Brain Res. **436:** 177-183.

28. SKREDE, K. K. & D. MALTHE-SORENSSEN. 1981. Brain Res. **208:** 436-441.
29. DOLPHIN, A. C., M. L. ERRINGTON & T. V. P. BLISS. 1982. Nature **297:** 496-498.
30. BLISS, T. V. P., R. M. DOUGLAS, M. L. ERRINGTON & M. A. LYNCH. 1986. J. Physiol. **377:** 391-408.
31. FEASEY, K. J., M. A. LYNCH & T. V. P. BLISS. 1986. Brain Res. **364:** 39-44.
32. MALINOW, R. & R. W. TSIEN. 1990. Nature. **346:** 177-180.
33. NORDEN, J., B. COSTELLO & J. A. FREEMAN. 1987. Invest. Ophthalmol. Vision Sci. **28**(Suppl.): 334.
34. NORDEN, J. J., B. COSTELLO & J. A. FREEMAN. 1987. Soc. Neurosci. Abstr. **13:** 1480.
35. FREEMAN, J. A., A. A. LETTES & B. COSTELLO. 1988. Soc. Neurosci. Abstr. **14:** 1126.
36. HUTTNER, W. B., W. SCHIEBLER, P. GREENGARD & P. DECAMILLI. 1983. J. Cell Biol. **96:** 1374-1388.
37. MORGAN, I. G., G. VINCENDON & G. GOMBOS. 1973. Biochim. Biophys. Acta **320:** 671-680.
38. MORGAN, I. G., W. C. BRECKENRIDGE, C. VINCENDON & G. GOMBOS. 1973. *In* Proteins of the Nervous System. D. J. Schneider, R. H. Angeletti, R. A. Bradshaw, A. Grasso & B. W. Moore, Eds.: 171-192. Raven Press. New York, NY.
39. TSIEN, R., T. RINK & M. POENIE. 1985. Cell Calcium **6:** 145-157.
40. CONNOR, J. A. 1986. Proc. Natl. Acad. Sci. USA **83:** 6179-6183.
41. GIRARD, P. R., G. J. MAZZEI, J. G. WOOD & J. F. KUO. 1985. Proc. Natl. Acad. Sci. USA **82:** 3030-3034.
42. DECAMILLI, P., T. UEDA, F. E. BLOOM, E. BATTENBERG & P. GREENGARD. 1979. Proc. Natl. Acad. Sci. USA **76:** 5977-5981.
43. BLOOM, F. E., T. UEDA, E. BATTENBERG & P. GREENGARD. 1979. Proc. Natl. Acad. Sci. USA **76:** 5982-5986.
44. MORGAN, I. G. 1976. Neuroscience **1:** 159-165.
45. WIEDENMANN, B. & W. W. FRANKE. 1985. Cell **41:** 1017-1028.
46. MEDZIHRADSKY, F., P. S. NANDHASRI, V. IDOYAGA-VARGAS & O. Z. SELLINGER. 1971. J. Neurochem. **18:** 1599-1603.
47. NOVIKOFF, A. B., P. M. NOVIKOFF, N. QUINTANA & C. DAVIS. 1972. J. Histochem. Cytochem. **20:** 745-749.
48. NOVIKOFF, A. B. 1980. J. Histochem. Cytochem. **28:** 1036-1038.
49. CIMLER, B. M., T. J. ANDREASEN, K. I. ANDREASEN & D. R. STORM. 1985. J. Biol. Chem. **260:** 10784-10788.
50. MOSKOWITZ, N., W. SCHOOK, K. BECKENSTEIN & S. PUSZKIN. 1983. Brain Res. **263:** 243-250.
51. DELORENZO, R. J., S. D. FREEDMAN, W. B. YOHE & S. C. MAURER. 1979. Proc. Natl. Acad. Sci. USA **76:** 1838-1842.
52. ALEXANDER, K. A., B. M. CIMLER, K. E. MEIER & D. R. STORM. 1987. J. Biol. Chem. **262:** 6108-6113.
53. VANLOOKEREN CAMPAGNE, M., A. B. OESTREICHER, P. M. P. VANBERGEN EN HENEGOUWEN & W. H. GISPEN. 1989. J. Neurocytol. **18:** 479-489.
54. SKENE, J. H. P. & I. VIRAG. 1989. J. Cell Biol. **108:** 613-624.
55. GORGELS, T. G. M. F., M. VANLOOKEREN CAMPAGNE, A. B. OESTREICHER, A. A. M. GRIBNAU & W. H. GISPEN. 1989. J. Neurosci. **9:** 3861-3869.
56. KARNS, L. R., S.-C. NG, J. A. FREEMAN & M. C. FISHMAN. 1987. Science **236:** 597-600.
57. COHAN, C. S., J. A. CONNOR & S. B. KATER. 1987. J. Neurosci. **7:** 3588-3599.
58. SILVER, R. A., A. G. LAMB & S. R. BOLSOVER. 1989. J. Neurosci. **9:** 4007-4020.
59. DEKKER, L. V., P. N. E. DEGRAAN, D. H. G. VERSTEEG, A. B. OESTREICHER & W. H. GISPEN. 1989. J. Neurochem. **52:** 24-30.
60. DEKKER, L. V., P. N. E. DEGRAAN, A. B. OESTREICHER, D. H. G. VERSTEEG & W. H. GISPEN. 1989. Nature **342:** 74-76.
61. KATZ, B. & R. MILEDI. 1967. J. Physiol. **189:** 535-544.
62. MILEDI, R. 1973. Proc. R. Soc. London **183:** 421-425.
63. DUNLAP, K. & G. D. FISCHBACH. 1981. J. Physiol. **317:** 519-535.
64. HOLZ, G. G., S. G. RANE & K. DUNLAP. 1986. Nature **319:** 670-672.
65. MCDONALD, J. R. & M. WALSH. 1985. Biochem. Biophys. Res. Commun. **129:** 603-610.
66. RANE, S. G. & K. DUNLAP. 1986. Proc. Natl. Acad. Sci. USA **83:** 184-188.

67. STRITTMATTER, S. M., D. VALENZUELA, T. E. KENNEDY, E. J. NEER & M. C. FISHMAN. 1990. Nature **344:** 836-841.
68. HESCHELER, J., W. ROSENTHAL, W. TRAUTWEIN & G. SCHULTZ. 1987. Nature **325:** 445-447.
69. DUNLAP, K., G. G. HOLZ & S. G. RANE. 1987. Trends Neurosci. **10:** 241-244.
70. PERNEY, T. M., L. D. HIRNING, S. E. LEEMAN & R. J. MILLER. 1986. Proc. Natl. Acad. Sci. USA **83:** 6656-6659.
71. KONGSAMUT, S. & R. J. MILLER. 1986. Proc. Natl. Acad. Sci. USA **83:** 2243-2247.
72. MILLER, R. J. 1987. Science **235:** 46-52.
73. HIRNING, L. D., A. P. FOX, E. W. MCCLESKEY, B. M. OLIVERA, S. A. THAYER, R. J. MILLER & R. W. TSIEN. 1988. Science **239:** 57-61.
74. ARMSTRONG, D. L. 1989. Trends Neurosci. **12:** 117-122.
75. LIU, Y. & D. R. STORM. 1989. J. Biol. Chem. **264:** 12800-12804.
76. CHEUNG, W. Y. 1982. Fed. Proc. **41:** 2253-2257.
77. KLEE, C. B., T. H. CROUCH & P. G. RICHMAN. 1980. Annu. Rev. Biochem. **49:** 489-515.
78. DELORENZO, R. J. 1982. Fed. Proc. **41:** 2265-2272.
79. RASMUSSEN, H., I. KOJIMA, K. KOJIMA, W. ZAWALICH & W. APFELDORF. 1984. *In* Advances in Cyclic Nucleotide and Protein Phosphorylation Research. P. Greengard & G. A. Robison, Eds. Vol. 18: 159-193. Raven Press. New York, NY.
80. SKENE, J. H. P. & M. WILLARD. 1981. J. Cell Biol. **89:** 86-95.
81. HSU, L., D. NATYZAK & J. D. LASKIN. 1984. Cancer Res. **44:** 4607-4614.
82. FREEMAN, J. A., P. B. MANIS, G. J. SNIPES, B. N. MAYES, P. C. SAMSON, J. P. WIKSWO & D. B. FREEMAN. 1985. J. Neurosci. Res. **13:** 257-283.
83. LIPSCOMBE, D., D. V. MADISON, M. POENIE, H. REUTER, R. Y. TSIEN & R. W. TSIEN. 1988. Proc. Natl. Acad. Sci. USA **85:** 2398-2402.
84. YOUNG, S. H. & M.-M. POO. 1983. Nature **305:** 634-637.
85. HUME, R. I., L. W. ROLE & G. D. FISCHBACH. 1983. Nature **305:** 632-634.
86. SUN, Y. & M.-M. POO. 1987. Proc. Natl. Acad. Sci. USA **84:** 2540-2544.
87. GARBER, S. S., T. HOSHI & R. W. ALDRICH. 1989. J. Neurosci. **9:** 3976-3987.
88. STREIT, J. & H. D. LUX. 1989. J. Neurosci. **9:** 4190-4199.
89. APPLEGATE, M. D., D. S. KERR & P. W. LANDFIELD. 1987. Brain Res. **401:** 401-406.
90. DESMOND, N. & W. B. LEVY. 1990. Synapse **5:** 139-143.

Ependymin, a Brain Extracellular Glycoprotein, and CNS Plasticity[a]

VICTOR E. SHASHOUA

Ralph Lowell Laboratories
McLean Hospital
Harvard Medical School
Belmont, Massachusetts 02178

INTRODUCTION

Ependymin is a major glycoprotein component of the extracellular fluid (ECF) of goldfish brain.[1] It has been implicated in the process of formation of synaptic changes during the consolidation step of learning[2,3] and in neuronal regeneration.[4] It is also present in mammalian brain.[5] Ependymin was first identified by double-labeling experiments, as a macromolecule that became more highly labeled after goldfish learned a new pattern of behavior.[1,6,7] Control experiments investigating the effects of stress and physical activity did not show such changes.[6,7] In addition, injections of ependymin antisera into the fourth ventricle of goldfish brain at 8-24 hr after the initiation of training were found to block long-term memory (LTM) but not short-term memory.[8] Antisera to ependymin were also found to block LTM in an avoidance conditioning-shuttle box type of experiment.[9]

CNS EPENDYMIN LEVELS CHANGE AFTER CLASSICAL CONDITIONING

Recent experiments indicate that the concentration of ependymin in ECF of goldfish brain changes after the animals learn a classical (Pavlovian) conditioning behavior.[10] Goldfish were trained in a sound-proof, light-tight chamber (FIG. 1) to associate the onset of a light stimulus (CS, or conditioned stimulus) with the arrival of a shock (US, or unconditioned stimulus) using the CS-US parameters established by Bitterman.[11] ELISA measurements of the concentrations of ependymin in brain extracts enriched with ECF proteins showed that the levels of ependymin decreased for the animals that received the CS-US in a paired presentation but not when the stimuli were delivered in an unpaired manner. TABLE 1 summarizes the data as a function of the time after initiation of training. The results for the paired stimulus group show an initial 19% decrease at 3.5 hr (p value of .05 in a t test) after the initiation of training, followed by a 20% increase at 5 hr (p value of .0005 in a t test). No significant changes were found for the unpaired stimulus group. These data are consistent with previous findings which indicated that increased ependymin synthesis and release into ECF occurs at 5-6 hr after the acquisition of a behavior,[1,2] a time span corresponding to the transcription and translation rate of messenger RNA in goldfish brain.[12] Similar ECF ependymin changes occurred when goldfish were trained to swim with a float.[13] The results suggest that, following a decrease of ECF ependymin levels, the brain is stimulated to synthesize

[a] This work was supported by the National Institute of Neurological and Communicative Disorders and Stroke (NS25748).

and secrete more of the protein into ECF. Thus in two separate experiments ECF ependymin was used up as a consequence of a training procedure. It was surprising to find that as much as 7 μg of the protein was used up in a 70 mg goldfish brain in the classical conditioning experiment. Such a result raises the possibility that ependymin might be involved in some general CNS function that takes place during the process.

Ependymin is also present in mammalian brain and appears to be involved in the consolidation step of learning.[5] In experiments using Balb-c mice, thirsty animals were trained to find water in one arm of a T-maze. Analysis of the pattern of protein synthesis by double-labeling methods showed that increased labeling of a protein with the same molecular weight, subcellular localization, and immunoreactivity properties as ependymin occurs in the brains of the trained mice in comparison with controls. Also, in studies of long-term potentiation of rat brain hippocampal slices the only specific

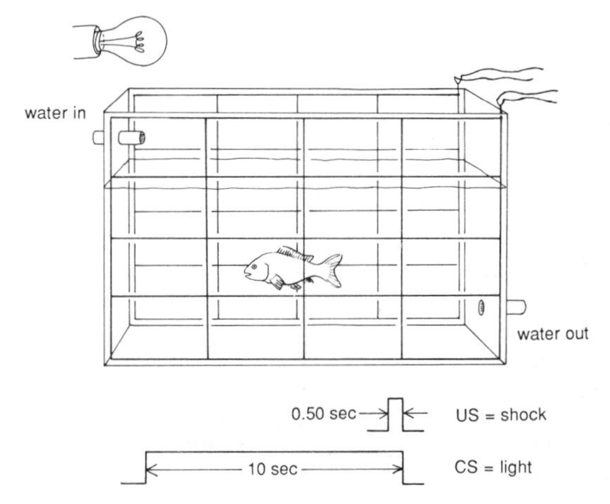

FIGURE 1. Apparatus for the classical conditioning experiment. Goldfish were trained to associate light as the CS with the onset of a shock (8 V at 60 Hz) as the US in a sound-proof, light-tight chamber. The paired stimulus group (experimentals) received 25 CS-US stimuli delivered randomly, with the stimuli being 90 or 180 sec apart. Controls (unpaired stimulus group) in a separate tank received CS or US (25 each) in a random sequence with the stimuli being 90 sec apart during a 90-min period.

protein changes observed were components of products that were released into the incubation medium.[14,15] Thus, three types of behavioral experiments in the goldfish (swimming with a float, avoidance conditioning, and classical conditioning) and two experiments in the mammalian brain (T-maze learning in the mouse and long-term potentiation in rat brain hippocampal slices) implicate ependymin in the consolidation step of learning and plasticity of the CNS.[16,17] These findings led to the hypothesis that ependymin may be involved in some aspect of synapse formation or modification that may occur during the consolidation step of LTM formation.[16] The data raise two important questions about the role of such a protein: 1) How can an extracellular protein participate in such a process? 2) Why is there so much of the protein used up as a result of the acquisition of a new pattern of behavior? In attempts to explore these and other aspects of ependymin function, we studied the molecular properties of ependymin.

TABLE 1. Ependymin Changes in ECF

	Paired CS-US Group[b]		Unpaired CS-US Group[c]			
Time[a] (hr)	Ependymin Level[d]	N	Ependymin Level[d]	N	Percentage Change (paired/unpaired)	Statistics[e]
3.5	115 ± 32	8	142 ± 24	7	−19	<.05, >.025
5.0	175 ± 18	24	139 ± 13	24	+20	<.0005
25.5	141 ± 11	8	145 ± 11	5	−3	NS

[a] Time after initiation of training.
[b] The paired groups received 25 stimuli in which the light was presented as the CS for 10 sec followed by a 0.5-sec shock (8 V at 60 Hz) as the US during the last 0.5 sec of the CS.
[c] The unpaired group received 25 presentations of the CS (10-sec light stimulus) and shock (0.5-sec duration) delivered separately 90 sec apart in a random order.
[d] Each entry below indicates a mean ± SEM for the level of ependymin in ECF in μg ependymin/mg ECF protein
[e] The entries below are p values derived from Student's t test. The p values in the first two lines indicate highly significant results.

MOLECULAR PROPERTIES OF EPENDYMIN

Ependymin is present in ECF as three types of disulfide-linked dimers of two acidic (α and β) polypeptide chains (37 kDa and 31 kDa).[16] About 50% of the ependymin content of goldfish brain is in the brain extracellular fluid, with 50% being in the cytosol fraction.[18] The protein contains at least 10% carbohydrate as an N-linked glycan.[19] Two-dimensional gel electrophoresis shows the presence of at least 7 α and 7 β subtypes of ependymin with isoelectric points in the range of pH 5 to 5.7, suggesting a microheterogeneity and a possible existence of 49 types of dimers.[20] The microheterogeneity of the molecule was found to persist even after removal of the N-linked glycan fragments by N-glycosidase F. This result was confirmed by analysis of primary amino acid sequence data[21,22] that showed the presence of several points in the sequence of the molecule that could have two or three alternate amino acid substituents. Investigations of the composition and properties[19] of the carbohydrate moiety of ependymin showed that it had the same antigenicity as the cell surface glycoprotein HNK-1. This has similar antigenic properties as the glycan epitope of the neural cell adhesion molecule (NCAM), J1, L1, the myelin-associated glycoprotein (MAG), and the sulfoglucuronyl glycolipid (SGGL)[23]; all of these contain a glucuronic acid sulfate.

The α and β polypeptide chains of ependymin have common antigenic sequences. Antibodies raised against α can recognize β and vice versa.[2,24] Treatment of β with the peptide N-glycosidase F removes the N-glycan fragment to yield a lower molecular weight form, γ-ependymin (26 kDa).[20] FIGURE 2 shows the primary amino acid sequence of the γ chain of ependymin determined by the cDNA method[21] in comparison to a partial sequence obtained by direct analysis of peptides isolated from proteolytic digests.[25] The two methods gave identical results.[25] The sequence shows no homology to any known polypeptides, indicating that ependymin may be a unique brain macromolecule. The molecule has an N-terminal leader sequence, consistent with the previous findings suggesting it was a secreted macromolecule.[7,26] An investigation of the phylogenetic distribution of ependymin-like macromolecules suggests that identical immunoreactive products were present in brains of vertebrates including goldfish, chick, mouse, rat, and human brain.[13,27] Immunoprecipitation data show that antisera to the goldfish brain protein can precipitate mouse brain ependymin, and those raised against the protein derived from mouse brain can precipitate goldfish brain ependymin,[28] suggesting that there may be some cross-reactivity. The mammalian form of the protein tends to remain as the dimer form in the presence of β-mercaptoethanol.

One surprising property of ependymin is its capacity to form aggregates of fibrous insoluble polymer (FIP). Solutions of pure ependymin form FIP[27,29] if the Ca^{2+} concentration in solution is suddenly reduced by about 50% by the addition of a Ca^{2+}-chelating agent such as EGTA. These aggregates, which may contain fibers as long as 1-2 mm (FIG. 3), once formed, could not be dissolved by restoring the Ca^{2+} to 10 mM or by using 2% sodium dodecylsulfate (SDS) in 5 M urea, 100% acetic acid, chloroform/methanol (2:1), saturated aqueous KCNS, and even 100% trifluoroacetic acid. FIP was partially soluble in 80% formic acid. Such solutions, however, did not yield water-soluble products when the formic acid was evaporated off or if it was removed by dialysis after neutralization with sodium hydroxide. These properties differentiate ependymin from other polymerizing macromolecules, such as tubulin, laminin, fibronectin, and actin, that are known to form fibers that dissolve in 2% SDS in 5 M urea. An investigation of the process of ependymin polymerization indicates that an activation step is necessary before the molecules become polymerizable, that is, sensitive to conditions that deplete Ca^{2+} from the medium. Preliminary studies[30] suggest that the activation process may be an ATP-dependent phosphorylation step that converts ependymin into its polymerizable form. Ependymin's capacity to form FIP aggregates, if it occurs *in vivo,* may be an important factor in its CNS function.

EPENDYMIN POLYMERIZATION AND NEURONAL FUNCTION: THE RULE OF TWO HYPOTHESIS

FIGURE 4 illustrates a proposed hypothesis for the role of ependymin in neuronal function.[27] The model shows two pathways converging on a single dendrite in a classical

```
S S H R Q P C H A P P L T S G T M K V V S T G G H D L E S G E F S Y D S K A N
S S H R Q P C H A P P L T S G T M K V V S T G                     Y D S K A N

K F R F V E D T A H A N K T S H M D V L I H F E E G V L Y E I D S K N E S C K
K F R F V E D T A H A N K T           I H F E E G V L Y

K E T L Q F R K H L M E I P D A T H E S E I Y M G S P S I T E Q G L R V R V W
  E T L Q F R               D A T H E S E I Y M G S P S I T E           V R V W

N G K F P E L H A H Y S M S T T S C G C L P V S G S Y H G E K K D L H F S F F
                                                            K K D L H F S F F

G V E T E V D D L Q V F V P P A Y C E G V A F E E A P D D H S F F D L F H D
G V E T E V D D L Q V F V P P A Y C E G V A F E E A P D D H S F F D L F H
```

FIGURE 2. Amino acid sequence of γ-ependymin. The total sequence data obtained by the cDNA method[21] is shown, as well as a partial sequence (underlined) determined by analysis of proteolytic digests of the γ peptide chain.[24]

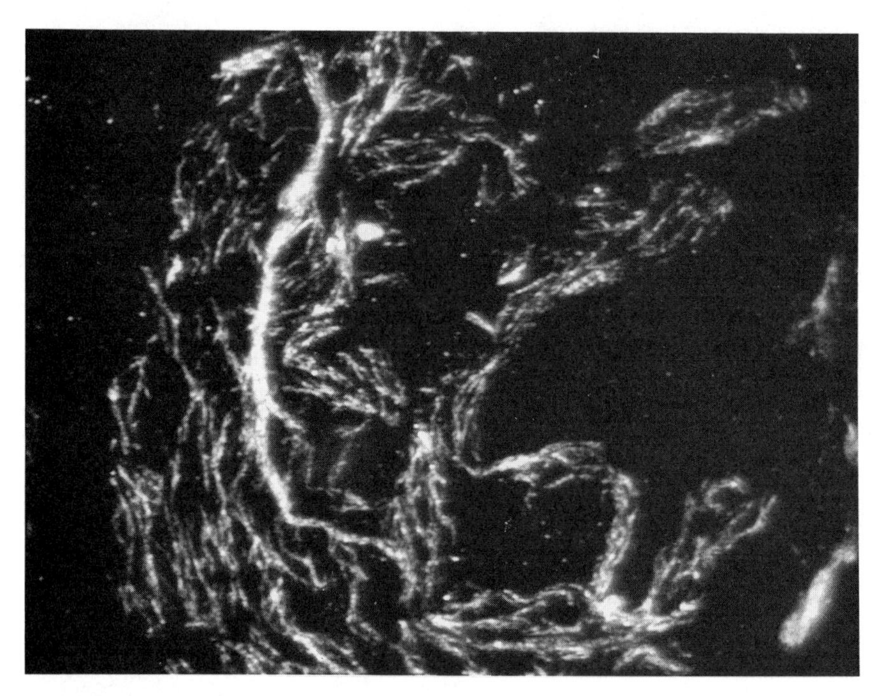

FIGURE 3. Polymerization properties of ependymin. Note the appearance of fibrous insoluble polymer (FIP) aggregates (immunofluorescence staining, dark-field, light microscopy; original magnification: 200×; reduced to 83% of original size).

conditioning paradigm. Simultaneous or a correct sequential firing of the strong synapse (large postsynaptic contact area) together with the weak (small contact area) synapse results in a local depletion of the extracellular Ca^{2+} (FIG. 4B). In this process the extracellular soluble ependymin becomes activated and polymerizes to form an extracellular matrix at the locus of the "converging associating" pathways. A crucial assumption for the model is the notion that the "associated" firing of two inputs results in a large depletion of extracellular Ca^{2+} (that is, $> 50\%$) as compared to smaller decreases that are normally obtained for the repetitive stimulation of a single pathway (that is, $< 10\%$). This step is essentially a modification of the Hebbian hypothesis[31] requiring the simultaneous activation of two pathways (Rule of Two) to a cell in order to initiate a change. The decrease in the local extracellular Ca^{2+} is accompanied by increased uptake of intracellular Ca^{2+}, activation of intracellular proteases, breakdown of cytoskeleton elements, and generation of spines (FIG. 4B).[32–34] The model also illustrates the possibility that soluble, locally activated ependymin in the ECF can enter into a spine, to immediately polymerize in the low Ca^{2+} environment to form part of the postsynaptic density. Thus an extracellular matrix and new cytoskeleton elements are formed. Such a structure essentially delineates the locus for the formation of new synapses, and initiates the modification of preexisting weak into strong synapses (FIG. 4C).

This Rule of Two hypothesis proposes that coincident stimulation as an activation mechanism may be generally employed by the nervous system for modifying neural circuits for events that follow learning, regeneration, and development.[32,35,36] Its application for the case of ependymin in the consolidation step of learning[27] has essentially four requirements: 1) there must be a steady supply of soluble ependymin in the extracellular medium; 2) the simultaneous firing of the two inputs at the locus of convergence must be able to generate a local extracellular calcium depletion signal which is large (that is, about 50%) as compared to the 10% that is normally observed for repetitive stimulation of a single pathway at frequencies below seizure thresholds; 3) an activation process must exist for converting the normally nonpolymerizable ependymin into its polymerizable state (see phosphorylation data on ependymin below); and 4) the polymerization initiation signal (that is, Ca^{2+} depletion) at a local convergence of two input sites must be rapid and transient. All these aspects of the Rule of Two hypothesis have some supporting evidence.

In the goldfish the supply of ependymin is well regulated in the brain ECF.[37] It is found that new synthesis and release of the molecule occurs after depletion of ECF levels following training.[6] Also, transient decreases (lasting about 2-3 min) of up to 70% in the extracellular Ca^{2+} levels have been reported following stimulation of multiple inputs in rat hippocampal slices *in vitro* as well as *in vivo*[38–40] and in goldfish optic tectum *in vitro.*[41] In addition, studies of associative learning[42] have demonstrated that calcium accumulation occurs in a postsynaptic cell of *Hermissenda* when two associating inputs are stimulated together, but not when each is stimulated separately. The uptake of calcium could presumably produce a depletion of extracellular Ca^{2+} at sites of converging inputs. This, however, cannot account for the total quantity of extracellular Ca^{2+} depletion that is required for ependymin polymerization. Presumably the major Ca^{2+} uptake system might involve anionic components of the bilipid layer of the cell membranes, such as an exchange of Na^+ for Ca^{2+} in the phosphate groups of lipids.

Recent studies of the polymerization of ependymin[30] suggest that its activation involves a phosphorylation of the molecule by protein kinase C in the presence of extracellular ATP. Both cholinergic[43] and adrenergic[44] systems have been found to release ATP into the extracellular space after stimulation. In addition, the existence of an ectokinase in the CNS has been reported.[45] Thus an energy source as well as an enzyme system capable of acting on an extracellular substrate, such as ependymin, are available in the CNS.

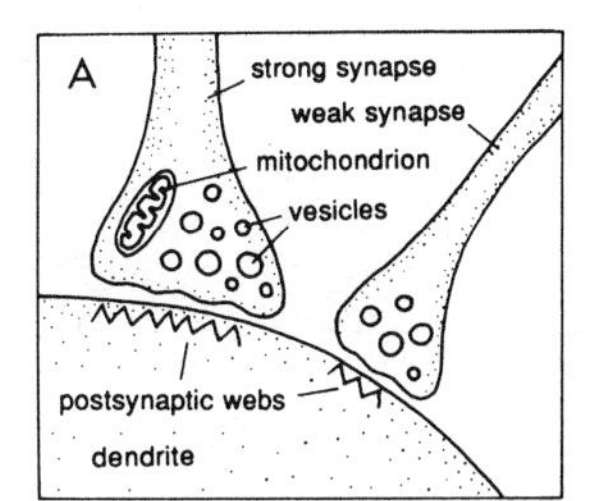

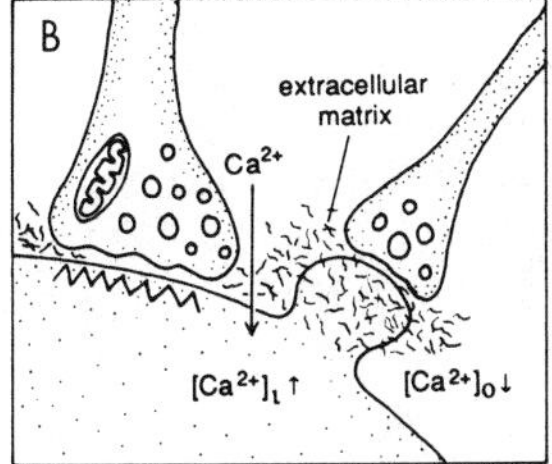

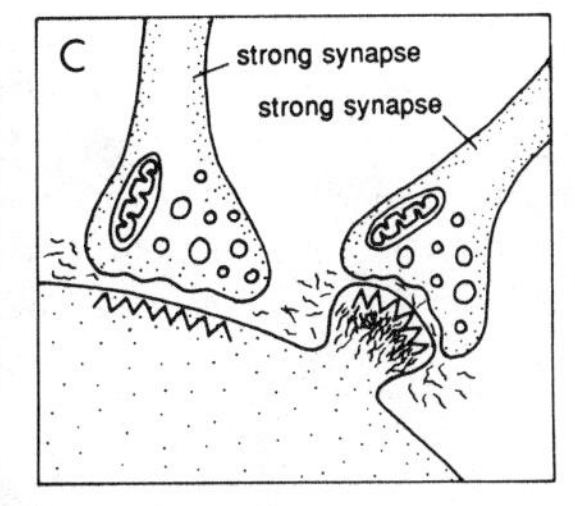

FIGURE 4. Diagram illustrating the ependymin polymerization (Rule of Two) hypothesis. In the initial state (**A**), two inputs—one strong (large contact area), one weak (small area of contact)—converge onto a dendrite. During associative learning a simultaneous or a sequentially correct firing of the two inputs results in a Ca^{2+} depletion in the local extracellular space and increases Ca^{2+} within the dendrite (**B**). Ependymin in the extracellular space polymerizes to form a fibrous extracellular matrix. The increased intracellular Ca^{2+} activates proteases that break down intracellular cytoskeletons to initiate the formation of a spine.[33,34] Entry of ependymin into the spine produces polymers that form a new cytoskeleton (**C**). These targeted sites are eventually converted to strong synapses that can independently fire the cell.

TESTS FOR THE RULE OF TWO HYPOTHESIS OF EPENDYMIN POLYMERIZATION

To test for the possibility that ependymin may be involved in the formation and modification of synapses, it is essential to select model systems in which one can know when and where to look for synaptic changes. The regenerating goldfish optic nerve is an example of such a system. Here a crushed optic nerve regenerates within a well-defined time course,[46] in a two-stage process in which the regenerating optic nerve first grows back to the optic tectum and establishes a broad, diffuse pattern of connections at 20-22 days postcrush.[47,48] In stage I, electrophysiological measurements show that the synapses do respond to visual stimulation, but the animal cannot see. In stage II, an activity-dependent refinement process occurs between day 22 and day 36 postcrush, during which the animal's behaviorally initiated movements promote connectivity changes and establish normal vision.[49] This process requires coincident visually driven inputs from neighboring retinal ganglion cells, to convert the initially diffuse pattern of connectivity to a precisely organized tectal map (FIG. 5). The synaptic contacts that do not receive coincident stimuli are then eliminated. This results in a reduction of the multiple unit receptive field (MURF) sizes from a large value (30-38°) to the smaller normal sizes (10-13°) by a process that eliminates inappropriate connections.[47,48] As shown in FIGURE 5, this phenomenon, in which the apparent coincident activation of overlapping inputs from neighboring cells results in neural circuit modifications, bears a formal resemblance to the requirements of the Rule of Two hypothesis as proposed for classical conditioning and ependymin polymerization (FIGS. 3 & 4). We therefore investigated the effects of antiependymin IgG on the regeneration pattern of crushed goldfish optic nerve.[4] The antiependymin IgG was delivered continuously by infusion via an osmotic minipump into the fourth ventricle of goldfish brain. The antibody had no effect on the pattern of the growth of the axons, that is, during stage I. In stage II, at days 22-35 postcrush, however, continuous infusion of the antibody produced a blockade of the sharpening of the retinotectal projections. Thus MURF sizes remained large (30-32°) for the animals receiving antiependymin IgG, whereas controls receiving either preimmune IgG, buffer alone, or another antibody (anti-S100 IgG) had no effect (TABLE 2). The MURF sizes in the control experiments (10.6-13°) followed a recovery time course comparable to that for untreated regenerating axons. The data suggest that antiependymin IgG may exert its effect by binding to the polymerized matrix that may be formed during the simultaneous stimulation of neighboring inputs, disrupting the subsequent steps that establish a precisely organized tectal map. The involvement of an extracellular matrix in this step may be similar to that reported for the reinnervation of the neuromuscular junction[50] and neuronal development in which growing or regenerating axons have been found to seek and attach to specialized extracellular matrix sites.

LTP AS A TEST SYSTEM FOR THE EPENDYMIN POLYMERIZATION HYPOTHESIS

Long-term potentiation (LTP) of the rat hippocampal slice preparation has been well established as a model system of neuroplasticity,[51–54] in which changes in the

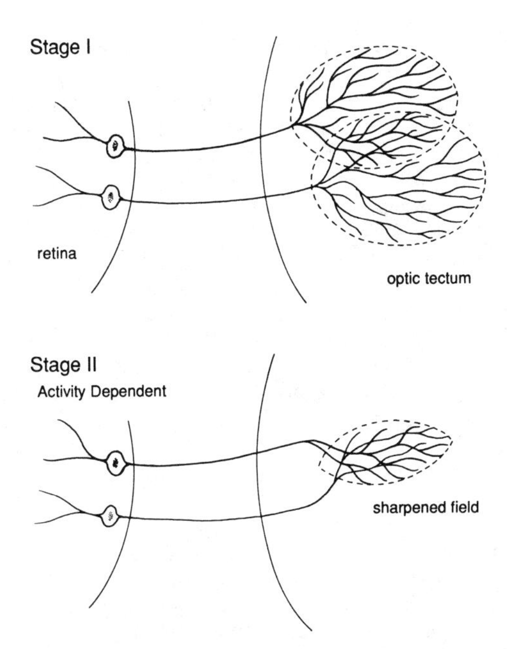

FIGURE 5. Stages in the regeneration of goldfish optic nerve axons. In stage I, the crushed optic nerve regenerates from retinal ganglion cells to establish a broad, diffuse pattern of tectal connections with overlapping regions from neighboring retinal ganglion cells. In stage II, an activity-dependent process results in a sharpening of the tectal projection field with elimination of inappropriate connections (nonoverlapping) to form smaller but stable receptive fields. The multiple unit receptive field size changes from about 32° to 10-13° during this process.

TABLE 2. MURF Sizes of Regenerating Goldfish Optic Nerves[a]

Experimentals		Controls	
Dosage in Antiependymin IgG Treatment (μg/ml)	MURF[b]	Treatment	MURF[b]
250	27.8 ± 4.3	Nonregenerating side	9.3 ± 0.4
250	32.8 ± 2.2	Nonregenerating side	10.1 ± 1.9
62	28.3 ± 1.9	Ringers	10.6 ± 0.1
62	30.8 ± 1.8	Goat antirabbit IgG	12.2 ± 0.5
62	35.5 ± 2.6	Goat antirabbit IgG	13.1 ± 0.8
62	32.4 ± 2.4	Goat antirabbit IgG	12.2 ± 0.4
62	34.1 ± 1.7	Anti-S100 IgG	12.2 ± 0.5
Average	31.8 ± 1.0		12.1 ± 0.4

[a] Control and experimental fish received an infusion of affinity-purified IgG (0.25 mg/ml) from an osmotic minipump on days 21-40 after optic nerve crush. The MURF measurements were carried out using extracellular electrodes on day 47 postcrush at an average of 17 sites for experimentals and 27 sites for controls. The pump flow rate was 3.5-4.4 μl/day. The results show that antiependymin IgG blocks sharpening of connections. The receptor field sizes for the experimentals remain large (31.8°) as compared to a variety of control treatments (12.1°).

[b] Each value below is a mean ± SEM. MURF values are expressed in degrees.

number of synapses[33,34] have been observed. The slice consists of a well-defined, three-synaptic circuit. Afferent stimulation of the Schaffer collateral pathway to CA1 dendrites by application of a brief tetanizing train (100 Hz for 1 sec) produces a long-lasting postsynaptic field potential response that is 2-10-fold larger than for unstimulated controls. LTP lasts for over 10 hr *in vitro* and several months *in vivo.*[54] Previous studies on the effects of LTP on the pattern of protein synthesis showed that the only changes detectable were those associated with the products released into the extracellular medium.[14,15] No protein changes were found in the cytoplasmic, synaptosomal, nuclear, and membrane subcellular fractions. Subsequent studies demonstrated that ependymin is a prominent component of the products released into the extracellular medium and that the CA1 pyramidal cells, as well CA2, CA3, and CA4, contain immunoreactive components that are similar to ependymin.[55-57] These findings raise the possibility that ependymin may be used up during LTP, perhaps by a process that generates FIP in stratum radiatum at sites where new synapse formation has been noted.[33,34]

We therefore used standard methods to obtain LTP at CA1 and developed an immunohistochemical method to look for the presence of polymerized ependymin in the slice. Each slice was stimulated via its Schaffer collateral pathway to give a 2-3-fold increased response that was stable for at least 1 hr[14,15] and then extracted at 0 °C for 3 hr in isotonic medium to remove any remaining soluble ependymin that may have been present in its extracellular space, prior to fixation for 15 min in 5% acrolein.[58] This step was found to be essential for eliminating nonspecific polymerization of ependymin during subsequent processing. After several washing steps, each slice was cut into 50-μm-thick sections in a vibratome and immunohistochemical studies were carried out on the second and third sections using rabbit antiependymin serum (HIROKO) raised against the mouse brain derived protein. FIGURE 6 shows the pattern of localization of the horseradish peroxidase reaction product at the light microscope level. Both the experimental and control unstimulated sections show the expected presence of intracellular ependymin in the hippocampal pyramidal and dentate granule cells. This is consistent with the known pattern of localization of the protein in hippocampal neurons.[56] In the potentiated sections, there was additional staining; a dot-like pattern was present in the stratum radiatum and stratum oriens of the CA1 region. This stopped abruptly at the border of the CA3 region of the slice. FIGURE 7 depicts the same fields at a higher magnification, illustrating the difference between the stimulated CA1 and the unstimulated CA3 regions, which serve as an internal control for the same slice. Additional control unstimulated slices (data not shown) that were treated with preimmune serum and those that received no primary antibody showed no such staining patterns. These data suggest that ependymin polymerization may occur at axodendridic regions of pyramidal cell dendrites where multiple pathways are stimulated during generation of LTP. This finding is implicated by the observed structural changes in previous studies[33,34] and is consistent with the Rule of Two hypothesis, which would predict the formation of FIP at sites of convergent inputs.

FIGURES 8 & 9 show the results of electron microscopic studies of these same slices after fixation with gluteraldehyde and osmium. Control and potentiated slices were cut into thin sections and processed by identical methods; no treatments with uranyl acetate or lead were used to allow for better contrast. The potentiated slices show the presence of a reaction product at extracellular sites localized in the synaptic cleft and at some postsynaptic sites. In control slices, the synaptic regions appear as faintly stained zones, suggesting the absence of an immunoreactive product. The results suggest that ependymin polymerization may have occurred in the potentiated slices, where it appears to be localized at synaptic extracellular regions and postsynaptic density regions, occasionally invading into some presynaptic sites.

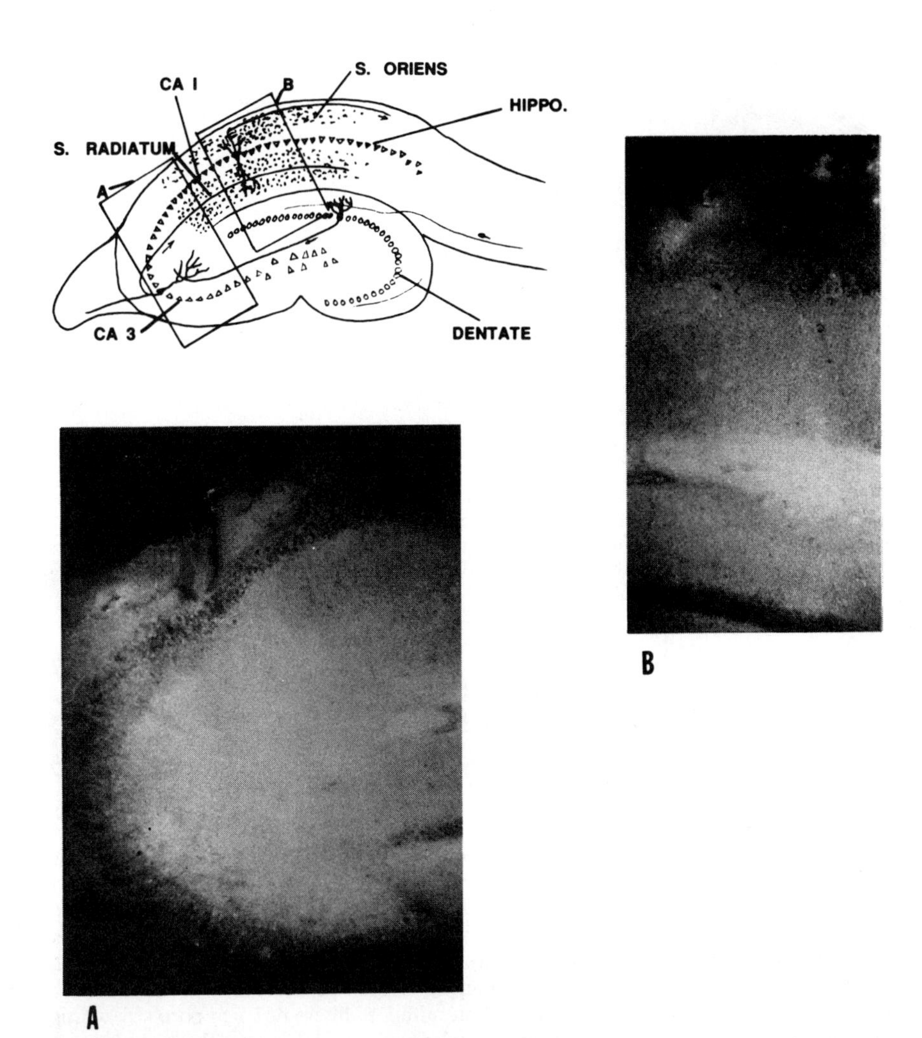

FIGURE 6. Immunohistochemical localization of antiependymin staining sites in a rat hippocampal slice after LTP. The diagram shows the distribution of the horseradish peroxidase reaction product in a slice potentiated via the Schaffer collaterals to the CA1 region (note the regions labeled **A** and **B** in the diagram; these regions correspond to the photographs labeled **A** and **B**). A dot-like staining pattern is found in stratum radiatum and stratum oriens at the CA1 pyramidal cell region. Photograph **A** shows the CA3 region and its border with CA1. Note the sudden appearance of dots at the demarcation zone between CA1 and CA3. The region below CA3 is essentially clear as compared to the CA1 stratum radiatium zone. Both CA1 and CA3 pyramidal cell somas contain ependymin and are stained by the antibody. (Magnification: 100×; reduced to 67% of original size).

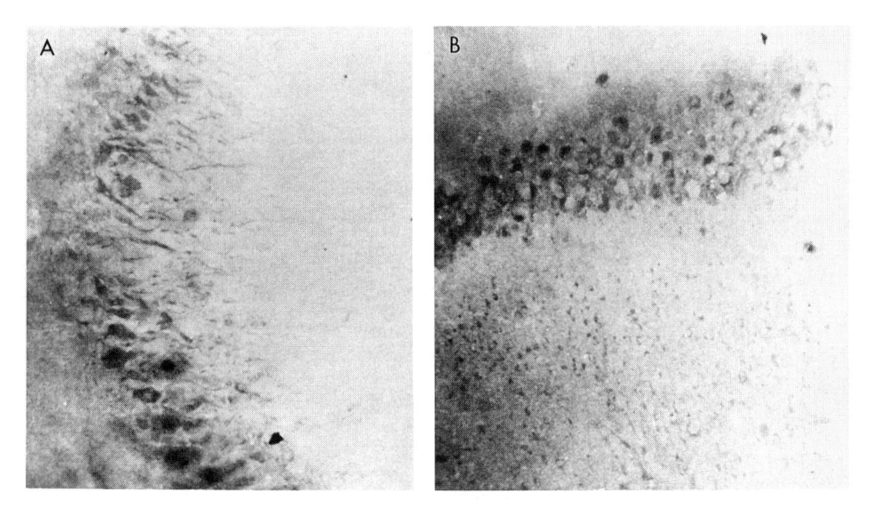

FIGURE 7. Comparisons of stratum radiatum staining patterns with antisera to ependymin below CA3 (**A**) and CA1 (**B**). Data are for the same slice as in FIGURE 6. Magnification: 400×; reduced to 49% of original size. The CA1 region has been stimulated; the CA3 region has not. Note the extra appearance of the dot-studded zone at CA1 but not at CA3.

EVIDENCE FOR THE POLYMERIZATION OF EPENDYMIN *IN VIVO*

The fact that ependymin polymerization may occur after LTP[59,60] and in regenerating optic nerve[4] led us to look for the possibility that it may also take place *in vivo* in the CNS. We therefore developed a method for isolating FIP from brain subcellular fractions.[29] The assay depends on the observation that once polymerized, ependymin fibers can no longer be dissolved in 2% SDS in 5 M urea (thus, with this solvent, one can dissolve away all the components of brain tissue to leave a residue of aggregates that may include polymerized ependymin). Such solutions, after sedimentation for 3 hr at 300,000 × g, yielded pellets that upon further purification by dialysis were found to contain fibers that were retained on millipore filters (0.2 μm pore size). This procedure showed that both goldfish and mouse brain contained about the same amount of FIP (6.7 μg/mg brain protein), and that this FIP was largely concentrated in the synaptosomal subcellular fraction.[29]

With the exception of ependymin, all known fibers formed from proteins such as tubulin, actin, neurofilament protein, laminin, and fibronectin are soluble in a 2% SDS-5 M urea solvent. Analysis of the immunoreactivity of FIP using a dot-blot assay showed that they readily interacted with antiependymin IgG but not with preimmune serum or with horseradish peroxidase-coupled secondary antibody alone. In addition, the aggregates of FIP were found to be cross-reactive toward antibodies raised against other fiber-forming components of the extracellular matrix such as laminin and fibronectin as well as the intracellular proteins tubulin and tau (a protein present in paired helical filaments[61]) (FIG. 10). They did not react with an antimouse neural cell adhesion molecule (NCAM) antibody directed against its polypeptide epitope. Immunoreactivity properties of the same type were found to be present in FIP prepared by the polymerization of pure ependymin *in vitro* (FIG. 10). The cross-reactivity properties were quite surprising because the primary amino acid sequence of ependymin did not show any significant homologies to any long segments of these proteins. However, if the sequence analyses were carried out comparing segments that were about five to seven amino acids long, then specific homologies were found. TABLE 3 shows that ependymin contains sequences that would be suitable epitopes for raising antisera that might recognize fibronectin, laminin, tubulin, as well as human NCAM (no analogy to mouse NCAM was found). This was confirmed by western blots.[29] FIGURE 11 depicts the γ polypeptide chain of ependymin showing the location of a fibronectin-related site near

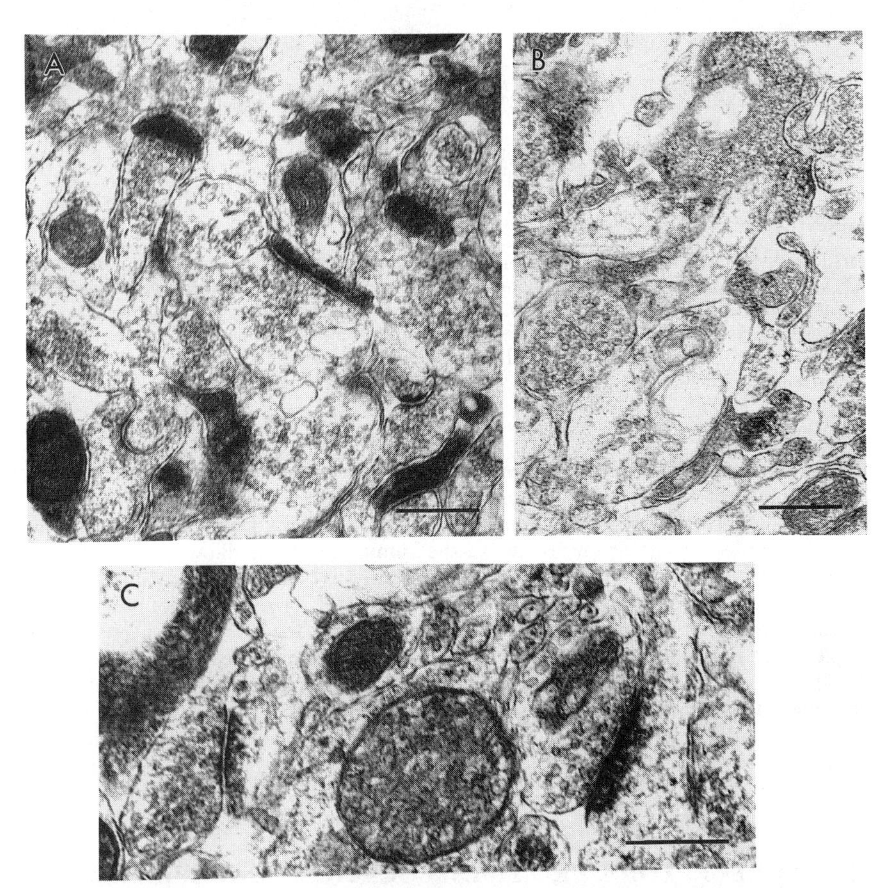

FIGURE 8. Electron micrographs of sections of rat brain hippocampal slices showing stratum radiatum regions stained with antisera to ependymin. **A** and **C** show sections from a potentiated slice (space bars: 0.6 μ); **B** is a control (space bar: 0.6 μ). All slices were extracted with isotonic media to remove residual soluble ECF ependymins prior to fixation in 5% acrolein. The sections were then stained first with $1/200$ rabbit antiependymin serum followed by $1/200$ ferritin-labeled goat antirabbit IgG. Note the staining pattern of extracellular and postsynaptic areas, together with an occasional invasion of a presynaptic site showing the locus of polymerized ependymin loci. No uranyl acetate or lead was used.

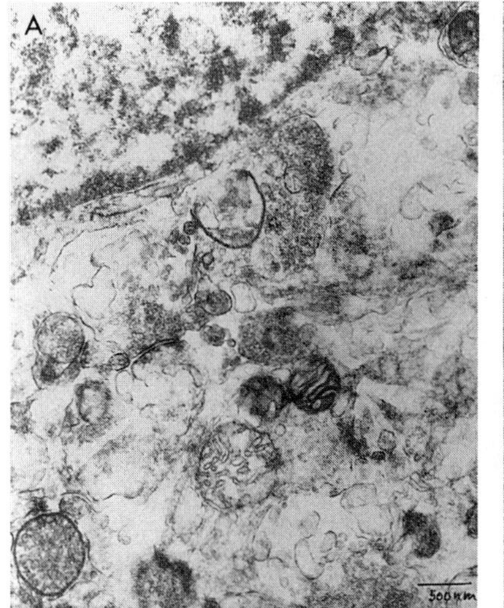

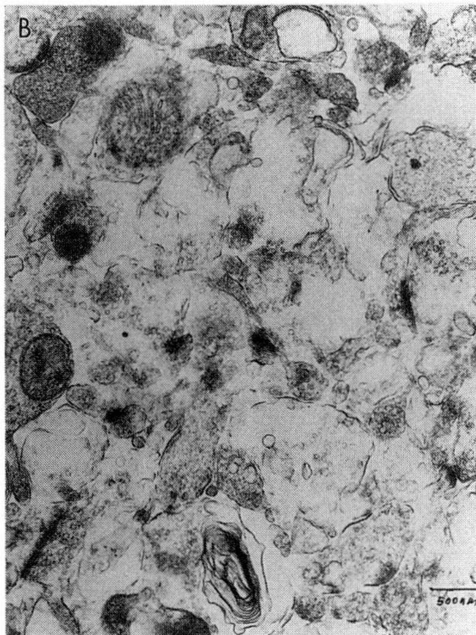

FIGURE 9. Electron micrographs of sections of stratum radiation regions of rat brain hippocampal slices. Both control (**A**) and potentiated (**B**) slices were first extracted with isotonic media at 0 °C for 3 hr to remove any soluble residual ECF ependymin prior to fixation with 5% acrolein. The slices were stained with $\frac{1}{200}$ rabbit antiependymin serum followed by $\frac{1}{200}$ goat antirabbit-horseradish peroxidase (avidin-biotin method). No uranyl acetate or lead salts were used. Note the presence of a dense stain at postsynaptic sites in the potentiated section.

the C-terminal, a laminin sequence near the N-terminal, and a tubulin segment near the midpoint of the molecule. Thus, although such short sequences are normally considered to be too close to a random correlation to be relevant, they should nevertheless be regarded as significant because of the immunoreactivity data.

Further evidence for the presence of FIP in brain was obtained from labeling studies with ^{35}S methionine.[29] These showed that about 0.23% of the incorporated counts, present in the synaptosomal fraction, were associated with the FIP fraction. The results suggest that FIP is formed *in vivo.* It appears to be a normal constituent of the CNS. The fact that it is formed in the brains of control animals suggests that it must be continuously generated. One must therefore consider the possibility that it has a turnover rate regulated by specific proteases that may exist in the brain for "FIP removal" after it has served its function.

NMDA RECEPTORS, PKC TRANSLOCATION, AND EPENDYMIN POLYMERIZATION

A number of studies have reported that the *N*-methyl-D-aspartate (NMDA) receptor channel complex plays a key role in molecular mechanisms of associative learning,[68,69] brain development,[70] and neuronal regeneration.[71] It has been found that following activation of NMDA receptors by glycine and glutamate, an intracellular Ca^{2+} influx takes place. This causes, among other changes, a release of intracellular diacylglycerol (DAG). DAG and Ca^{2+} then activate protein kinase C (PKC)[68,72] and promote its translocation to the cell membrane. After classical conditioning, PKC was found[73] to translocate to cell membrane sites in the stratum radiatum and stratum oriens regions of the rat hippocampal slice. These sites are essentially identical to those obtained by staining with antiependymin sera after LTP (FIGS. 6 & 7), and raise the

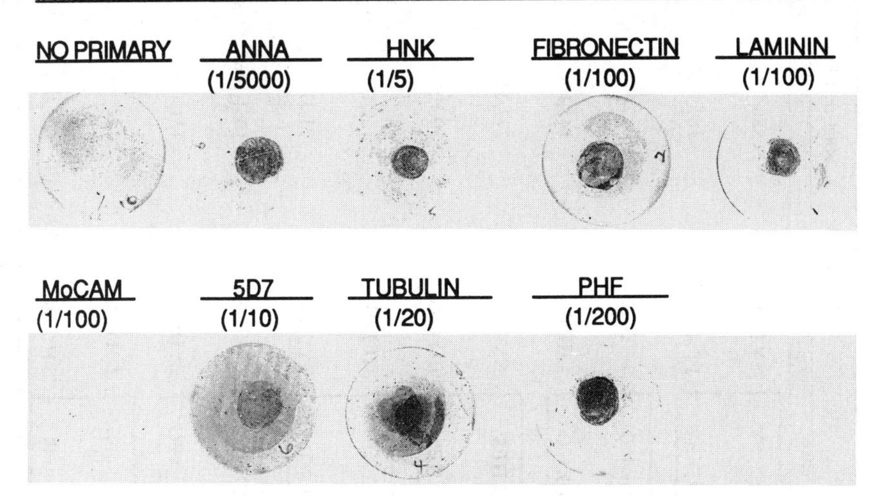

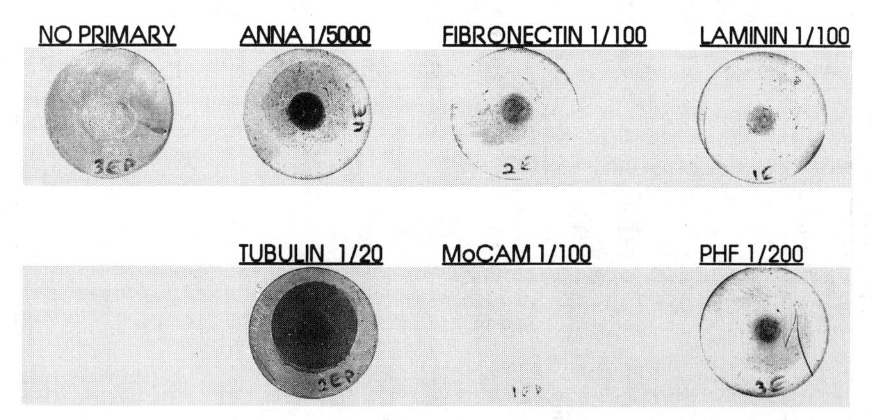

FIGURE 10. Comparisons of dot-blot assay patterns of the immunoreactivity properties of FIP isolated from goldfish brain synaptosomes with FIP prepared by polymerization of pure ependymin. Note that both types have similar staining patterns. Control filters loaded with equivalent amounts of FIP but not treated with any primary antibody and those treated with antimouse NCAM did not stain. ANNA is a rabbit polyclonal antigoldfish ependymin antibody; HNK and 5D7 are monoclonal antigoldfish antibodies directed against the carbohydrate and polypeptide epitopes of ependymin; MoCAM and PHF are polyclonal antibodies to polypeptide epitopes of mouse NCAM and tau.[61]

TABLE 3. Comparison of Sequence Homologies

Protein	Reference	Sequence			Start and End of Sequence	Percentage Homology[a]
Human fibronectin	62	VPSPR	DLQ_FV	EVTDV	886,890	83.3
Ependymin		ETEVD	DLQVFV	PPAYC	160,175	
Fibronectin receptor	63	RAFNK	GEKKD	TCTQE	660,674	100.0
Human β-ependymin		SGSYH	GEKKD	LHFSF	142,156	
Tubulin yeast 2-3	64	LSKKR	ATHES	NSVSE	277,291	100.0
Ependymin		EIPPD	ATHES	EIYMG	90,104	
Human and mouse β_1-laminin	65,66	VTIQL	DLE_AEF	HFTHL	121,136	71.4
Ependymin		STGGH	DLESGEF	SYDSK	21,37	
Human β_1-laminin	65,66	TGQCS	CLPNVIG	QNCDR	1051,1067	71.4
Ependymin		TTSCG	LP__VSG	SYHGE	131,148	
NCAM	66	KPEIR	LP_SGS	DHVML	192,206	83.3
Ependymin		TSCGC	LPVSGS	YHGEK	134,149	

[a] With the exception of laminin, only the homologies with values above the 80% level are shown. The percentage homology data are calculated for those portions of the sequences shown in brackets.

possibility that the formation of FIP from activated ependymin, occurring at such sites, may be related to the presence of translocated PKC. (PKC has also been implicated in LTP.[74,75]) Because the activation of ependymin molecules to their polymerizable state requires a phosphorylation step, one must postulate the existence of a mechanism that generates an ectoenzyme at the sites of "associating inputs." One possible candidate for such a function may be the translocated PKC, which, as a result of its migration to the cell membrane, becomes available to phosphorylate an extracellular substrate. This speculative hypothesis is illustrated in FIGURE 12. Here the enzyme acts as an ectokinase, using the extracellularly released ATP to phosphorylate extracellular ependymin at the locus of associating inputs. The phosphorylated ependymin can then polymerize to form an extracellular matrix at the site where NMDA activation occurs, provided that the extracellular Ca^{2+} depletion is sufficiently large. The possibility that the phosphorylated ependymin could actually enter the cell is supported by recent findings showing that phosphorylated macromolecules do enter into cells.[76] This mechanism assigns a function to the PKC translocation phenomenon; that is, this mechanism suggests that PKC translocation is a device for targeting an enzyme to work on an

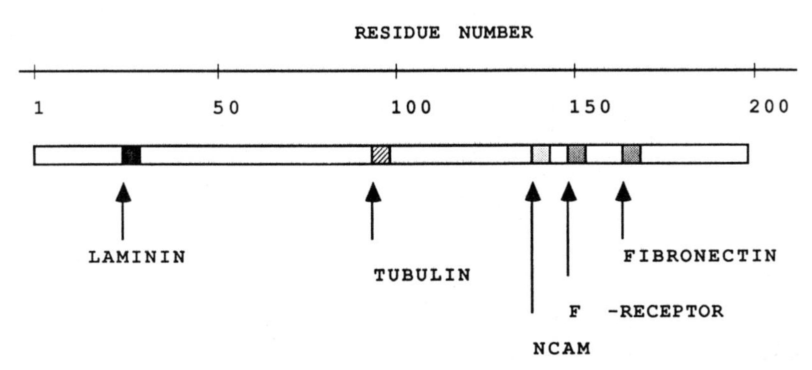

FIGURE 11. Diagram of the loci of cross-reactive regions in the primary amino acid sequence of the ependymin γ polypeptide chain.

extracellular substrate at a specific locus. One consequence of this notion is that it provides a means for coupling events that have been postulated[68,74] for short-term memory to a long-term memory related aspect, that is, the polymerization of ependymin and its participation in the consolidation step of learning.[29,30] We do not yet know whether the hypothesis illustrated in FIGURES 4 & 12 will prove to be valid. Any proposed mechanism, however, will not only have to account for the depletion of ependymin from ECF and a local decrease of extracellular Ca^{2+} at the convergence of associating inputs, but also to provide a purpose for the translocation of PKC to cell membranes and the release of ATP into ECF.

SUMMARY

Ependymin, a glycoprotein of the brain ECF, has been implicated in the neurochemistry of memory and neuronal regeneration. Three behavioral experiments (swimming with a float, avoidance conditioning, and classical conditioning) in the goldfish and one in the mouse (T-maze learning) indicate that ependymin has a role in the synaptic changes that take place in the consolidation step of memory formation and the activity-dependent phase of sharpening of goldfish retinotectal connections during neuronal regeneration. The ECF concentration of the protein was found to decrease after the goldfish learned to associate a light stimulus (CS) with the subsequent arrival of a shock (US): paired CS-US gave changes whereas an unpaired presentation of CS-US gave no changes relative to the unstimulated controls.

Ependymin is present in ECF as a mixture of three disulfide-linked dimers of two acidic (α and β) polypeptide chains (37 kDa and 31 kDa). Upon removal of its N-linked

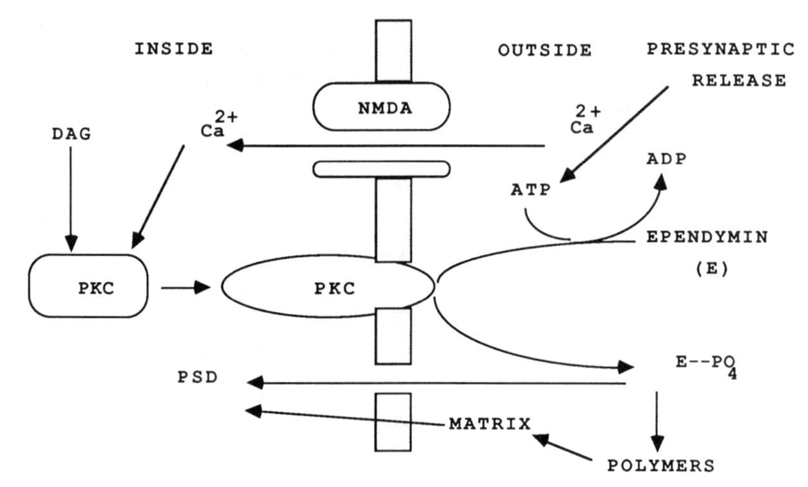

FIGURE 12. Diagram illustrating a possible role of protein kinase C (PKC) in the phosphorylation and polymerization of ependymin. The hypothesis shows that following the activation of the NMDA receptor, an influx of extracellular Ca^{2+} together with diacylglycerol (DAG) results in the activation and translocation of PKC to the cell membrane. The PKC is inserted into the membrane with its enzymatic active site protruding into the extracellular space. This then uses the extracellular presynaptically released ATP to phosphorylate ependymin (E). The E-PO_4 may then enter the postsynaptic site and polymerize in the cytosol to form part of the postsynaptic density (PSD) complex as well as polymerize extracellularly to form an extracellular matrix. Such a sequence of biochemical changes is proposed for initiating the modification of preexisting synapses or the formation of new synapses (see FIG. 4).

glycan fragment by N-glycosidase F, the β chain yields γ-ependymin (26 kDa). Determinations of the amino acid sequence of γ-ependymin indicate that it is a unique protein with no long sequence homologies to any known polypeptide. There are, however, small segments (5-7 amino acids long) with homologies to fibronectin, laminin, and tubulin.

Ependymin has the capacity to polymerize into FIP (after activation by phosphorylation) in response to events that deplete ECF calcium. FIP is insoluble in 2% SDS in 6 M urea, 10 mM $Ca^{2+}Ac_2$, 100% acetic acid, chloroform/methanol (2/1), saturated KCNS, and even 100% trifluoroacetic acid. FIP was found to be present in goldfish brain and to be formed as a labeled product *in vivo.* Ependymin's FIP-forming property was used to propose a molecular hypothesis for generating synaptic changes in response to local extracellular depletions of calcium at sites of "associating inputs." The model assumes that, following NMDA receptor stimulation, the translocated PKC that is generated activates extracellular ependymin by converting it to its phosphorylated form using presynaptically released ATP. The hypothesis was tested in studies of LTP of rat hippocampal slices at CA1. After LTP, new sites that stained with antisera to ependymin, visible at 100×, were obtained in its potentiated radiatum in the CA1 region but not in the unpotentiated CA3. Electron microscopic studies showed that the horseradish peroxidase reaction product obtained was localized at synaptic clefts and postsynaptic regions. The results suggest that FIP may be formed at extracellular and postsynaptic loci where multiple associating inputs interact at CA1.

REFERENCES

1. SHASHOUA, V. E. 1976. Brain metabolism and the acquisition of new behaviors. I. Evidence for specific changes in the pattern of protein synthesis. Brain Res. **111:** 347-364.
2. SHASHOUA, V. E. 1977. Brain protein metabolism and the acquisition of new behaviors. II. Immunological studies of the α-, β-, and γ-proteins of goldfish brain. Brain Res. **122:** 113-124.
3. SHASHOUA, V. E. 1982. Molecular and cell biological aspects of learning: Towards a theory of memory. *In* Advances in Cellular Neurobiology. S. Fedoroff & L. Hertz, Eds. Vol. 3: 97-141. Academic Press. New York, NY.
4. SCHMIDT, J. T. & V. E. SHASHOUA. 1988. Antibodies to ependymin block the sharpening of the regenerating retinotectal projection in goldfish. Brain Res. **446:** 269-284.
5. SHASHOUA, V. E. 1983. The role of specific brain protein in long-term memory formation. *In* Changing Concepts of the Nervous System. A. R. Morrison & P. L. Strick, Eds.: 681-716. Academic Press. New York, NY.
6. SHASHOUA, V. E. 1977. Brain protein metabolism and the acquisition of new patterns of behavior. Proc. Natl. Acad. Sci. USA **74:** 1743-1747.
7. SHASHOUA, V. E. 1979. Brain metabolism and the acquisition of new behaviors. III. Evidence for secretion of two proteins into the brain extracellular fluid after training. Brain Res. **166:** 349-358.
8. SHASHOUA V. E. & M. E. MOORE. 1978. Effect of antisera to β and γ goldfish brain proteins on the retention of a newly acquired behavior. Brain Res. **148:** 441-449.
9. PIRONT, M. L. & R. SCHMIDT. 1988. Inhibition of long-term memory formation by anti-pendymin antisera after classical shock avoidance conditioning in goldfish. Brain Res. **442:** 53-62.
10. SHASHOUA, V. E. & G. W. HESSE. 1989. Classical conditioning leads to changes in extracellular concentrations of ependymin in goldfish brain. Brain Res. **484:** 333-339.
11. BITTERMEN, M. E. 1964. Classical conditioning in the goldfish as a function of the CS-US interval. J. Comp. Physiol. Psychol. **58:** 359-366.
12. SHASHOUA, V. E. 1970. RNA metabolism in goldfish brain during acquisition of new behavioral patterns. Proc. Natl. Acad. Sci. USA **65:** 160-167.
13. SCHMIDT, R. 1986. Biochemical participation of glycoproteins in memory consolidation after two different training paradigms in goldfish. *In* Learning and Memory: Mechanisms of Information Storage in the Nervous System, Advances in Bioscience. H. Mattheis, Ed. Vol. 59: 213-222. Pergamon. Oxford.
14. DUFFY, C., T. J. TEYLER & V. E. SHASHOUA. 1981. Long-term potentiation in the hippocampal slice: Evidence for stimulated secretion of newly synthesized proteins. Science **212:** 1148-1151.
15. HESSE, G. W., R. HOFSTEIN & V. E. SHASHOUA. 1984. Protein release from hippocampus *in vitro.* Brain Res. **305:** 61-66.
16. SHASHOUA, V. E. 1985. The role of brain extracellular proteins in neuroplasticity and learning. Cell. Mol. Neurobiol. **5:** 183-207.

17. SHASHOUA, V. E. 1988-1989. The role of ependymin in the development of long-lasting synaptic changes. J. Physiol. (Paris) **83:** 232-239.
18. SCHMIDT, R. & V. E. SHASHOUA. 1981. A radioimmunoassay for ependymins β and γ: Two goldfish brain proteins involved in behavioral plasticity. J. Neurochem. **36:** 1368-1377.
19. SHASHOUA, V. E., P. F. DANIEL, M. E. MOORE & F. B. JUNGALWALA. 1986. Demonstration of glucuronic acid on brain glycoproteins which react with HNK-1 antibody. Biochem. Biophys. Res. Commun. **138:** 902-909.
20. SHASHOUA, V. E. 1988. Monomeric and polymeric forms of ependymin: A brain extracellular glycoprotein implicated in memory consolidation processes. Neurochem. Res. **13:** 649-655.
21. KÖNIGSTORFER, A., S. STERRERS, C. ECKERSKORN, F. LOTTSPEICH, R. SCHMIDT & W. HOFFMANN. 1989. Molecular characterization of an ependymin precursor from goldfish brain. J. Neurochem. **52:** 310-312.
22. KÖNIGSTORFER, A., S. STERRERS & W. HOFFMAN. 1989. Biosynthesis of ependymins from goldfish brain. J. Biol. Chem. **264:** 13689-13692.
23. CHAN, D. K. H., A. A. ILYAS, J. E. EVANS, C. COSTELLO, R. H. QUARALES & F. B. JUNGALWALA. 1986. Structure of sulfated glucoronyl SGGl glycolipids in the nervous system reacting with HNK-1 antibody and some IgH paraproteins in neuropathy. J. Biol. Chem. **261:** 11717-11725.
24. SCHMIDT, R. & V. E. SHASHOUA. 1983. Structural and metabolic relationships between goldfish brain glycoproteins participating in functional plasticity of the central nervous system. J. Neurochem. **40:** 652-660.
25. SHASHOUA, V. E. & G. W. HESSE. 1990. Primary amino acid sequence of γ-ependymin by analysis of proteolytic digests. Submitted for publication.
26. MAJOCHA, R. E., R. SCHMIDT & V. E. SHASHOUA. 1982. Cultures of zona ependyma cells of goldfish brain: An immunological study of the synthesis and release of ependymins. J. Neurosci. Res. **8:** 331-342.
27. SHASHOUA, V. E. 1985. The role of brain extracellular proteins in neuroplasticity and learning. Cell. Mol. Neurobiol. **5:** 183-207.
28. SHASHOUA, V. E., H. EPSTEIN & M. E. MOORE. 1990. Extracellular protein changes during periods of rapid postnatal mouse brain development. Submitted for publication.
29. SHASHOUA, V. E., G. W. HESSE & B. MILINAZZO. 1990. Evidence for the *in vivo* polymerization of ependymin: A brain extracellular protein. Brain Res. **522:** 181-190.
30. NOLAN, P. M. & V. E. SHASHOUA. 1989. Mechanisms of ependymin phosphorylation in goldfish brain ECF. Soc. Neurosci. Abstr. **15**(1): 961.
31. HEBB, D. O. 1949. The Organization of Behavior. J. Wiley & Sons. New York, NY.
32. SHASHOUA, V. E. 1985. The role of extracellular proteins in learning and memory. Am. Sci. **73:** 364-370.
33. LEE, K. S., F. SCHOTTLER, M. OLIVER & G. LYNCH. 1980. Brief bursts of high-frequency stimulation produce two types of structural change in rat hippocampus. J. Neurophysiol. **44:** 274-258.
34. CHANG, F. L. & W. T. GREENOUGH. 1984. Transient and enduring morphological correlates of synaptic activity and efficacy change in rat hippocampal slice. Brain Res. **309:** 35-46.
35. BEAR, M. F., L. N. COOPER. & F. F. EBNER. 1988. A physiological basis for a theory of synapse modification. Science **237:** 42-48.
36. ALKON, D. L. 1984. Calcium-inactivated potassium currents: A biophysical memory trace. Science **276:** 1037-1045.
37. SHASHOUA, V. E. 1981. Extracellular fluid proteins of goldfish brain: Studies of concentration and labeling patterns. Neurochem. Res. **6:** 1129-1147.
38. KRNJEVIC, K., M. E. MORRIS & R. J. REIFFENSTEIN. 1982. Stimulation-evoked potentiation changes in extracellular K^+ and Ca^{2+} in pyramidal layers of the rat's hippocampus. Can. J. Physiol. Pharmacol. **60:** 1643-1657.
39. KRNJEVIC, K., M. E. MORRIS, R. J. REIFFENSTEIN & N. ROPERT. 1982. Depth distribution and mechanism of changes in extracellular K^+ and Ca^{2+} concentrations in the hippocampus. Can. J. Physiol. Pharmacol. **60:** 1658-1671.
40. SOMJEN, G. G. 1984. Acidification of interstitial fluid in hippocampal formation caused by seizures and by spreading depression. Brain Res. **311:** 186-188.
41. MORRIS, M. E., N. ROPERT & V. E. SHASHOUA. 1986. Stimulus-evoked changes in extracel-

lular calcium in optic tectum of goldfish: Possible role in neuroplasticity. Ann. N.Y. Acad. Sci. **481:** 375-377.

42. Connor, J. A. & D. L. Alkon. 1984. Light- and voltage-dependent increases of calcium ion concentration in molluscan photoreceptors. J. Neurophysiol. **51:** 745-752.
43. White, T. D. 1977. Direct detection of depolarization-induced release of ATP from a synaptosomal preparation. Nature **267:** 67-68.
44. Richardson, P. J. & S. J. Brown. 1987. ATP release from affinity-purified rat cholinergic nerve terminals. J. Neurochem. **48:** 622-630.
45. Zhang, J., E. Kornecki, J. Jackman & Y. H. Ehrlich. 1988. ATP secretion and extracellular protein phosphorylation by CNS neurons in primary culture. Brain Res. Bull. **21:** 459-464.
46. Grafstein, B. 1969. Changes in the morphology and amino acid incorporation of regenerating goldfish optic neurons. Exp. Neurol. **23:** 544-560.
47. Schmidt, J. T., D. L. Edwards & C. A. O. Stuermer. 1983. The reestablishment of synaptic transmission of regenerating optic axons in goldfish: Time course and effects of blocking activity by intraocular injection and tetrodotoxin. Brain Res. **269:** 15-27.
48. Schmidt, J. T. & D. L. Edwards. 1983. Activity sharpens the map during regeneration of the retinotectal projection in goldfish. Brain Res. **269:** 29-39.
49. Edwards, D. L. & B. Grafstein. 1983. Intraocular tetrodotoxin in goldfish hinders optic nerve regeneration. Brain Res. **269:** 1-14.
50. Sanes, J. R. 1983. Roles of extracellular matrix in neural development. Annu. Rev. Physiol. **45:** 581-600.
51. Anderson, P., S. H. Sundberg, J. N. Swann & H. Wigstrom. 1980. Possible mechanisms for long-lasting potentiation of synaptic transmission in hippocampal slices from guinea pigs. J. Physiol. **302:** 463-482.
52. Bliss, T. V. P. & T. Lomo. 1973. Long-lasting potentiation of synaptic transmission in the dentate area of the anaesthetized rabbit following stimulation of the perforant path. J. Physiol. **232L:** 1-356.
53. Lynch, G. & J. Schubert. 1980. The use of *in vitro* brain slices for multidisciplinary studies of synaptic functions. Annu. Rev. Neurobiol. **3:** 1-22.
54. Teyler, T. J. 1980. Brain slice preparation: Hippocampus. Brain Res. Bull. **5:** 391-403.
55. Shashoua, V. E. & G. W. Hesse. 1985. Role of brain extracellular proteins in the mechanism of long-term potentiation in rat brain hippocampus. Neurosci. Abstr. **11:** 782.
56. Schmidt, R., F. Loffler, H. W. Muller & W. Seifert. 1986. Immunological cross-reactivity of cultured rat hippocampal neurons with goldfish brain protein synthesized during memory consolidation. Brain Res. **386:** 245-257.
57. Fazeli, M. S., M. L. Errington, A. C. Dolphin & T. V. P. Bliss. 1988. Long-term potentiation in the dentate gyrus of the anesthetized rat is accompanied by an increase in protein efflux into push-pull cannula perfusates. Brain Res. **473:** 51-59.
58. King, J. C., R. M. Lechan, G. Kugel & E. L. P. Anthony. 1983. Acrolein: A fixative for immunocytochemical localization of peptides in the central nervous system. J. Histochem. Cytochem. **31:** 62-68.
59. Shashoua, V. E., G. W. Hesse & P. Paskevich. 1989. Localization of ependymin in potentiated rat brain hippocampal slices. Submitted for publication.
60. Shashoua, V. E. 1989. The role of ependymin in the development of long-lasting synaptic changes. J. Physiol. (Paris) **83:** 232-239.
61. Kosik, K. S., L. D. Orecchio, L. Binder, J. Q. Trojanowski, V. M.-Y. Lee & G. Lee. 1988. Epitopes that span the tau molecule are shared with paired helical filaments. Neuron **1:** 817-825.
62. Kornblith, A. R., K. Umezawa, K. Vibe-Pedersen & F. E. Baradle. 1985. Fibronectin-human sequences translated from mRNA. EMBO J. **4:** 1755-1759.
63. Argravos, W. S., S. Suzuki, H. Arai, K. Thompson, M. D. Pierschbacher & E. Ruoslahti. 1987. Amino acid sequence of the human fibronectin receptor. J. Cell Biology **105:** 1183-1190.
64. Schatz, P. J., L. Pillus, P. Grisafi, F. Salomon & D. Botstein. 1986. Two functional α-tubulin genes of the yeast *Saccharomyces cervisiae* encode divergent proteins. Mol. Cell Biol. **6:** 3711-3721.
65. Barlow, D. P., N. M. Green, M. Kurkinen & B. L. M. Hogan. 1984. Sequencing of

laminin B chain cDNAs reveals C-terminal regions of coiled-coil alpha-helix. EMBO J. **3:** 2355-2362.
66. SASAKI, M., S. KATO, K. KOHNO & G. R. MARTIN. 1987. Sequence of the cDNA encoding the laminin B_1 chain reveals a multidomain protein containing cysteine-rich repeats. Proc. Natl. Acad. Sci. USA **84:** 935-939.
67. DICKSON, G., H. J. GOWER, C. H. BARTON, H. M. PRENTICE, V. L. ELSOM, S. E. MOORE, R. D. COX, C. QUINN, W. PUTT & F. S. WALSH. 1987. Human muscle neural cell adhesion molecule (N-CAM): Identification of a muscle-specific sequence in the extracellular domain. Cell **50:** 1119-1130.
68. ALKON, D. L. & T. J. NELSON. 1990. Specificity of molecular changes in neurons involved in memory storage. FASEB J. **4:** 1567-1576.
69. HARRIS, E. W., A. H. GANONG & C. W. COTMAN. 1984. Long-term potentiation in the hippocampus involves activation of *N*-methyl-D-aspartate receptors. Brain Res. **323:** 132-137.
70. SCHERER, W. S. & S. B. UDIN. 1988. The role of NMDA receptors in the development of binocular maps in *Xenopus tectum.* Soc. Neurosci. Abstr. **14:** 675.
71. SCHMIDT, J. T. 1990. Long-term potentiation and activity-dependent retinotopic sharpening in the regenerating retinotectal projection of goldfish: Common sensitive period and sensitivity to NMDA blockers. J. Neurosci. **10:** 233-246.
72. NISHIZUKA, Y. 1984. Turnover of inositol phospholipids and signal transduction. Science **225:** 1365-1370.
73. OLDS, J. L., M. L. ANDERSON, D. L. MCPHIE, L. D. STATEN & D. L. ALKON. 1989. Imaging and memory-specific changes in the distribution of protein kinase C in the hippocampus. Science **245:** 866-869.
74. ACKERS, R. F., D. M. LOVINGER, P. A. COLLEY, D. J. LINDEN & A. ROUTTENBERG. 1986. Translocation of protein kinase C activity may mediate hippocampal long-term potentiation. Science **231:** 587-589.
75. MEFFERT, M. K., K. D. PARFITT, V. A. DOZE, G. A. COHEN & D. V. MADISON. 1990. Protein kinases and long-term potentiation. Ann. N.Y. Acad. Sci. This volume.
76. CASANOVA, J. E., P. P. BREITFELD, S. A. ROSS & K. E. MOSTOV. 1990. Phosphorylation of the polymeric immunoglobulin receptor is required for its efficient transcytosis. Science **248:** 742-744.

Fosvergnügen

The Excitement of Immediate-Early Genes

KARL SCHILLING,[a,b] TOM CURRAN,[c] AND
JAMES I. MORGAN[b]

[b]*Department of Neurosciences*
and
[c]*Department of Molecular Oncology and Virology*
Roche Institute of Molecular Biology
Roche Research Center
Nutley, New Jersey 07110

INTRODUCTION

Neurons typically respond to external stimuli by short-lived changes in membrane conductance and excitability. It is now well established, however, that neurons are also able to transduce these short-term stimuli into long-lasting changes. These include functional and morphological alterations that are believed to underlie such fundamental phenomena as learning and memory. Moreover, neuronal activity has been recognized to play a major regulatory role during neural development. This implies that neuronal excitation can also control gene expression in the maturing nervous system. Indeed, it may be that similar biological responses underlie neuronal development in the embryo and neuronal plasticity in the adult.

A fundamental quest of current neurobiological research is to understand the molecular mechanisms linking ephemeral changes of membrane excitability to persistent adaptive modifications. While the physiology of stimulus-excitation-coupling and the molecular structures of the components subserving these phenomena have become increasingly clear over recent years, our knowledge of the mechanisms mediating long-term changes remains marginal, at best. Thus, while it has been postulated that long-lasting functional and morphological reorganization of the nervous system ultimately requires changes in gene expression (see, for example, reference 1), the molecular mechanisms responsible for such alterations are still unclear.

New avenues to approach these problems opened with the observation that neuronal stimulation rapidly activates the transcription of several protooncogenes (see, for example, references 2-4; for a recent review, see reference 5). Several of these genes encode DNA-binding proteins[6] that were subsequently shown to have the properties of transcription factors. Thus, they have the potential to regulate, in a hierarchical fashion, the transcription of genes that ultimately determine cellular behavior and phenotype. Because of their rapid and transient induction, even in the absence of protein synthesis, these genes have been classed together as cellular immediate-early genes (IEGs).[7,8] This is by analogy to the IEGs of viruses that are promptly expressed upon infection of a cell even if protein synthesis is inhibited. Furthermore, the analogy has been extended in the sense that since viral IEGs regulate later aspects of the replicative cycle, their

[a]Karl Schilling is supported by a research fellowship from the Deutsche Forschungsgemeinschaft (Schi 271/2-2).

cellular analogues are thought to regulate expression of genes involved in a coordinated phenotypic response to stimulation. It should be emphasized that not all cellular IEGs are transcriptional factors. Indeed, some encode soluble, secreted proteins. Therefore, it is suggested that stimulation elicits a cellular immediate-early response that includes, as one of its facets, regulation of target gene transcription.

To date, the best studied cellular IEG involved in signal transduction is c-*fos*. Consequently, this gene has recently attracted the attention of numerous scientists working in different fields of neurobiology.

The present communication is intended to give a short introduction to the role of c-*fos* in signal transduction and its application to the neurosciences. Several extensive reviews covering both the molecular biology of c-*fos* and its utility in neurobiological research have appeared recently,[5,9–13] and the interested reader is referred to these publications for further details.

THE c-*fos* GENE AND ITS PROTEIN

The structure of c-*fos,* the cellular homologue of the viral v-*fos* gene, has been extensively characterized (for reviews, see references 9, 12, and 14). The c-*fos* gene encodes a nuclear protein, Fos (apparent molecular weight: 62 kDa), that is subject to extensive posttranslational modification, notably phosphorylation. In cooperation with other transcription factors encoded by the *jun* gene family, Fos contributes to the transcriptional regulator known as activator protein-1 (AP-1) (reviewed in references 6 and 12; see also below). The interaction of Fos and its various partners is mediated by α-helical protein domains consisting of heptad repeats of leucine residues, termed the leucine zipper.[15] Upon dimerization, these domains form a coiled coil that is held together by a strong hydrophobic interaction.[16–18] The DNA-binding regions of Fos and its related proteins, as well as those of the *jun* family, are located N-terminal to the leucine zipper region.[16] The potential to bind to DNA and to regulate the transcription of other genes is a key requirement for any switch that can translate short-term environmental signals into lasting changes in cellular phenotype and behavior.

To date, several regulatory sequences have been shown to regulate transcription of c-*fos* in response to a plethora of extracellular stimuli.[11,19,20] For example, the serum response element (SRE) (also known as the dyad-symmetry element (DSE)), located 300 bases upstream of the transcription start site, mediates c-*fos* induction by serum and several other agents, including activators of protein kinase C. The SRE participates also in the autoregulation of c-*fos*[21] (for a critical discussion of this issue, see reference 20). The *sis*/conditioned medium response element (SCME), at −346, confers inducibility by platelet-derived growth factor (PDGF) onto the c-*fos* promoter and functions in an additive manner to the SRE, which is also responsive to PDGF.[22] At −60, c-*fos* has a cyclic AMP/calcium response element (CRE/CaRE) that mediates regulation by the second messengers cAMP and calcium. All of these *cis*-acting regulatory sequences are located in the 5′-untranslated region of c-*fos*. In contrast, the *fos* intragenic regulatory element[23] (FIRE) sequence lies at the 3′-end of the first exon. This element affects transcriptional elongation as opposed to initiation.

Although such a strict definition of response elements is helpful in experimental approaches to the characterization of c-*fos* regulation, increasing evidence suggests that it does not adequately account for the regulation of c-*fos* expression. Rather, there seems to be extensive cross-talk between various response elements (see references 5 and 10 for further details; see also references 24-27).

In summary, extracellular stimuli that activate "classical" second messenger systems as well as polypeptide growth factors induce c-*fos* expression. The c-*fos* promoter is capable not only of sensing these various stimuli, but also of integrating them. Consequently, it seems appropriate to view c-*fos* not simply as a passive member in the signal-transduction chain, but rather as an active signal-processing element that participates in the integration of multiple stimuli into a unified (though by no means simple) cellular response.

c-*fos* IN NEUROBIOLOGY

The protooncogene c-*fos* caught the attention of neurobiologists with the observation that it is expressed by PC12 cells following treatment with nerve growth factor.[2–4] These observations were extended by experiments indicating that agents causing the activation of classical neurotransmitter receptors,[28] membrane depolarization,[3,4,29] and direct activation of second messenger pathways[3,4] also lead to c-*fos* induction in PC12 cells. Together, these findings prompted investigators to look for a role for c-*fos* in the nervous system.

This was done initially in two paradigms. First, Morgan *et al.*[30] used the Metrazole-seizure model to follow c-*fos* expression after a general, rather synchronous depolarization of neurons and compared it to expression of c-*fos* in nonseizured brain. Although there was some c-*fos* expression and Fos-like immunoreactivity (FLI—see below) in control brains, seizures resulted in a massive induction of c-*fos* and FLI in the dentate gyrus and the pyriform and cingulate cortex, which became positive 90 min after stimulation. Subsequently, FLI appeared in many more areas, including the limbic system, hippocampus, and entire cerebral cortex. Second, Hunt *et al.*[31] studied induction of FLI in spinal cord neurons after peripheral sensory stimulation. These studies led to the concept that Fos (or rather, FLI) could be used to map metabolic activity in the brain with single-cell resolution. A number of studies[32–34] subsequently compared induction of FLI by several physiologic stimuli to changes in [^{14}C]2-deoxyglucose uptake, an established technique to monitor metabolic activity in the brain.[35] While a considerable overlap between the neuronal populations labeled by both techniques was observed, there were also distinct and important differences (see references 32 and 33, reviewed in reference 5). Thus while both c-*fos* (FLI) and deoxyglucose are valuable methods to monitor neuronal activity, they clearly reveal different facets of neural signaling processes.

Subsequently, a wealth of stimuli has been shown to induce expression of c-*fos* and/or FLI in the intact nervous system and in primary cultures derived thereof (see, for example, references 36-38, reviewed in reference 5).

While c-*fos* has proved to be a very helpful tool for neurobiologists, neurobiology in turn contributed to our understanding of the functional importance of c-*fos*. In particular, the seizure model[30] of c-*fos* induction in the nervous system helped to elucidate the relationship between c-*fos* and its related antigens, that is, Fos-related antigens (Fras). Picked up initially by their cross-reactivity with antibodies to Fos,[39] these molecules have now been characterized as DNA-binding proteins that, together with Fos and proteins encoded by the *jun* gene family, form transcription factor AP-1 (for review, see references 40 and 41). Analysis of expression of Fos and Fras in seizured brain provided a detailed temporal and quantitative map of the expression of Fos and Fras following a massive, synchronous stimulation. These results were pivotal in formulating the concept that AP-1 is not a static entity, but that its composition changes over time (reviewed in reference 10). Consequently, the binding preference of the AP-1 protein complex for different variations of the AP-1 recognition motif may change in a time-dependent manner. A second possibility is that variation in the composition of AP-1 affects the gene regulatory activity of this complex. Finally, one may envision a combination of both these possibilities as the most likely effect that a change in AP-1 composition may have on its function.

From a practical point of view, the expression of Fras in concert with Fos means that proper identification of Fos by immunocytochemistry is not possible, unless an absolutely monospecific antibody is used. As the history of the detection of Fras suggests, such specific antibodies are rare, and it is not always possible to predict specificity *a priori.* For this reason, it seems sound to speak of Fos-like immunoreactivity (FLI), rather than Fos itself, whenever immunocytochemistry is used to map "Fos."

Whenever it is vital to be certain of the molecular entity recognized by a given antibody, the combined application of immunocytochemical and biochemical (immunoprecipitation and/or immunoblotting) methods is essential. Unfortunately, biochemical studies cannot match the cellular resolution obtained by immunocytochemistry. Thus, novel approaches have to be taken if one is interested in mapping c-*fos* expression with high anatomical resolution, as would be desirable in the central nervous system.

As outlined above, several response elements have been identified in the c-*fos* promoter. Thus, the question arises which elements are involved in c-*fos* induction in particular neurons following a specific type of stimulus. Could, for example, the same stimulus mediate c-*fos* induction by different elements in different neuronal populations or subsets? Alternatively, can different stimuli elicit c-*fos* induction through distinct elements in the same set of neurons? Such information could point to the second messengers mediating c-*fos* induction in specific neurons and ultimately help to unravel the neuronal circuits being activated following a specific type of stimulation (for example, seizure). But how could such a map be generated?

FUTURE DIRECTIONS

One strategy to approach these questions is to tag Fos with an additional, unique label. To this end, we recently constructed a c-*fos-lac*Z fusion gene and stably transfected it into B104 neuroblastoma cells.[42] Expression of the fusion gene can be measured by assessing β-galactosidase activity either histochemically or with a soluble substrate. Induction and regulation of the fusion gene by both external stimuli and direct activation of second messenger pathways is essentially comparable to that of the cognate c-*fos.* This cell line thus provides a convenient tool to investigate, in a quantitative manner, induction of c-*fos* expression (FIG. 1). It should be helpful in the further characterization of interactions between various response elements in the c-*fos* promoter.

The results obtained with this cell line also encouraged us to introduce the c-*fos-lac*Z fusion gene into the germ line of mice, thus generating transgenic animals in which expression of c-*fos* can be monitored with absolute specificity by simple histochemistry for β-galactosidase.[43]

We have also used these transgenic animals as a source of primary neural cultures. Such cultures provide a model system that is intermediate in complexity between the *in vivo* situation and (monoclonal) cell lines derived from neural tumors. They allow stringent control of experimental conditions, and are amenable to both morphological and biochemical analysis (see, for example, reference 44).

As may be presumed from the heterogeneity of cells found in these cultures, a large variety of treatments is capable of inducing c-*fos.*[43] These include potassium-induced membrane depolarization, direct stimulators of second messenger pathways (dbcAMP, phorbol esters), excitatory amino acids, adrenergic and muscarinic agonists, and growth factors. An unexpected finding is that many of these substances can be grouped into one of two categories: stimuli that induce c-*fos* expression in neurons, and those that result in c-*fos* expression by nonneuronal, mostly glial, cells. Prototypic representatives for these two types of stimuli are depolarization by elevated levels of potassium, and the phorbol ester, TPA (FIG. 2). These findings nurture the hope that experimental paradigms can be designed in which c-*fos* can be used to monitor neuron-glia interactions. Such interactions might be of particular interest in developmental neurobiology, an area in which c-*fos* is just coming of age.[5,37,45,46]

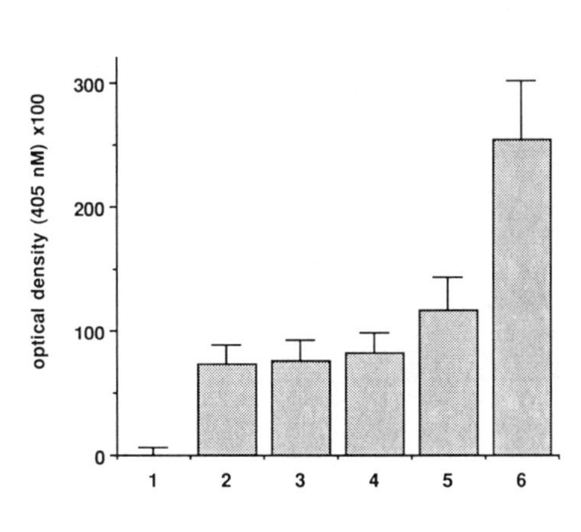

FIGURE 1. Induction of β-galactosidase activity in B104 neuroblastoma cells stably transfected with a c-*fos-lac*Z fusion gene.[42] The control (column 1) has been set to 0. Treatment of cells for 3 hr with 15% serum (column 2), 8 nM TPA (column 3), or 1 mM dbcAMP (column 4) results in about equal levels of β-galactosidase activity. However, β-galactosidase activity induced by simultaneous treatment with serum and TPA (column 5) is lower than would be expected from values obtained with these substances individually (columns 2 and 3). In contrast, simultaneous treatment with serum and dbcAMP (column 6) induces β-galactosidase activity to levels higher than those calculated by adding the levels of induction caused by the individual substances (columns 2 and 4). β-Galactosidase activity was measured using the soluble substrate ONPG. Each value is a mean $\pm$ 1 SD.

Although this short review has concentrated on c-*fos,* there is now ample evidence that other protooncogenes and cellular IEGs are also expressed in the nervous system. As mentioned above, these include the Fos-related antigens and members of the *jun* gene family. Other immediate-early genes expressed in the CNS include *egr*-1, *egr*-2, and *nur*-77 (for references and review, see references 5 and 11). The cellular IEG c-*fos* can thus be regarded a prototype of a set of cellular genes that are responsive to extracellular signals and regulate gene expression in response to these signals. Understanding the function and interactions of these proteins may provide a key to solve some of the most prominent questions of current neurobiological research.

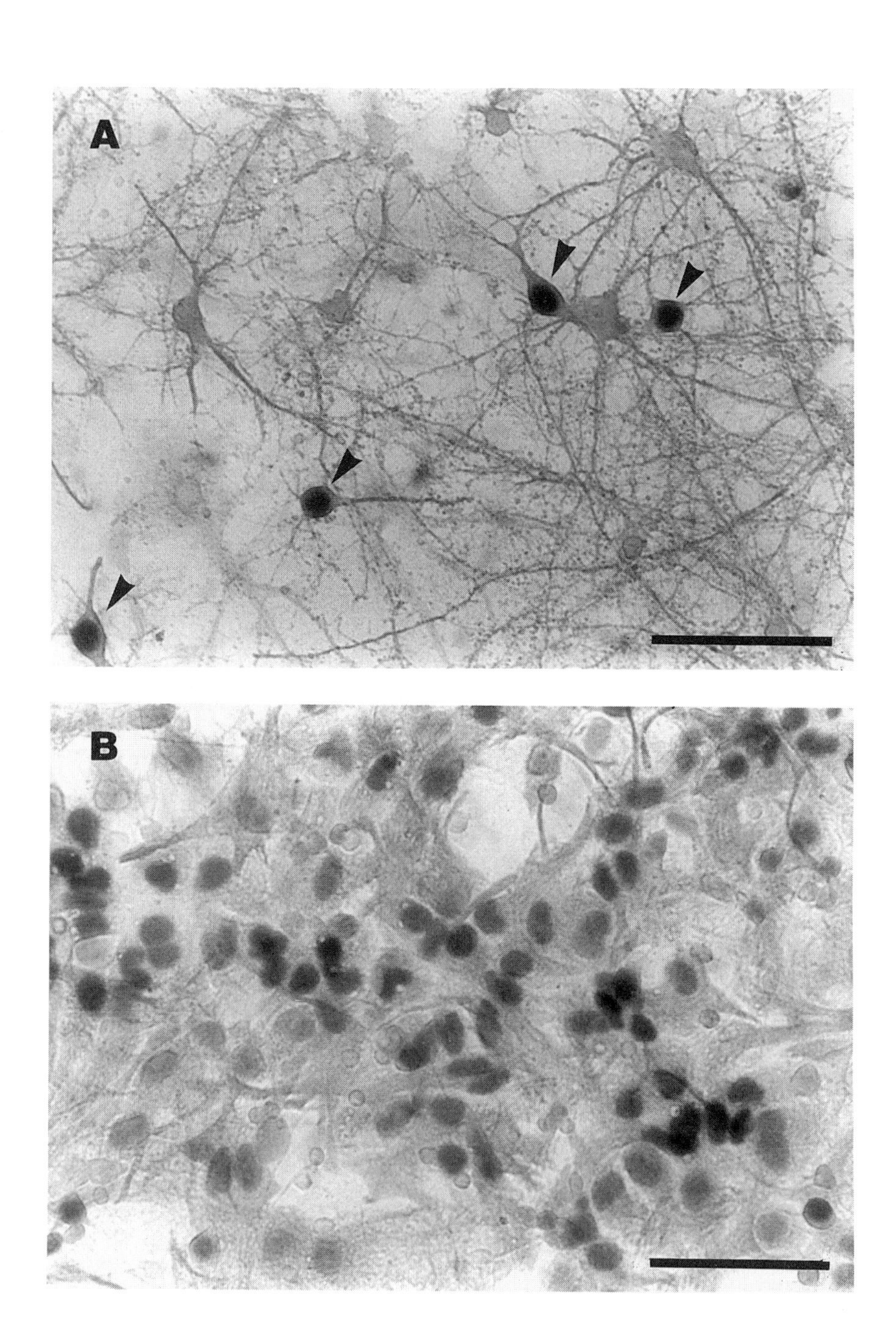
A
B

FIGURE 2. Expression of β-galactosidase activity in dissociated primary cortical cultures derived from c-*fos-lacZ* transgenic animals.[43] Eight-day-old cultures were stimulated (**A**) by depolarization with 30 mM K^+ or (**B**) by treatment with TPA. Three hours later, they were fixed and processed to demonstrate β-galactosidase activity with X-gal.[47] This yielded a blue reaction product. Subsequently, cultures were immunostained for (**A**) microtubule-associated protein 2 (MAP2) or (**B**) glial fibrillary acidic protein (GFAP). Primary antibody binding was visualized with alkaline phosphatase using the ABC technique. This resulted in a pinkish-red staining of immunopositive cells. (**A**) Depolarization resulted in c-*fos-lacZ* induction in neurons (arrowheads), but not in underlying (unstained) glial cells. Note also that not all neurons express the fusion gene. (**B**) The phorbol ester, TPA, resulted in a massive induction of c-*fos-lacZ* in GFAP-positive glial cells. Staining of TPA-stimulated cultures with MAP2 showed that only a very minor population of neurons expressed the transgene following this stimulus (data not shown). Bars = 100 μm.

REFERENCES

1. KANDEL, E. R. & J. H. SCHWARTZ. 1982. Molecular biology of learning: Modulation of transmitter release. Science **218:** 433-443.
2. CURRAN, T. & J. I. MORGAN. 1985. Superinduction of c-*fos* by nerve growth factor in the presence of peripherally active benzodiazepines. Science **229:** 1265-1268.
3. GREENBERG, M. E., L. A. GREENE & E. B. ZIFF. 1985. Nerve growth factor and epidermal growth factor induce rapid transient changes in proto-oncogene transcription in PC 12 cells. J. Biol. Chem. **260:** 14101-14110.
4. KRUIJER, W., D. SCHUBERT & I. M. VERMA. 1985. Induction of the proto-oncogene *fos* by nerve growth factor. Proc. Natl. Acad. Sci. USA **82:** 7330-7334.
5. MORGAN, J. I. & T. CURRAN. 1991. Stimulus-transcription coupling in the nervous system: Involvement of the inducible proto-oncogenes *fos* and *jun*. Annu. Rev. Neurosci. **14:** 421-451.
6. CURRAN, T. & B. R. FRANZA, JR. 1988. Fos and Jun: The AP-1 connection. Cell **55:** 395-397.
7. LAU, L. F. & D. NATHANS. 1987. Expression of a set of growth-related immediate-early genes in BALB/c 3T3 cells: Coordinate regulation with c-*fos* or c-*myc*. Proc. Natl. Acad. Sci. USA **84:** 1182-1186.
8. CURRAN, T. & J. I. MORGAN. 1987. Memories of *fos*. BioEssays **7:** 255-258.
9. CURRAN, T. 1988. The *fos* oncogene. *In* The Oncogene Handbook. E. P. Reddy, A. M. Skalka & T. Curran, Eds.: 307-325. Elsevier. Amsterdam.
10. MORGAN, J. I. & T. CURRAN. 1989. Stimulus-transcription coupling in neurons: Role of cellular immediate-early genes. TINS **12:** 459-462.
11. SHENG, M. & M. E. GREENBERG. 1990. The regulation and function of c-*fos* and other immediate-early genes in the nervous system. Neuron **4:** 477-485.
12. COHEN, D. R. & T. CURRAN. 1989. The structure and function of the *fos* proto-oncogene. Crit. Rev. Oncogen. **1:** 65-88.
13. RIVERA, V. M. & M. E. GREENBERG. 1990. Growth factor-induced gene expression: The ups and downs of c-*fos* regulation. New Biol. **2:** 751-758.
14. CURRAN, T. 1988. Recent developments concerning the *fos* oncogene. *In* The Oncogene Handbook. E. P. Reddy, A. M. Skalka & T. Curran, Eds.: 551-554. Elsevier. Amsterdam.
15. LANDSCHULZ, W. H., P. F. JOHNSON & S. L. MCKNIGHT. 1988. The leucine zipper: A hypothetical structure common to a new class of DNA-binding proteins. Science **240:** 1759-1764.
16. GENTZ, R., F. J. RAUSCHER III, C. ABATE & T. CURRAN. 1989. Parallel association of Fos and Jun leucine zippers juxtaposes DNA-binding domains. Science **243:** 1695-1699.
17. O'SHEA, E. K., R. RUTKOWSKI & P. S. KIM. 1989. Evidence that the leucine zipper is a coiled coil. Science **243:** 538-542.
18. PATEL, L., C. ABATE & T. CURRAN. 1990. Altered protein conformation on DNA binding by Fos and Jun. Nature **347:** 572-575.

19. Treisman, R. 1990. The SRE: A growth-factor-responsive transcriptional regulator. *In* Seminars in Cancer Biology, Transcription Factors, Differentiation and Cancer. N. C. Jones, Ed. Vol. 1: 47-58. Saunders. London.
20. Curran, T. 1991. Fos and Jun: Intermediary transcription factors. *In* The Hormonal Control Regulation of Gene Transcription. P. Cohen & J. G. Foulkes, Eds.: 371-384. Elsevier. Amsterdam.
21. Gius, D., X. Cao, F. J. Rauscher III, D. R. Cohen, T. Curran & V. P. Sukhatme. 1990. Transcriptional activation and repression by Fos are independent functions: The C-terminus represses immediate-early gene expression via CArG elements. Mol. Cell. Biol. **10:** 4243-4255.
22. Wagner, B. J., T. E. Hayes, C. J. Hoban & B. H. Cochran. 1990. The SIF binding element confers *sis*/PDGF inducibility onto the c-*fos* promoter. EMBO J. **9:** 4477-4484.
23. Lamb, N. J. C., A. Fernandez, N. Tourkine, P. Jeanteur & J.-M. Blanchard. 1990. Demonstration in living cells of an intragenic negative regulatory element within the rodent c-*fos* gene. Cell **61:** 485-496.
24. Diamond, M. I., J. N. Miner, S. K. Yoshinaga & K. R. Yamamoto. 1990. Transcription factor interactions: Selectors of positive or negative regulation from a single DNA element. Science **249:** 1266.
25. Gaub, M.-P., M. Bellard, I. Scheuer, P. Chambon & P. Sassone-Corsi. 1990. Activation of the ovalbumin gene by the estrogen receptor involves the Fos-Jun complex. Cell **63:** 1267-1276.
26. Schüle, R., P. Rangarajan, S. Kliewer, L. J. Ransone, J. Bolado, N. Yang, I. M. Verma & R. M. Evans. 1990. Functional antagonism between oncoprotein c-*jun* and the glucocorticoid receptor. Cell **62:** 1217-1226.
27. Owen, T. A., R. Bortell, S. A. Yocum, S. L. Smock, M. Zhang, C. Abate, V. Shalhoub, N. Aronin, K. L. Wright, A. J. van Wijnen, J. L. Stein, T. Curran, J. B. Lian & G. S. Stein. 1990. Coordinate occupancy of AP-1 sites in the vitamin D-responsive and CCAAT box elements by Fos-Jun in the osteocalcin gene: Model for phenotype suppression of transcription. Proc. Natl. Acad. Sci. USA **87:** 9990-9994.
28. Greenberg, M. E., E. B. Ziff & L. A. Greene. 1986. Stimulation of neuronal acetylcholine receptors induces rapid gene transcription. Science **234:** 80-83.
29. Morgan, J. I. & T. Curran. 1986. Role of ion flux in the control of c-*fos* expression. Nature **322:** 552-555.
30. Morgan, J. I., D. R. Cohen, J. L. Hempstead & T. Curran. 1987. Mapping patterns of c-*fos* expression in the central nervous system after seizure. Science **237:** 192-197.
31. Hunt, S. P., A. Pini & G. Evan. 1987. Induction of c-*fos*-like protein in spinal cord neurons following sensory stimulation. Nature **328:** 632-634.
32. Jorgensen, M. B., J. Deckert, D. C. Wright & D. R. Gehlert. 1989. Delayed c-*fos* proto-oncogene expression in the rat hippocampus induced by transient global cerebral ischemia: An *in situ* hybridization study. Brain Res. **484:** 393-398.
33. Sagar, S. M., F. R. Sharp & T. Curran. 1988. Expression of c-*fos* protein in brain: Metabolic mapping at the cellular level. Science **240:** 1328-1331.
34. Sharp, F. R., M. F. Gonzalez, J. W. Sharp & S. M. Sagar. 1989. c-*fos* expression and [^{14}C]2-deoxyglucose uptake in the caudal cerebellum of the rat during motor/sensory cortex stimulation. J. Comp. Neurol. **284:** 621-636.
35. Sokoloff, L., M. Reivich, C. Kennedy, M. H. Des Rosiers, C. S. Patlak, K. D. Pettigrew, O. Sakurada & M. Shinogara. 1977. The [^{14}C]deoxyglucose method for the measurement of local cerebral glucose utilization: Theory, procedure, and normal values in the conscious and anesthetized albino rat. J. Neurochem. **28:** 897-916.
36. Dragunow, M., G. S. Robertson, R. L. M. Faull, H. A. Robertson & K. Jansen. 1990. D2 dopamine receptor antagonists induce Fos and related proteins in rat striatal neurons. Neuroscience **37:** 287-294.
37. Hisanaga, K., S. M. Sagar, K. J. Hicks, R. A. Swanson & F. R. Sharp. 1990. c-*fos* proto-oncogene expression in astrocytes associated with differentiation or proliferation but not depolarization. Mol. Brain Res. **8:** 69-75.
38. Jacobson, L., F. R. Sharp & M. F. Dallman. 1990. Induction of Fos-like immunoreactivity in hypothalamic corticotropin-releasing factor neurons after adrenalectomy in the rat. Endocrinology **126:** 1709-1719.

39. COHEN, D. R. & T. CURRAN. 1988. *fra*-1: A serum-inducible, cellular immediate-early gene that encodes a Fos-related antigen. Mol. Cell. Biol. **8:** 2063-2069.
40. COHEN, D. R., P. C. P. FERREIRA, R. GENTZ, B. R. FRANZA, JR. & T. CURRAN. 1989. The product of a Fos-related gene, *fra*-1, binds cooperatively to the AP-1 site with Jun: Transcription factor AP-1 consists of multiple protein complexes. Genes Dev. **3:** 173-184.
41. ABATE, C. & T. CURRAN. 1990. Encounters with Fos and Jun on the road to AP-1. Semin. Cancer Biol. **1:** 19-26.
42. SCHILLING, K., D. LUK, T. CURRAN & J. I. MORGAN. 1991. Regulation of *fos-lac*Z: A paradigm for quantitative molecular pharmacology. Submitted for publication.
43. SMEYNE, R. J., K. SCHILLING, L. M. ROBERTSON, D. LUK, J. OBERDICK, T. CURRAN & J. I. MORGAN. 1991. Activity-dependent expression of a *fos-lac*Z fusion gene in transgenic mice. In preparation.
44. SCHILLING, K., P. OEDING, H. SCHMALE & C. PILGRIM. 1991. Regulation of vasopressin expression in cultured diencephalic neurons by glucocorticoids. Neuroendocrinology **53:** 528-535.
45. CAUBERT, J. 1989. c-*fos* proto-oncogene expression in the nervous system during mouse development. Mol. Cell. Biol. **9:** 2269-2272.
46. WILKINSON, D. G., S. BHATT, R.-P. RYSECK & R. BRAVO. 1990. Tissue-specific expression of c-*jun* and *jun*B during organogenesis in the mouse. Development **106:** 465-471.
47. OBERDICK, J., R. J. SMEYNE, J. R. MANN, S. ZACKSON & J. I. MORGAN. 1990. A promoter that drives transgene expression in cerebellar Purkinje and retinal bipolar neurons. Science **248:** 223-226.

KEYNOTE ADDRESS

Neural and Molecular Bases of Nonassociative and Associative Learning in *Aplysia*[a]

JOHN H. BYRNE, DOUGLAS A. BAXTER, DEAN V. BUONOMANO, LEONARD J. CLEARY, ARNOLD ESKIN,[b] JASON R. GOLDSMITH, EV McCLENDON, FIDELMA A. NAZIF, FLORENCE NOEL, AND KENNETH P. SCHOLZ[c]

Department of Neurobiology and Anatomy
University of Texas Medical School at Houston
Houston, Texas 77225

INTRODUCTION

One of the major unresolved problems in neurobiology is to explain the anatomical, physiological, and biochemical bases of learning and memory. Specifically, what parts of the nervous system are critical for learning? How is information about a learned event acquired and encoded in neuronal terms? How is the information stored, and once stored how is it retrieved or read out? Most neuroscientists believe that answers to these questions lie in understanding how the properties of individual nerve cells in general, and synapses in particular, are changed when learning occurs.

As a result of research on several vertebrate and invertebrate model systems, several rather general principles have emerged. A list of these principles might include the following: 1) the neural representation of learning and memory is distributed throughout the nervous system; 2) at least short-term forms of learning and memory require changes in existing neural circuits; 3) these changes may involve multiple cellular mechanisms within single neurons; 4) second messenger systems appear to play a key role in mediating cellular changes; 5) changes in the properties of membrane channels are commonly correlated with learning and memory; and 6) long-term memory requires new protein synthesis whereas short-term memory does not (for a review see reference 1). Although these principles have been placed in the context of learning and memory,

[a] This work was supported by fellowships from the National Institute of Mental Health (F31 MH09895, to D. V. B.; F31 MH09956, to F. A. N.), a training grant from the National Institutes of Health (T32 HD07324, to J. R. G.), a fellowship from the Foundation Fyssen (to F. N.), a grant from the Texas Higher Education Coordinating Board (1945, to L. J. C.), an award from the National Institute of Mental Health (K02 MH00649, to J. H. B.), grants from the Air Force Office of Scientific Research (87-0274 and 91-0027, to J. H. B.), a grant from the National Institute of Mental Health (R01 MH41979, to A. E.), and a grant from the National Institutes of Health (R01 NS19895, to J. H. B.).

[b] Present address: Department of Biochemistry, University of Houston, Houston, Texas 77204.

[c] Present address: Department of Pharmacological and Physiological Sciences, University of Chicago, Chicago, Illinois 60637.

it is fair to say that many of them also apply equally well to general forms of adult plasticity.

In this chapter, we illustrate some of these principles by reviewing work on the changes that occur in identified neurons of the marine mollusc *Aplysia*, changes that contribute to simple forms of nonassociative and associative learning. To begin, we describe some of the features of the particular behavioral system in *Aplysia* that we have analyzed and some of its underlying neurocircuitry. An understanding of the neural circuit for a behavior is a necessary prerequisite for the analysis of neuronal changes that underly a behavioral modification. This may sound trivial, but it has been one of the major obstacles hindering progress in this field. Next, we describe sensitization of the behavior and the biochemical and biophysical mechanisms contributing to both the short-term form, lasting minutes, and the long-term form, lasting days, of this simple example of nonassociative learning. Finally, we review our attempts to understand the mechanisms underlying a simple example of associative learning, namely classical or Pavlovian conditioning.

DEFENSIVE REFLEX RESPONSES

FIGURE 1 illustrates two behaviors of *Aplysia* that have been used to analyze the neuronal mechanisms contributing to nonassociative and associative learning.[1,2] Within the mantle cavity is the respiratory organ of the animal, the gill, and protruding from the mantle cavity is the siphon. One behavior is the siphon-gill withdrawal reflex (FIG. 1A), which has been studied extensively by Eric Kandel and his colleagues.[2] A tactile or electrical stimulus delivered to the siphon elicits a reflex withdrawal of the siphon and gill. Many of the neurons that control the siphon-gill withdrawal reflex are located in the abdominal ganglion. The other behavior, the tail-siphon withdrawal reflex, is illustrated in FIGURE 1B. Tactile or electrical stimulation of the tail elicits a coordinated set of defensive responses, two components of which are illustrated. The stimulus produces a reflex withdrawal of the tail and the siphon. Neurons that control the tail withdrawal component of the reflex are located in the pleural and pedal ganglia.

SENSITIZATION OF THE TAIL-SIPHON WITHDRAWAL REFLEX

We have used the tail-siphon withdrawal reflex to analyze the neural and molecular mechanisms of a simple form of nonassociative learning, sensitization. Sensitization refers to an enhancement in the behavioral response to a test stimulus produced by application of a strong, usually noxious stimulus to the animal. In contrast to associative learning (see below), this type of learning occurs without any specific temporal relationship between the test stimulus and the noxious reinforcing stimulus.

The basic experimental procedure used to induce sensitization of the tail-siphon withdrawal reflex is illustrated in FIGURE 2A. For behavioral analysis, each animal was initially tested by the delivery of a weak electric shock to the tail via implanted electrodes. The test stimuli produced a coordinated set of defensive responses, including reflex withdrawal of the tail and siphon. Changes in the duration of withdrawal of the siphon were used as a measure of sensitization. After five test stimuli delivered at 5-min intervals, the animal was subjected to sensitization training, which consisted of a train of 10 shocks diffusely applied to the body wall via a hand-held electrode over a 10-sec period (FIG. 2B). The intensity of the sensitizing shocks was initially adjusted so that they caused release of ink from the ink gland (an indication that the animal has received a noxious stimulus). After sensitization training, at least five additional test stimuli were presented. As indicated in FIGURE 2C, the sensitizing stimuli led to an abrupt and prolonged enhancement of the siphon withdrawal.

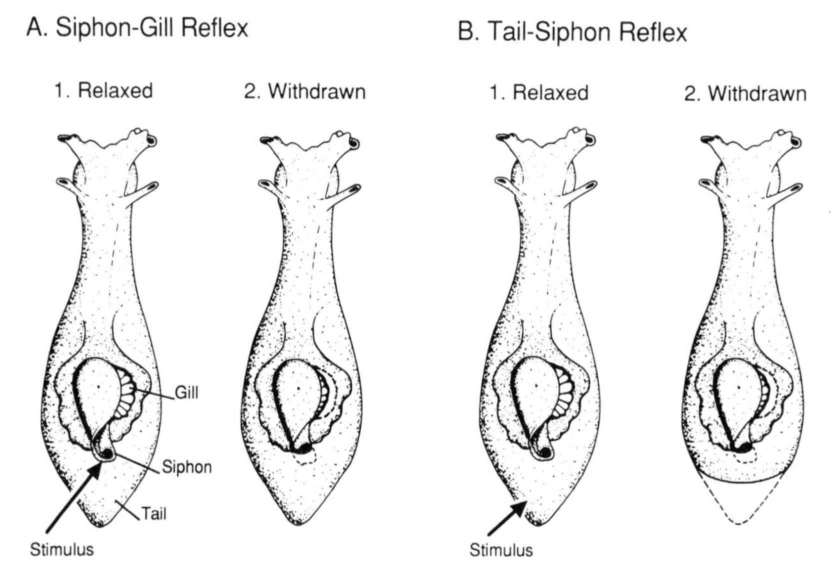

FIGURE 1. Siphon-gill and tail-siphon withdrawal reflexes of *Aplysia*. (**A**) Siphon-gill withdrawal reflex. 1: Dorsal view of *Aplysia*, relaxed position; 2: a stimulus (such as a water jet, brief touch, or weak electric shock) applied to the siphon causes the siphon and the gill to withdraw into the mantle cavity. (**B**) Tail-siphon withdrawal reflex. 1: Relaxed position; 2: a stimulus (such as a brief touch or weak electric shock) applied to the tail region elicits a coordinated set of defensive responses including reflex withdrawal of the tail and siphon.

NEURAL CIRCUIT FOR THE TAIL-SIPHON WITHDRAWAL REFLEX

Some of the known elements of the circuit that mediate the tail-siphon withdrawal reflex are indicated schematically in FIGURE 3. The cell bodies of the sensory neurons (SN) that innervate the tail are located in symmetric clusters in the left and right pleural ganglia.[3] When a test stimulus is applied to the tail, the sensory neurons are excited. The sensory neurons make monosynaptic excitatory connections with motor neurons (MN) in the adjacent pedal ganglion, which produce a withdrawal of the tail. In addition to their connections with motor neurons, the sensory neurons form synapses with various identified interneurons (I).[4] Some of these interneurons[5] activate motor neurons in the abdominal ganglion, which control reflex withdrawal of the siphon.

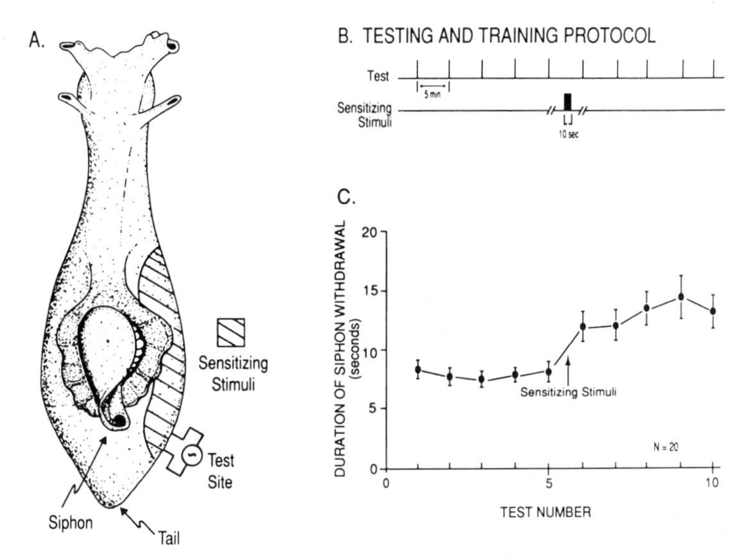

FIGURE 2. Sensitization of the tail-siphon withdrawal reflex. (**A**) Sites for delivering test and sensitizing stimuli. (**B**) Testing and training protocol. Testing consisted of 20-msec AC shocks delivered to one side of the animal's tail via implanted electrodes (at an interstimulus interval of 5 min). The stimulus intensity was twice the threshold value necessary to elicit a siphon withdrawal. Sensitizing stimuli consisted of a 10-sec train of 10 500-msec, 60-mA AC shocks delivered via a handheld bipolar electrode to a diffuse area of the body wall on the same side as that receiving the test stimuli. (**C**) Behavioral results. A single, brief train (10 sec) of sensitizing stimuli leads to an enhancement of siphon withdrawal elicited by stimuli to the tail. Error bars represent the SEM.

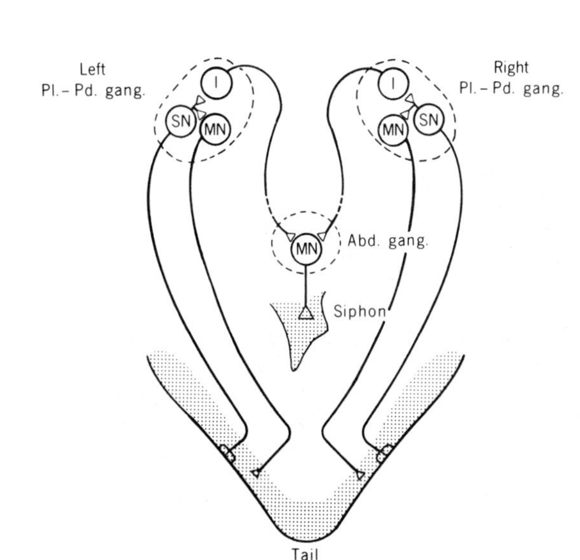

FIGURE 3. Simplified schematic of the neural circuit controlling the tail-siphon withdrawal reflex.

Moreover, several additional interneurons (not shown) modulate the tail-siphon withdrawal reflex.

The sensory neurons appear to be key plastic elements in the neural circuit. Changes in their membrane properties and synaptic efficacy are associated with sensitization and the effects of procedures that mimic short- and long-term sensitization training (see below).

SEROTONIN: A PUTATIVE MODULATORY TRANSMITTER RELEASED BY SENSITIZING STIMULI

Several lines of evidence indicate that serotonin (5-HT) is one of the modulatory transmitters that induces sensitization. First, serotonergic circuitry is available in the nervous system to mediate sensitization. Second, 5-HT mimics many of the short- and long-term effects of sensitization training. Third, depleting the nervous system of 5-HT by injection of a serotonergic neurotoxin significantly reduces sensitization of the siphon-gill withdrawal reflex.[6]

Evidence supporting a direct action of 5-HT comes from the observation that serotonergic fibers are in close proximity to pleural sensory neurons. Although there are no serotonergic neurons in the pleural ganglion itself,[7–10] there are many serotonergic axons within the neuropil of the pleural ganglion that originate from neurons in other ganglia.[11] As illustrated in FIGURE 4, many axons with 5-HT immunoreactivity send fine processes to surround cell bodies of pleural sensory neurons. Numerous varicosities are found along these fine processes. Subsequent examination using electron microscopic immunocytochemistry has shown that these varicosities directly appose the plasma membrane of the sensory neurons; "contact" sites with no intervening glial cells have been observed.[12] Moreover, synaptic contacts may occur at either the cell body or the axon hillock. These results suggest that 5-HT is released onto cell bodies of pleural sensory neurons.

Although neurons that give rise to the serotonergic processes surrounding the pleural sensory neurons have not yet been identified, there are several candidates.[13] Most promising among these are the serotonergic cells in the B cluster of neurons in the cerebral ganglion (reference 14; Cleary and Byrne, unpublished results). These cells appear to send axons through the cerebral-pleural connectives into the pleural ganglia and from there continue into the abdominal ganglion via the pleural-abdominal connectives. Stimulation of the serotonergic cerebral B cells produces both spike broadening in siphon sensory neurons in the abdominal ganglion and facilitation of the excitatory postsynaptic potentials (EPSPs) produced by siphon sensory neurons in their follower motor neurons,[14] effects that could contribute to sensitization of the siphon-gill withdrawal reflex.

Serotonin mimics several of the biochemical and electrophysiological correlates of sensitization. This has been studied by isolating the pleural-pedal ganglia in a recording chamber while peripheral nerves to the tail remained intact. The selective application of 5-HT to the isolated ganglia mimicked the effects of sensitizing stimuli to the tail; 5-HT increased the intensity of the tail withdrawal reflex, increased the activation of motor neurons, and enhanced the strength of the connections between sensory neurons and motor neurons.[15,16]

Recordings from sensory and motor neurons revealed that 5-HT has four distinct, but probably related, effects on tail sensory neurons and their synaptic connections.[15–19] First, 5-HT increases the size of the monosynaptic EPSPs in motor neurons.

Second, 5-HT produces a slow and long-lasting depolarization of the sensory neuron. This depolarization is associated with a decrease in membrane input conductance. Third, 5-HT increases the excitability of sensory neurons (an index of excitability is the number of spikes that are elicited by a single depolarizing current pulse of fixed duration and amplitude). Fourth, 5-HT enhances the duration of action potentials in sensory neurons. An example of the effects of 5-HT on spike duration and excitability is illustrated in FIGURE 5. Application of 5-HT to sensory neurons doubled both the duration of the action potential and the number of spikes elicited by a 1-sec constant

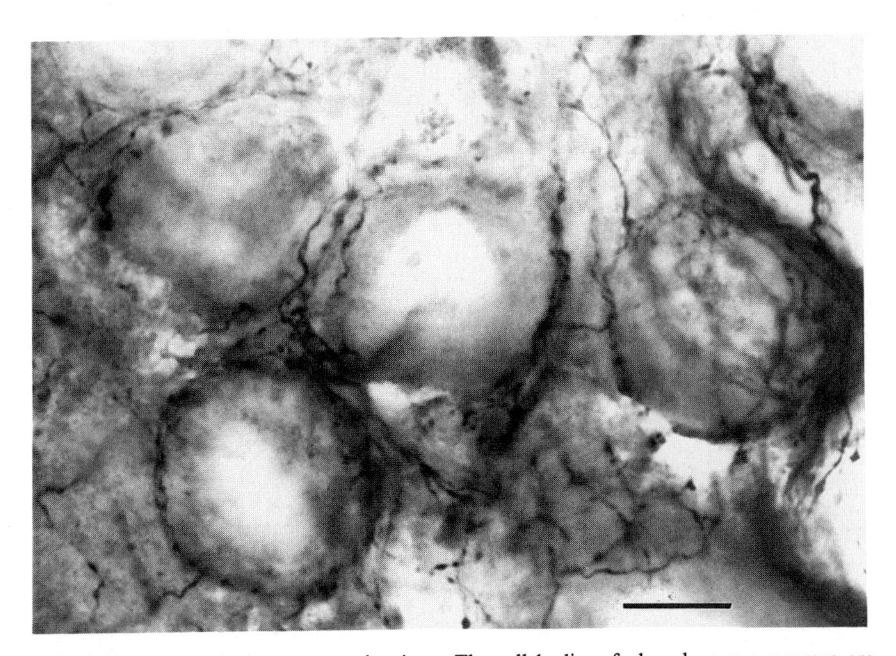

FIGURE 4. Serotonin immunocytochemistry. The cell bodies of pleural sensory neurons are surrounded by serotonergic fibers that frequently expand into varicosities of varying size. Subsequent electron microscopic examination has shown that direct contacts occur between serotonergic processes and the plasma membrane of sensory neurons. Pleural ganglia were fixed, and clusters of sensory neurons were isolated from them. The clusters were incubated in primary antibody for 1 week. Distribution of the primary antibody was revealed by an avidin-peroxidase technique. The tissue was then osmicated, embedded in plastic, and cut into 30-μm-thick sections. Scale bar: 25 μm.

current depolarizing pulse.[18,20] Voltage-clamp experiments indicate that the effects of 5-HT on the resting membrane potential, duration of the action potential, and excitability of sensory neurons are due to the modulation of at least three K^+ currents: the S-K^+ current ($I_{K,S}$), a slow component of the Ca^{2+}-activated K^+ current ($I_{K,Ca}$), and the delayed or voltage-dependent K^+ current ($I_{K,V}$).[17,18,21,22] The reduction of K^+ currents that are normally active leads to the slow depolarization and decreased membrane input conductance, and it also contributes to the enhanced excitability. Moreover, the duration of the action potential is prolonged because of the reduction in K^+ currents that repolarize the neuron.

ROLE OF cAMP IN THE INDUCTION OF SHORT-TERM SENSITIZATION

Cyclic AMP appears to have an important role in mediating the effects of sensitizing stimuli and 5-HT. Exposure of isolated sensory neuron clusters to 5-HT significantly enhanced their cAMP content compared to that of the contralateral control sensory neurons.[19,23] The half-maximal 5-HT concentration for elevation of cAMP was about 15 μM. Other studies have found that the levels of cAMP were elevated for about 15 min after a transient application of 5-HT.[24] This time-course roughly corresponds to the duration of short-term sensitization.

If cAMP is involved in presynaptic facilitation of the tail sensory neurons, one would predict that sensitizing stimuli delivered to the animal should lead to changes

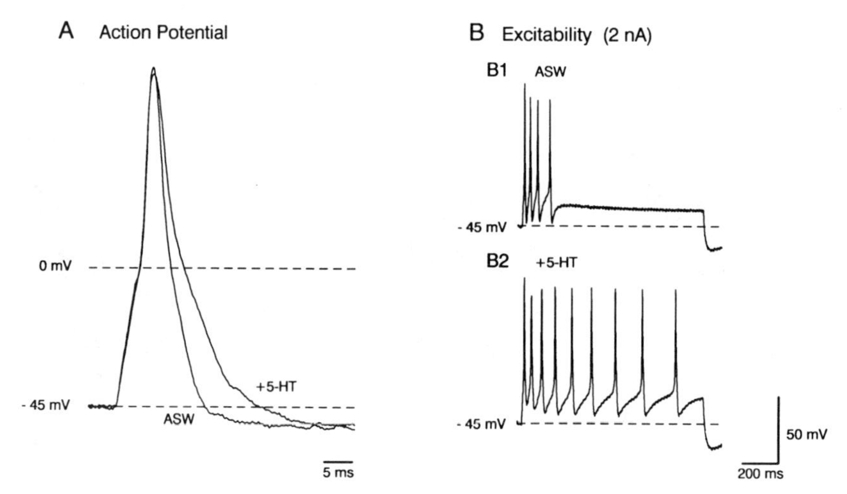

FIGURE 5. Effects of 5-HT on spike duration and excitability. **(A)** The addition of 5-HT (50 μM) to the bath increased the duration of the action potential. **(B)** In the same cell, the addition of 5-HT increased the number of action potentials elicited by a 1-sec, 2-nA depolarizing current pulse from four to nine. Thus, the application of 5-HT increased both the duration of the action potential and the excitability in the somata of sensory neurons. From reference 18.

in the levels of cAMP in these cells. This hypothesis was tested by bisecting an animal and its nervous system and then presenting a sensitizing stimulus to *one* side of the animal. Both sets of sensory neuron clusters were then removed and assayed for cAMP. The level of cAMP was significantly increased in the sensory neuron cluster from the stimulated side of the animal compared to the control cluster from the unstimulated side of the animal.[25] These results demonstrate that the same stimuli that produce sensitization in *Aplysia* also produce an elevation of cAMP in the tail sensory neurons. In addition, cAMP affects membrane currents, excitability, release of transmitter, and levels of phosphorylation of specific proteins in sensory neurons.[2,17,18,20,26–32] Not all of the short-term modulatory effects of 5-HT are mediated by cAMP, however. For example, modulation of $I_{K,V}$ is not mediated by cAMP,[17,18] and mobilization of transmitter (see below) is modulated by both cAMP-dependent protein kinase and protein kinase C[26] (see also reference 33).

LONG-TERM SENSITIZATION

Sensitization also exists in a long-term form. Whereas short-term sensitization can be produced by a single brief shock or train of shocks, the induction of long-term sensitization seems to require a more extended training period.[34] The basic experimental design used to produce long-term sensitization is similar to that illustrated in FIGURE 2. In the long-term studies both sides of each animal were initially tested by the delivery of weak electric shocks to the posterior part of the body wall (tail). Immediately after the initial testing, one randomly chosen side of the animal was subjected to sensitization training, which consisted of four blocks of diffusely applied shocks delivered over an interval of 1.5 hr. The other side of the animal served as the untrained control. Twenty-four hours after sensitization training, the animal was tested again. The percent change in the test score after training, relative to before training, was determined for each side. In contrast to the control side in which no enhancement was produced, the reflex elicited by stimulation of the tail on the side of the animal that received the training procedure was significantly enhanced for at least 24 hr after training.[35] These results indicate that the memory for sensitization persists for at least 24 hr and that the effects of sensitization training are lateralized.

The fact that the behavioral change persists for at least a day raises three questions. First, what is the locus for the long-term memory? Is it the same as for the short-term memory, namely the sensory neurons? Second, what is the mechanism of the long-term memory? Third, is there a mechanistic relationship between short- and long-term memory?

LOCUS FOR LONG-TERM MEMORY: BIOPHYSICAL CORRELATES OF LONG-TERM SENSITIZATION IN THE SENSORY NEURONS

Because the behavioral effects of long-term sensitization training were lateralized, we were able to compare the membrane properties of "sensitized" neurons with the membrane properties of the "control" neurons on the opposite side of the animal.[35] Twenty-four hours after sensitization training, the left and right pleural-pedal ganglia were removed from the animal, so that the biophysical properties of tail sensory neurons could be analyzed. The net outward membrane currents of sensory neurons from the side receiving the sensitization training were reduced as a consequence of long-term sensitization training. There are similarities between one of the currents ($I_{K,S}$) modulated in the short-term by 5-HT and cAMP[17,18,21,36,37] and the current altered in the long-term by sensitization training. The net membrane current that is reduced as a consequence of long-term sensitization training has a mild voltage dependence, is slow to activate, and shows little or no inactivation.

These results indicate that the sensory neurons are not only a cellular locus for short-term sensitization but that they are also a cellular locus for long-term sensitization. Therefore, it does not seem that different neurons are necessary to store short- and long-term memories; the same neurons can do both. Furthermore, at least one cellular correlate of both short- and long-term sensitization is a reduction of net outward currents in the sensory neurons. The functional consequences of this reduction in net outward currents would be a larger or broader action potential and an enhanced release of transmitter. In addition, a reduction of net outward currents would increase the excitability of the sensory neurons and thus enhance the probability of initiating subsequent action potentials after a single test stimulus.

ROLE OF cAMP IN THE INDUCTION OF LONG-TERM SENSITIZATION

Electrophysiological Changes Induced by cAMP

In view of the fact that the sensory neurons are a locus for both short- and long-term memory, do the two processes involve common mechanisms? What is the nature of the intracellular signal(s) that leads to the induction of the long-term electrophysiological changes? One possibility is that cAMP leads to the induction of the long-term changes as it does the short-term changes. Consequently, we have investigated the hypothesis that cAMP triggers long-term sensitization. To test this hypothesis, cAMP was injected into sensory neurons, and its effects on outward currents were measured 24 hr later.[38] Individual sensory neurons were injected with either cAMP or a metabolite of cAMP, 5′-adenosine monophosphate (5′-AMP). The injection electrodes also contained fast green so that the same neurons could be reimpaled 24 hr later. Twenty-four hours after injection, sensory neurons were reimpaled with two microelectrodes and voltage-clamped. Cells that had been injected with cAMP had significantly smaller net outward currents than sensory neurons that had been injected with 5′-AMP. This current has a mild voltage dependence, is slow to activate, and shows little or no inactivation. The membrane current that was altered 24 hr after injection of cAMP was nearly identical to the membrane current that was reduced 24 hr after sensitization training of the animal, as well as to the membrane current ($I_{K,S}$) that was affected in the short-term by both 5-HT and cAMP.

These results indicate that the same second messenger that is capable of inducing the short-term correlates of sensitization is also capable of inducing at least one long-term correlate of sensitization.

Morphological Changes Induced by cAMP

For decades, neurobiologists have been intrigued by the possibility that learning and memory are associated with changes in the shape of neurons. Perhaps most provocative is the idea that outgrowth of neuronal processes establishes new circuits that encode learned information. Bailey and Chen[39–41] found that long-term sensitization was correlated with changes in the structure of sensory neurons in the abdominal ganglion mediating the siphon-gill withdrawal reflex (see also reference 42). Because the intracellular injection of cAMP mimics the effects of sensitization on membrane currents (see above and reference 38), we were interested in determining whether it altered cell morphology as well.[43,44] One sensory neuron in each pleural ganglion was impaled with a microelectrode containing a solution of fast green dye and either cAMP or 5'-AMP. Approximately 22 hr after the iontophoretic injection of the nucleotides, a solution of horseradish peroxidase (HRP) was pressure injected into each cell. The ganglia were then incubated for 2-4 hr to allow HRP to fill the axon and its branches.

A camera lucida drawing of a pair of sensory neurons is shown in FIGURE 6. The soma, the main axon that projects to the pedal ganglion, and many fine branches and varicosities are seen in both the control and the cAMP-injected neuron. Two morphological features of sensory neurons were altered 24 hr after the injection of cAMP. There were approximately twice as many varicosities and 50% more branch points in cAMP-filled cells compared to 5'-AMP-filled cells. Summary data are shown in FIGURE 7.

These morphological effects of cAMP in the pleural sensory neurons are similar to those produced in abdominal sensory neurons by long-term behavioral sensitization training.[40–42] For example, in both studies the number of varicosities was approximately doubled by the experimental treatment. Moreover, 5-HT increases the number of varicosities in cultured sensory neurons.[45] In our studies, the increased number of varicosities and branch points suggests that the sensory neurons are making more synapses with interneurons in the pleural ganglion. As a result, existing connections between sensory neurons and their followers would be strengthened. Additional follower neurons might also be recruited by the formation of new connections. Both of these alterations would contribute to the enhanced amplitude and duration of the withdrawal reflex.

The electrophysiological and morphological results, together with those showing that sensitizing stimuli elevate cAMP in sensory cells,[25] suggest that cAMP is one of the critical intracellular signals that trigger long-term sensitization. Additional support for this conclusion comes from the work of Schacher *et al.*[46] showing that transient bath application of permeable analogues of cAMP leads to an enhancement of the synaptic connections between cultured sensory and motor neurons 24 hr after treatment.

An important remaining question is how cAMP mediates these long-term effects. The induction of long-term sensitization appears to require protein synthesis.[47] Several studies[46,48,49] point to a critical role for new protein synthesis in producing long-term synaptic facilitation and increases in excitability of sensory neurons in culture. In addition, it would seem likely that new protein synthesis is required for the long-term morphological changes. Indeed, for many years it has been suggested that short- and long-term memory can be distinguished by the requirement of new protein synthesis for long-term memory.[50–53] Therefore, it is reasonable to propose that cAMP mediates long-term effects by altering the synthesis of a number of proteins.

IN VITRO ANALOGUE OF SENSITIZATION TRAINING

Ideally, in order to examine the role of protein synthesis in long-term sensitization, one would like to examine changes in synthesis of proteins in the sensory neurons during and at various times after behavioral training. Because of technical difficulties in the treatment of intact animals, we decided to address this issue in a reduced

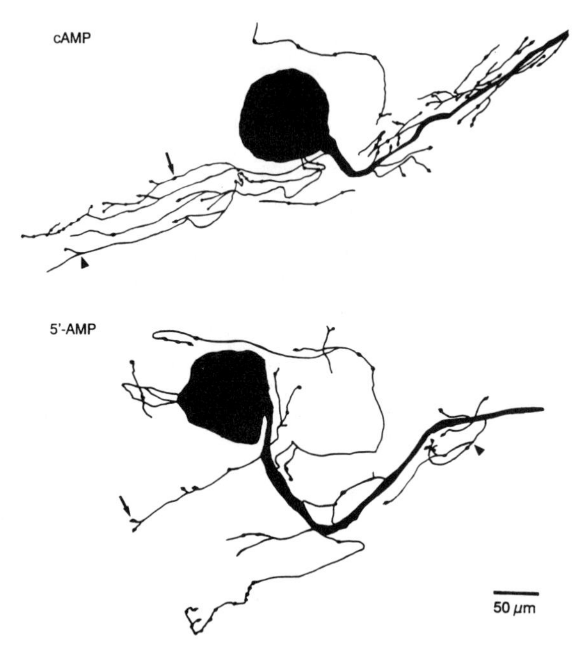

FIGURE 6. A camera lucida drawing of the axonal arborization of individual sensory neurons. One pair of injected cells from the same animal is shown. The cell in the top panel was injected with cAMP and that in the lower panel was injected with 5′-AMP 22 hr prior to being injected with HRP. In each cell a single large axon extends from the cell body to the pleural-pedal connective. Because we analyzed only their number, varicosities (arrows) were drawn slightly larger than scale to enhance their visibility. In the cAMP-filled cell, 78 varicosities were counted; in the 5′-AMP-filled cell, 46. In the cAMP-filled cell, 32 branch points were counted (arrowheads); in the 5′-AMP-filled cell, 24. Not all the branch crossings in these drawings are branch points because these are three-dimensional objects drawn in two dimensions. A few of the branches appear disconnected from the rest of the neuron. This artifact presumably occurred after fixation, as a result of shrinkage of the tissue during processing. From reference 44.

preparation. Specifically, we developed an *in vitro* analogue of the training procedure to investigate the role of protein synthesis in long-term sensitization.

Pleural-pedal ganglia were isolated from the animal. Peripheral nerves were stimulated with four blocks of shocks (FIG. 8) mimicking the behavioral training protocol.[35] Because sensitizing stimuli delivered to the animal lead to the release of ink from the ink gland, the effectiveness of the shocks to the peripheral nerves was estimated by making intracellular recordings from the L14 ink motor neurons.[54] The *in vitro* training procedure reliably produced brisk discharges in the ink motor neurons.

Electrophysiological Changes Produced by the in Vitro *Training Analogue*

Although only one pleural-pedal ganglion is illustrated in FIGURE 8, both pairs of ganglia were removed. One pair received the training, whereas the contralateral pair did not. Twenty-four hours after stimulation of the peripheral nerves, the sensory neurons were voltage-clamped and their membrane currents were examined. Nerve stimulation led to a significant long-term reduction of net outward currents in sensory neurons from the stimulated side compared to those from the contralateral (unstimulated) control side (FIG. 9).[55] The reduction of outward currents was particularly evident toward the end of the pulse. These effects on outward currents were nearly identical to those produced by behavioral sensitization training and the injection of cAMP.[35,38]

We have also used the *in vitro* training analogue to examine the connections between sensory neurons and their follower cells.[56] The connections between the sensory and motor neurons were examined before, 5 min after, and again 24 hr after the training procedure by firing a sensory neuron with a brief intracellular depolarizing pulse and recording the monosynaptic EPSPs in a motor neuron (FIG. 10A). After the 5-min test, electrodes were removed and cells adjacent to the sensory and motor neurons were injected with fast green dye to aid in later identification of the neurons of interest. The ganglia were maintained in organ culture for the 24-hr period between recordings. The mean amplitude of the EPSPs recorded 5 min after training was significantly larger than control. Furthermore, the mean amplitude of the EPSPs recorded 24 hr after training was also significantly larger than control. Summary data are shown in FIGURE 10B.

These results indicate that an *in vitro* analogue of behavioral sensitization training is capable of producing both significant short- and long-term enhancement of the sensory-motor connections responsible for mediating the tail withdrawal reflex as well as long-term modulation of membrane currents in the sensory neurons. This simplified system therefore seems to be a useful model with which to explore further the mechanisms of long-term facilitation, and in particular, the role of protein synthesis.

Protein Changes Produced by the in Vitro *Analogue*

In order to study proteins that may be responsible for the induction of long-term sensitization, we studied the effects of nerve stimulation on the incorporation of amino acids into proteins of the sensory neurons.[55,57,58] Both experimental and control ganglia

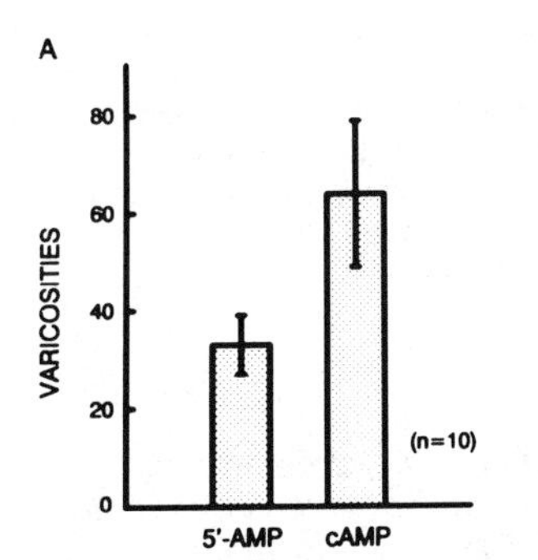

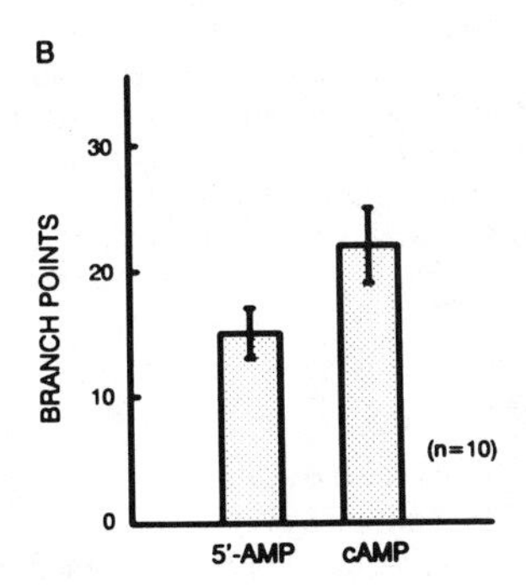

FIGURE 7. Total number of varicosities (**A**) and branch points (**B**) per sensory neuron in pairs of control and cAMP-injected neurons. Error bars represent the SEM. From reference 44.

were exposed to [^{35}S]methionine for a 2-hr period. Both groups of ganglia were treated identically except that the peripheral nerves of the experimental ganglia were stimulated with four blocks of shocks beginning 0.5 hr after the label was applied. At the end of stimulation, experimental and matched control clusters of sensory cells were removed from the pleural ganglia and incorporation of amino acids into their proteins was

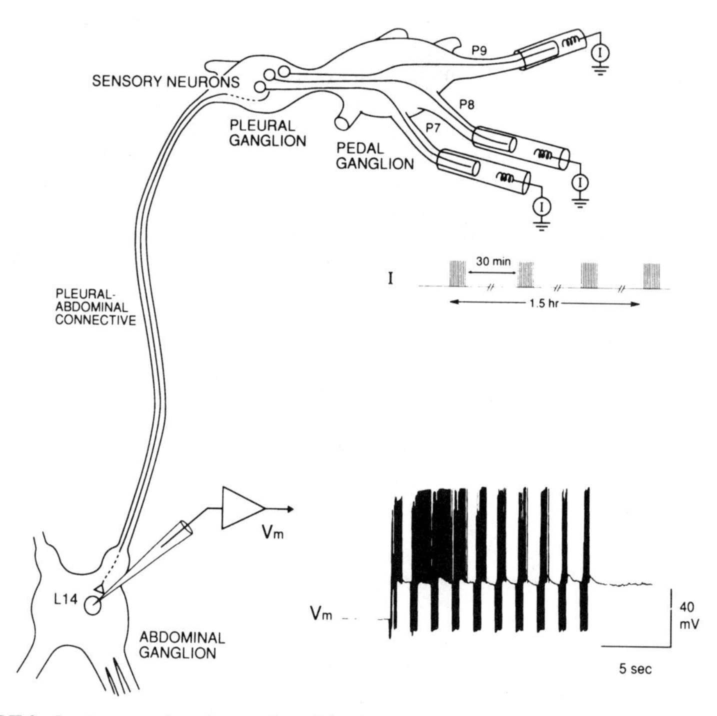

FIGURE 8. *In vitro* neural analogue of sensitization training. Left and right pleural-pedal ganglia were isolated from an anesthetized animal and placed into separate chambers (only one pleural-pedal ganglion is illustrated). Peripheral nerves of both ganglia were drawn into suction electrodes. After equilibration for 2 hr, the peripheral nerves of one randomly selected ganglion were stimulated. The contralateral ganglion served as the unstimulated control. Stimulation consisted of four blocks of shocks over a 1.5-hr period, a procedure designed to mimic that used for long-term sensitization training in the intact animal.[35] Because sensitizing stimuli delivered to the animal lead to the release of ink from the ink gland, control experiments with the abdominal ganglia attached were performed and revealed that this training procedure was sufficient to strongly activate the L14 ink motor neurons in the abdominal ganglion.

analyzed by autoradiography of two-dimensional polyacrylamide gels. Several proteins appeared to be changed reliably by electrical stimulation. Incorporation of amino acid was decreased in some proteins and increased in others. In addition, we compared the effects of electrical stimulation with those of 1.5-hr applications of 5-HT or analogues of cAMP. A number of the proteins affected by electrical stimulation were also changed

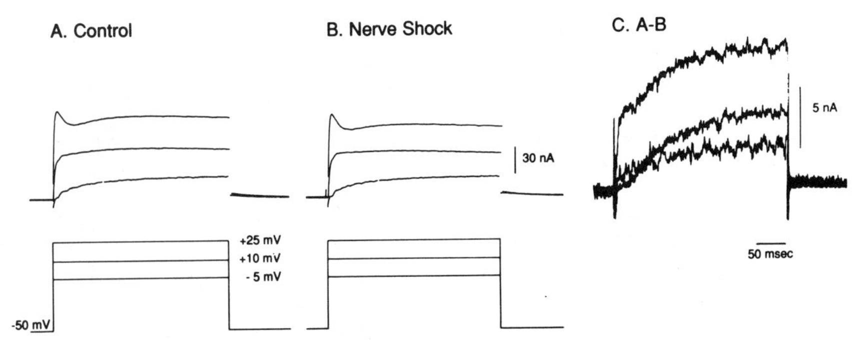

FIGURE 9. Long-term changes in membrane currents produced by the neuronal analogue of sensitization training. Twenty-four hours after the stimulation procedure, sensory neurons were voltage-clamped, and membrane currents elicited in response to a series of different voltage pulses were recorded and digitized. Thus, for each pleural sensory cell, a response family was generated. The overall average of control (**A**) and stimulated (**B**) cells was then obtained by averaging the characteristic response families of the respective clusters ($N = 10$). The data in **C** were obtained by subtracting the response family of the stimulated cells from its corresponding control family. The results for all the experiments were then averaged. Thus, the family of traces in **C** represents the net outward currents that were reduced 24 hr after the analogue of sensitization training.

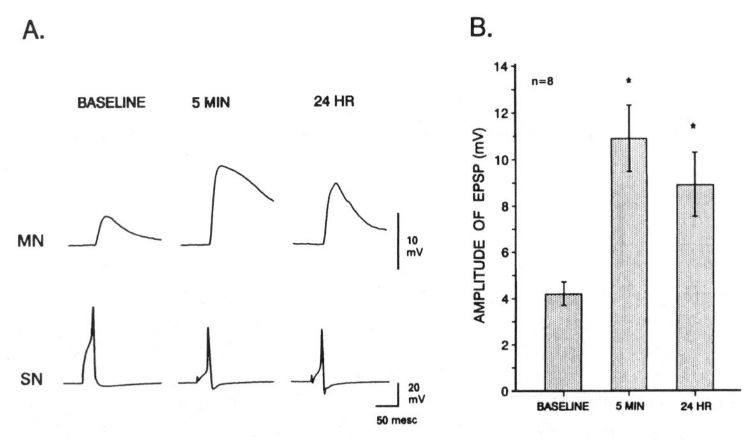

FIGURE 10. Short- and long-term enhancement of synaptic connections produced by a neuronal analogue of sensitization training. (**A**) Recordings from a representative electrophysiological experiment. Across the bottom panel of the figure are recordings of the action potentials elicited in the sensory neuron (SN) during the baseline test, the 5-min posttraining test, and the 24-hr posttraining test. Above each of these traces are the EPSPs recorded in the follower motor neuron (MN) during the respective tests. The training protocol leads to both short- and long-term enhancement of the amplitude of the EPSPs. (**B**) Mean amplitude of EPSPs. The mean amplitude ($\pm$SEM) of EPSPs recorded in the motor neurons is shown for each test. The mean amplitude of the EPSPs recorded 5 min after training was significantly larger than baseline. Moreover, the mean amplitude of the EPSPs recorded 24 hr after training was still significantly larger than baseline. These data indicate that the *in vitro* training procedure is capable of producing significant short- and long-term enhancement of the sensory-to-motor connection.

in a similar way by both 5-HT and analogues of cAMP.[55,57] These results demonstrate that the electrical stimulation of peripheral nerves, which leads to long-term electrophysiological changes in the sensory neurons and their synaptic connections, also leads to the alteration of incorporation of amino acids into specific proteins *during* the period of stimulation, a time when protein synthesis is required to produce long-term effects. Moreover, the effects of electrical stimulation may be mediated at least in part by 5-HT and cAMP.

To extend these studies we investigated changes of proteins that occurred 24 hr after stimulation and compared these changes with those observed just after stimulation.[58] The ganglia were exposed to [^{35}S]methionine for a 2-hr period as with the previous experiment, but now the label was first applied 22 hr after stimulation. The incorporation of label into several proteins was increased both at the end of stimulation and 24 hr after stimulation. These proteins may be involved in events common to the induction and maintenance of long-term sensitization. Several proteins were affected just after the stimulation but not 24 hr later. These proteins may be involved in the induction of sensitization. On the other hand, one protein was increased 24 hr after the stimulation but was unchanged just after stimulation. This protein may be involved in the maintenance or expression of long-term sensitization. Thus, training appears to produce a temporal spectrum of effects on proteins; some proteins are only affected early, some only late, and others both early and late after stimulation (see also references 59 and 60). It will be important to further identify and characterize the proteins affected by the *in vitro* training procedure as well as to determine their cellular locations and functions. Amino acid sequencing of these proteins will help determine their specific roles in the induction, expression, and maintenance of sensitization.

ASSOCIATIVE MODIFICATIONS OF SENSORY NEURONS

We have also used the neural circuit mediating the tail-siphon withdrawal reflex to gain insights into possible mechanisms of associative learning such as classical or Pavlovian conditioning. Previously, we suggested that a cellular mechanism called activity-dependent neuromodulation could mediate short-term classical conditioning.[61–63] A cellular model of activity-dependent neuromodulation and how it might be related to classical conditioning is shown in FIGURE 11. Assume that two neurons (sensory cells in this case) make weak subthreshold connections to a response system (for example, a motor neuron). A reinforcing stimulus has two effects. First, it directly activates the motor neuron and produces the unconditioned response (UR) (FIG. 11A). Second, it activates a diffuse modulatory or facilitatory system that nonspecifically enhances the connections of all the sensory neurons onto their target cells. Such nonspecific enhancement would lead to sensitization. The temporal specificity characteristic of associative learning is achieved by spike activity in one of the sensory neurons (SN1) paired with the reinforcing stimulus. This contiguity produces a selective amplification of the modulatory effects in that specific sensory neuron. Unpaired activity or no activity (SN2) fails to produce the amplification. The amplification of the modulatory effects in the paired sensory neuron leads to an enhancement of the ability of the paired sensory neuron to activate the response system and produce the conditioned response (CR) (FIG. 11B).

TESTING THE ACTIVITY-DEPENDENT NEUROMODULATION HYPOTHESIS

In order to test the activity-dependent neuromodulation hypothesis we used a neural analogue of a classical conditioning protocol. In this neural analogue of classical conditioning, activation of a sensory neuron represents the conditioned stimulus (CS), electrical stimulation of the tail or of a peripheral nerve represents the reinforcing or unconditioned stimulus (US), and the EPSPs in the motor neuron produced by stimulation of a sensory neuron represent the conditioned response (FIG. 12). Simultaneous recordings were made from a motor neuron (MN) and two sensory neurons (SN1 and SN2), each of which made a monosynaptic connection with the motor neuron.[64] One

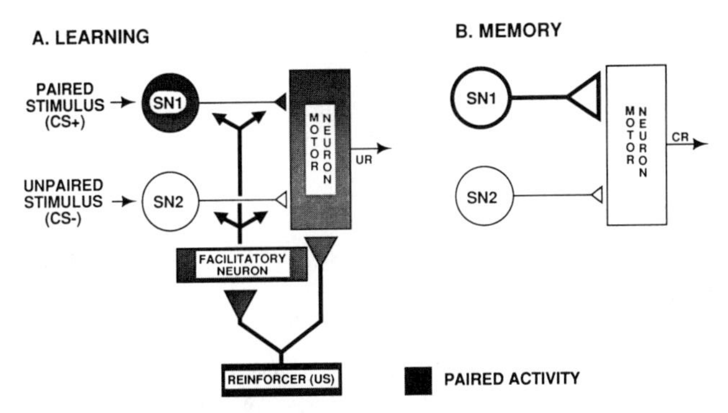

FIGURE 11. Model of associative learning. (**A**) Learning. A motivationally potent reinforcing stimulus (unconditioned stimulus, US) activates the motor neuron to produce an unconditioned response (UR) as well as modulatory or facilitatory neurons that regulate the efficacy of diffuse sensory afferents that make synaptic connections with the motor neurons. Increased spike activity in the paired afferent (CS+, SN1) immediately before the modulatory signal amplifies the degree and duration of the modulatory effects. The unpaired afferent neuron (CS−, SN2) does not show an amplification of the modulatory effects. (**B**) Memory. The amplified modulatory effects cause long-term increases in transmitter release and/or excitability of the paired neuron (SN1), which in turn strengthens the functional connection between the paired neuron and the response system to produce the conditioned response (CR). Modified from reference 63.

cell (the CS+) was artificially fired to produce a high-frequency train of spikes just prior to the reinforcing stimulus (FIG. 12A). This CS+ cell was used to test the hypothesis that paired activity in a sensory neuron can enhance the modulatory effects in that neuron. The second cell, the CS− cell, was also artificially activated (producing a similar train of spikes as in the CS+ cell), but the CS− cell was stimulated 2.5 min after the reinforcing stimulus was delivered. Thus, the CS+ and CS− cells both were stimulated to produce a train of spikes and both received the modulatory effects of the reinforcing stimulus. The key difference is that activity in the CS+ cell was paired with the reinforcement, whereas activity in the CS− cell was not paired with the reinforcement.

As indicated in FIGURE 12B, individual action potentials artificially produced in either of the two sensory neurons (SN1 and SN2) produced small monosynaptic EPSPs

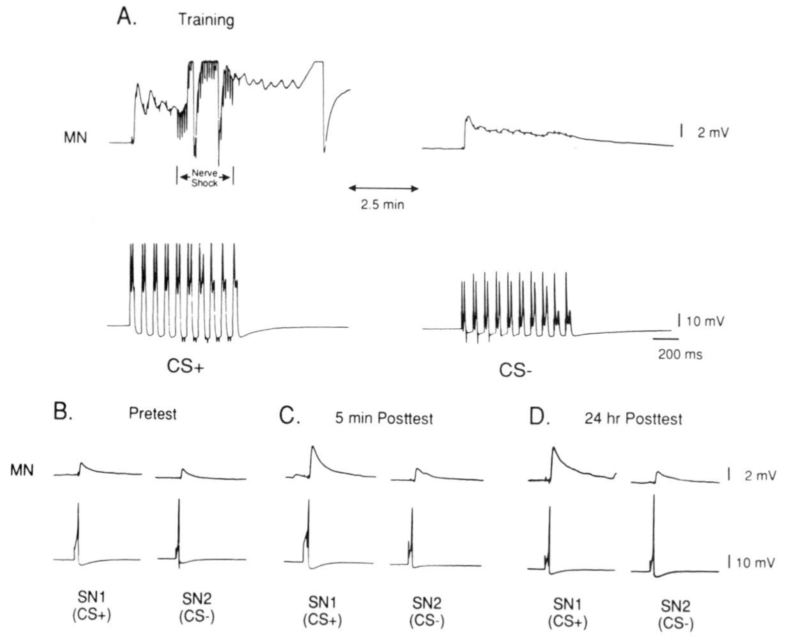

FIGURE 12. Training procedure for short- and long-term activity-dependent neuromodulation. **(A)** Training procedure. Both sensory neurons were activated with a train of 10 depolarizing pulses, eliciting on average 17 spikes (lower traces). One sensory neuron (SN1) was activated 400 msec before (CS+) a train of electric shocks to the nerve. The other sensory neuron (SN2) was activated 2.5 min after (CS−) stimulation of the nerve. **(B)** Pretest. Amplitudes of the EPSPs produced by both sensory neurons before training. **(C)** Amplitude of the EPSPs 5 min after training. The increase in the EPSP amplitude produced by SN1 was larger than that produced by SN2, representing short-term associative plasticity. **(D)** Amplitude of the EPSPs produced by the same sensory neurons 24 hr after training. Long-term associative plasticity is indicated by the observation that 24 hr after training the EPSP produced by SN1 still displayed a larger increase than that produced by SN2. During each test phase three action potentials were elicited in each sensory neuron (the third EPSP of each test is shown). Modified from reference 64.

in the motor neuron. Each of these sensory neurons was then randomly assigned to receive either the paired spike activity with the modulatory stimulus, the CS+, or the unpaired activity, the CS−. The onset of a high-frequency train of spikes produced artificially in SN1 (CS+) preceded the onset of the shock. The train of action potentials in the sensory neuron produced summating EPSPs in the motor neuron. The shock did not fire the sensory neuron, but as one might expect, it produced EPSPs and spikes in the motor neuron. This represents the neural analogue of the unconditioned response. SN2 was activated in a specifically unpaired fashion (CS−). Two and a half minutes after the shock, a high-frequency burst of spikes was produced in the CS− cell. This sequence (FIG. 12A) was repeated five times at 5-min intervals. The EPSPs produced by individual spikes were measured at three time points: before, 5 min after, and 24 hr after training. As shown in FIGURES 12C & 12D, the EPSP produced by the sensory neuron receiving paired stimulation (CS+) is enhanced dramatically from its pretest levels. The EPSP produced by the sensory neuron receiving unpaired stimulation (CS−) is also enhanced, but to a lesser extent. This enhancement of the EPSP produced by the CS− cell is a reflection of the heterosynaptic facilitation associated with sensitization.

Summary data from these experiments are shown in FIGURE 13. There was a significant short-term associative effect, determined by comparing the EPSPs produced by the paired (CS+) group to those produced by the unpaired (CS−) group 5 min after training. Moreover, 24 hr after training there was still a significant enhancement of the EPSPs produced by the paired (CS+) group compared to those produced by the unpaired (CS−) group. Thus, it appears that the effectiveness of modulatory input to a sensory neuron is regulated by prior spike activity in that sensory neuron. Hence, this phenomenon is called activity-dependent neuromodulation. This study establishes for the first time the existence of long-term (24 hr) associative plasticity of synaptic connections in *Aplysia*. Further experiments on the mechanisms underlying the induction, expression, and maintenance of long-term associative memories are now feasible using this neural analogue of a classical conditioning protocol.

MECHANISM OF THE ASSOCIATIVE NEURONAL MODIFICATION AND ACTIVITY-DEPENDENT NEUROMODULATION

Given that there is an enhancement of synaptic transmission produced by activity-dependent neuromodulation, what are the mechanisms that might account for this

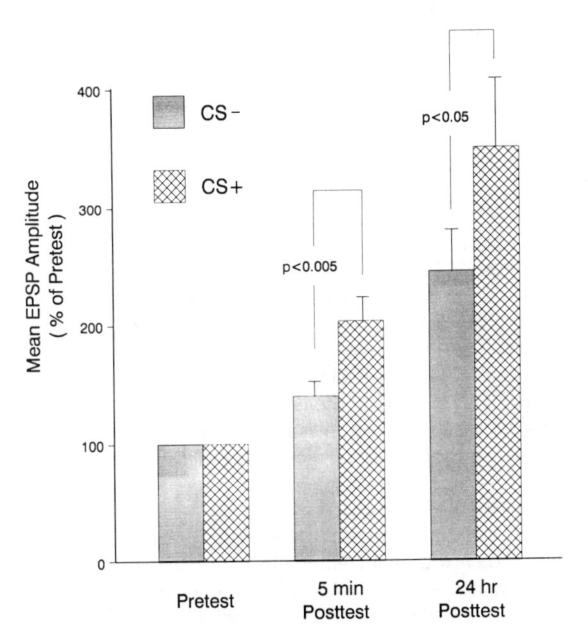

FIGURE 13. Summary data on short- and long-term associative plasticity. There was a significant difference in the amplitudes of the EPSPs produced by the CS− and CS+ groups, both at the 5-min and the 24-hr posttraining test. Error bars represent the SEM. Modified from reference 64.

associative phenomenon? Because activity seems to amplify the effects of sensitization, and because previous work has shown that the effects of sensitizing stimuli appear to be linked to a serotonin-mediated increase in the levels of the second messenger cAMP,[24,25] we predicted that pairing spike activity with a sensitizing stimulus would lead to specific enhancement of cAMP levels.

To test this hypothesis, we developed a biochemical analogue of the pairing procedure in which levels of cAMP were measured in the entire population of sensory neurons.[65] This experiment was possible because the sensory neurons can be readily identified and removed from the ganglia. Clusters of sensory neurons were removed from the left and right pleural ganglia. One of the isolated clusters received a paired presentation of a high concentration of K^+ (to depolarize all the cells and thus mimic spike activity) and an application of 5-HT (to mimic the effects of the unconditioned stimulus). The contralateral cluster received the same exposure to high K^+ and 5-HT, but the treatments were separated by 2 min. The paired group had a significantly greater elevation of cAMP relative to that of the unpaired group from the contralateral cluster.[65] These results indicate that a single pairing of high K^+ with 5-HT leads to a significant enhancement of levels of cAMP.

CHANGES IN PROTEINS PRODUCED BY AN ANALOGUE OF ASSOCIATIVE CONDITIONING

One important question is whether long-term associative plasticity involves pairing-specific changes in protein synthesis. A secondary question is whether long-term associative plasticity is simply an elaboration of the mechanisms for long-term sensitization. To begin to answer these questions, we are investigating the effects on protein synthesis of pairing high K^+ and 5-HT to determine potential sites of interactions of second messenger systems upon specific proteins and to set the stage for studying proteins involved in associative learning.[66]

Abdominal ganglia were used in these experiments because they are an abundant source of neuronal tissue. Ganglia were subjected to four different conditions: 1) 1 hr of high K^+ (80 mM) alone; 2) 2 hr of 5-HT (5 μM) alone; 3) 1 hr of high K^+ paired with 5-HT, with the high K^+ overlapping by 0.5 hr the 2 hr of 5-HT; 4) no treatment. The ganglia were incubated in [^{3}H]leucine for 2 hr beginning at a time corresponding to 5 hr after the end of the 5-HT treatment. After incubation, ganglia were processed for two-dimensional polyacrylamide gel electrophoresis. Several proteins were affected by the treatments with high K^+ or 5-HT alone but were affected to a lesser extent by the combined treatment of high K^+ plus 5-HT. In addition, several proteins seemed to show effects specific to the paired treatment. These proteins were not affected by high K^+ or 5-HT alone but were affected by the combined treatment of high K^+ plus 5-HT.

These experiments should be repeated using sensory neurons rather than abdominal ganglia and with the necessary unpaired controls. Nevertheless, the results are intriguing because they indicate that an analogue of a classical conditioning procedure produces what appears to be pairing-specific changes in the incorporation of label into proteins. Furthermore, the results provide insights into the relationship between long-term associative and nonassociative plasticity. The mechanisms responsible for long-term associative plasticity may be similar to those responsible for long-term nonassociative plasticity. In this case, one would expect to see only quantitative differences in the changes in proteins between the paired and unpaired 5-HT-high K^+ groups. We have, however, observed qualitative differences; that is, different proteins are altered in the paired and unpaired groups, implying that long-term associative plasticity may rely on some mechanisms different from those involved in long-term nonassociative plasticity. Such effects could occur if the modulatory transmitter and Ca^{2+} pathways (activated by the depolarizing stimulus) interacted synergistically downstream from the adenylate cyclase. Cyclic AMP and Ca^{2+} pathways could interact at the genomic level,[67,68] regulating the expression of genes specific for long-term associative plasticity.

SUMMARY AND CONCLUSION

Model of Short- and Long-Term Sensitization

A model that summarizes some of the neural and molecular mechanisms contributing to short- and long-term sensitization is shown in FIGURE 14. Sensitizing stimuli lead to the release of a modulatory transmitter such as 5-HT. Both serotonin and sensitizing stimuli lead to an increase in the synthesis of cAMP and the modulation of a number of K^+ currents through protein phosphorylation. Closure of these K^+ channels leads to membrane depolarization and the enhancement of excitability. An additional consequence of the modulation of the K^+ currents is a reduction of current during the repolarization of the action potential, which leads to an increase in its duration. As a result, Ca^{2+} flows into the cell for a correspondingly longer period of time, and additional transmitter is released from the cell. Modulation of the pool of transmitter available for release (mobilization) also appears to occur as a result of sensitizing stimuli.[69–75] Recent evidence indicates that the mobilization process can be activated by both cAMP-dependent protein kinase and protein kinase C.[26] Thus, release of transmitter is enhanced not only because of the greater influx of Ca^{2+} but also because more transmitter is made available for release by mobilization. The enhanced release of transmitter leads to enhanced activation of motor neurons and an enhanced behavioral response.

Just as the regulation of membrane currents is used as a read out of the memory for short-term sensitization, it also is used as a read out of the memory for long-term sensitization. But long-term sensitization differs from short-term sensitization in that morphological changes are associated with it, and long-term sensitization requires new protein synthesis. The mechanisms that induce and maintain the long-term changes are not yet fully understood (see the dashed lines in FIG. 14) although they are likely to be due to direct interactions with the translation apparatus and perhaps also to events occurring in the cell nucleus. Nevertheless, it appears that the same intracellular messenger, cAMP, that contributes to the expression of the short-term changes, also triggers cellular processes that lead to the long-term changes. One possible mechanism for the action of cAMP is through its regulation of the synthesis of membrane modulatory proteins or key effector proteins (for example, membrane channels).[46,49,76] It is also possible that long-term changes in membrane currents could be due in part to enhanced activity of the cAMP-dependent protein kinase so that there is a persistent phosphorylation of target proteins.[31,77] This model suggests that the analysis of long-term sensitization in *Aplysia* reflects a general problem in cell biology, which encompasses many aspects of neuronal plasticity and development. Specifically, how is protein synthesis regulated by environmental stimuli and how do changes in protein synthesis lead to changes in cell function?

Analysis of the tail-siphon withdrawal reflex and the analogous siphon-gill withdrawal reflex may allow for a fairly complete understanding of the mechanisms underlying the induction, expression, and maintenance of a form of long-term memory as well as help address fundamental questions regarding the relationship between short- and long-term memory.

Model of Associative Plasticity

A model summarizing the electrophysiological and biochemical results that are related to associative plasticity is shown in FIGURE 15 (some of the cellular detail has been omitted to highlight the key points of convergence). Part A shows some aspects of the action of the modulator alone, an effect called heterosynaptic facilitation. The associative mechanism shown in part B depends upon mechanisms already in place for nonassociative effects. Specifically, spike activity paired with the modulatory or reinforcing stimulus, in addition to causing release, leads to a further increase in the levels of cAMP compared to that produced by modulatory stimuli alone. The enhanced cAMP levels in turn would produce greater closure of K^+ channels, greater spike broadening, and greater transmitter release. How might this pairing-specific increase in cAMP levels take place? Pairing might activate an additional second messenger system that could then lead to synergistic effects of the second messengers. One attractive possibility is that Ca^{2+} entering during spike activity might act through calmodulin to prime the adenylate cyclase so that the effects of 5-HT would be amplified, enhancing the subsequent production of cAMP.[65,78,79] Synergistic effects could occur at other cellular loci as well.

An important conclusion is that this model of the mechanisms for short-term associative learning is simply an elaboration of mechanisms already in place that mediate a simpler form of nonassociative learning, sensitization. If true, this conclusion raises the interesting possibility that even more complex forms of learning may be achieved by using simple forms of learning as building blocks, an idea that has been suggested by some physiologists for many years, but one that up until now has not been testable at the cellular level.

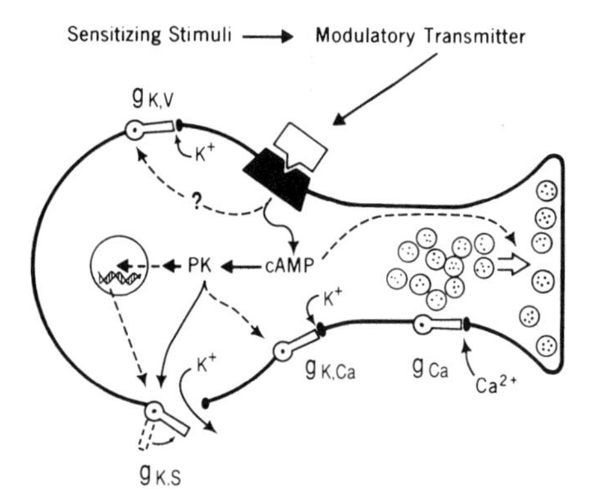

FIGURE 14. Model of heterosynaptic facilitation that contributes to short- and long-term sensitization in *Aplysia.* Dashed lines represent pathways that require further investigation (see text for details).

A. HETEROSYNAPTIC FACILITATION

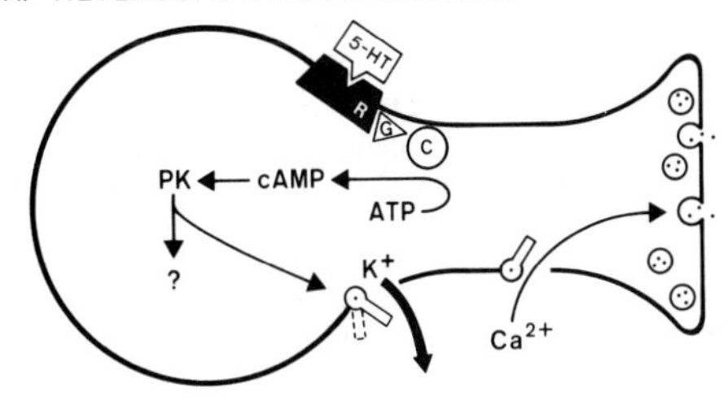

B. ACTIVITY-DEPENDENT NEUROMODULATION

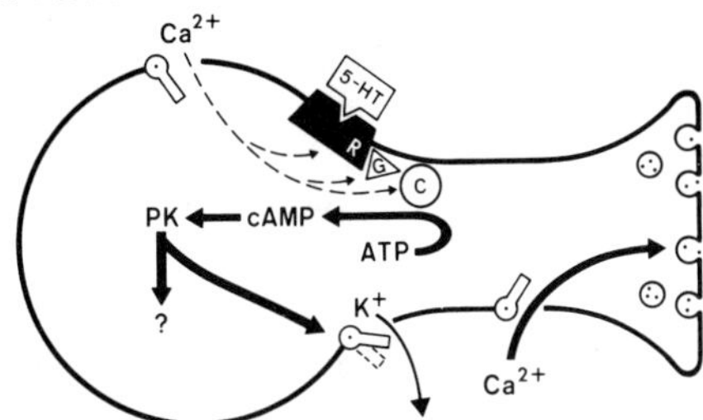

FIGURE 15. Model of molecular events contributing to short-term sensitization and short-term activity-dependent neuromodulation. **(A)** Neurotransmitter binding to receptor (R) activates adenylate cyclase (C) via a regulatory subunit (G). The cAMP that is produced activates one or more protein kinases (PK) whose action(s) includes closure of steady state K^+ channels. Closing K^+ channels results, indirectly, in an increased Ca^{2+} influx and an increase in transmitter release during subsequent action potentials. **(B)** If Ca^{2+} entering during spike activity (CS) interacts synergistically with one or more of the components of the adenylate cyclase complex, subsequent activation by 5-HT (US) would result in an amplification of cAMP levels. As a consequence of the amplified cAMP levels, transmitter release would also be enhanced. Modified from reference 65.

The results from these studies on *Aplysia* are generally consistent with those from other groups studying examples of simple forms of learning and synaptic plasticity in both invertebrates and vertebrates. Although much progress has been made, the sobering news is that much work needs to be done both in *Aplysia* and in other model systems. For the near future, a major experimental question that needs to be answered is to what extent mechanisms for learning are common both within any one animal and among different species. Although many common principles are emerging, work so far on several systems indicates that there are some specific differences in the details. It will be important to determine to what extent different mechanisms are used selectively for different examples of learning and cellular plasticity. More ambitious questions include the analyses of more complex phenomena such as learning to recognize a face or telephone number and whether such learning involves neuronal processes and mechanisms similar to or fundamentally different from those underlying simpler forms of learning, such as sensitization and classical conditioning.

REFERENCES

1. Byrne, J. H. 1987. Cellular analysis of associative learning. Physiol. Rev. **67:** 329-439.
2. Kandel, E. R. & J. H. Schwartz. 1982. Molecular biology of learning: Modulation of transmitter release. Science **218:** 433-442.
3. Walters, E. T., J. H. Byrne, T. J. Carew & E. R. Kandel. 1983. Mechanoafferent neurons innervating tail of *Aplysia*. I. Response properties and synaptic connections. J. Neurophysiol. **50:** 1522-1542.
4. Cleary, L. J. & J. H. Byrne. 1985. Interneurons contributing to the mediation and modulation of the tail withdrawal reflex in *Aplysia*. Soc. Neurosci. Abstr. **11:** 642.

5. CLEARY, L. J. & J. H. BYRNE. 1986. Associative learning of the gill and siphon withdrawal reflex in *Aplysia*: Interneurons mediating the unconditioned response. Soc. Neurosci. Abstr. **12:** 397.
6. GLANZMAN, D. L., S. L. MACKEY, R. D. HAWKINS, A. M. DYKE, P. E. LLOYD & E. R. KANDEL. 1989. Depletion of serotonin in the nervous system of *Aplysia* reduces the behavioral enhancement of gill withdrawal as well as the heterosynaptic facilitation produced by tail shock. J. Neurosci. **9:** 4200-4213.
7. GOLDSTEIN, R., H. B. KISTLER, JR., H. W. STEINBUSCH & J. H. SCHWARTZ. 1984. Distribution of serotonin-immunoreactivity in juvenile *Aplysia*. Neuroscience **11:** 535-547.
8. LONGLEY, R. D. & A. J. LONGLEY. 1986. Serotonin immunoreactivity of neurons in the gastropod *Aplysia californica*. J. Neurobiol. **17:** 339-358.
9. ONO, J. K. & R. E. MCCAMAN. 1986. Immunocytochemical localization and direct assays of serotonin-containing neurons in *Aplysia*. Neuroscience **11:** 549-560.
10. TRITT, S. H., I. P. LOWE & J. H. BYRNE. 1983. A modification of the glyoxylic acid-induced histofluorescence technique for demonstration of catecholamines and serotonin in tissues of *Aplysia californica*. Brain Res. **259:** 159-162.
11. LO, L.-T., J. H. BYRNE & L. J. CLEARY. 1987. Distribution of three modulatory transmitters within the pleural ganglion of *Aplysia*. Soc. Neurosci. Abstr. **13:** 1440.
12. ZHANG, Z., L. J. CLEARY, D. W. MARSHAK & J. H. BYRNE. 1988. Serotonergic varicosities make apparent synaptic contacts with pleural sensory neurons of *Aplysia*. Soc. Neurosci. Abstr. **14:** 841.
13. HAWKINS, R. D. 1989. Localization of potential serotonergic facilitator neurons in *Aplysia* by glyoxylic acid histofluorescence combined with retrograde fluorescent labeling. J. Neurosci. **9:** 4214-4226.
14. MACKEY, S. L., E. R. KANDEL & R. D. HAWKINS. 1989. Identified serotonergic neurons LCB1 and RCB1 in the cerebral ganglia of *Aplysia* produce presynaptic facilitation of siphon sensory neurons. J. Neurosci. **9:** 4227-4235.
15. HAMMER, M., L. J. CLEARY & J. H. BYRNE. 1989. Serotonin acts in the synaptic region of sensory neurons in *Aplysia* to enhance transmitter release. Neurosci. Lett. **104:** 235-240.
16. WALTERS, E. T., J. H. BYRNE, T. J. CAREW & E. R. KANDEL. 1983. Mechanoafferent neurons innervating tail of *Aplysia*. II. Modulation by sensitizing stimulation. J. Neurophysiol. **50:** 1543-1559.
17. BAXTER, D. A. & J. H. BYRNE. 1989. Serotonergic modulation of two potassium currents in the pleural sensory neurons of *Aplysia*. J. Neurophysiol. **62:** 665-679.
18. BAXTER, D. A. & J. H. BYRNE. 1990. Differential effects of cAMP and serotonin on membrane current, action potential duration, and excitability in somata of pleural sensory neurons of *Aplysia*. J. Neurophysiol. **64:** 978-990.
19. OCORR, K. A. & J. H. BYRNE. 1985. Membrane responses and changes in cAMP levels in *Aplysia* sensory neurons produced by 5-HT, tryptamine, FMRFamide and SCP_B. Neurosci. Lett. **55:** 113-118.
20. KLEIN, M., B. HOCHNER & E. R. KANDEL. 1986. Facilitatory transmitters and cAMP can modulate accommodation as well as transmitter release in *Aplysia* sensory neurons: Evidence for parallel processing in a single cell. Proc. Natl. Acad. Sci. USA **83:** 7994-7998.
21. KLEIN, M., J. S. CAMARDO & E. R. KANDEL. 1982. Serotonin modulates a new potassium current in the sensory neurons that show presynaptic facilitation in *Aplysia*. Proc. Natl. Acad. Sci. USA **79:** 5713-5717.
22. WALSH, J. P. & J. H. BYRNE. 1989. Modulation of a steady state, Ca^{2+}-activated K^+ current in tail sensory neurons of *Aplysia*: Role of serotonin and cAMP. J. Neurophysiol. **61:** 32-44.
23. OCORR, K. A. & J. H. BYRNE. 1986. Evidence for separate receptors that mediate parallel effects of serotonin and small cardioactive peptide B (SCP_B) on adenylate cyclase in *Aplysia californica*. Neurosci. Lett. **70:** 283-288.
24. BERNIER, L., V. F. CASTELLUCCI, E. R. KANDEL & J. H. SCHWARTZ. 1982. Facilitatory transmitter causes a selective and prolonged increase in adenosine 3′:5′-monophosphate in sensory neurons mediating the gill and siphon withdrawal reflex in *Aplysia*. J. Neurosci. **2:** 1682-1691.
25. OCORR, K. A., M. TABATA & J. H. BYRNE. 1986. Stimuli that produce sensitization lead

to an elevation of cyclic AMP levels in tail sensory neurons of *Aplysia*. Brain Res. **371:** 190-192.

26. BRAHA, O., N. DALE, B. HOCHNER, M. KLEIN, T. W. ABRAMS & E. R. KANDEL. 1990. Second messengers involved in the two processes of presynaptic facilitation that contribute to sensitization and dishabituation in *Aplysia* sensory neurons. Proc. Natl. Acad. Sci. USA **87:** 2040-2044.
27. ESKIN, A., J. H. BYRNE & K. S. GARCIA. 1988. Identification of proteins that may mediate sensitization. Soc. Neurosci. Abstr. **14:** 840.
28. ICHINOSE, M., S. ENDO, S. D. CRITZ, S. SHENOLIKAR & J. H. BYRNE. 1990. Microcystin-LR, a potent protein phosphatase inhibitor, prolongs the serotonin- and cAMP-induced currents in sensory neurons of *Aplysia californica*. Brain Res. **533:** 137-140.
29. SHUSTER, M. J., J. CAMARDO, S. A. SIEGELBAUM & E. R. KANDEL. 1985. Cyclic AMP-dependent protein kinase closes the serotonin-sensitive K^+ channels of *Aplysia* sensory neurons in cell-free membrane patches. Nature (London) **313:** 392-395.
30. SIEGELBAUM, M. J., J. S. CAMARDO & E. R. KANDEL. 1982. Serotonin and cyclic AMP close single K^+ channels in *Aplysia* sensory neurons. Nature (London) **299:** 413-417.
31. SWEATT, J. D. & E. R. KANDEL. 1989. Persistent and transcriptionally dependent increase in protein phosphorylation in long-term facilitation of *Aplysia* sensory neurons. Nature **339:** 51-54.
32. SWEATT, J. D., A. VOLTERRA, B. EDMONDS, K. A. KARL, A. SIEGELBAUM & E. R. KANDEL. 1989. FMRFamide reverses protein phosphorylation produced by 5-HT and cAMP in *Aplysia* sensory neurons. Nature **342:** 275-278.
33. SACKTOR, T. C. & J. H. SCHWARTZ. 1990. Sensitizing stimuli cause translocation of protein kinase C in *Aplysia* sensory neurons. Proc. Natl. Acad. Sci. USA **87:** 2036-2039.
34. FROST, W. N., V. F. CASTELLUCCI, R. D. HAWKINS & E. R. KANDEL. 1985. Monosynaptic connections made by the sensory neurons of the gill- and siphon-withdrawal reflex in *Aplysia* participate in the storage of long-term memory for sensitization. Proc. Natl. Acad. Sci. USA **82:** 8266-8269.
35. SCHOLZ, K. P. & J. H. BYRNE. 1987. Long-term sensitization in *Aplysia*: Biophysical correlates in tail sensory neurons. Science **235:** 685-687.
36. KLEIN, M. & E. R. KANDEL. 1980. Mechanism of calcium current modulation underlying presynaptic facilitation and behavioral sensitization in *Aplysia*. Proc. Natl. Acad. Sci. USA **77:** 6912-6916.
37. POLLOCK, J. D., L. BERNIER & J. S. CAMARDO. 1985. Serotonin and cyclic adenosine 3′:5′-monophosphate modulate the potassium current in tail sensory neurons in the pleural ganglion of *Aplysia*. J. Neurosci. **5:** 1862-1871.
38. SCHOLZ, K. P. & J. H. BYRNE. 1988. Intracellular injection of cAMP induces a long-term reduction of neuronal K^+ currents. Science **240:** 1664-1666.
39. BAILEY, C. H. & M. CHEN. 1983. Morphological basis of long-term habituation and sensitization in *Aplysia*. Science **220:** 91-93.
40. BAILEY, C. H. & M. CHEN. 1988. Long-term memory in *Aplysia* modulates the total number of varicosities of single identified sensory neurons. Proc. Natl. Acad. Sci. USA **85:** 2373-2377.
41. BAILEY, C. H. & M. CHEN. 1989. Time course of structural changes at identified sensory neuron synapses during long-term sensitization in *Aplysia*. J. Neurosci. **9:** 1774-1780.
42. BAILEY, C. H. & M. CHEN. 1991. Morphological aspects of synaptic plasticity in *Aplysia*: An anatomical substrate for long-term memory. Ann. N.Y. Acad. Sci. This volume.
43. NAZIF, F., J. H. BYRNE & L. J. CLEARY. 1989. Intracellular injection of cAMP produces a long-term (24 hr) increase in the number of varicosities in pleural sensory neurons of *Aplysia*. Soc. Neurosci. Abstr. **15:** 1283.
44. NAZIF, F., J. H. BYRNE & L. J. CLEARY. 1991. cAMP induces long-term morphological changes in sensory neurons of *Aplysia*. Brain Res. **539:** 324-327.
45. GLANZMAN, D. L., E. R. KANDEL & S. SCHACHER. 1990. Target-dependent structural changes accompanying long-term facilitation in *Aplysia* neurons. Science **249:** 799-802.
46. SCHACHER, S., V. F. CASTELLUCCI & E. R. KANDEL. 1988. cAMP evokes long-term facilitation in *Aplysia* sensory neurons that requires new protein synthesis. Science **240:** 1667-1669.

47. CASTELLUCCI, V. F., H. BLUMENFELD, P. GOELET & E. R. KANDEL. 1989. Inhibitor of protein synthesis blocks long-term behavioral sensitization in isolated gill-withdrawal reflex of *Aplysia.* J. Neurobiol. **20:** 1-9.
48. DALE, N., E. R. KANDEL & S. SCHACHER. 1987. Serotonin produces long-term changes in the excitability of *Aplysia* sensory neurons in culture that depend on new protein synthesis. J. Neurosci. **7:** 2232-2238.
49. MONTAROLO, P. G., P. GOELET, V. F. CASTELLUCCI, R. MORGAN, E. R. KANDEL & S. SCHACHER. 1986. A critical period for macromolecular synthesis in long-term heterosynaptic facilitation in *Aplysia.* Science **234:** 1249-1254.
50. AGRANOFF, B. W., ED. 1972. The Chemistry of Mood Activation and Memory. Plenum. New York, NY.
51. BARONDES, S. H. 1975. Protein synthesis-dependent and protein synthesis-independent memory storage processes. *In* Short-Term Memory. D. Deutsh & J. A. Deutsh, Eds.: 379-390. Academic Press. New York, NY.
52. DAVIS, H. P. & L. R. SQUIRE. 1984. Protein synthesis and memory: A review. Physiol. Bull. **96:** 518-559.
53. MCCABE, N. & S. P. R. ROSE. 1987. Increased fucosylation of chick brain proteins following training: Effects of cycloheximide. J. Neurochem. **48:** 538-542.
54. CAREW, T. J. & E. R. KANDEL. 1977. Inking in *Aplysia californica.* I. Neural circuit of an all-or-none behavioral response. J. Neurophysiol. **40:** 692-707.
55. NOEL, F., K. P. SCHOLZ, A. ESKIN & J. H. BYRNE. 1989. Biophysical and protein changes produced in pleural sensory neurons of *Aplysia* by electrical stimulation, an *in vitro* analogue of sensitization training. Soc. Neurosci. Abstr. **15:** 1283.
56. GOLDSMITH, J. R. & J. H. BYRNE. 1989. Long-term (24 hr) enhancement of the sensory-motor connection mediating tail withdrawal reflex in *Aplysia* is produced by nerve stimulation, an *in vitro* analogue of sensitization training. Soc. Neurosci. Abstr. **15:** 1283.
57. NOEL, F., K. P. SCHOLZ, A. ESKIN & J. H. BYRNE. 1991. Common set of proteins in *Aplysia* sensory neurons affected by an *in vitro* analogue of long-term sensitization training, 5-HT and cAMP. In preparation.
58. NOEL, F., A. ESKIN, R. HOMAYONI & J. H. BYRNE. 1990. Short-term and long-term changes of proteins in sensory neurons of *Aplysia* produced by an *in vitro* analogue of sensitization training. Soc. Neurosci. Abstr. **16:** 596.
59. BARZILAI, A., T. E. KENNEDY, J. D. SWEATT & E. R. KANDEL. 1989. 5-HT modulates protein synthesis and the expression of specific proteins during long-term facilitation in *Aplysia* sensory neurons. Neuron **2:** 1577-1586.
60. ESKIN, A., K. S. GARCIA & J. H. BYRNE. 1989. Information storage in the nervous system of *Aplysia*: Specific proteins affected by serotonin and cAMP. Proc. Natl. Acad. Sci. USA **86:** 2458-2462.
61. CAREW, T. J., R. D. HAWKINS & E. R. KANDEL. 1983. Differential classical conditioning of a defensive withdrawal reflex in *Aplysia californica.* Science **219:** 397-400.
62. HAWKINS, R. D., T. W. ABRAMS, T. J. CAREW & E. R. KANDEL. 1983. A cellular mechanism of classical conditioning in *Aplysia*: Activity-dependent amplification of presynaptic facilitation. Science **219:** 400-405.
63. WALTERS, E. T. & J. H. BYRNE. 1983. Associative conditioning of single sensory neurons suggests a cellular mechanism for learning. Science **219:** 405-408.
64. BUONOMANO, D. V. & J. H. BYRNE. 1990. Long-term synaptic changes produced by a cellular analogue of classical conditioning in *Aplysia.* Science **249:** 420-423.
65. OCORR, K. A., E. T. WALTERS & J. H. BYRNE. 1985. Associative conditioning analogue selectively increases cAMP levels of tail sensory neurons in *Aplysia.* Proc. Natl. Acad. Sci. USA **82:** 2548-2552.
66. ESKIN, A., F. NOEL, U. RAJU, R. G. COOK, M. NUÑEZ-REGUERIO & J. H. BYRNE. 1990. Novel effects on proteins produced by pairing serotonin with depolarization. Soc. Neurosci. Abstr. **16:** 596.
67. HYMAN, S. E., J. COMB & T. V. NGUYEN. 1989. Synergistic activation of proenkephalin gene expression by cAMP, phorbol esters, and depolarization. Soc. Neurosci. Abstr. **15:** 645.

68. SHENG, M. G., G. MCFADDEN & M. E. GREENBERG. 1990. Membrane depolarization and calcium induces c-fos transcription via phosphorylation of transcription factor CREB. Neuron **4:** 571-582.
69. GINGRICH, K. J., D. A. BAXTER & J. H. BYRNE. 1988. Mathematical model of cellular mechanisms contributing to presynaptic facilitation. Brain Res. Bull. **21:** 513-520.
70. GINGRICH, K. J. & J. H. BYRNE. 1985. Stimulation of synaptic depression, posttetanic potentiation, and presynaptic facilitation of synaptic potentials from sensory neurons mediating gill-withdrawal reflex in *Aplysia*. J. Neurophysiol. **53:** 652-669.
71. GINGRICH, K. J. & J. H. BYRNE. 1987. Single-cell neuronal model for associative learning. J. Neurophysiol. **57:** 1705-1715.
72. HOCHNER, B., M. KLEIN, S. SCHACHER & E. R. KANDEL. 1986. Action potential duration and the modulation of transmitter release from sensory neurons of *Aplysia* in presynaptic facilitation and behavioral sensitization. Proc. Natl. Acad. Sci. USA **83:** 8410-8414.
73. HOCHNER, B., M. KLEIN, S. SCHACHER & E. R. KANDEL. 1986. Additional component in the cellular mechanism of presynaptic facilitation contributes to behavioral dishabituation in *Aplysia*. Proc. Natl. Acad. Sci. USA **83:** 8794-8798.
74. PIERONI, J. P. & J. H. BYRNE. 1989. Differential effects of serotonin, SCP_B, and FMRFamide on processes contributing to presynaptic facilitation in sensory neurons of *Aplysia*. Soc. Neurosci. Abstr. **15:** 1284.
75. PIERONI, J. P. & J. H. BYRNE. 1990. The mobilization process of facilitation produced by serotonin is inhibited by FMRFamide in sensory neurons of *Aplysia*. Soc. Neurosci. Abstr. **16:** 596.
76. DASH, P. K., B. HOCHNER & E. R. KANDEL. 1990. Injection of the cAMP-responsive element into the nucleus of *Aplysia* sensory neurons blocks long-term facilitation. Nature **345:** 718-721.
77. GREENBERG, S. M., V. F. CASTELLUCCI, H. BAILEY & J. H. SCHWARTZ. 1987. A molecular mechanism for long-term sensitization in *Aplysia*. Nature **329:** 63-65.
78. KANDEL, E. R., T. W. ABRAMS, L. BERNIER, T. J. CAREW, R. D. HAWKINS & J. H. SCHWARTZ. 1983. Classical conditioning and sensitization share aspects of the same molecular cascade in *Aplysia*. Cold Spring Harbor Symp. Quant. Biol. **48:** 821-830.
79. ELLIOT, L. S., Y. DUDAI, E. R. KANDEL & T. W. ABRAMS. 1989. Ca^{2+}/calmodulin sensitivity may be common to all forms of neural adenylate cyclase. Proc. Natl. Acad. Sci. USA **86:** 9564-9568.

Introduction

Recent evidence in several systems supports the existence of synapses that are nontransmitting but are capable of being recruited to an active state.

In the goldfish Mauthner cell, Faber *et al.* report that each synaptic bouton, whether inhibitory or excitatory, has a single active zone and behaves as an all-or-none unit during transmission. Presynaptic spike broadening recruits additional excitatory boutons (previously silent synaptic contacts). Pairing of impulses in two inhibitory inputs or injection of cyclic AMP into the postsynaptic neuron increases the number of active inhibitory boutons. Thus, in this system both presynaptic and postsynaptic mechanisms can recruit silent synaptic contacts.

Crustacean neuromuscular junctions possess numerous individual boutons and active zones of variable morphology, but Wojtowicz *et al.* find far fewer physiologically active units during transmission. Changes in probability of quantal release and in the number of active units can be detected during and after stimulation, and the probability of release is not uniform among the active units. Presynaptic second messenger systems appear to play a role in modulating this activity-dependent recruitment of silent synapses.

In the cat spinal cord, Henneman describes a marked increase following tetanic stimulation in the excitatory postsynaptic potentials that individual Ia afferents produce in motoneurons. He suggests that tetanic stimulation reduces the failure of impulse conduction at branch points in Ia afferent terminal arbors.

These presentations indicate the widespread occurrence and importance of activity-driven recruitment of otherwise silent synaptic contacts. They also indicate that both the mechanisms by which these synapses are recruited and the functional significance of their recruitment appear to differ across, and even within, systems.

Silent Synaptic Connections and Their Modifiability[a]

DONALD S. FABER,[b] JEN-WEI LIN,[c] AND HENRI KORN[d]

[b]*Neurobiology Laboratory*
Department of Physiology
State University of New York at Buffalo
Buffalo, New York 14214

[c]*Department of Physiology and Biophysics*
New York University Medical Center
New York, New York 10016

[d]*Laboratoire de Neurobiologie Cellulaire (INSERM U261)*
Department des Biotechnologies
Institut Pasteur
Paris 15, France

INTRODUCTION

Two general categories encompass ways in which plasticity of synaptic transmission can be achieved in neural networks: 1) structural remodeling, such as the proliferation and retraction of neuronal connections, and 2) modifications in the efficacy or strength of existing connections, which are particularly important once wiring diagrams are established. In the second case, an intriguing possibility that has been the subject of much speculation[1–5] is that there are silent synapses or connections that may become functional under the appropriate conditions, with the corollary that the opposite may also occur. Because studies of silent connections necessarily are concerned with negative results, it has been difficult to demonstrate their existence or the possibility that they may serve as templates for plasticity. More recently, however, combined morphophysiological investigations of pairs of cells have shed some light on this issue and have, at the same time, raised additional questions. In this paper we briefly review evidence that has been generated with two types of synaptic inputs to the goldfish Mauthner cell and demonstrates silent synaptic connections that are potentially functional.

FUNCTIONAL DEFINITIONS: QUANTAL ANALYSIS

A major advance in the study of synaptic transmission in the central nervous system has been the development of techniques that allow quantal analysis to be used in situations where a postsynaptic cell receives inputs from a large number of different

[a]The authors' research reviewed here was supported in part by grants from the National Institutes of Health (NS-15335 and NS-26539) and by the Institut National de la Santé et de la Recherche Médicale.

sources. The results of such investigations in a number of systems are reviewed in detail elsewhere.[6,7] For the purpose of the material discussed here, it is important to establish certain operative definitions that have emerged. These definitions are based upon comparisons of the results of the quantal analysis of individual inhibitory connections onto the Mauthner cell with morphological reconstructions of the synaptic relations between that target cell and the studied presynaptic neurons. The major point is that it is important to distinguish between a *synapse* or synaptic unit and a *synaptic connection.* The term synapse has been used to refer to structures that are very complex, such as the squid giant synapse, and to those that seem much simpler, for example, the contact between a bouton and the postsynaptic membrane. In order to establish a definition that can be generalized and has the same implications for a variety of connections, we propose that synapse be used to refer to the unit consisting of a presynaptic active zone or release site, with its complement of vesicles containing neurotransmitter, and the apposed postsynaptic receptor matrix. Quantal analysis further suggests that when it is sufficiently depolarized one synapse releases the contents of a single vesicle, with a finite probability, p_i. As shown in FIGURE 1, a terminal that contains more than one active zone may then be characterized as establishing multiple synapses with its target, with each of these synapses operating independently but according to the same rules.

The nomenclature above takes into consideration the structural and functional variability of presynaptic terminals, which may have one or more release sites. Consequently, n boutons, each with one active zone, are operationally equivalent to one larger bouton with n active zones. Then, it follows that a synaptic connection is the collection of the total number of synapses between two cells, with each synapse being an independent release unit. Furthermore, because transmitter release is probabilistic, it can be expected that evoked responses will fluctuate in amplitude from one trial to the next, unless n and/or p are very large. This latter point is very important for understanding the operation of central synaptic connections, most of which exhibit large moment-to-moment variations in efficacy. A notable exception is the connection between a climbing fiber and a Purkinje cell in the cerebellum, which resembles the squid giant synapse and the vertebrate neuromuscular junction in that it contains a large number of synapses and transmits quite reliably.[8] It should be pointed out, however, that this type of connection actually processes little information and rather functions as a relatively simple and secure relay.[6]

The concept that a connection consists of n quantal units releasing transmitter probabilistically can be used to generate a statistical description of response fluctuations, for comparison with experimental observations. The two quantal models that have been used most extensively are the simple and compound binomials, with the former assuming that the average p_i is essentially the same for all synapses within a connection, and the latter allowing this parameter to vary from one to the next.[6,7] The distinctions between the two can be very important in considering the differences between silent synapses and silent connections. As shown in FIGURE 2, the term silent synapse refers to the situation whereby some, but not all, synapses within a connection are nonfunctional. That is, those synapses consistently fail to evoke a response when the presynaptic cell is activated. From the perspective of a quantal analysis, the number of quantal release units is then less than the number of morphologically identified active zones. In contrast, a silent connection is one in which all the synapses are nontransmitting and there is no response when the presynaptic cell is stimulated. Its existence is difficult to confirm in the absence of direct evidence for physical contacts between the two neurons, although in a number of systems experimental manipulations have produced changes in connectivity on a rapid time scale considered to rule out morphological restructuring.[2,3]

FIGURE 2 also lists possible mechanisms or sites at which failure of transmission may occur, in either case. They include conditions where 1) the synaptic ending is depolarized but the synapse fails to release transmitter ($p_i = 0$), 2) there is secretion but the postsynaptic receptors are missing or fail to respond, and 3) the presynaptic impulse fails to sufficiently depolarize the terminals (for example, the branch point failure hypothesis of Hennemann and colleagues[9,10]). In the two types of Mauthner cell afferents reviewed here, we have evidence for silent connections, but not for silent synapses, and the failure is not related to the issue of impulse propagation but rather occurs at the other loci. We also describe experimental conditions that can convert silent connections to transmitting ones.

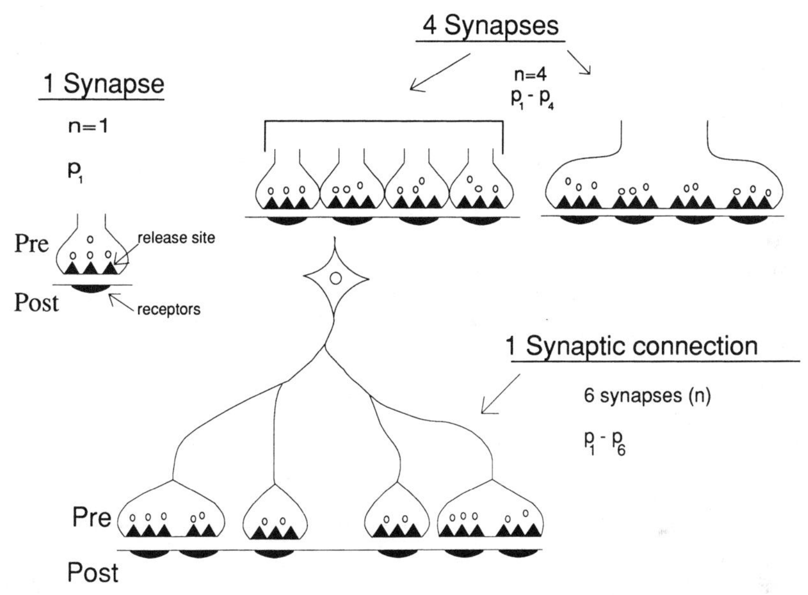

FIGURE 1. Operational definitions of a synapse and a synaptic connection. Upper diagrams show different combinations of synapses. To the left is a single synapse characterized by one active zone in terminal ending. The synapse consists of pre- and postsynaptic specializations, namely, the release site and synaptic vesicles, and the transmitter receptor, respectively. When the ending is activated, it releases the contents of one vesicle, with a certain probability, p_i, and it is, therefore, referred to as a quantal release unit. Center and right: two examples of four synapses on a target cell. One case is characterized by separate boutons, each with one release site (center), whereas the other illustrates one large ending with four independent release sites. In binomial models these two situations are equivalent, with binomial $n = 4$. In a simple binomial all average p_i values are the same, whereas in a compound model they are not. Lower diagram: example of a synaptic connection between pre- and postsynaptic neurons, with binomial n being equal to the total number of release sites. Note that although there are four terminals, binomial $n = 6$, since two endings contain two release sites each.

MAUTHNER CELL INHIBITORY AFFERENTS

The Mauthner cell is a large paired midbrain neuron found in many teleosts and premetamorphic amphibians.[11] It is an ideal model for studies of synaptic physiology because it and certain afferents can be identified both physiologically and morphologically.[12–14] Much of the work providing the background for this report has been reviewed

Silent Synapses or Connections?

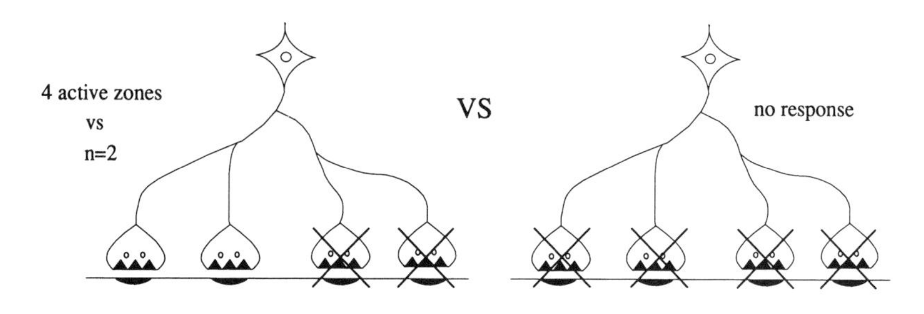

Mechanisms?

- → $p_i = 0$ (no release)
- → receptors do not respond to transmitter
- → failure of impulse propagation

FIGURE 2. Contrast between silent synapses within a single connection and a silent connection. The diagram on the left illustrates a transmitting synaptic connection where two of the four synapses are not functional. The number of quantal release units identified with a binomial analysis of the fluctuations in amplitude of the postsynaptic responses would be less than the number of active zones. That is, there would be a mismatch between physiology and morphology. A chemically silent synaptic connection is shown to the right; in this instance there is no postsynaptic response to a presynaptic element, unless the two units are electrotonically coupled (not shown). Listed below are possible mechanisms for silent synapses or connections.

recently[14,15] and will not be repeated in detail here. Briefly, quantal analysis has been most successfully applied to the connections between an identifiable class of inhibitory interneurons and the Mauthner cell.[16–20] In experiments on these connections, simultaneous pre- and postsynaptic recordings were used and the interneurons were injected with dye to allow morphological reconstruction of individual connections. A simple binomial model was applied to the analysis of the amplitude fluctuations of the evoked inhibitory postsynaptic potentials (IPSPs), which are glycinergic and due to an increased Cl^- conductance.[21–23] This analysis used a deconvolution technique to distinguish statistically the contributions of noise and the probabilistic release process to the amplitude histograms. When the morphological and physiological results were compared, we found that for each connection the number of mathematically defined quantal release units, binomial n, equalled the number of presynaptic terminals.[16–17] Furthermore, each inhibitory ending contains only one active zone.[18] Because the postsynaptic response to the transmitter released from one site (that is, quantal size, q) was relatively small and corresponded to the opening of about 1000 Cl^- channels, the hypothesis that emerged is that following a presynaptic action potential each active zone either does or does not release the contents of a single vesicle, with a certain probability (FIG. 3). Finally, it should be noted that in this system the predicted quantal size was confirmed with direct measurements of spontaneous or miniature events in synaptic noise.[24,25] This direct comparison has not yet been achieved at other central junctions where quantal analysis has been attempted.

Two additional studies of the inhibitory projections to the Mauthner cell demonstrated that p is a modifiable variable. First, when the frequency of stimulation is progressively increased, from 1 or 2 Hz to between 50 and 100 Hz, the averaged evoked responses decrease in amplitude. According to the binomial model, the mean response equals the $n \cdot p \cdot q$ product, and quantal analysis, which can be used to distinguish changes in individual parameters, indicated that a reduction in p alone occurred during this form of synaptic depression.[19] Indeed, at even the highest frequencies there was no indication that some release units dropped out; that is, no sites became silent. Second, synaptic strength and the binomial parameters were compared for different connections, as within this homogeneous population of afferents there is a large range in the number of presynaptic terminals. It turned out that the synaptic strength was not scaled in proportion to n and that interneurons with markedly different complements of boutons could be equipotent, in terms of the average IPSPs evoked by their impulses. That is, innervation density does not provide a weighting function for synaptic strength. Furthermore, the quantal analysis indicated that this apparent leveling process was due to the fact that p decreases as n increases.[20] This finding is consistent with similar reports for the neuromuscular junction, where it appears that transmitter output remains constant when motor units are made smaller experimentally[26] or when more contacts are being established.[27]

As mentioned, analysis of inhibitory connections requires simultaneous pre- and postsynaptic recordings, with both neurons being identified on the basis of electrophysiological criteria. In the course of these studies it was not uncommon to encounter a presynaptic neuron that was expected to evoke a response in the Mauthner cell upon stimulation but did not. As already noted, it is difficult to interpret such negative results in the absence of dye injections. We do have evidence, however, from two types of experiments supporting the notion that there are silent connections and that the site of the block is postsynaptic.

The first indication of silent connections comes from studies where we paired two sources of inhibitory inputs to the Mauthner cell, while recording the evoked responses in voltage-clamp. One input was activated by extracellular stimulation of the contralateral eighth nerve, which projects to the inhibitory interneurons.[28–30] Stimulus strength was adjusted to excite a few presynaptic cells at most, as judged by the amplitude of the inhibitory postsynaptic current (IPSC) (FIG. 4A). The second input was produced by direct intracellular stimulation of an inhibitory interneuron not activated by the weak eighth nerve stimulus. We previously reported that when the two inputs are activated simultaneously the postsynaptic current and its underlying conductance changes are greater than predicted from algebraic summation of the two responses evoked separately.[31] The explanation advanced for this finding is based on the fact that terminals of different inhibitory interneurons synapse on the Mauthner cell soma and proximal dendrites and are intermingled there.[18,32] In addition, channel activation requires the binding of at least two glycine molecules to a postsynaptic receptor (re-

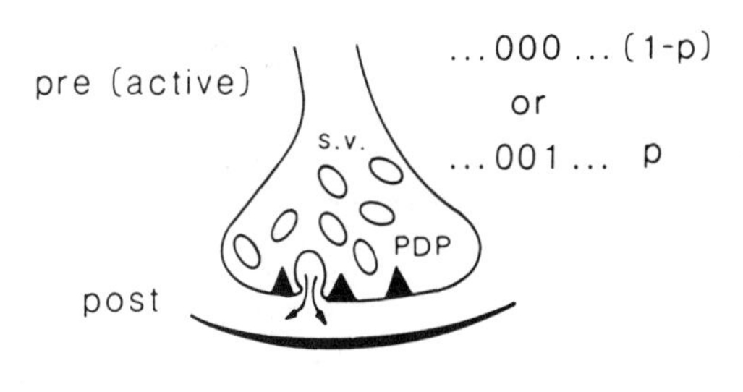

FIGURE 3. Diagram illustrating the notion that after a presynaptic impulse no more than one vesicle releases its contents toward the postsynaptic receptors. PDP: presynaptic dense projections; s.v.: synaptic vesicles. The binary notation emphasizes the hypothesis that, even though there are many vesicles, a maximum of one can be released after each presynaptic impulse (from reference 50).

viewed in references 12 and 15). We therefore suggested that the receptors underlying a single synapse are not saturated by the contents of a single vesicle, and this notion was confirmed by demonstrating that the IPSC evoked by the eighth nerve was enhanced in the presence of iontophoretically applied glycine. Consequently, we proposed that the synergism found when the two physiological sources of inhibition were activated synchronously was due to the lateral diffusion of transmitter to the fringe of a synapse, with the concentration in that region being higher, and with more channels being open, when the adjacent synapse was also active.

In this experimental series, we again encountered presynaptic units that when excited did not produce a measurable response in the Mauthner cell (FIG. 4B). When this stimulus was paired with one to the eighth nerve, however, there was an additional synaptic current, with the appropriate monosynaptic delay (FIGS. 4C & 4D). This finding led us to suggest that at these silent synaptic connections, transmitter is released, but the postsynaptic receptors at those synapses are unresponsive.[15] The response observed with the paired stimulation paradigm would then be due to spread of the transmitter to an adjacent activated synaptic region.

A second, more indirect result supporting the above conclusion comes from experiments in which postsynaptic injections of cAMP were shown to enhance the inhibitory responses of the Mauthner cell.[33] We have argued that this effect is postsynaptic, because cAMP and its immediate metabolic by-products are restricted to the injection site. In these experiments, we found that a weak stimulus to the contralateral eighth nerve, though failing to evoke any inhibition in the Mauthner cell in the control condition, succeeded in doing so after cAMP injection. That is, the second messenger, applied postsynaptically, appeared to unblock silent connections. The most parsimonious explanation of this result is that the protein kinase activated by cAMP produced a modification in the glycine receptor's structure, from an unresponsive form to a functional one. These two sets of observations raise the possibility that patches of receptors scattered over the surface membrane of a neuron and corresponding to postsynaptic sites for a given neuron may be collectively switched on or off (see reference 34 for a similar notion considered in the context of hippocampal long-term potentiation). This is a particularly intriguing idea because it also suggests that while receptor patches at one connection are switched "off," others that are intermingled with them may be "on." That is, this sort of block, if it occurs, may require postsynaptic recognition of the synapses established by a given presynaptic cell. An alternative explanation of this result with cAMP would involve feedback from the Mauthner cell to the presynaptic cells.

MAUTHNER CELL EXCITATORY AFFERENTS

The second Mauthner cell input that has been studied with paired intracellular recordings is the excitatory projection from the so-called large myelinated club endings of eighth nerve afferents. These terminals have morphologically mixed synapses with the Mauthner cell lateral dendrite. That is, they exhibit both chemical synapses and gap junctions,[35,36] the latter mediating electrotonic transmission. The structural organization of these connections is quite different from that of the inhibitory inputs. Specifically, intracellular dye injections have shown that each excitatory afferent sends only one axonal branch to the Mauthner cell,[37] and the presynaptic terminal contains multiple active zones,[35,36,38,39] apparently at least 10. This system has a number of advantages for the issues being considered here. Because the afferents are electrotonically coupled to the Mauthner cell, the presence of a coupling potential produced by a presynaptic spike could be taken as evidence of a connection, once initial dye injection experiments had provided confirmation.[40] Also, the electrotonic potential is a representation of the afferent impulse, and it could therefore be used to assess the branch point failure hypothesis directly by comparing fluctuations in the two modes of transmission.

As already indicated, a single presynaptic impulse may evoke a two-component response in the Mauthner cell, with a brief electrotonic coupling potential being followed by a chemically mediated excitatory postsynaptic potential (EPSP) (FIG. 5).[37] The size of the latter was typically quite small, on the order of a few hundred microvolts,

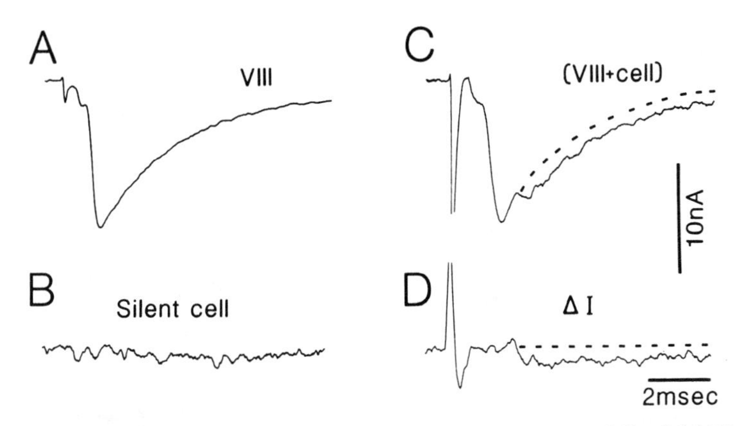

FIGURE 4. Unmasking transmitter release at the connection between a "silent" inhibitory interneuron and the Mauthner cell. **(A-D)** Inhibitory postsynaptic currents (IPSCs) recorded in the Mauthner cell (single-electrode voltage-clamp) in one experiment. **(A)** IPSC evoked by intracellular stimulation of the contralateral eighth nerve, which excites the commissural interneuron. In this experiment, about three presynaptic cells were activated. **(B)** Lack of Mauthner cell response to intracellular stimulation of an identified interneuron. **(C & D)** Synergism observed by pairing the two inputs at a short interval. **(C)** The composite response, with the baseline of the IPSC evoked by the control eighth nerve being indicated by the dashed curve. Note the added synaptic current. **(D)** Subtraction of the record in **A** (VIII) from that in **C** (VIII + cell) revealed the amplitude and time course of the incremental current (ΔI). All records are averages of 20 individual sweeps. As described in the text, this effect is probably of postsynaptic origin, suggesting that at these silent connections the presynaptic impulse nevertheless does release transmitter (from reference 15).

and, as discussed below, the majority of the connections appeared to be chemically silent. Because of the small EPSP size, a direct quantal analysis similar to that used for the inhibitory inputs could not be applied in this case. Rather, an indirect approach was necessary, one that took advantage of the fact that there is a pronounced facilitation of the EPSP when presynaptic stimuli are paired at short intervals (FIGS. 6A & 6B). A number of tests indicated that, for single-fiber stimulation, the locus of the paired pulse facilitation was presynaptic, and the shift in the statistical distribution of the evoked responses could be fit by a simple binomial model if p increased, but not if the change was in n.[40] Thus, these results indicated once more that the probability of release is activity dependent. The fact that with repetitive activity p is enhanced at the excitatory junctions and depressed at the inhibitory ones may reflect the findings that the low-frequency release probability is small in the first case[40] and large in the second[17,20] (see discussion in reference 6).

The results of the indirect quantal analysis at single excitatory connections indicated that binomial n was 10 or greater, which is consistent with the proposal that each active zone is a presynaptic quantal release unit, and not the terminal bouton itself, as proposed by others.[5,7,41,42] This conclusion was further tested directly by using the deconvolution method, applied successfully at the inhibitory connections,[16,17] and by setting binomial n equal to one, since, as mentioned above, there is only one ending per connection. The resulting best-fitting binomial descriptions of the amplitude histograms of the evoked responses were statistically unsatisfactory; however, the fits generated with parameters derived from the paired-pulse paradigm were satisfactory. Thus, the hypothesis that n simply equals the number of endings could be rejected for the case of one ending with multiple release sites.[40] Finally, it should be noted that fluctuations in

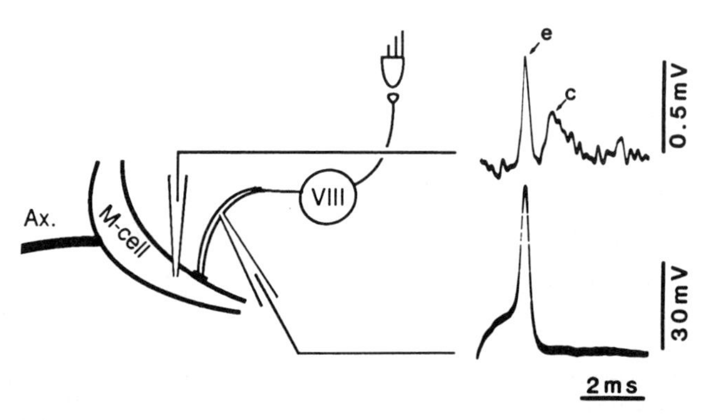

FIGURE 5. Mixed synaptic transmission between single eighth nerve afferents and the Mauthner cell lateral dendrite. The diagram illustrates the circuit studied and the locations of the pre- and postsynaptic intracellular microelectrodes. To the right are sample recordings from the indicated sites. Note that in this case a presynaptic action potential in the eighth nerve fiber produced an electrotonic (e) coupling potential followed by a chemical (c) excitatory postsynaptic potential (modified from reference 37).

the amplitudes of the electrotonic coupling potential and the EPSP were uncorrelated, thereby providing direct evidence that variations in chemical transmission are not due to changes in action potential propagation along the presynaptic axon. Indeed, in this system we did not obtain any evidence for intermittent failure of impulse propagation.

A striking feature of the excitatory connections was that about 80% of those studied were chemically silent.[37,40] That is, in the majority of the fibers studied, presynaptic impulses produced electrotonic coupling potentials in the Mauthner cell lateral dendrite, indicating the two cells were connected, but EPSPs could not be detected, even when signal averaging was used to improve the signal-to-noise ratio (FIGS. 7A & 7B). The average amplitudes of the coupling potentials recorded at transmitting and silent connections were the same, and injections of horseradish peroxidase revealed that in both conditions the afferents were club endings. Also, as shown in FIGURE 7, repetitive stimulation of single eighth nerve fibers, at frequencies that produced facilitated transmitter release when EPSPs were evident, did not unmask chemical transmission at the silent connections.

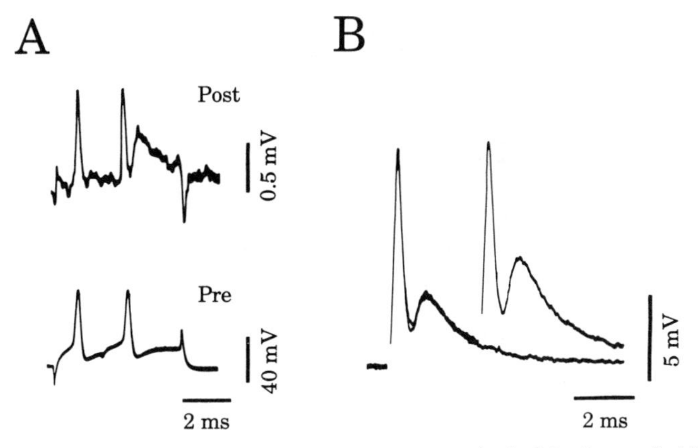

FIGURE 6. Paired-pulse facilitation of eighth nerve responses in the Mauthner cell. (**A**) Simultaneous pre- and postsynaptic recordings from a single eighth nerve fiber and the Mauthner cell lateral dendrite, illustrating facilitation of the second EPSP. (**B**) Similar facilitation is observed when recording Mauthner cell responses to extracellular stimulation of the eighth nerve. Two superimposed records show the response to the first stimulus alone, and to the pair, to indicate the actual amplitude of the second EPSP (modified from reference 40).

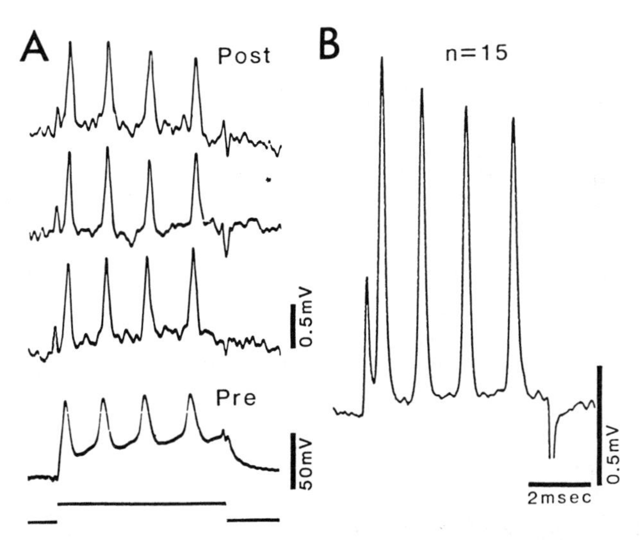

FIGURE 7. Demonstration of silent synaptic connections between a single eighth nerve club ending and the Mauthner cell lateral dendrite. (**A & B**) Same recording conditions as in FIGURE 6. (**A**) Individual postsynaptic responses (upper three traces, labeled Post) produced by repetitive presynaptic firing (bottom trace, labeled Pre). Note that there are coupling potentials, but no chemically mediated EPSPs. (**B**) The average of 15 postsynaptic records from the same experiment, with an improved signal-to-noise level, confirms the absence of chemical transmission. These records also illustrate that repetitive firing at intervals producing facilitation of chemically transmitting connections does not unmask any EPSPs (from reference 40).

Another experimental approach, however, did convert silent connections to transmitting ones. As shown in FIGURE 8, injection of a K^+ channel blocker into the eighth nerve fiber produced small but measurable increases in the duration of the presynaptic action potential, and this prolongation was transmitted to the synaptic ending, as indicated by the coupling potential, which also became broader (A_1 versus A_2). In the experiment of FIGURE 8, the potassium blocker was Cs and there was no clear EPSP prior to injection (A_1). After three Cs injections, the half-width of the coupling potential had increased by 112 μsec (from 244 to 356 μsec) and there was an obvious EPSP, of about 400 μV in amplitude. The plot in FIGURE 8B shows that the EPSP amplitude

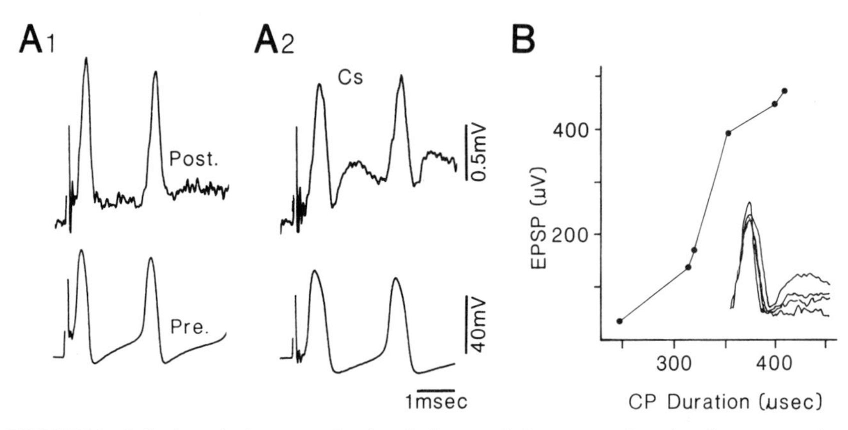

FIGURE 8. Spike broadening unmasks chemical transmission at an otherwise silent connection between an eighth nerve fiber and the Mauthner cell lateral dendrite. (**A_1, A_2 & B**) Records from one experiment, using the same recording conditions as in FIGURE 6, with the exception that the presynaptic electrode contained 2M CsCl rather than KCl. Spike broadening was produced by Cs injection. (**A_1 & A_2**) Average traces of pre- and postsynaptic recordings obtained before (n = 9) and after (n = 17) three periods of Cs injection. In the control (**A_1**) the half-width of the first coupling potential was 244 μsec and there was virtually no EPSP following it. After the Cs injections (**A_2**) the coupling potential was broader (half-width = 356 μsec) and there were clear EPSPs. (**B**) Plot of the first EPSP amplitudes versus coupling potential (CP) half-width. Data points are averages based on sample sizes of 25 to 54 traces. The slope in the steepest region is 7.1 μV/μsec. Inset: superimposed averaged traces of the first coupling potentials and EPSPs. The smallest and largest EPSPs correspond to those in **A_1** and **A_2**, respectively, whereas the intermediate traces are from responses collected following the first and second Cs injections (from reference 40).

(and, hence, transmitter output) was a graded, but steep, function of coupling potential half-width, with a slope of about 700 μV per 100 μsec. In another experimental series, in which the presynaptic electrode contained 4-AP, spike width typically increased immediately after penetration of the axon, without injection of the blocker, and 56% of the connections transmitted chemically. Again the difference between the durations of the coupling potentials at transmitting and nontransmitting connections was small, 320 ± 92 versus 242 ± 16 μsec. Thus, these connections operate at the low end of the input-output curve relating transmitter release to presynaptic spike width. The slope of this relation is similar to that described for the squid giant synapse[43] and *Aplysia* sensory-motor connections.[44]

The large number of chemically silent excitatory connections and the evidence for at least one mechanism to unblock them suggests that they might function as a reserve pool, available to be recruited in certain physiological or behavioral conditions. That is, they may provide an important substrate for certain types of neural plasticity. It should not be assumed that spike broadening is the only means to unmask chemical transmission at these connections, because in control conditions we did not find a significant difference in this parameter when comparing those connections which transmitted in both modes with those which only exhibited electrotonic coupling. Furthermore, we have recently described long-term potentiation of the Mauthner cell responses to extracellular stimulation of the eighth nerve[45] but could not detect a measurable change in the duration of the coupling potential. Nevertheless, it is tempting to consider the possibility that this use-dependent modification in synaptic efficacy might involve recruitment of previously silent connections.

SUMMARY

Comparison of the two afferent systems illustrates certain features common to synaptic transmission as well as differences that might be important for synaptic plasticity. Transmission at both the inhibitory and excitatory connections is satisfactorily described by a simple binomial model that considers the average probability of release to be the same at each active site, although it should be stressed that the best evidence derives from the first set of afferents. Another similarity between the two systems is that short-term changes in synaptic efficacy, namely, facilitation and depression, appear to be due to changes in p. We previously suggested that both phenomena occur during repetitive stimulation, with the dominant effect depending upon the initial probability of release.[6] It remains to be seen if depression dominates at other inhibitory connections, although it is already clear that one cannot generalize about excitation, because some excitatory junctions have an initial high p and exhibit a marked depression[46] rather than the facilitation described here.

We have found no evidence for the notion that some synapses within a connection may be silent. That idea has been proposed, but not proven, for other synaptic connections in the vertebrate central nervous system. Indeed, it will be difficult to assess as long as quantal release cannot be reliably detected at these junctions (see references 6 and 7), and morphological confirmation at the ultrastructural level will also be required. On the other hand, evidence from a few peripheral junctions where one presynaptic afferent establishes hundreds of contacts with its target cell, does suggest the possibility of silent synapses, or at least an extremely low probability of release in those cases.[47–49] These situations may correspond to extremes of our finding that as the number of release sites increases, p decreases.[20] Regardless, the inverse relation between n and p suggests caution should be exercised in interpreting data indicating that synaptic plasticity is associated with increased numbers of synapses between two cells.

Although we have not detected silent synapses within a transmitting connection, we have observed chemically silent connections between neurons, and the evidence reviewed here suggests transmission may be blocked postsynaptically, as with the inhibitory connections, or presynaptically, as with the excitatory ones. Although the underlying mechanisms are only partially elucidated, it is also clear that such connections can be switched into a transmitting mode. Consequently, they may provide a significant reserve that might well become functional in different behavioral states or in response to certain patterns of activity. Conversely, it is also possible that the efficacy of a given input pathway to a neuron or set of neurons may be down-regulated by turning off a subset of the connections.

REFERENCES

1. WALL, P. D. & R. WERMAN. 1976. The physiology and anatomy of long-range afferent fibers within the spinal cord. J. Physiol. **255:** 321-334.
2. DOVSTROVSKY, J. O., J. MILLAR & P. D. WALL. 1976. The immediate shift of afferent drive of dorsal column nucleus cells following deafferentation: A comparison of acute and chronic deafferentation in gracile nucleus and spinal cord. Exp. Neurol. **52:** 480-495.
3. NELSON, S. C., T. C. COLLATOS, A. NIECHAJ & L. M. MENDELL. 1979. Immediate increase in Ia-motoneuron synaptic transmission caudal to spinal cord transection. J. Neurophysiol. **42:** 655-664.
4. MENDELL, L. M. 1984. Modifiability of spinal synapses. Physiol. Rev. **64:** 260-324.
5. REDMAN, S. J. & B. WALMSLEY. 1983. Amplitude fluctuations in synaptic potentials evoked in cat spinal motoneurons at identified group Ia synapses. J. Physiol. **343:** 135-145.
6. KORN, H. & D. S. FABER. 1987. Regulation and significance of probabilistic release mechanisms at central synapses. *In* New Insights into Synaptic Function. G. M. Edelman, W. E. Gall & W. M. Cowan, Eds.: 57-108. Neurosciences Research Foundation/John Wiley & Sons. New York, NY.
7. REDMAN, S. 1990. Quantal analysis of synaptic potentials in neurons of the central nervous system. Physiol. Rev. **70:** 165-198.
8. ECCLES, J. C., M. ITO & J. SZENTAGOTHAI. 1968. The Cerebellum as a Neuronal Machine. Springer. New York, NY.
9. HENNEMAN, E., H.-R. LUSCHER & J. MATHIS. 1984. Simultaneously active and inactive synapses of single Ia fibres on cat spinal motoneurones. J. Physiol. **352:** 147-161.
10. LUSCHER, H.-R., P. RUENZEL & E. HENNEMAN. 1982. Composite EPSPs in motoneurons of different sizes before and during PTP: Implications for transmission failure and its relief in Ia projections. J. Neurophysiol. **49:** 269-289.
11. ZOTTOLI, S. J. 1978. Comparative morphology of the Mauthner cell in fish and amphibians. *In* Neurobiology of the Mauthner Cell. D. S. Faber & H. Korn, Eds.: 13-45. Raven Press. New York, NY.
12. FABER, D. S. & H. KORN. 1978. Electrophysiology of the Mauthner cell: Basic properties, synaptic mechanisms, and associated networks. *In* Neurobiology of the Mauthner Cell. D. S. Faber & H. Korn, Eds.: 47-131. Raven Press. New York, NY.
13. FABER, D. S., J. R. FETCHO & H. KORN. 1989. Neuronal networks underlying the escape response in goldfish. Ann. N.Y. Acad. Sci. **563:** 11-33.
14. KORN, H., D. S. FABER & A. TRILLER. 1990. Convergence of morphological, physiological, and immunocytochemical techniques for the study of single Mauthner cells. *In* Handbook of Chemical Neuroanatomy. Vol. 8: Analysis of Neuronal Microcircuits and Synaptic Interactions. A. Björklund, T. Hökfelt, F. G. Wouterloud & A. N. VanderPol, Eds.: 403-480. Elsevier. Amsterdam.
15. KORN, H. & D. S. FABER. 1990. Quantitative electrophysiological studies and modelling of central glycinergic synapses. *In* Glycine Neurotransmission. O. P. Ottersen & J. Storm-Mathisen, Eds.: 139-170. John Wiley & Sons. New York, NY.
16. KORN, H., A. TRILLER, H. MALLET & D. S. FABER. 1981. Fluctuating responses at a central synapse: *n* of binomial fit predicts number of stained presynaptic boutons. Science **213:** 898-901.
17. KORN, H., A. MALLET, A. TRILLER & D. S. FABER. 1982. Transmission at a central inhibitory synapse. II. Quantal description of release, with a physical correlate for the binomial *n*. J. Neurophysiol. **48:** 679-707.
18. TRILLER, A. & H. KORN. 1982. Transmission at a central inhibitory synapse. III. Ultrastructure of physiologically identified and stained terminals. J. Neurophysiol. **48:** 708-736.

19. KORN, H., D. S. FABER, A. BURNOD & A. TRILLER. 1984. Regulation of efficacy at central synapses. J. Neurosci. **4:** 125-130.
20. KORN, H., D. S. FABER & A. TRILLER. 1986. Probabilistic determination of synaptic strength. J. Neurophysiol. **55:** 402-421.
21. FABER, D. S. & H. KORN. 1982. Transmission at a central inhibitory synapse. I. Magnitude of the unitary postsynaptic conductance change and kinetics of channel activation. J. Neurophysiol. **48:** 654-678.
22. FABER, D. S. & H. KORN. 1988. Unitary conductance changes at teleost Mauthner cell glycinergic synapses: A voltage-clamp and pharmacological analysis. J. Neurophysiol. **60:** 1982-1999.
23. SEITANIDOU, T., A. TRILLER & H. KORN. 1988. Distribution of glycine receptors on the membrane of a central neuron: An immunoelectron microscopy study. J. Neurosci. **8:** 4319-4333.
24. KORN, H., Y. BURNOD & D. S. FABER. 1987. Spontaneous quantal currents in a central neuron match predictions from binomial analysis of evoked responses. Proc. Natl. Acad. Sci. USA **84:** 5981-5985.
25. KORN, H. & D. S. FABER. 1990. Transmission at a central inhibitory synapse. IV. Quantal structure of synaptic noise. J. Neurophysiol. **63:** 198-222.
26. HERRERA, A. A. & A. D. GRINNELL. 1980. Transmitter release from frog motor nerve terminals depends upon motor unit size. Nature **287:** 649-651.
27. TRUSSELL, L. O. & A. D. GRINNELL. 1985. The regulation of synaptic strength within motor units of the frog cutaneous pectoris muscle. J. Neurosci. **5:** 243-254.
28. TRILLER, A. & H. KORN. 1978. Mise en evidence electrophysiologique et anatomique de neurones vestibulaires inhibiteurs commissuraux chez la tanche. (*Tinca tinca.*) C.R. Acad. Sci., Ser. D (Paris) **286:** 89-92.
29. ZOTTOLI, S. J. & D. S. FABER. 1980. An identifiable class of statoacoustic interneurons with bilateral projections in the goldfish medulla. Neuroscience **5:** 1287-1302.
30. TRILLER, A. & H. KORN. 1981. Morphologically distinct classes of inhibitory synapses arise from the same neurons: Ultrastructural identification from crossed vestibular interneurons intracellularly stained with HRP. J. Comp. Neurol. **203:** 131-155.
31. FABER, D. S. & H. KORN. 1988. Synergism at central synapses due to lateral diffusion of transmitter. Proc. Natl. Acad. Sci. USA **85:** 8708-8712.
32. TRILLER, A. & H. KORN. 1986. Variability of axonal arborizations hides simple rules of construction: A topological study from HRP intracellular injections. J. Comp. Neurol. **253:** 500-513.
33. WOLSZON, L. R. & D. S. FABER. 1989. The effects of postsynaptic levels of cyclic AMP on excitatory and inhibitory responses of an identified central neuron. J. Neurosci. **9:** 784-797.
34. BEKKER, J. M. & C. F. STEVENS. 1990. Presynaptic mechanism for long-term potentiation in the hippocampus. Nature **346:** 724-729.
35. NAKAJIMA, Y. 1974. Fine structure of the synaptic endings on the Mauthner cell of the goldfish. J. Comp. Neurol. **156:** 375-402.
36. NAKAJIMA, Y. & K. KOHNO. 1978. Fine structure of the Mauthner cell: Synaptic topography and comparative study. *In* Neurobiology of the Mauthner Cell. D. Faber & H. Korn, Eds.: 133-166. Raven Press. New York, NY.
37. LIN, J.-W. & D. S. FABER. 1988. Synaptic transmission mediated by single club endings on the goldfish Mauthner cell. I. Characteristics of electrotonic and chemical postsynaptic potentials. J. Neurosci. **8:** 1302-1312.
38. TUTTLE, R., S. MASUKO & Y. NAKAJIMA. 1986. Freeze-fracture study of the large myelinated club ending synapse on the goldfish Mauthner cell: Special reference to the quantitative analysis of gap junctions. J. Comp. Neurol. **246:** 202-211.
39. KOHN, K. & N. NOGUCHI. 1986. Large myelinated club endings on the Mauthner cell in the goldfish: A study with thin sectioning and freeze-fracturing. Anat. Embryol. **173:** 361-370.
40. LIN, J.-W. & D. S. FABER. 1988. Synaptic transmission mediated by single club endings on the goldfish Mauthner cell. II. Plasticity of excitatory postsynaptic potentials. J. Neurosci. **8:** 1313-1325.
41. NELSON, P. G., K. C. MARSHALL, R. Y. K. PUN, C. N. CHRISTIAN, W. H. SHERIFF, JR., R. L. MACDONALD & E. A. NEALE. 1983. Synaptic interactions between mammalian

central neurons in cell culture. II. Quantal analysis of EPSPs. J. Neurophysiol. **49:** 1442-1458.
42. NEALE, E. A., P. G. NELSON, R. L. MACDONALD, C. N. CHRISTIAN & L. M. BOWERS. 1983. Synaptic interactions between mammalian central neurons in cell culture. III. Morphological correlates of quantal synaptic transmission. J. Neurophysiol. **49:** 1459-1468.
43. LLINAS, R., I. Z. STEINBERG & K. WALTON. 1981. Relationship between presynaptic calcium current and postsynaptic potential in squid giant synapse. Biophys. J. **33:** 323-352.
44. HOCHNER, B., M. KLEIN, S. SCHACHER & E. R. KANDEL. 1986. Additional component in the cellular mechanism of presynaptic facilitation contributes to behavioral dishabituation in *Aplysia.* Proc. Natl. Acad. Sci. USA **83:** 8794-8798.
45. YANG, Y.-D., H. KORN & D. S. FABER. 1990. Long-term potentiation of electrotonic coupling at mixed synapses. Nature **348:** 542-545.
46. HIGHSTEIN, S. M. & M. V. L. BENNETT. 1975. Fatigue and recovery of transmission at the Mauthner fiber and giant fiber synapse of the hatchetfish. Brain Res. **98:** 229-242.
47. WOJTOWICZ, J. M. & H. L. ATWOOD. 1986. Long-term facilitation alters transmitter releasing properties at the crayfish neuromuscular junction. J. Neurophysiol. **55:** 484-498.
48. ASTRAND, P. & L. STJARNE. 1989. On the secretory activity of single varicosities in the sympathetic nerves innervating the rat tail artery. J. Physiol. **409:** 207-220.
49. WOJTOWICZ, J. M., B. R. SMITH & H. L. ATWOOD. 1991. Activity dependent recruitment of silent synapses. Ann. N.Y. Acad. Sci. This volume.
50. KORN, H. 1984. What central inhibitory pathways tell us about mechanisms of transmitter release. Exp. Brain. Res. **9**(Suppl.): 201-224.

The Size Principle and Its Relation to Transmission Failure in Ia Projections to Spinal Motoneurons

ELWOOD HENNEMAN

Department of Cellular and Molecular Physiology
Harvard Medical School
Boston, Massachusetts 02115

One of the more obvious things about the central nervous system (CNS) is how widely nerve cells differ in size. Some of their properties also differ with size, for example—conduction velocity. The systematic nature of these relations led us to formulate what I have called the Size Principle. This basically simple notion has led to some very unexpected findings. It has generally been assumed that impulses in a parent nerve fiber are conducted into every branch or collateral and invade every one of its terminals, finally to release transmitter from every synapse it forms. As I hope to show, this notion, for which there was never any experimental evidence, is clearly not correct for the system I will be discussing.

Approximately 300 motor neurons (MNs) innervate the muscle fibers in the cat's medial gastrocnemius (MG) muscle. The largest MNs in this pool have about 10 times the surface area of the smallest in the same pool. Largely because of the greater surface areas, their input resistance is only about 1/10th that of the smallest ones. As a result, the potential, E, elicited by an afferent nerve impulse, is much smaller in a large MN because $E = IR$ (ionic current $\times$ input resistance).

As Cajal showed, everywhere in the nervous system large nerve cells have correspondingly thicker axons than small ones. When they reach their destinations (in this case the MG muscle) they give off proportionally more terminals. Each terminal innervates just one muscle fiber in the MG. So, impulses from large MNs with many terminals can produce strong contractions (as much as 120 g for a single motor unit). Impulses from smaller MNs with few terminals produce much smaller tensions (as little as 0.5 g). Already it is obvious why we began to refer to a size principle. This is only the beginning, however. Impulses in large MNs are conducted very rapidly to their muscle. Although the contractions they elicit in their motor units are rapid and powerful, the muscle fibers in these units fatigue rapidly. Such big units are anaerobic. For energy they depend on stored glycogen, which is used up rapidly. This is why maximum efforts cannot be maintained very long. The smaller, slower contracting motor units are aerobic and relatively nonfatiguing. They play a major role in activity requiring endurance.

Once we had recognized this 240-fold range of strength and the great differences in contractile speed, endurance, and metabolic economy, an intriguing question arose. How does the nervous system make use of the amazing assortment of motor units in every muscle? And how can this be done at no undue cost in contractile energy, which is vital for survival in the wild? Faced with these questions, what solution was reached in the course of evolution?

As Francis Crick recently pointed out, "Natural selection is not a clean designer. Evolution is a tinkerer. It is opportunistic: anything will do as long as it works. Its mechanisms may not embody deep general principles. It may prefer a series of slick tricks. Only a close inspection of the gadgetry will tell."[1]

Let us consider the gadgetry that controls muscles. The pool of 300 MNs controlling the cat's MG might be regarded as a large, neural keyboard. This analogy suggests that the CNS fingers the 300 keys selectively as a pianist does. If a strong, fast contraction is urgently required, the CNS might activate only large, fast motor units to produce the necessary tension quickly. Many believe in highly selective activation of this type.

However, consider the problems posed by such a method of control. It would require an enormous number of inputs to each pool. To achieve highly selective action of individual motor units, each afferent would necessarily go to just one, or a small group of MNs, so that only motor units with the necessary speed, power, endurance, and metabolic economy would be activated. Each MN would have to receive a sufficiently strong input from limited sources to be discharged readily. Because the MG may be called upon to develop any tension up to about 12,000 g quite precisely, a great many different *combinations* of motor units would be needed, depending upon the tensions and contractile properties required. "Combination" is one of the stumbling blocks in the notion of selective activation.

There are more than 10^{90} possible combinations in a pool of 300 MNs[2] and, of course, more than 10^{90} combinations of reflex and voluntary inputs would be needed to activate them. Instead of the unimaginable complexity that "selective activation" would require, evolution arrived at a far simpler solution, namely "orderly recruitment of MNs by size."

When fine filaments of appropriate ventral roots are placed on recording electrodes, and progressively stronger excitatory inputs are applied by stretching the MG in a decerebrate cat, MNs begin to discharge in an orderly sequence strictly determined by their size, small to large. Inhibitory inputs silence the largest MNs first, intermediate next, and smallest last. Output is precisely graded by input. If a strong, fast contraction is needed, the entire pool is activated and the required speed and power are obtained. If less powerful contractions are desired, only MNs up to a certain size are fired.[3]

What does orderly recruitment tell us? 1) It reveals that selective activation does not ordinarily occur. 2) It indicates that in a pool of 300 MNs there is not an infinite number of possible combinations of active units, but only 300! 3) It relieves the CNS of the impossible burden of providing an enormous number of selective inputs, so enormous that to accommodate them, the brain would have to be much larger.

How does this simple system work? Not an easy question. The organization of the Ia input from the stretch receptors in muscle to the MN pool provides a quite unexpected clue. All of the 60 Ia's in the MG project directly to the MN pool. To determine how the inputs in this projection are distributed, we recorded intracellularly from the MNs of the MG. Without going into the details of this technique,[4] we found that each Ia fiber from the MG projected directly to all of its 300 MNs. This, of course, was just the opposite of the highly selective distribution that many had proposed.

It has been shown anatomically[5] that, regardless of its size, each MN in a pool has an equal density of synapses. As a result, although MNs with 10 times the surface area of the smallest cells have only about 1/10th of the input resistance, they apparently have 10 times as many synapses. It was conjectured that the larger number of Ia synapses on big MNs might compensate for their lower input resistance. However, when all of the Ia fibers in the MG nerve were stimulated with a single shock, the aggregate or total excitatory postsynaptic potential (EPSP) elicited in a large MN with a low input resistance was small, whereas the EPSP in a small MN with an input resistance 10 times greater was much larger.[6] These findings indicated clearly that the greater number of Ia synapses on large cells did not *necessarily* compensate for their lower input resistances.

As many have learned, findings that do not fit are often the most interesting. When we recorded the aggregate EPSP in the same MNs after a 10-sec conditioning tetanus to the muscle nerve at 500 per sec, the amplitude of the response in the largest MN increased by 122%. In intermediate-sized cells it increased by 92%, and in the smallest cell only by 11%. Clearly, this posttetanic potentiation, as it is called, varies greatly with cell size.[6] In a longer series of observations on many MNs it was apparent that percent potentiation was always greatest in the large cells with small EPSPs and was least in the small MNs with large EPSPs.[6]

These results led us to understand why cell size was so important. In order to supply 10 times as many terminals to a very large MN, a Ia fiber must branch and rebranch repeatedly. It has been shown[7] that the safety factor for axonal conduction is lower at branch points in peripheral nerve fibers, and may result in conduction into only one of two branches. The successive branch points an impulse encounters in the terminal arborizations to MNs apparently results in similar failures. Because there are more branch points in the projections to large cells, there should be more failures in them, and, as a result, smaller EPSPs in them, as we found.

If this is so, there must be more failures that could be relieved during potentiation in large MNs than in small ones. This is the case.[6] Transmission failure *and its relief* during potentiation are greater in projections to large MNs and least in projections to small ones.

The sequence of reflex changes in the MN pools of newborn kittens illustrates some of these points nicely.[8] Shortly after birth, when the motoneurons are small, an afferent volley in all the Ia fibers in a muscle nerve discharges all the MNs in its pool. As the kittens grow older and their motoneurons enlarge, the amplitude of this monosynaptic reflex gradually decreases, indicating that some of the MNs in the pool are no longer being discharged by the afferent volley. This sequence suggests that the greater surface areas of the growing MNs induce more branching of the Ia fibers approaching them, more transmission failure at the branch points, less release of transmitter, and a progressively larger subliminal fringe of unfired cells attributable to the failures.

Consider the developmental implications of this sequence. If MN size determines the degree of branching in its own afferents, and if susceptibility to failure is a function of this branching, it follows that the size of an MN exerts a strong influence on its own excitability through its effect on the input it receives. This is a newly recognized type of interaction between neurons, which suggests how a size principle may develop. The main alternative explanation to branch-point transmission failure is change in transmitter release or effectiveness at active boutons.[6,9,10]

Just a note about clinical implications. It has been reported recently that anything that reduces input to eighth nerve nuclei leads to reduction of the sizes of the nerve cells there. Restoration of input to these nuclei causes a rapid regrowth in cell size. According to the authors,[11] this regulation in cell size by afferent activity is truly bidirectional.

From this one may infer that any trauma or disease that cuts off important inputs to CNS neurons may cause equally rapid changes in postsynaptic cell size and excitability—changes from which there might be little recovery. Measures to minimize major effects of this kind might greatly reduce the impact of disease or trauma on the nervous system.

REFERENCES

1. Crick, F. 1989. The recent excitement about neural networks. Nature **337:** 129-132.
2. Henneman, E. & L. Mendell. 1981. Functional organization of a motoneuron pool and its inputs. *In* Handbook of Physiology—The Nervous System II. Sect. 1, vol. II, part 1: 423-507. American Physiological Society.

3. HENNEMAN, E., G. SOMJEN & D. O. CARPENTER. 1965. Excitability and inhibitability of motoneurons of different sizes. J. Neurophysiol. **28:** 599-620.
4. MENDELL, L. & E. HENNEMAN. 1971. Terminals of single Ia fibers: Location, density and distribution within a pool of 300 homonymous motoneurons. J. Neurophysiol. **34:** 171-187.
5. GELFAN, S. & A. F. RAPISARDA. 1964. Synaptic density on spinal neurons of normal dogs and dogs with experimental hind limb rigidity. J. Comp. Neurol. **123:** 73-96.
6. LÜSCHER, H.-R., P. RUENZEL & E. HENNEMAN. 1983. Composite E.P.S.P.s in motoneurons of different sizes before and during P.T.P.: Implications for transmission failure and its relief in Ia projections. J. Neurophysiol. **49:** 269-289.
7. KRNJEVIC, K. & R. MILEDI. 1959. Presynaptic failure of neuromuscular propagation in rats. J. Physiol. (London) **149:** 1-22.
8. BAWA, P. 1986. Posttetanic changes in the afferent and efferent activity of monosynaptic reflexes in kittens. Can. J. Physiol. Pharmacol. **65:** 328-336.
9. CALMANN, P. H., M.-S. RIOULT-PEDOTTI & H.-R. LÜSCHER. 1991. The influence of noise on quantal EPSP size obtained by deconvolution in spinal motoneurons in the cat. J. Neurophysiol. **65:** 67-75.
10. REDMAN, S. 1990. Quantal analysis of synaptic potentials in neurons of the central nervous system. Physiol. Rev. **70:** 165-198.
11. BORN, D. E. & E. W. RUBEL. 1985. Afferent influences on brainstem auditory nuclei of the chicken: Neuron number and size following cochlear removal. J. Comp. Neurol. **231:** 435-445.

Activity-Dependent Recruitment of Silent Synapses[a]

J. M. WOJTOWICZ,[b] B. R. SMITH,[c] AND
H. L. ATWOOD [b]

[b]*Department of Physiology*
University of Toronto
Toronto, Ontario, Canada M5S 1A8

[c]*Department of Statistics*
Dalhousie University
Halifax, Nova Scotia, Canada B3H 4J1

INTRODUCTION

There have been numerous reports in recent years indicating the possibility of "silent" synapses that can be brought into an active condition by imposed neural activity. In the mammalian spinal cord, for example, Wall and co-workers[1] found that synchronous stimulation of sensory inputs extended a cell's receptive field beyond the boundaries discerned during more normal natural stimulation. A statistical analysis of synaptic responses suggests that treatment of spinal cord preparations by agents that enhance transmitter release leads to appearance of additional transmitting synapses between Ia afferents and motor neurons.[2] This class of effect has been found in other parts of the nervous system. In general, activity-dependent strengthening of synaptic transmission seems to be a widespread phenomenon, and one of the possible mechanisms for such strengthening involves recruitment of weak or unresponsive synapses.

A good model for investigating the hypothesis of activity-dependent recruitment of unresponsive synapses is provided by crustacean motor neurons. The axons of these neurons ramify extensively within their target muscles and make thousands of individual synaptic contacts with the muscle fibers.[3] Each muscle fiber has at least 50 points of synaptic transmission, according to estimates in the well-studied opener muscle of the crayfish.[4] Electron microscopic observations suggest that there are numerous small synapses on each nerve terminal making contact with a muscle fiber.[5] There is at present no firm evidence at the morphological level that can be used to discriminate "active" and "inactive" synapses. However, the individual morphologically defined synapses show heterogenous features: some lack the presynaptic dense bars and associated collections of vesicles that conventional wisdom associates with an actively transmitting synapse.[6] Therefore, it is conceivable that some of these numerous synapses are undeveloped or inactive.

Many crustacean motor axons of the fatigue-resistant "tonic" type show activity-dependent enhancement of transmission, both long-term and short-term. These plastic phenomena have been reviewed recently.[7,8] Single impulses or short-lasting trains of impulses enhance transmitter release dramatically but transiently (short-term facilita-

[a]The authors received research grants from the Medical Research Council of Canada, the Natural Sciences and Engineering Research Council of Canada, and the Whitehall Foundation, Inc.

tion), whereas long-lasting trains of stimuli at moderate frequencies enhance transmission progressively and persistently (long-term facilitation, as described initially by Sherman and Atwood[9]) (FIG. 1). Following stimulation at 10-20 Hz for several minutes, long-lasting enhancement of transmission ensues. At individual terminals where active transmission can be measured, the quantal content of transmission is higher than normal during the long-lasting phase, while no change in amplitude or time course of the individual quantal currents can be demonstrated.[10,11] Thus, the enhancement of transmission appears to be due exclusively to changes in the presynaptic terminals induced by the extra stimulation.

Several types of experiments have been conducted to clarify the mechanism of the persistent enhancement of transmission. There is no change in amplitude or width of the preterminal action potential during the long-lasting enhancement.[10] During and shortly after the train of stimulation, the potentiation of the excitatory postsynaptic potential is linked to build-up of Na^+ and Ca^{2+} in the nerve terminal[12]; about half the tetanic enhancement is Na^+ dependent.[7,13] However, the persistent after-effect of stimulation can be induced even with negligible entry of Na^+ and Ca^{2+}. It is possible

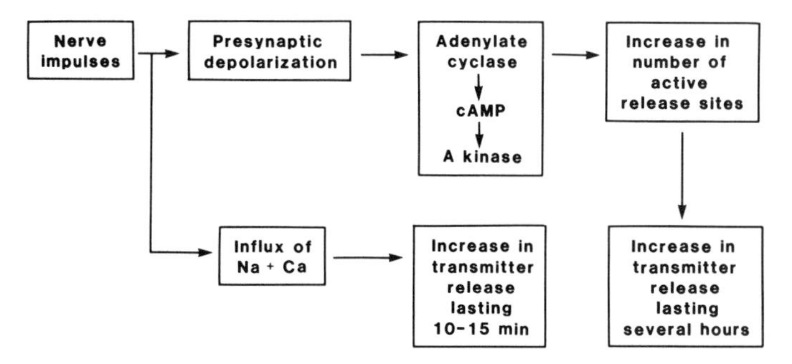

FIGURE 1. Summary of working hypothesis for short-term and long-term facilitation. Ionic changes (Na^+ and Ca^{2+}) modulate short-term facilitation; depolarization can produce long-term facilitation via second messenger systems without major ionic buildup in the nerve terminal.

that an as yet undefined process initiated by depolarization but not requiring influx of Na^+ or Ca^{2+} leads to the long-lasting potentiation.

The initial event apparently results in activation of second messenger systems and protein phosphorylation. The long-lasting potentiation (but not tetanic enhancement) can be blocked selectively by injecting agents into the presynaptic nerve terminal to block adenylate cyclase or the cyclic AMP-activated protein kinase.[14] Thus, adenylate cyclase and reactions dependent on it appear to be necessary for the long-lasting potentiation. The protein substrates of these reactions are presently not known.

Evidence that protein phosphorylation leads in turn to recruitment of synapses comes from physiological and morphological studies. The physiological studies have provided data that can be analyzed statistically using methods of quantal counting and the binomial model applied earlier in other systems.[10,11] It is relatively easy to count quantal release at crustacean neuromuscular junctions using macropatch electrodes for quantal current recording.[11] It is even possible to determine quantal events from intracellular recordings from single muscle fibers,[10] though the analysis is then somewhat more complex. Analyses of quantal events at the crayfish neuromuscular junction have shown that many data sets can be satisfactorily described by the binomial model, in which quantal content is the product of parameters n and p.[10,11,15-17] Several authors have suggested that n may represent the number of active synapses, and p the mean probability of transmitter release per impulse at an active synapse.[18,19] Strong evidence for this type of model has been obtained in studies of the Mauthner cell of goldfish, in which an exact correspondence between the binomial n and the number of individual morphologically identified boutons has been demonstrated.[19]

In the data sets taken from crayfish terminals, the binomial n usually increases during long-lasting potentiation, while p changes less.[10,11] The increase in n can be taken as partial evidence for addition of new active synapses to the responsive pool.

Parallel ultrastructural evidence has been presented to show that the number of presynaptic dense bars may increase during the long-lasting potentiation,[20,21] or with repeated bouts of stimulation.[22] This evidence, however, is still preliminary.

One problem that has always plagued the analyses based on the simple binomial model is the possibility, or even likelihood, of nonuniform response probability at the different synapses sampled by electrophysiological measurements. Calculations have shown that errors introduced by nonuniform probability are potentially serious.[23] This problem has been dealt with and reviewed by Redman.[24] Previous studies by Hatt and Smith[17] have provided evidence for nonuniform release at crayfish neuromuscular junctions. In other crustacean neuromuscular junctions, both morphological and physiological evidence suggest nonuniformity of synapses, and hence the likelihood of nonuniform probability of release.[25] In order to determine whether the result of increased n in long-term potentiation obtained with the simple binomial model is realistic, it is necessary to apply statistical methods allowing for nonuniform probability to the data sets.

The present study outlines a statistical approach that we have developed and employed for data sets obtained with macropatch electrodes for terminals in the crayfish opener muscle. The approach described here is more general than that adopted earlier for the crayfish neuromuscular junction by Hatt and Smith,[17] though the intent is the same. The results of the present analysis indicate that the finding of increased n for long-lasting potentiation holds when nonuniform probability is incorporated into the statistical treatment. Reassuringly, better fits to the data are provided with nonuniform probability, and the values of n thus obtained are not greatly different from those found with the simple binomial model. Thus, the more serious errors foreseen for the simple binomial model are not present, and it is still reasonable to think that altered n may represent changes in the number of active synapses.

METHODS

Electrophysiology

The experiments were performed on an *in vitro* preparation of the crayfish opener muscle. The dissecting and intracellular recording techniques were described in detail by Wojtowicz and Atwood.[13,26] The macropatch recording technique is similar to that used by Dudel,[27] except that in our laboratory the macropatch pipettes were manufactured from the thin-walled omega-dot tubing (1.5 mm outside diameter). The tips of these pipettes had lumens measuring 20-40 μm. The pipettes filled with the standard physiological solution had resistances of 200-500 kΩ. When placed on the surface of the nerve terminal, the pipettes formed a seal of about 1 MΩ thus allowing high-resolution recording of the postsynaptic current. Single quantal currents were easily measured.

Quantal Analysis

A new procedure has been developed to identify the number of available quanta for synaptic transmission and the probabilities of their release. Since we assume that a single release site releases at most one quantum in response to a single presynaptic action potential,[19] the analysis yields the number of active (or responding) sites.

The procedure consists of fitting the observed counts of released quanta over a large number of stimuli to three hypothetical models. These are the binomial, Poisson, and nonuniform probability models.

Suppose that the observed counts are $N_0, N_1, \ldots, N_m$; that is, the largest response corresponds to m released quanta, and we observe N_0 zeros, N_1 singles, and so on. Let $N = \sum_{i=0}^{m} N_i$, and let E_i be the expected number of "i quantum" releases. Then $E_i = N\pi_i$, where π_i is the probability of i releases, and so E_i depends on the model considered (binomial, Poisson, or nonuniform probability).

$$X^2 = \sum_{i=o}^{m} (N_i - E_i)^2/E_i$$

For each model under consideration, we minimize the statistic with respect to the parameters. This provides parameter estimates for each competing model. The criterion we use to select a model is $MAIC = \chi^2 + k$, where k is a penalty term related to the number of estimated parameters and χ^2 is the minimized value of X^2. We refer to this as a modified Akaike information criterion (*AIC*). Akaike[28] introduced the *AIC* criterion as a means of choosing from among several competing models, and the theoretical basis for the *MAIC* is given in Donoho.[29] For Poisson models $k = 1$ (the mean), for binomial models $k = 2$ (n and p), and for the nonuniform release model $k = (1 +$ number of distinct $p_i)$, where the extra 1 makes the comparison of binomial and nonuniform models valid. For example, if a nonuniform model with $n = 2$ is fitted under the constraint $p_1 = p_2$, then the penalty term is 2 (that is, 1 + the number of distinct release probabilities). Because $p_1 = p_2$, the model is really binomial, and again the penalty term is 2.

It is easy to see that this procedure should work for large sample sizes. As $N \to \infty$, $N_i \to \pi_i N$, and E_i will converge to $\pi_i N$ if and only if the model chosen is either the correct model, or includes the correct model as a particular case. For example, if the true model is binomial with $n = 2$, then χ^2 will be near 0 for that binomial model, but not for the Poisson model, because for the Poisson, E_i is non-0 for any i. Nor will χ^2 be close to 0 for any binomial model with $n \neq 2$. χ^2 will be near 0 for any nonuniform model with $n \geq 2$, because each such a model contains the true binomial model as a submodel. Because we are penalizing by the number of estimated parameters, the nonuniform models with $n > 2$ will be rejected, and so the final comparison will be between the binomial $(2, p)$ and the nonuniform $(2, p_1 < p_2)$ models. As the binomial $(2, p)$ model is correct in this example, the comparison of these two competitors will lead to acceptance of the binomial, whose penalty term is smaller by one.

Model parameters are estimated by finding those values that minimize the X^2. We use an iterative, quasi-Newton algorithm to perform the minimization (see Fletcher[30]).

We wish to find the global minimizer, and therefore must begin the iteration in a suitably small neighborhood of the true global minimizer. Our method is to generate random starting values according to a uniform distribution on the parameter space, and, for each starting value, to iterate the minimization procedure until convergence is achieved (convergence is assumed to have occurred when successive iterates vary by less than 10^{-4} for each parameter). The iterative solution providing the smallest value of χ^2 is taken to be the global minimum of X^2. In practice the method appears to be independent of the starting iterate, indicating that X^2 might be unimodal. We have been unable to show this analytically.

The adequacy of sample sizes of 1000-2000 has been determined in pilot experiments where large (6000 stimuli) samples were obtained and subsequently broken down into smaller fragments. It was found that samples of 1000 give values of n that are consistent with the n of the whole sample (± 1). However, estimates of p_i (in case of the nonuniform model) evidently require larger samples.

The fit of the data to a model was judged satisfactory when the expected counts (E_i) differed from the observed values (N_i) by 1 or less. All data presented in this paper fulfilled this criterion.

It should be noted, however, that a perfect fit of the data to a model does not guarantee the correctness of the physical interpretation of parameters n and p_i. Further work is needed in order to determine if n indeed stands for the number of release sites and if probabilities of release at individual sites are independent of each other.

RESULTS AND DISCUSSION

The results of this study are primarily concerned with our current search for the mechanism of long-term facilitation expression.

The macropatch recording technique permits detection of quantal transmitter release from circumscribed regions of the axonal arbor (FIG. 2). The average number of synapses between the nerve and the muscle in small crayfish opener muscle preparations is 1.7 ± 0.2 (mean $\pm$ SD) per 1 μm of the terminal (unpublished data based on serial sectioning and reconstruction of electron micrographs from three control preparations). In other crustacean preparations, larger values have been reported.[25,31] Therefore, in small crayfish opener muscle preparations, one terminal crossing the whole diameter of the pipette may possess as many as 45 synapses. Physiological recordings revealed that up to 10 quanta can be released approximately simultaneously if the axon is stimulated at high frequency or when release is enhanced by treatment with 3,4-diaminopyridine. Under normal circumstances, however, only up to five quanta are released per stimulus when the axon is stimulated at 1-10 Hz, and the usual mean quantal content varies between 0.05 and 0.2 (FIG. 3).

In principle the maximal number of quanta released per stimulus provides a measure of the number of active release sites. In practice this approach is not always feasible because, when the probability of release from individual release sites is low, the chance that they will release simultaneously is too low to be seen in reasonable sample sizes. For example, three sites with probabilities of 0.05 each can be expected to release together in one out of 10,000 stimuli!

Therefore we employed a statistical model as outlined above to match the observed and expected release patterns and thus identify the number of release sites and their individual probabilities from smaller, experimentally feasible sample sizes of 1000-2000 responses.

We applied the new analytical procedure to data from four experiments on long-term facilitation. Representative records are shown in FIGURE 4. In these experiments the excitatory axon was stimulated at 2 Hz either by evoking action potentials or applying depolarizing pulses directly to the terminal through an intraaxonal electrode in the presence of tetrodotoxin. The stimulation produced release of transmitter quanta that were counted and analyzed as described above.

Long-term facilitation induced in these four preparations by 20-Hz stimulation increased the quantal content (the average number of quanta released per stimulus) measured at 2 Hz after establishment of the plateau phase by 112 ± 53% (mean ±

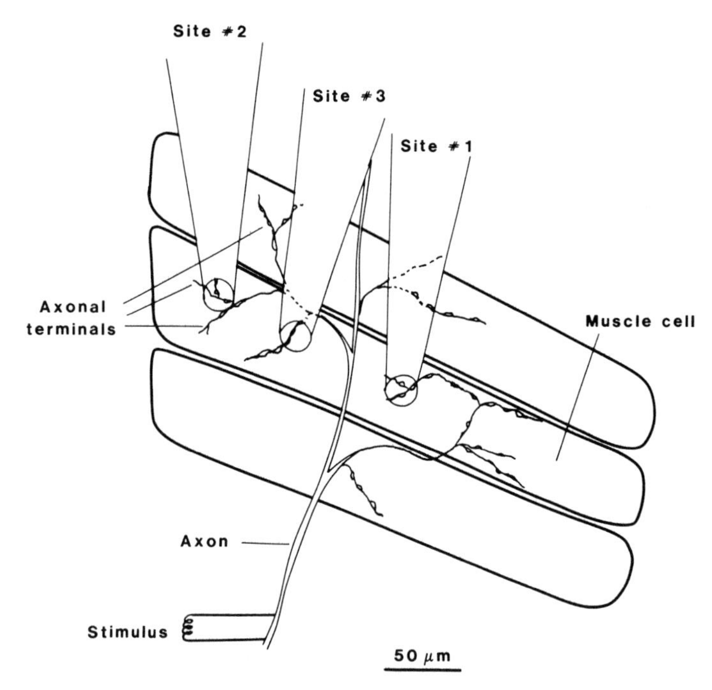

FIGURE 2. Experimental arrangement for macropatch recordings of quantal currents. Macropatch pipettes were placed on the surface of the muscle fibers over terminal arborizations of the excitatory axon.

SD). A large part of this enhancement can be accounted for by the increase (83%) in the number of active sites (TABLE 1). Thus the analysis confirms earlier results estimated from the simple binomial model[10,11] and gives credence to the hypothesis that synaptic connections can be strengthened by recruitment of inactive (or dormant) release sites.

Our previous efforts were hampered by a simplification in the data analysis which assumed that the probabilities of release at all responding sites are equal. Although in theory such a simplification could introduce serious errors into the analysis,[20] in practice the errors in the estimates of the parameter n are not significant and therefore the results

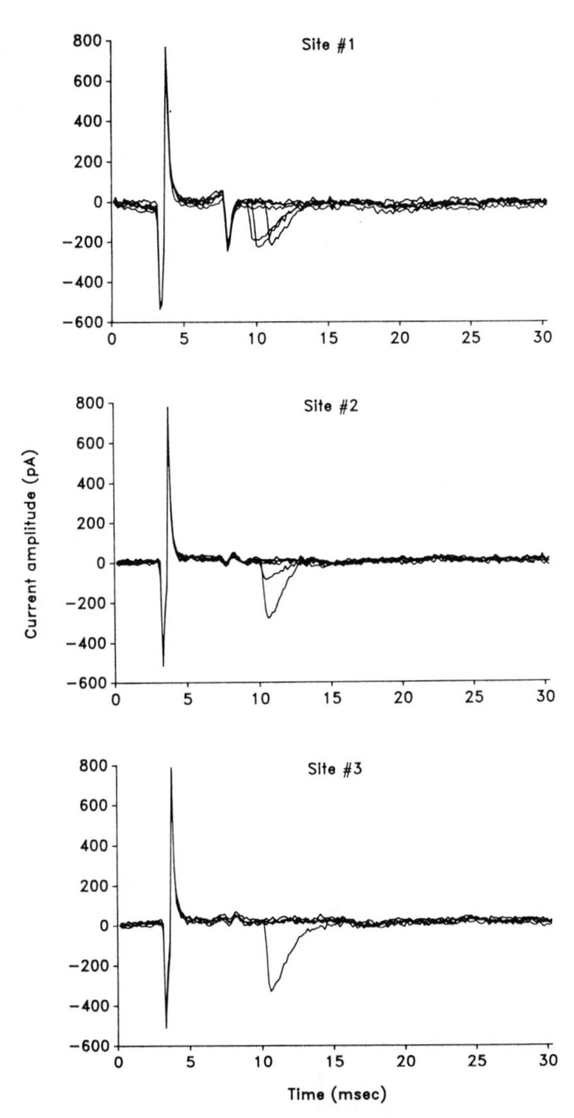

FIGURE 3. Sample responses obtained by placing the same pipette on three different "hot spots" of release on one muscle fiber. The average number of quantal releases was different in the three locations, but in all cases there were many failures of release. The low quantal content at stimulation frequencies of 1-10 Hz is the usual finding in this preparation.

presented here are not very different from those reported previously for long-term facilitation.[11]

A comparison of the estimates of parameters n and p on a single data set for short-term facilitation is given in TABLE 2. In this experiment the axon was stimulated at four different frequencies ranging between 1 and 10 Hz. Such an increase in frequency results in about a 15-fold enhancement of transmitter release with corresponding changes in the binomial parameters. It can be seen that in this type of short-term facilitation the main factor that accounts for the increase in transmitter release is the probability p. Moreover, assuming equal probabilities of release among the release sites

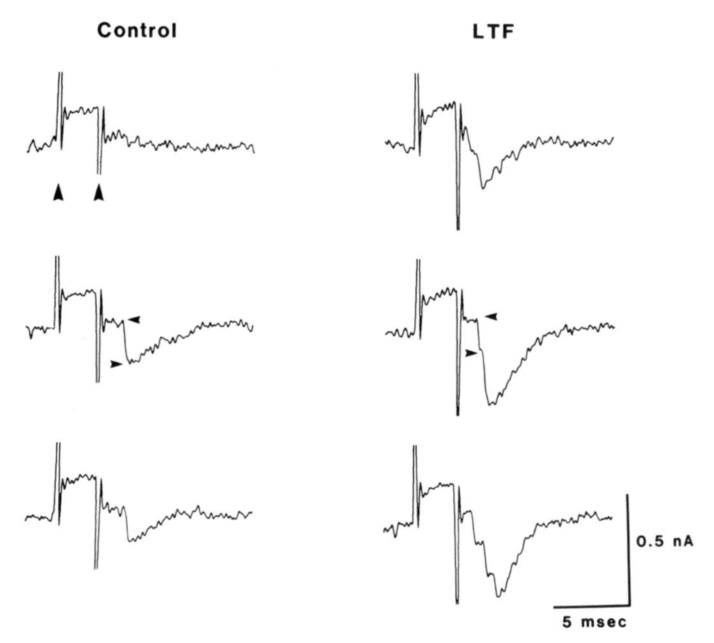

FIGURE 4. Sample data obtained by means of macropatch recording demonstrates long-term facilitation. In each trace the coupling artifact (vertical arrowheads) between the presynaptic depolarizing current pulse and the macropatch pipette is followed by synaptic responses. Horizontal arrowheads point to single quanta evoked by the pulse. Typically more quanta were evoked during long-term facilitation than in the control period.

does not cause serious distortions in the estimates of the parameter n, except at high frequencies when probabilities vary widely.

The results for short-term facilitation are roughly comparable to those reported previously for crayfish neuromuscular junctions.[15–18]

In conclusion, we have found that the number of responding release sites is considerably smaller than the number of available synapses so that there is a pool of dormant synapses that can be recruited during activity. The threshold for this recruitment is apparently reduced during long-term facilitation, as evidenced by larger values for the estimated parameter n.

TABLE 1. Changes in Binomial Parameters during Long-Term Facilitation[a]

Experiment	Control	Long-Term Facilitation
1	m = 0.24 Nonuniform n = 3 p_1 = .05 p_2 = .10 p_3 = .10	m = 0.38 Nonuniform n = 3 p_1 = .03 p_2 = .16 p_3 = .19
2	m = 0.23 Nonuniform n = 3 p_1 = .03 p_2 = .06 p_3 = .14	m = 0.58 Nonuniform n = 4 p_1 = .02 p_3 = .17 p_2 = .10 p_4 = .29
3	m = 0.61 Uniform n = 4 p_{1-4} = .15	m = 1.07 Nonuniform n = 6 p_1 = .09 p_4 = .13 p_2 = .12 p_5 = .17 p_3 = .13 p_6 = .44
4	m = 0.21 Uniform n = 2 p_{1-2} = .10	m = 0.55 Nonuniform n = 5 p_1 = .05 p_4 = .13 p_2 = .10 p_5 = .14 p_3 = .13

[a] Results of four experiments where long-term facilitation was induced by a 10-min stimulation at 20 Hz. The quantal release was sampled at 2 Hz (1200 responses) before and after the high-frequency burst and showed the increase of quantal content (m). In experiment 1, the increase in m was accounted for by the increase in the probabilities of release at three release sites. In the remaining experiments, there was a change in the number of responding quantal units as well as in the probabilities of release. Although in general the data were best described by nonuniform binomial distributions, the control responses in experiments 3 and 4 were best described by simple binomial models.

TABLE 2. Estimates of Binomial Parameters during Frequency Facilitation[a]

Frequency (Hz)	m	Binomial n & p	Nonuniform n & p
1	0.07	2 & .03	2 & p_1 = .03, p_2 = .04
2	0.14	2 & .07	2 & p_1 = .02, p_2 = .13
5	0.34	3 & .11	3 & p_1 = .06, p_2 = .11 p_3 = .17
10	1.08	No satisfactory fit	4 & p_1 = .08, p_2 = .22 p_3 = .23, p_4 = .55

[a] Results from a single experiment in which the axonal terminal was stimulated at 1-10 Hz and quantal content (m) measured in 1000 successive sweeps. Increased m indicates short-term facilitation of release. Although the data were best described by nonuniform binomial models, the simple binomial models provided accurate estimates of n at lower frequencies.

SUMMARY

At the crayfish neuromuscular junction, a long-lasting enhancement of synaptic transmission can be induced by tetanic stimulation of 10-20 Hz for several minutes. The long-lasting enhancement is presynaptic in origin, because quantal content increases but not quantal size, and is not dependent upon broadening or enlargement of the presynaptic action potential. The enhancement can be selectively blocked by presynaptic injection of agents that inhibit adenylate cyclase or the cyclic AMP-dependent protein kinase. Entry of calcium may not be sufficient in itself to produce the enhancement. Analyses of quantal events using both a simple binomial statistical method, and a more refined method that takes into account the possibility of unequal probabilities of responding units, have shown that the number of responding units increases during the long-lasting enhancement. In addition, there is an increase in the probability of transmitter release at preexisting units. In contrast, during short-term facilitation accompanying repetitive stimulation, response probability increases greatly whereas the number of responding units increases only moderately with frequencies of activation up to 20 Hz, which increase quantal output severalfold. These results indicate that responding units, hypothesized to be transmitting synapses, can be recruited to active transmission from an unresponsive pool by tetanic activity, and that protein phosphorylation is required for long-lasting changes to occur. The existence of an excess of synapses on crustacean nerve terminals is indicated by ultrastructural studies, which invariably show many synapses on the terminals. The number of morphologically defined synapses is always greater than the number of responding units seen in statistical analyses of quantal release for the same recording location.

ACKNOWLEDGMENTS

The authors wish to thank Marianne Hegström-Wojtowicz for her assistance in preparing the manuscript. H. L. A. and J. M. W. are members of the Medical Research Council Group for Nerve Cells and Synapses.

REFERENCES

1. WALL, P. D. 1977. The presence of ineffective synapses and the circumstances which unmask them. Philos. Trans. R. Soc. London, Ser. B **278:** 361-372.
2. JACK, J. J. B., S. J. REDMAN & K. WONG. 1981. Modifications to synaptic transmission at group Ia synapses on cat spinal motoneurones by 4-aminopyridine. J. Physiol. (London) **321:** 111-126.
3. BITTNER, G. D. 1968. Differentiation of nerve terminals in the crayfish opener muscle and its functional significance. J. Gen. Physiol. **51:** 731-758.
4. DUDEL, J. & S. W. KUFFLER. 1961. The quantal nature of transmission and spontaneous miniature potentials at the crayfish neuromuscular junction. J. Physiol. (London) **155:** 514-529.
5. ATWOOD, H. L. 1982. Synapses and neurotransmitters. *In* The Biology of Crustacea. H. L. Atwood & D. C. Sandeman, Eds. Vol. 3: 105-150. Academic Press. New York, NY.
6. JAHROMI, S. S. & H. L. ATWOOD. 1974. Three-dimensional ultrastructure of the crayfish neuromuscular apparatus. J. Cell Biol. **63:** 599-613.

7. Atwood, H. L. & J. M. Wojtowicz. 1986. Short-term and long-term plasticity and physiological differentiation of crustacean motor synapses. Int. Rev. Neurobiol. **28:** 275-362.
8. Zucker, R. S. 1989. Short-term synaptic plasticity. Annu. Rev. Neurosci. **12:** 13-31.
9. Sherman, R. G. & H. L. Atwood. 1971. Synaptic facilitation: Long-term neuromuscular facilitation in crustaceans. Science **171:** 1248-1250.
10. Wojtowicz, J. M. & H. L. Atwood. 1986. Long-term facilitation alters transmitter releasing properties at the crayfish neuromuscular junction. J. Neurophysiol. **55:** 484-498.
11. Wojtowicz, J. M., I. Parnas, H. Parnas & H. L. Atwood. 1988. Long-term facilitation of synaptic transmission demonstrated with macro-patch recording at the crayfish neuromuscular junction. Neurosci. Lett. **90:** 152-158.
12. Delaney, K. R., R. S. Zucker & D. W. Tank. 1989. Calcium in motor nerve terminals associated with posttetanic potentiation. J. Neurosci. **9:** 3558-3567.
13. Wojtowicz, J. M. & H. L. Atwood. 1988. Presynaptic long-term facilitation at the crayfish neuromuscular junction: Voltage-dependent and ion-dependent phases. J. Neurosci. **8:** 4667-4774.
14. Dixon, D. & H. L. Atwood. 1989. Adenylate cyclase system is essential for long-term facilitation at the crayfish neuromuscular junction. J. Neurosci. **9:** 4246-4252.
15. Johnson, I. W. & A. Wernig. 1971. The binomial nature of transmitter release at the crayfish neuromuscular junction. J. Physiol. (London) **218:** 757-767.
16. Wernig, A. 1975. Estimates of statistical release parameters from crayfish and frog neuromuscular junctions. J. Physiol. (London) **244:** 207-221.
17. Hatt, H. & D. O. Smith. 1976. Non-uniform probabilities of quantal release at the crayfish neuromuscular junction. J. Physiol. (London) **259:** 395-404.
18. Zucker, R. S. 1973. Changes in the statistics of transmitter release during facilitation. J. Physiol. (London) **229:** 787-810.
19. Korn, H., A. Mallet, A. Triller & D. S. Faber. 1982. Transmission at a central inhibitory synapse. II. Quantal description of release with a physical correlate for binomial *n*. J. Neurophysiol. **48:** 679-707.
20. Chiang, R. G. & C. K. Govind. 1986. Reorganization of synaptic ultrastructure at facilitated lobster neuromuscular terminals. J. Neurocytol. **15:** 63-74.
21. Wojtowicz, J. M., L. Marin & H. L. Atwood. 1989. Synaptic restructuring during long-term facilitation at the crayfish neuromuscular junction. Can. J. Physiol. Pharmacol. **61:** 167-171.
22. Mearow, K. M. & C. K. Govind. 1989. Stimulation-induced changes at crayfish (*Procambarus clarkii*) neuromuscular terminals. Cell Tissue Res. **256:** 119-123.
23. Brown, T. H., D. H. Perkel & M. W. Feldman. 1976. Evoked neurotransmitter release—Statistical effects of nonuniformity and nonstationarity. Proc. Natl. Acad. Sci. USA **73:** 2913-2917.
24. Redman, S. 1990. Quantal analysis of synaptic potentials in neurons of the central nervous system. Physiol. Rev. **70:** 165-198.
25. Atwood, H. L. & F. W. Tse. 1988. Changes in binomial parameters of quantal release at crustacean motor axon terminals during presynaptic inhibition. J. Physiol. (London) **402:** 177-193.
26. Wojtowicz, J. M. & H. L. Atwood. 1984. Presynaptic membrane potential and transmitter release at the crayfish neuromuscular junction. J. Neurophysiol. **52:** 99-113.
27. Dudel, J. 1981. The effect of reduced calcium on quantal unit current and release at the crayfish neuromuscular junction. Pflügers Arch. **391:** 35-40.
28. Akaike, H. 1974. A new look at the statistical model identification. IEEE Trans. Autom. Control **19:** 716-723.
29. Donoho, D. L. 1988. One-sided inference about functionals of a density. Ann. Math. Stat. **16:** 1390-1420.
30. Fletcher, R. 1970. A new approach to variable metric algorithms. Comput. J. **13:** 317-322.
31. Atwood, H. L. & L. Marin. 1983. Ultrastructure of synapses with different transmitter-releasing characteristics on motor axon terminals of a crab, *Hyas areneas.* Cell Tissue Res. **231:** 103-115.

Introduction

This section examines activity-driven structural changes in the nervous system during development and during learning in the adult. Examples are drawn from both vertebrate and invertebrate nervous systems.

New synaptic contacts form during long-term memory storage in adults. In the mammalian CNS, Greenough and Anderson report formation of new synapses on hippocampal CA1 neurons after long-term potentiation and on Purkinje cells after complex motor learning. In *Aplysia*, Bailey and Chen describe a similar increase in synapse number accompanying learning that leads to long-term memory. Sensory neuron synaptic terminals that show increased transmitter release with learning undergo an increase in the number of synaptic varicosities and active zones, and this increase persists in parallel with the long-term memory.

Altered activity during development can greatly change synaptic organization and structure. Greenough and Anderson find that rats raised in a complex environment have larger Purkinje cell dendritic fields and a greater number of synapses per neuron than those raised in a deprived environment. The effect of early activity on development is particularly dramatic in the mammalian visual system, where competition occurs between the afferents from the two eyes. After monocular deprivation in the cat, Tieman finds that deprived terminals in visual cortex are smaller and contain fewer synapses than experienced terminals. In addition, the deprived terminals contain lower levels of the putative neurotransmitter, *N*-acetylaspartylglutamate. Lnenicka reports that crayfish phasic motoneurons, which develop under very low activity levels, have smaller motor terminals containing smaller synapses and fewer mitochondria than their more experienced tonic counterparts. This difference is largely attributable to the difference in activity levels, since chronic stimulation of the phasic motoneurons in young animals produces synaptic varicosities, enlarges synapses and mitochondria, and increases capacity for transmitter release.

These papers permit comparison of the activity-driven morphological changes seen during development and during adult learning, and of the various types of neuronal activity that produce these morphological changes.

Morphological Aspects of Synaptic Plasticity in *Aplysia*

An Anatomical Substrate for Long-Term Memory[a]

CRAIG H. BAILEY[b] AND MARY CHEN

Center for Neurobiology and Behavior
Departments of Anatomy and Cell Biology and Psychiatry
College of Physicians and Surgeons
Columbia University
and
New York State Psychiatric Institute
New York, New York

INTRODUCTION

In the last few years it has become increasingly apparent that the mature central nervous system can exhibit a surprising degree of synaptic remodeling.[1] For example, in vertebrates and invertebrates behavioral training, changes in the complexity of the environment, as well as neuronal activity have been shown to alter both the number of synapses and aspects of their structure.[2] The specific role such structural changes may play during learning and memory has been a question central to the study of behavior but one that has been difficult to address in the more complex nervous systems of vertebrates where the role of individual synapses in learning and memory is not yet well defined. To bridge this gap, the more tractable central nervous systems of several higher invertebrates have proven useful for correlating changes in cellular and molecular function with learning.[3-5] We shall focus here on one such model system, the gill-and-siphon withdrawal reflex of the marine mollusc, *Aplysia californica*, to review recent morphological studies of elementary forms of learning and explore the role structural changes at identified synapses may play in the expression of a long-term memory trace.

ORGANIZATION OF THE CENTRAL NERVOUS SYSTEM SYNAPSE IN *APLYSIA*

A synaptic contact in *Aplysia* consists of a single presynaptic component, which is usually an irregularly shaped varicose expansion occurring along or at the end of a

[a] This work was supported by the National Institute of Mental Health (MH37134), the National Institutes of Health (GM32099), and the McKnight Endowment Fund for Neuroscience.

[b] Address for correspondence: Center for Neurobiology and Behavior, 722 West 168th Street, New York, New York 10032.

fine neurite, and a single postsynaptic element, which is typically a small diameter spine.[6-8] The functional architecture of these sites is similar to the synaptic morphology described in other higher invertebrates as well as in the vertebrate central nervous system.[9-14]

As is the case in other species, only a restricted portion of the contact in *Aplysia* appears to be modified for synaptic transmission.[15-17] These focal and modified sites appear analogous to the active zones described at other central synapses and also consist of a presynaptic component, which is thought to participate in the translocation of vesicles prior to their fusion and release, and a postsynaptic component, which is modified for the reception of this chemical information.

The organization of the axoplasm contiguous to the presynaptic active zone is replete with fine-diameter filaments and small electron-lucent synaptic vesicles. In fact, the number of vesicles at these specialized sites is roughly 12 times greater than at unspecialized regions of the synaptolemma. Clumps of electron-dense fibrillar material can be found adherent to the cytoplasmic leaflets of both the pre- and postsynaptic membrane. These paramembranous components of the active zone can be isolated and studied in detail by the application of a variety of selective cytochemical stains.[17,18] If such approaches are utilized, the presynaptic specialization at *Aplysia* synapses can be seen to consist of a series of small, truncated, dense projections that probably represent a condensed form of the filamentous material that is present near the active zone membrane in conventionally fixed material.[19] The postsynaptic specialization is typically more modest and consists of a thin, fairly continuous sheet-like density.

Based upon these cytochemical and ultrastructural studies, synaptic connections in *Aplysia* appear to have clearly differentiated, discrete active zones similar to those described in other species. Our next goal was to determine what effect, if any, learning and memory might have on the structure of these specialized release sites.

MORPHOLOGICAL BASIS OF LEARNING AND MEMORY IN *APLYSIA*

Toward that end, we have utilized a simple behavioral system—the gill-and-siphon withdrawal reflex of *Aplysia*. This reflex is analogous to vertebrate defensive escape and withdrawal reflexes and like them can be modified by a variety of nonassociative and associative forms of learning.[20] We will focus here on an elementary type of nonassociative learning—sensitization.

Sensitization is a process by which an animal learns about the properties of a novel, usually noxious stimulus. Through it an animal learns to strengthen its defensive reflexes and to respond vigorously to what was previously a neutral or indifferent stimulus. In both invertebrates and vertebrates, sensitization can exist in both a short-term form and a long-term form. The duration of the memory is dependent upon the number of training trials; that is, the memory is graded. In *Aplysia*, for example, a single training trial produces short-term sensitization that lasts from minutes to hours.[21] Repeated training trials can produce a long-term memory that persists for at least several weeks.[22,23]

Among the great advantages of the invertebrate brain for the cellular analysis of behavior and its modification is a greatly reduced cell number and the presence of large, invariant, identified nerve cells. In *Aplysia*, certain elementary behaviors that can be modified by learning may use fewer than 100 cells. This neural parsimony makes it possible to delineate in detail the wiring diagram of the behavior and thus pinpoint the contribution of individual nerve cells to the behavior in which they participate.

Exploiting this reductionist approach, Eric Kandel and his colleagues have analyzed the neuronal circuits responsible for the gill-and-siphon withdrawal.[20] A well-studied component of this reflex is the monosynaptic connection between mechanoreceptor sensory neurons that innervate the siphon skin and the motor neurons that supply the gill and the siphon. There are six motor cells to the gill that receive information from the siphon skin by means of 24 sensory neurons as well as several identified interneurons. Once this wiring diagram had been delineated, Kandel and his colleagues began an analysis of the cellular and molecular events that underlie sensitization. They found that the mechanisms that underlie both short- and long-term sensitization involve an enhancement in the effectiveness of the synapses made by the sensory neurons onto their follower cells. When a sensitizing stimulus is applied to the neck or tail of *Aplysia*, facilitating neurons are activated that in turn act on the sensory neurons to enhance transmitter release. This heterosynaptic facilitation involves activation of the enzyme, adenylate cyclase, resulting in an increase in the level of cyclic adenosine monophosphate (cAMP) within the sensory neurons (for a review, see reference 3). Serotonin and a small interrelated peptide can mimic the physiological effects of sensitization and are candidates for the facilitating transmitter. The increased cAMP in turn activates a second enzyme, the cAMP-dependent protein kinase, which leads to the phosphorylation of a number of proteins. One of these is a K^+ channel protein or associated protein. In this case, the addition of a phosphate group reduces one of the K^+ currents that normally repolarizes the action potential, thus allowing more Ca^{2+} to flow into the terminals and increasing transmitter release.[24–27] Serotonin can also enhance mobilization of transmitter at these sensory neuron synapses.[28] Electrophysiological experiments examining the strength of the monosynaptic connection between sensory neurons and the gill motor neuron L7 have shown that the same synaptic loci are involved in the storage of the long-term memory for sensitization.[23]

Although several aspects of the biophysical and biochemical changes that underlie sensitization are well understood, less is known about the morphological mechanisms and in particular the role that structural changes at these identified sensory neuron synapses might play in the induction and maintenance of the long-term form.

STRUCTURAL PLASTICITY AT IDENTIFIED SENSORY NEURON SYNAPSES DURING LONG-TERM MEMORY

To address these questions we have begun to examine the family of structural changes at sensory neuron synapses that may accompany long-term memory. We have combined selective intracellular labeling techniques (horseradish peroxidase (HRP)) with the analysis of serial sections to quantitatively study complete reconstructions of sensory neuron synapses in naive and behaviorally modified animals.

The results of our initial ultrastructural studies[29] examining the effects of long-term training on active zone morphology are illustrated in FIGURE 1. Here we compared the number, size, and vesicle complement of completely reconstructed sensory neuron active zones in each of three groups of animals: control, long-term habituated, and long-term sensitized. We found a consistent trend; that is, all three active zone parame-

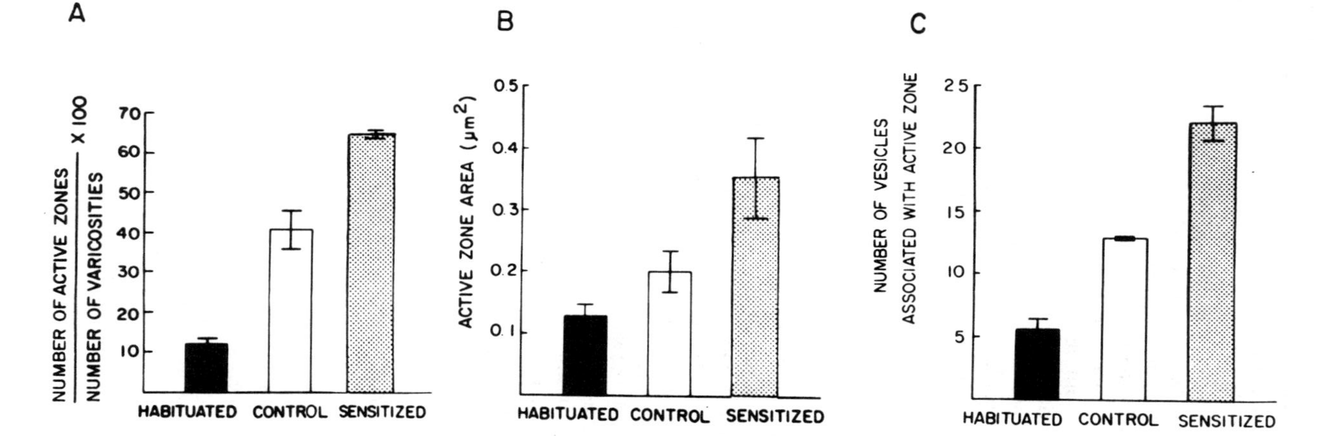

FIGURE 1. Effects of long-term memory on active zone morphology at identified sensory neuron synapses. (**A**) Incidence of active zones. (**B**) Active zone area. (**C**) Vesicle complement. Each bar represents a mean ± SEM (two animals in each behavioral group). (Data taken from reference 29; figure from reference 53.)

ters were larger in sensitized animals than in controls and smaller in habituated animals. This study indicated that clear structural changes could accompany long-term memory in *Aplysia* and demonstrated for the first time that these changes could be detected at the level of identified synapses known to be critically involved in the behavior. Moreover, these results suggested a considerable amount of synaptic remodeling could be induced by the learning process. In an attempt to determine the limits of this structural plasticity during long-term memory, we turned to a slightly different experimental approach.

LONG-TERM SENSITIZATION INVOLVES GROWTH AND AN INCREASE IN THE NUMBER OF SENSORY NEURON SYNAPSES

To examine if long-term training had any effect on the total number of synapses per sensory neuron, we quantitatively analyzed the total axonal arbor of single HRP-injected sensory neurons from control and behaviorally modified animals.[30] The results of this study are illustrated in FIGURE 2 and reveal a trend similar to that described above for active zone morphology but expressed here in a much more global fashion. Sensory neurons from control animals had on average 1300 varicose expansions per sensory neuron. This number was reduced to a mean of 850 varicosities per sensory neuron in long-term habituated animals and dramatically increased to 2700 in long-term sensitized animals. FIGURE 3 illustrates the correlation between these changes in varicosity number and each behavioral modification.

In addition to changes in the total number of presynaptic varicosities, sensory neurons from long-term sensitized animals demonstrate additional evidence of growth; that is, they display enlarged neuropil arbors (FIG. 4). Quantitative estimates indicate that the total neuropil arbor of sensory neurons from long-term sensitized animals is 2.7 times greater in linear extent than that from controls.

To determine which of these anatomical changes at sensory neuron synapses are necessary for the onset of long-term sensitization[31] and which are required for its maintenance,[32] we have begun to examine the relationship between the time course of each morphological alteration and the behavioral duration of the memory (FIG. 5). Our results suggest that not all of the structural changes persist as long as the memory. For example, the increase in the size and vesicle complement of sensory neuron active zones, which is present 24-48 hr following the completion of training, eventually is reversed, control levels being present one week later (FIG. 6). By contrast, the increase in active zone and varicosity number persists unchanged for at least one week and is only partially reversed at the end of three weeks (FIG. 7). The relative permanence of these changes and their similarity in time course to the behavioral duration of the memory suggest that an alteration in the number of sensory neuron synapses is the most likely of the structural candidates to contribute to the retention of long-term sensitization. Additional support for this notion comes from a recent study examining the effects of long-term sensitization on the structure of an identified postsynaptic target of the sensory neurons—the gill motor neuron L7.[33] Quantitative ultrastructural analysis revealed a striking increase in the frequency of presynaptic contacts onto L7 processes in sensitized compared to control animals. The results from this study are consistent with our earlier observations that long-term sensitization produces an in-

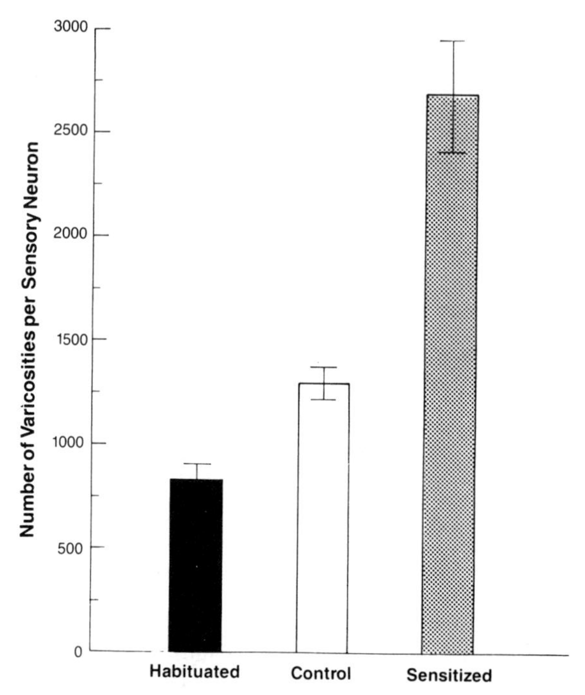

FIGURE 2. Total number of varicosities per sensory neuron in control and long-term behaviorally modified animals. Each bar represents the mean ± SEM. Control groups from both experiments were not significantly different from one another and have been combined: 1306 ± 82 varicosities, mean ± SEM, $N = 16$. The total number of varicosities per sensory neuron from each group was as follows: for long-term habituation animals, 836 ± 675 varicosities, mean ± SEM, $N = 10$ (for their controls, 1291 ± 138 varicosities, mean ± SEM, $N = 8$), and for long-term sensitized animals, 2697 ± 277 varicosities, mean ± SEM, $N = 10$ (for their controls, 1320 ± 99 varicosities, mean ± SEM, $N = 8$). The mean number of varicosities for long-term habituated animals is significantly less than the mean for their controls ($t = 3.05$; $p < .01$), and the mean for long-term sensitized animals is significantly larger than the mean for their controls ($t = 4.25$; $p < .01$). (From reference 30.)

crease in the number of synapses that sensory neurons make onto their target cells and strengthen the argument that synapse number changes may contribute to the maintenance of a long-term memory.

AN OVERALL VIEW

How information is stored in the brain is a question central to both neurobiology and psychology. In the last few years, studies on several higher invertebrates have enhanced our knowledge about the specific synaptic loci and mechanisms that are involved in the acquisition and retention of various elementary forms of learning and memory.[3-5] One such model system, the gill-and-siphon withdrawal reflex in *Aplysia,* has proven particularly advantageous for examining the cellular and molecular events that underlie both short- and long-term memory.[20] We have exploited the cellular specificity of this reflex and have begun to examine the functional relationship between synaptic structure and the prolonged changes in synaptic effectiveness that accompany long-term behavioral modifications.

Toward that end we have combined a variety of morphological techniques with biophysical and behavioral approaches to determine the nature, extent, and time course of structural changes at identified sensory neuron synapses during long-term memory. Because aspects of the memory for sensitization can be trapped at this specific synaptic locus, we have been able to analyze, in a set of identified neurons that are causally related to the behavior, the mechanisms underlying the acquisition and retention of a long-term memory trace. Central issues include the following: Can a long-term memory trace be specified in morphological terms? Can morphological approaches be used directly as probes for examining the mechanisms that underlie learning and memory? What is the time course of the structural synaptic changes? Which are necessary for the onset of the memory and which are required for its maintenance?

Our results indicate that learning in *Aplysia* produces morphological as well as functional changes at specific synaptic loci. Long-term memory (lasting several weeks) is accompanied by a profound degree of structural plasticity at the level of identified sensory neuron synapses. These structural changes occur at two different levels of synaptic organization: 1) remodeling of sensory neuron active zones—which for a long-term sensitization is reflected by an increase in the number, size, and vesicle complement of these release sites—and 2) a parallel but even more widespread change involving modulation of the total number of presynaptic varicosities per sensory neuron

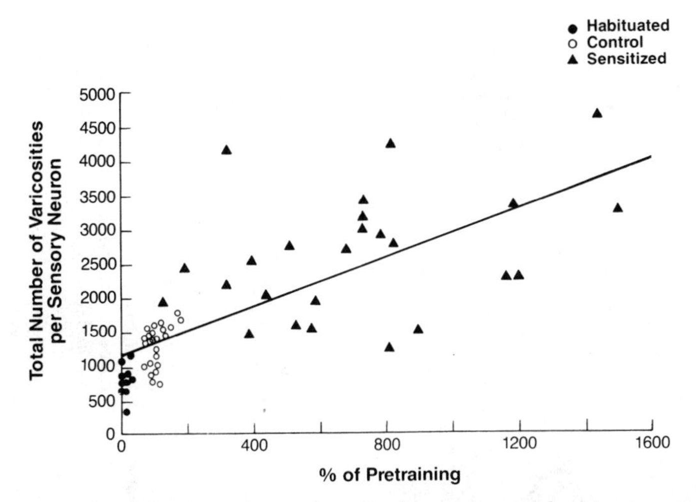

FIGURE 3. Correlation between changes in varicosity number and long-term memory. Data points represent animals trained for long-term habituation, animals trained for long-term sensitization, and untrained (control) animals. Structural changes (total number of varicosities per sensory neuron) are plotted against changes in behavioral efficacy. The responses of each 10-trial session, 24-48 hr following the completion of training, have been summed and expressed as a single median score compared to each animal's own pretraining score (Spearman's rank correlative coefficient: 0.825; $p < .001$). (From reference 32.).

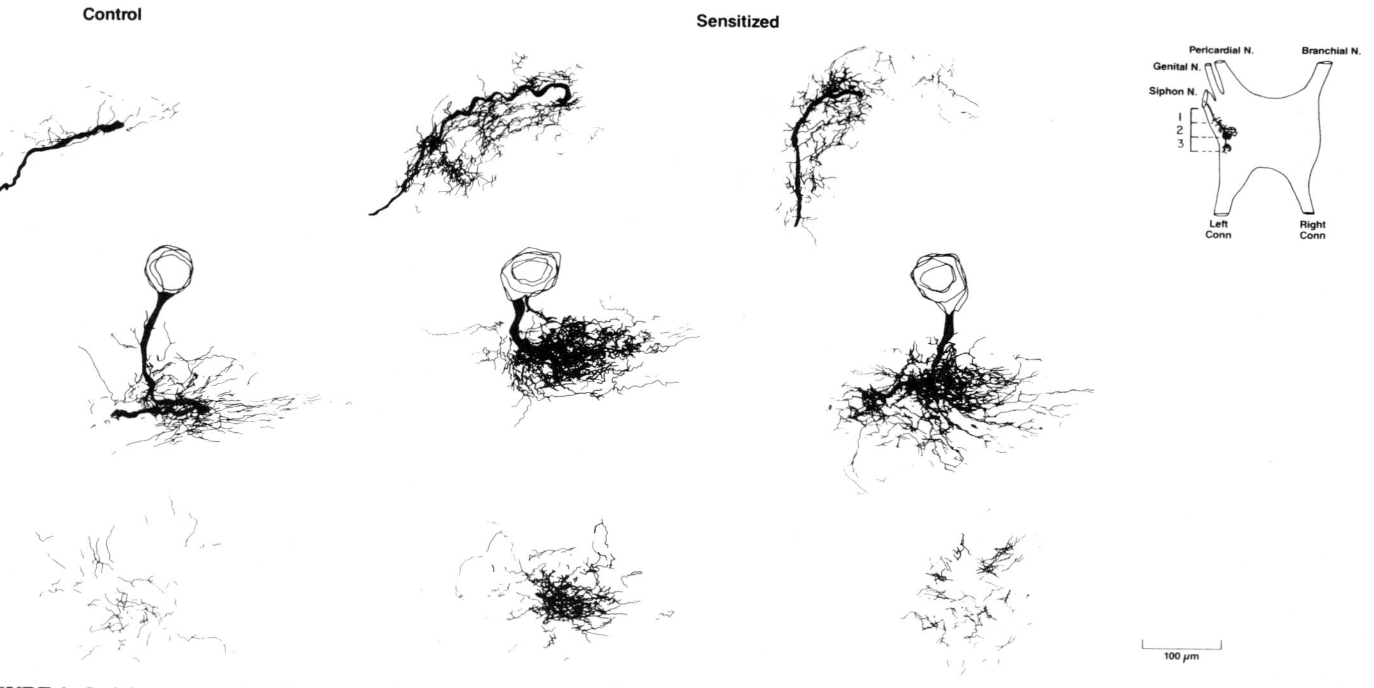

FIGURE 4. Serial reconstruction of sensory neurons from long-term sensitized and control animals. Total extent of the neuropil arbors of sensory neurons from one control animal and two long-term sensitized animals are shown. In each case the rostral (row 3) to caudal (row 1) extent of the arbor is divided roughly into thirds. Each panel was produced by the superimposition of camera lucida tracings of all HRP-labeled processes present in 17 consecutive slab-thick sections and represents a linear segment through the ganglion of roughly 340 μm. For each composite, ventral is up, dorsal is down, lateral is to the left, and medial is to the right. By examining images across each row (rows 1, 2, and 3), the viewer is comparing similar regions of each sensory neuron. In all cases, the arbor of long-term sensitized cells is markedly increased compared to control. Quantitative estimates indicate that the total neuropil arbor of sensory neurons from long-term sensitized animals is 2.7 times greater in extent (22,254 $\pm$ 2415 μm, mean $\pm$ SEM, $N = 6$) than that from control animals (8415 $\pm$ 1640 μm, mean $\pm$ SEM, $N = 3$) ($t = 3.746$; $p < .01$, two-tailed test). Siphon N., Genital N., Pericardial N., and Branchial N. are various peripheral nerves of the abdominal ganglion; Left and Right Conn are left and right connectives, fiber tracts connecting the abdominal ganglion with other ganglia. (From reference 30.)

(FIG. 8). Quantitative analysis of the time course over which these anatomical changes occur has further demonstrated that only alterations in the number of sensory neuron synapses persist in parallel with the behavioral retention of the memory. These results directly link a change in synaptic structure to long-lasting behavioral memory and provide evidence for a fundamental notion—that varicosities and their active zones are plastic rather than immutable components of the nervous system. Such morphological changes could represent an anatomical substrate for memory consolidation.

The nature and extent of the structural alterations at mechanoreceptor sensory neurons are consistent with changes in the known behavioral effectiveness of habitua-

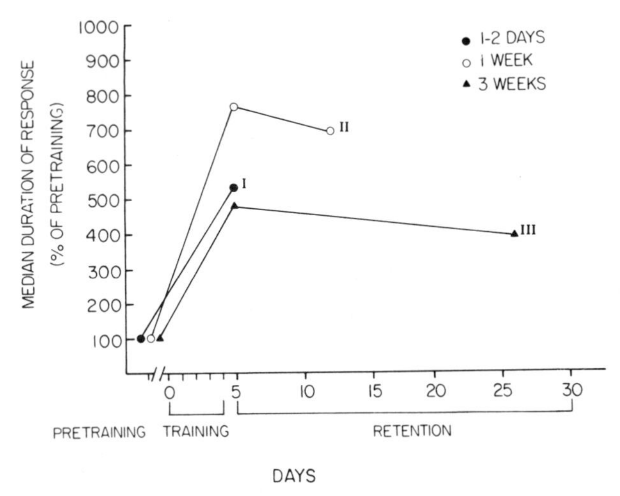

FIGURE 5. Behavioral time course of long-term sensitization. Retention of sensitization was tested at different intervals following the completion of 4 days of long-term training. Behavioral performance of animals in each group was estimated by comparing their behavioral scores with their pretraining scores. This was achieved by summing the responses for each daily session (10-trial block; intertrial interval: 30 sec) and expressing each as a single score compared with their own pretraining score. Animals in behavioral experiments were killed at 1-2 days (I: 529 ± 121, $N = 6$), 1 week (II: 700 ± 173, $N = 4$), and 3 weeks (III: 414 ± 86, $N = 5$) following training, in preparation for the analysis of the structure of sensory neuron synapses in each group. (From reference 32.)

tion and sensitization[22,34] as well as with the demonstration of an enduring alteration in the amplitude of the sensory-to-follower cell connection following long-term training.[23,35] These findings suggest the interesting possibility that learning may modulate the structure of the synapse to alter both the strength of synaptic connections as well as behavioral efficacy. The alterations exhibited by sensory neuron release sites following learning and memory are consistent in many respects with the changes in active zone structure reported at the neuromuscular junction following various forms of stimulation (see references 36-40; for a review, see references 41 and 42). Moreover, the changes in synapse number following long-term sensitization in *Aplysia* are remarkably similar to the reports in the vertebrate CNS of alterations in the number and/or pattern of

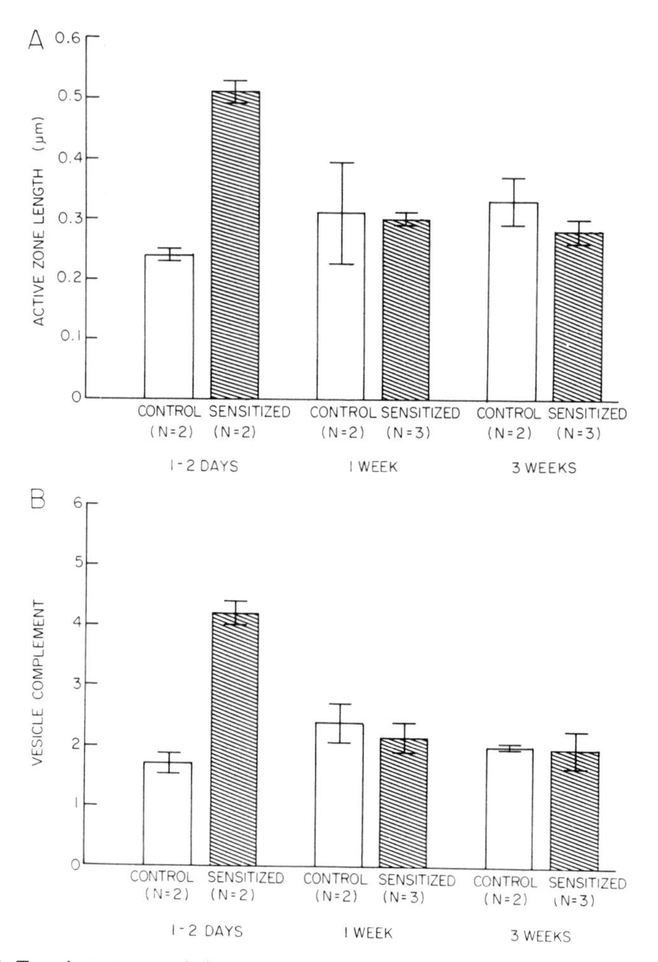

FIGURE 6. Transient structural changes at sensory neuron synapses during long-term sensitization. (**A**) Duration of changes in active zone size. (**B**) Duration of changes in the vesicle complement of active zones. Unlike the changes in varicosity and active zone number, the increases in active zone size and associated vesicles peak at 1-2 days following the completion of training and then return to control levels at 1 week and 3 weeks, making them unlikely candidates to contribute to the maintenance of long-term sensitization. Each bar represents the mean ± SEM. (From reference 32.)

synaptic connections following experimental manipulation and training (for a review, see references 43 and 44).

What is the role of this structural plasticity during learning and memory? One clue comes from the results of studies on both vertebrates and invertebrates which indicate that long-term memory depends upon the synthesis of proteins and RNA.[45–47] In *Aplysia* these studies indicate that long-term sensitization (lasting days to weeks) requires gene products that are not required for short-term sensitization (lasting minutes to hours).

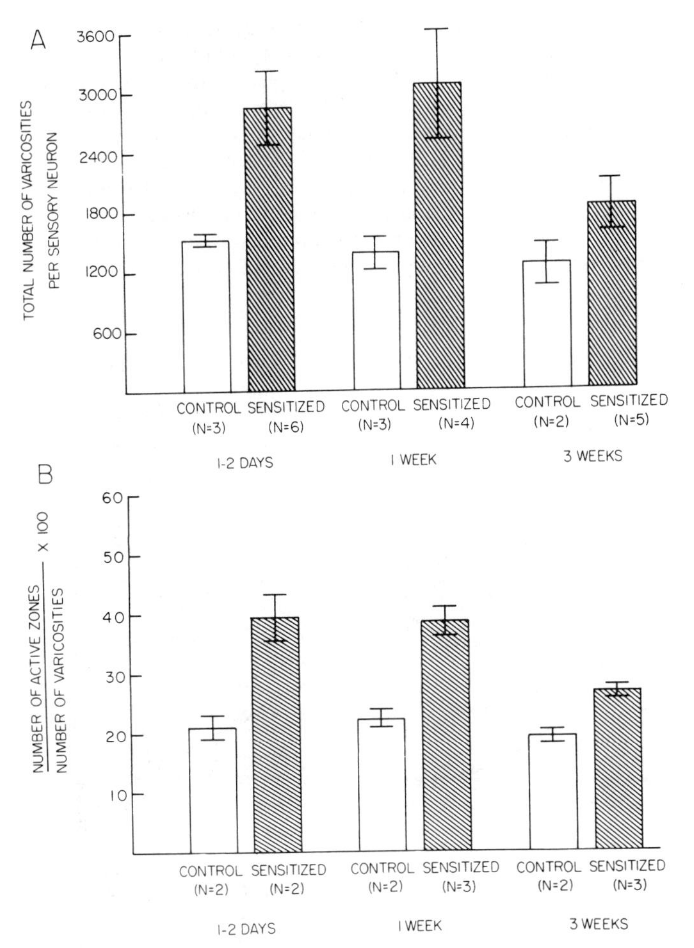

FIGURE 7. Enduring structural changes at sensory neuron synapses during long-term sensitization. **(A)** Duration of changes in the total number of varicosities per sensory neuron. **(B)** Duration of changes in the incidence of sensory neuron active zones. To determine which class of alterations in the functional architecture of identified sensory neuron synapses might underlie long-term sensitization, we have compared their duration to the persistence of the memory. By examining the structure of sensory neuron synapses at different intervals following the completion of training—1-2 days, 1 week, and 3 weeks—we have found that only the duration of changes in the number of varicosities and their active zones parallels the behavioral time course for the retention of long-term sensitization. Each bar represents the mean ± SEM. (From reference 32.)

From a molecular perspective this means that the gene products required for short-term memory are either preexisting or else turned over slowly whereas the gene products required for long-term memory must be newly synthesized.

In examining the possible role of these newly synthesized proteins, one is forced to consider the evidence that associates structural changes in synapses with behavioral training. In fact, the ample evidence that long-term memory involves a process of cell growth provides a rationale for thinking about the types of intracellular changes that protein synthesis might bring about.[48] For example, in *Aplysia* at least some of the newly synthesized proteins induced during long-term sensitization training must lead not only to functional changes but must contribute as well to the structural changes—growth of new synaptic contacts—that occur in the sensory neurons.

Barzilai, Kennedy, Sweatt, and Kandel[49] have begun to determine the specific gene products that underlie the acquisition of long-term sensitization by studying the incorporation of ^{35}S-labeled methionine into proteins in the sensory neurons. They found that long-term training initiates a large increase in overall protein synthesis. In

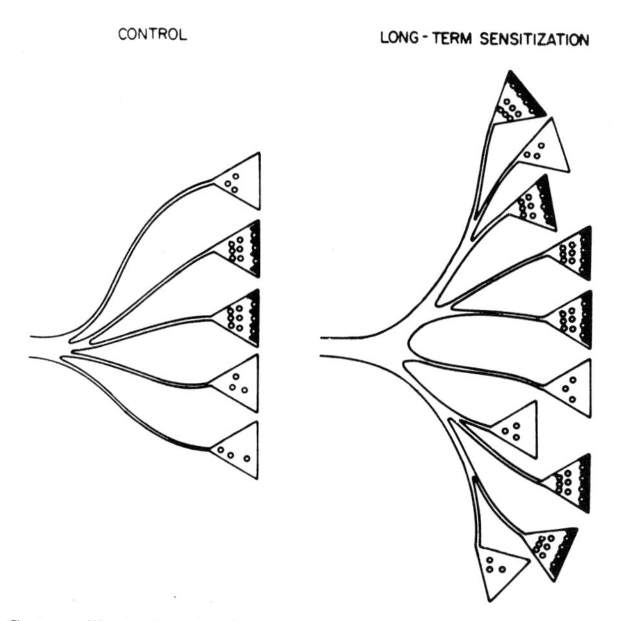

FIGURE 8. Cartoon illustrating the family of structural alterations produced by long-term sensitization. These involve a doubling in the number of sensory neuron varicosities, an increase in the incidence of their active zones, and an increase in the neuropil arbor of the sensory neurons. (Modified from reference 30.)

addition, they found that long-term training produced rapid and transient alterations in gene expression suggesting the possibility of a gene cascade, whereby early proteins that are activated during learning might be regulators for the expression of late effector genes.[48] The structural changes at sensory neuron synapses are likely to require the increased expression of a substantial number of such late effector proteins.

Some additional studies on *Aplysia* into the macromolecular candidates that may underlie these structural changes have begun to clarify the nature of the cytoplasmic signals involved in the flow of information from membrane receptors to the genome during learning.[50] Studies by Glanzman, Kandel, and Schacher, who used co-cultured sensory neurons and the gill motor neuron L7,[51] have established that bath application of serotonin, in accordance with a paradigm that produces long-term facilitation in the dish, also produces an increase in the number of sensory neuron varicosities. These results, combined with the molecular studies by Kandel and his colleagues described previously, suggest the intriguing notion that modulatory transmitters like serotonin may also function as growth factors exerting additional control over the gene expression required for neuronal growth during learning.[48] In addition to the effects of repeated or prolonged application of serotonin on sensory neuron varicosities, the work of Glanzman *et al.*[51] suggest a role for the postsynaptic neuron in these structural changes; that is, only sensory neuron branches in contact with L7's main neurite displayed a varicosity increase. Results from the work of Nazif, Byrne, and Cleary[52] have indicated a second mechanistic candidate for the structural changes in sensory neurons seen during long-term sensitization. They have shown that the intracellular injection of cAMP into pleural sensory neurons can mimic the increase in varicosity number and enlarged neuropil arbor that we have described following behavioral sensitization. These studies further define the family of cytoplasmic signals that are involved in the structural changes during learning and clearly implicate a role for the cAMP cascade.

The precise relationship between long-term memory and structural plasticity at the synapse will ultimately depend upon the increased convergence of behavioral, molecular, and morphological approaches. The striking similarities in the response of both vertebrate and invertebrate central neurons to behavioral training indicate that learning produces changes in neuronal architecture across a broad segment of the animal kingdom and suggest synapse formation as a highly conserved mechanism accompanying long-term memory.

SUMMARY

The morphological basis of long-term sensitization of the gill-and-siphon withdrawal reflex in *Aplysia* was explored by examining the structure of identified sensory neuron synapses in control and behaviorally modified animals. Following long-term training, sensitized animals displayed an increase in the number of sensory neuron synapses compared to control animals. The relative permanence of these structural changes and their similarity in time course to the behavioral duration of sensitization suggest a role for synapse number changes during long-term memory.

REFERENCES

1. Purves, D. & J. T. Voyvodic. 1987. Imaging mammalian nerve cells and their connections over time in living animals. Trends Neurosci. **10:** 398-404.
2. Greenough, W. T. & C. H. Bailey. 1988. The anatomy of memory: Convergence of results across a diversity of tests. Trends Neurosci. **11:** 142-147.
3. Kandel, E. R. & J. H. Schwartz. 1982. Molecular biology of an elementary form of learning: Modulation of transmitter release by cyclic AMP. Science **218:** 433-443.
4. Alkon, D. L. 1984. Calcium-mediated reduction of ionic currents: A biophysical memory trace. Science **226:** 1037-1045.
5. Byrne, J. H. 1987. Cellular analysis of associative learning. Physiol. Rev. **67:** 329-439.
6. Coggeshall, R. E. 1967. A light and electron microscope study of the abdominal ganglion of *Aplysia californica*. J. Neurophysiol. **30:** 1263-1287.

7. Bailey, C. H., E. B. Thompson, V. F. Castellucci & E. R. Kandel. 1979. Ultrastructure of the synapses of sensory neurons that mediate the gill-withdrawal reflex in *Aplysia*. J. Neurocytol. **8:** 415-444.
8. Tremblay, J. P., M. Colonnier & H. McLennan. 1979. An electron microscopic study of synaptic contacts in the abdominal ganglion of *Aplysia californica*. J. Comp. Neurol. **188:** 367-396.
9. Palay, S. L. 1958. The morphology of synapses in the central nervous system. Exp. Cell Res. **5:** 275-293.
10. Gray, E. G. 1959. Axosomatic and axodendritic synapses of the cerebral cortex: An electron microscopic study. J. Anat. **83:** 420-433.
11. Muller, K. J. & U. J. McMahan. 1976. The shapes of sensory and motor neurons and the distribution of their synapses in ganglia of the leech: A study using intracellular injection of horseradish peroxidase. Proc. R. Soc. London, Ser. B **194:** 481-499.
12. Peters, A., S. L. Palay & H. deF. Webster. 1976. The Fine Structure of the Nervous System: The Neurons and Supporting Cells. Saunders. Philadelphia, PA.
13. Wood, M. R., K. H. Pfenninger & M. J. Cohen. 1977. Two types of presynaptic configurations in insect central synapses: An ultrastructural analysis. Brain Res. **130:** 25-45.
14. Vrensen, G. & J. Nunes Cardozo. 1981. Changes in size and shape of synaptic connections after visual training: An ultrastructural approach of synaptic plasticity. Brain Res. **218:** 79-97.
15. Couteaux, R. & M. Pecot-Dechavassine. 1970. Vesicules synaptiques et poches au niveau des zones actives de la jonction neuromusculaire. C. R. Acad. Sci. **271:** 2346-2349.
16. Heuser, J. E., T. S. Reese, M. J. Dennis, Y. Jan, L. Jan & L. Evans. 1979. Synaptic vesicle exocytosis captured by quick freezing and correlated with quantal transmitter release. J. Cell Biol. **81:** 275-300.
17. Bailey, C. H., P. Kandel & M. Chen. 1981. The active zone at *Aplysia* synapses: Organization of presynaptic dense projection. J. Neurophysiol. **46:** 356-368.
18. Bloom, F. E. & G. K. Aghajanian. 1978. Fine structural and cytochemical analysis of the staining of synaptic junctions with phosphotungstic acid. J. Ultrastruct. Res. **22:** 361-375.
19. Landis, D. M. D., A. K. Hall, L. A. Weinstein & T. S. Reese. 1988. The organization of cytoplasm at the presynaptic active zone of a central nervous system synapse. Neuron **1:** 201-209.
20. Kandel, E. R. 1976. Cellular Basis of Behavior: An Introduction to Behavioral Neurobiology. W. H. Freeman. San Francisco, CA.
21. Pinsker, H. M., I. Kupfermann, V. F. Castellucci & E. R. Kandel. 1970. Habituation and dishabituation of the gill-withdrawal reflex in *Aplysia*. Science **167:** 1740-1742.
22. Pinsker, H. M., W. A. Hening, T. J. Carew & E. R. Kandel. 1973. Long-term sensitization of a defensive withdrawal reflex in *Aplysia*. Science **182:** 1039-1042.
23. Frost, W. N., V. F. Castellucci, R. D. Hawkins & E. R. Kandel. 1985. Monosynaptic connections made by the sensory neurons of the gill- and siphon-withdrawal reflex in *Aplysia* participate in the storage of long-term memory for sensitization. Proc. Natl. Acad. Sci. USA **82:** 8266-8269.
24. Klein, M. & E. R. Kandel. 1978. Presynaptic modulation of voltage-dependent Ca^{2+} current: Mechanism for behavioral sensitization in *Aplysia californica*. Proc. Natl. Acad. Sci. USA **75**(7): 3512-3516.
25. Klein, M. & E. R. Kandel. 1978. Mechanism of calcium current modulation underlying presynaptic facilitation and behavioral sensitization in *Aplysia californica*. Proc. Natl. Acad. Sci. USA **77:** 6912-6916.
26. Castellucci, V. F., E. R. Kandel, J. H. Schwartz, F. D. Wilson, A. C. Nairn & P. Greengard. 1980. Intracellular injection of the catalytic subunit of cyclic AMP-dependent protein kinase simulates facilitation of transmitter release underlying behavioral sensitization in *Aplysia*. Proc. Natl. Acad. Sci. USA **77:** 7492-7496.
27. Siegelbaum, S. A., J. S. Camardo & E. R. Kandel. 1982. Serotonin and cyclic AMP close single K^+ channels in *Aplysia* sensory neurons. Nature **299:** 413-417.
28. Hochner, B., M. Klein, S. Schacher & E. R. Kandel. 1986. Additional components in the cellular mechanism of presynaptic facilitation contributes to behavioral dishabituation in *Aplysia*. Proc. Natl. Acad. Sci. USA **83:** 8794-8798.

29. Bailey, C. H. & M. Chen. 1983. Morphological basis of long-term habituation and sensitization in *Aplysia*. Science **220:** 91-93.
30. Bailey, C. H. & M. Chen. 1988. Long-term memory in *Aplysia* modulates the total number of varicosities of single identified sensory neurons. Proc. Natl. Acad. Sci. USA **85:** 2373-2377.
31. Bailey, C. H. & M. Chen. 1989. Onset of structural changes at identified sensory neuron synapses and the acquisition of long-term sensitization in *Aplysia*. Soc. Neurosci. Abstr. **15:** 1285.
32. Bailey, C. H. & M. Chen. 1989. Time course of structural changes at identified sensory neuron synapses during long-term sensitization in *Aplysia*. J. Neurosci. **9**(5): 1774-1780.
33. Bailey, C. H. & M. Chen. 1988. Long-term sensitization in *Aplysia* increases the number of presynaptic contacts onto the identified gill motor neuron L7. Proc. Natl. Acad. Sci. USA **85:** 9356-9359.
34. Carew, T. J., H. M. Pinsker & E. R. Kandel. 1972. Long-term habituation of a defensive withdrawal reflex in *Aplysia*. Science **175:** 451-454.
35. Castellucci, V. F., T. J. Carew & E. R. Kandel. 1978. Cellular analysis of long-term habituation of the gill-withdrawal reflex of *Aplysia californica*. Science **202:** 1306-1308.
36. Atwood, H. L. & L. Marin. 1983. Ultrastructure of synapses with different transmitter-releasing characteristics on motor axon terminals of a crab, *Hyas areneas*. Cell Tissue Res. **231:** 103-115.
37. Herrera, A. A., A. P. Grinnell & B. Wolowske. 1985. Ultrastructural correlates of naturally occurring differences in transmitter release efficacy in frog motor nerve terminals. J. Neurocytol. **14:** 193-202.
38. Rheuben, M. D. 1985. Quantitative comparison of the structural features of slow and fast neuromuscular junctions in *Manduca*. J. Neurosci. **5:** 1704-1716.
39. Walrond, J. P. & T. S. Reese. 1985. Structure of axon terminals and active zones at synapses on lizard twitch and tonic muscle fibers. J. Neurosci. **5:** 1118-1131.
40. Chiang, R. G. & C. K. Govind. 1986. Reorganization of synaptic ultrastructure at facilitated lobster neuromuscular terminals. J. Neurocytol. **15:** 63-74.
41. Atwood, H. L. & J. M. Wojtowicz. 1986. Short-term and long-term plasticity and physiological differentiation of crustacean motor systems. Int. Rev. Neurobiol. **28:** 275-362.
42. Atwood, H. L. & G. A. Lnenicka. 1986. Structure and function in synapses: Emerging correlations. Trends Neurosci. **9:** 248-250.
43. Greenough, W. T. 1984. Structural correlates of information storage in the mammalian brain: A review and hypothesis. Trends Neurosci. **7:** 229-233.
44. Greenough, W. T. & F.-L. F. Chang. 1985. Synaptic structural correlates of information storage in mammalian nervous systems. *In* Synaptic Plasticity. C. W. Cotman, Ed.: 335-372. Guilford Press. New York, NY.
45. Agranoff, B. W., R. E. Davies & J. J. Brink. 1967. Actinomycin D blocks formation of memory of shock avoidance in goldfish. Science **158:** 1600-1601.
46. Davis, H. P. & L. R. Squire. 1984. Protein synthesis and memory: A review. Psychol. Bull. **96:** 518-559.
47. Montarolo, P. G., P. Goelet, V. F. Castellucci, J. Morgan, E. R. Kandel & S. Schacher. 1986. A critical period of macromolecular synthesis in long-term heterosynaptic facilitation in *Aplysia*. Science **234:** 1249-1254.
48. Kandel, E. R. 1989. Genes, nerve cells and the remembrance of things past. J. Neuropsychiatry **1:** 103-125.
49. Barzilai, A., T. E. Kennedy, J. D. Sweatt & E. R. Kandel. 1989. 5-HT modulates protein synthesis and the expression of specific proteins during long-term facilitation in *Aplysia* sensory neurons. Neuron **2:** 1577-1586.
50. Goelet, P. & E. R. Kandel. 1986. Tracking the flow of learned information from membrane receptors to genome. Nature **9:** 472-499.
51. Glanzman, D. L., E. K. Kandel & S. Schacher. 1990. Target-dependent structural changes accompanying long-term synaptic facilitation in *Aplysia* neurons. Science **249:** 799-802.

52. NAZIF, F., J. H. BYRNE & L. J. CLEARY. 1989. Intracellular injection of cAMP produces a long-term (24 hr) increase in the number of varicosities in pleural sensory neurons of *Aplysia*. Soc. Neurosci. Abstr. **15:** 1283.
53. BAILEY, C. H. & E. R. KANDEL. 1985. Molecular approaches to the study of short- and long-term memory. *In* Functions of the Brain. C. W. Coen, Ed.: 98-129. Clarendon Press. New York, NY.

The Role of Activity in the Development of Phasic and Tonic Synaptic Terminals[a]

GREGORY A. LNENICKA

Neurobiology Research Center
Department of Biological Sciences
State University of New York
Albany, New York 12222

INTRODUCTION

It is well established that changes in activity during development can result in long-term, if not permanent, changes in the physiology and morphology of synaptic connections. In the vertebrate CNS, dramatic changes in synapses have been reported to result from altered early experience. These include changes in synaptic morphology and physiology in specific neural pathways resulting from well-defined changes in sensory input. For example, stroboscopic illumination or dark rearing results in the abnormal formation of retinotectal connections in the goldfish,[1] and monocular deprivation produces abnormal synapses in the cat visual cortex,[2] whereas changes in synapses of the cerebral cortex result from rearing rats in "enriched" environments.[3]

One problem that has been of interest is whether activity plays a role in the development of the simpler, invertebrate nervous system.[4] This interest stems from both the technical advantages of using invertebrates in this area of study and questions regarding the generality of the role of activity in the development of nervous systems. The crustacean neuromuscular system offers a group of well-characterized, identified synapses with diverse properties. The development of a variety of well-defined synaptic properties can be studied at these relatively accessible synapses. In this paper, I will discuss the differences in the synaptic properties of phasically and tonically active crustacean neuromuscular synapses, and present evidence for the role of activity in the development of these synaptic properties. These results will be compared with findings in other invertebrate nervous systems, as well as the vertebrate nervous system, and will provide evidence for the generality of the contributions of activity to neural development, across a variety of species.

DIFFERENCES BETWEEN TONIC AND PHASIC SYNAPSES

Physiology

The system used in these studies includes crustacean phasic and tonic motoneurons where neuromuscular synaptic properties are matched to motoneuron impulse activity

[a] This work was supported by a grant from the National Science Foundation (BNS-8720135).

levels. Phasic motoneurons in crustaceans are characterized by low impulse activity levels and neuromuscular synapses that release large amounts of transmitter during initial and low-frequency stimulation. These neuromuscular synapses show fatigue during repetitive stimulation (FIG. 1).

Tonic motoneurons are active for prolonged periods, often at high frequencies. Although they release only small amounts of transmitter during initial activation, transmitter release generally increases during repetitive activation because of synaptic facilitation, and is maintained during prolonged high-frequency stimulation (FIG. 1). Even though phasic terminals release more transmitter during initial or low-frequency stimulation, in this case low-frequency stimulation is not a good measure of "synaptic strength," as higher frequency stimulation shows that the tonic terminals have the capacity to release much more transmitter than phasic terminals.

Vertebrate phasic and tonic motoneurons show similar, though less dramatic, differences in impulse activity levels[5] and neuromuscular synaptic properties.[6] Although data on the properties of phasically and tonically active central synapses are incomplete, they appear to show similar properties. For example, the properties of some of the central synapses in the phasically active circuits underlying defensive reflexes have been studied. Many of these synapses when repeatedly activated show synaptic depression resulting in behavioral habituation.[7,8]

It has been proposed that these differences in initial transmitter release for crustacean phasic and tonic motor terminals are due to differences in terminal excitability. Dudel *et al.*[9] have shown that most of the terminals of the phasically active fast extensor motoneurons of the spiny lobster are electrically excitable, whereas the terminals of the tonically active opener excitor are electrically unexcitable.[10,11] Greater invasion of the action potential into the terminals will result in a large depolarization and calcium influx at the synapses. Differences in motor terminal excitability may also exist in

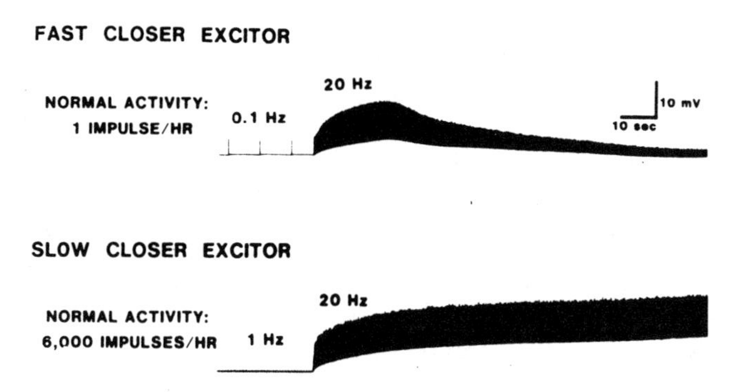

FIGURE 1. Physiology of phasic and tonic synapses. The crayfish claw closer muscle is innervated by the phasically active Fast Closer Excitor (FCE), and the tonically active Slow Closer Excitor (SCE).[66] *In vivo* recording has shown that the phasic axon fires approximately 1 impulse/hr while the tonic axon fires 6000 impulses/hr. Excitatory postsynaptic potentials were recorded from muscle fibers on the dorsal surface of the closer muscle during stimulation of the phasic or tonic axon. The phasic axon (FCE) produces large EPSPs at low frequencies of stimulation (0.1 Hz), which depress rapidly during high-frequency stimulation (20 Hz). The tonic axon (SCE) produces small EPSPs (<0.5 mV) at low stimulation frequencies (1 Hz); however, during high-frequency stimulation facilitation occurs, and no synaptic depression is observed. (Reproduced with permission from reference 67.)

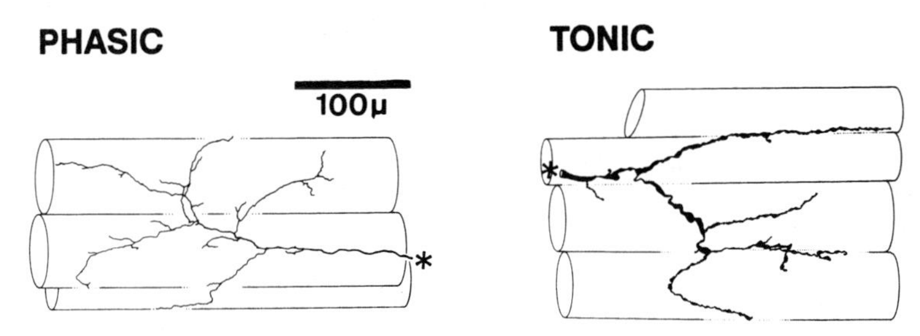

FIGURE 2. Morphology of phasic and tonic motor terminals in the crayfish claw closer muscle. Representative camera lucida drawings of tonic and phasic terminal arbors after injection of HRP. Differences in the shape of the phasic and tonic terminals are apparent, with the tonic terminals having many more synaptic varicosities. Approximate muscle fiber borders are represented schematically. The proximal end of the arbor is indicated by an asterisk. (After reference 42.)

vertebrates,[12,13] but it is not known whether excitability differs in phasic and tonic systems. Morphological differences that may contribute to differences in initial transmitter release are discussed below.

Morphology

Observations of phasic and tonic motor terminals in the crayfish reveal obvious differences in shape (FIG. 2). The greater transmitter releasing capability of the tonic terminals is correlated with a more robust morphology. The tonic terminals are varicose, most terminals having a series of well-differentiated synaptic varicosities and bottlenecks. In contrast, the phasic motor terminals are generally thinner in diameter and lack the large synaptic varicosities found in the tonic terminals. The large synaptic varicosities found along tonic terminals contain large synapses, and a high mitochondrial and synaptic vesicle content, as indicated by studies in crayfish,[14] lobster,[15] crab,[16,17] locust,[18] and moth.[19] The larger population of synaptic vesicles and large mitochondria found in the tonic synaptic varicosities may be responsible for their ability to release large amounts of transmitter for prolonged periods.

In vertebrates, tonic motor terminals are generally thicker than phasic motor terminals.[20–22] Although there is currently no study comparing mitochondria in vertebrate phasic and tonic motor terminals, greater mitochondrial content in the cell bodies of tonic motoneurons, compared to phasic, has been demonstrated.[23]

Differences in the size and structure of active zones have been reported for phasic and tonic motor terminals and appear to be related to the differences in initial transmitter release. Active zone area per length of terminal is greater in moth phasic terminals than in the tonic terminals.[19] In crustacea, greater active zone size has been reported for terminals that have greater initial transmitter release.[24] However, even where differences in active zone size exist, they may be insufficient to totally account for differences in transmitter release.[25]

Differences in the density or number of calcium channels at the active zone may account for these differences in initial transmitter release.[26] Freeze-fracture techniques reveal large intramembranous particles in the presynaptic membrane at the active zone.

These particles are thought to be calcium channels based upon their location and the fact that they are present in appropriate numbers.[27] A greater density of these putative calcium channels has been found at active zones of lizard phasic motor terminals than at tonic motor terminals.[28] In addition, moth phasic terminals have a greater number of active zone particles than tonic motoneurons.[19] Intramembranous particles have also been examined at the active zones of crustacean terminals[29] but have not been compared at phasic and tonic synapses.

ROLE OF ACTIVITY IN THE DIFFERENTIATION OF PHASIC AND TONIC SYNAPTIC TERMINALS

Physiology

Experiments performed on juvenile crayfish showed that *in vivo* electrical stimulation of the phasic axon to the crayfish claw closer muscle at 5 Hz for 2 hr/day for 14 days transformed its neuromuscular synaptic properties so that they became more similar those of a tonic motoneuron.[30] After conditioning, there was a 44% decrease in initial excitatory postsynaptic potential (EPSP) amplitude and an 11-fold decrease in synaptic fatigue measured during 30 min of 5-Hz stimulation (FIG. 3). These changes in EPSP amplitude during initial and repetitive stimulation are due to changes in transmitter release[30,31] and have been demonstrated in other crustacean phasic motoneurons.[31,32] This transformation, termed long-term adaptation,[30] is long-lasting; little or no decrease in the effect of chronic stimulation was observed when animals were rested for 2 weeks after 2 weeks of stimulation,[30] or 1 week after 3 days of stimulation.[33] A decrease in motoneuron electrical activity produces synaptic changes consistent with the synaptic changes produced by chronic stimulation. Claw immobilization in juvenile crayfish decreases the electrical activity of the phasic axon, and induces more phasic-like synaptic properties: an increase in the initial EPSP amplitude and greater fatiguability of the neuromuscular synapses.[34]

There is evidence that activity may also be playing a role in the differentiation of vertebrate phasic and tonic neuromuscular synapses. In the adult, immobilization of a tonically active vertebrate muscle, which produces a more phasic firing pattern in the motoneurons supplying the muscle, results in a greater initial release of transmitter from the neuromuscular synapses and greater synaptic fatigue.[35] Thus, the synaptic phenotype appears to be transformed to a more phasic type. In addition, recent evidence suggests that *in vivo* tonic stimulation of frog motor axons transforms the neuromuscular synapses to a more tonic type.[36]

Vertebrate and invertebrate central synapses appear to undergo similar activity-dependent changes. Increased activation of a cricket auditory sensory pathway during postembryonic development results in decreased initial responsiveness of the pathway and decreased habituation in the adult.[37] Visual deprivation during locust development results in increased habituation within a visual pathway.[38] In both cases, the changes in the response properties of the pathway appear to be due to changes in central synapses. A chronic decrease in impulse activity produces similar effects at mammalian autonomic and central synapses. A chronic block of impulse activity at synapses in an autonomic ganglion[39] or in the spinal cord[40,41] produces an increase in initial EPSP amplitude. It is not known whether synaptic fatigue increased, because these measurements were not performed.

Morphology

The role of impulse activity in the morphological differentiation of crayfish phasic and tonic motor terminals was examined by conditioning the phasic axon in a manner previously shown to be sufficient to transform the physiology of the synaptic terminals to a more tonic type. Conditioning of the phasic terminals resulted in a significant increase in the number of synaptic varicosities seen along the terminals (FIGS. 4 & 5). Tonic conditioning resulted in approximately a 3-fold increase in synaptic varicosity frequency, making the phasic terminals more similar to that of tonic terminals.[42] Serial-section electron microscopy has shown that these newly formed varicosities contain enlarged mitochondria and synapses.[14] These changes transform the synaptic terminal phenotype to one more similar to a tonic motoneuron, and may be responsible for the observed physiological transformation, particularly the increased fatigue resistance.

The evidence indicates that the synaptic varicosities form uniformly along preexisting terminals.[42] Because there is no evidence for a change in terminal length, there

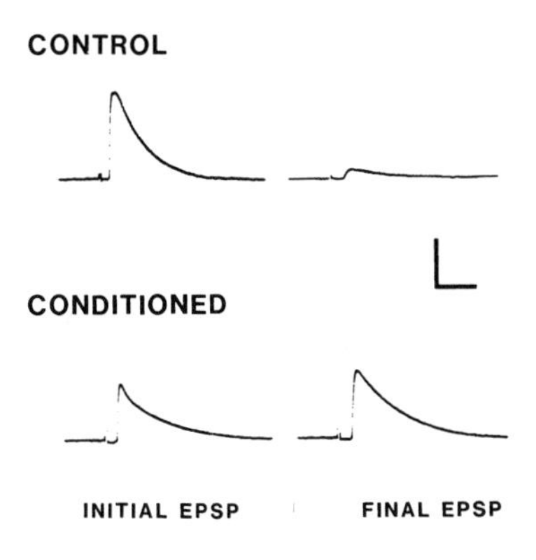

FIGURE 3. Representative EPSPs recorded from claw closer muscle fibers during stimulation of a control or conditioned phasic axon. The phasic axon was conditioned in one claw for two weeks by applying 5 Hz of electrical stimulation for 2 hr/day. One day after the final conditioning period both conditioned and contralateral control claws were removed and the physiological properties of the synapses were tested. The *initial* EPSP was measured during 0.01-Hz stimulation. The *final* EPSP was measured after a 30-min period of 5-Hz stimulation. For the conditioned phasic axon, initial EPSP is smaller in amplitude and there is less synaptic fatigue. Calibration: 20 msec; 4 mV. (After reference 30.)

is an increase in the total number of synaptic varicosities per terminal. This change in terminal shape represents a dramatic change in the location of neuromuscular synapses. Varicosities in conditioned phasic terminals contained a concentration of synapses, whereas in control phasic terminals synapses were evenly distributed along the terminal.[14] Thus, the morphological changes involve the movement of synaptic membrane, or the loss and reformation of synapses. Evidence from studies of the hippocampus indicate that synapses can form within minutes,[43,44] and there is evidence at crustacean neuromuscular terminals that active zones can form or disappear within minutes.[45]

Thus far, there are no observed morphological changes that could account for the decrease in initial transmitter release. Possible changes in the size and structure of the active zone, however, have not been fully explored. Nor have we examined the possibility that there are biochemical changes, such as phosphorylation of channel or synaptic proteins underlying the reduction in initial transmitter release.

Studies of the influence of activity upon neuromuscular development in other species are consistent with our results. Tonic stimulation of a vertebrate fast nerve during ectopic neuromuscular junction formation results in a partial transformation to a slow junction type, suggesting that the development of motor terminal morphology is

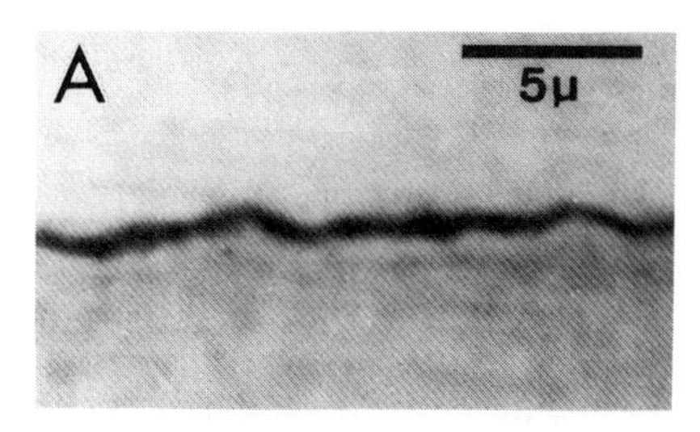

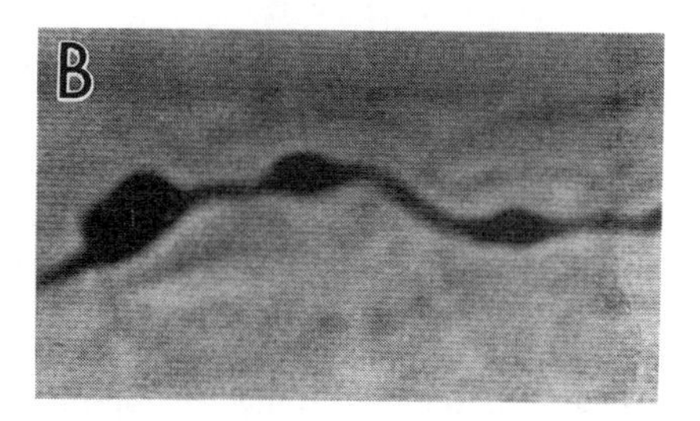

FIGURE 4. Morphological effects of conditioning. Conditioning was performed on the phasic motoneuron in one of the paired claws, and the nonstimulated claw was used as a control. One to six days after the final *in vivo* stimulation period, claws were removed, and the terminals were injected with HRP. In order to examine these differences in synaptic terminal morphology, terminals were examined using video microscopy and contrast enhancement. The morphology of the (**A**) control and (**B**) conditioned phasic synaptic terminals was compared by examining distal terminal branches, that is, those that underwent no further branching. Phasic distal terminal branches were typically slender and uniform in diameter and contained few or no synaptic varicosities. Conditioned phasic distal terminal branches contained numerous synaptic varicosities compared to the control. (After reference 42.)

partially dependent upon electrical activity.[46] Recent evidence in *Drosophila* indicates that the frequency of synaptic varicosities along motor terminals in the larva is greater if impulse activity is increased during development.[47]

Studies of the vertebrate CNS provide convincing evidence that the neuron's history of impulse activity plays a role in the morphological maturation of synaptic terminals. A reduction in impulse activity in the cat visual pathway during early postnatal development results in a reduction in the terminal cross-sectional area and mitochondria size at retinogeniculate[48] and geniculocortical synaptic terminals.[2] These morphological changes are accompanied by increased fatigability of responses in the visual cortex to visual stimuli.[49,50] Thus, these terminals appear to be similar to the fatigable crustacean phasic motor terminals, which normally develop under conditions of low impulse activity; they are smaller in cross-sectional area and have a lower mitochondrial content than their more active tonic counterparts, as discussed previously. However, the extent to which the mechanisms of these activity-dependent changes are the same is presently unclear (see below).

MECHANISMS OF ACTIVITY-INDUCED SYNAPTIC CHANGE

Initial Transmitter Release

Previous studies of the reduction in initial transmitter release produced by the *in vivo* stimulation of a crayfish phasic motoneuron have demonstrated these results: 1) The reduction in initial EPSP amplitude can be induced by an increase in the central impulse activity of the motoneuron (FIG. 6). Thus, changes in postsynaptic electrical activity or transmitter release are not necessary to induce this effect. In addition, selective stimulation of the distal axon and its terminals also induces the effect, demonstrating that impulse activity in any region of the motoneuron is sufficient (FIG. 7). 2) The effect can be produced by subthreshold synaptic potentials at the central processes of the motoneuron in the absence of changes in impulse activity.[51] These results suggest that ionic flux or enzyme activity resulting from depolarization of any region of the motoneuron is the primary molecular event responsible for triggering this synaptic change. We have recently reproduced these results *in vitro* and have found that a reduction in initial EPSP amplitude can be seen within 5 hr after an application of depolarizing pulses to the cell body.[52] Preliminary results also indicate that depolarizing pulses applied to the axon in the presence of tetrodotoxin also induce this effect. Thus, Na^+ influx is probably not essential to induce this effect. Because Ca^{2+} channels are found along crustacean axon,[53,54] the role of Ca^{2+} needs to be explored.

These results obtained at crustacean neuromuscular synapses appear consistent with recent findings in the mammalian CNS and sympathetic ganglia. Chronic reduction in Ia sensory afferent impulse activity in the cat results in enhancement of initial EPSP amplitude at the Ia-motoneuron synapse.[40,41] This change is confined to those synapses on the motoneuron formed by the chronically blocked Ia sensory fibers and therefore is likely to be the result of changes in presynaptic activity.[40] Blockage of impulses in a collateral branch of preganglionic fibers in the sympathetic nervous system results in increased transmitter release from both blocked and unblocked branches, even though they synapse on different neurons.[39] These results can be explained if the synapses are responding to the overall change in impulse activity of the presynaptic neuron as shown in crayfish motoneurons.

At the mammalian neuromuscular junction, blocking the motor nerve with tetrodotoxin produces both sprouting[55] and increased initial transmitter release.[56] Although

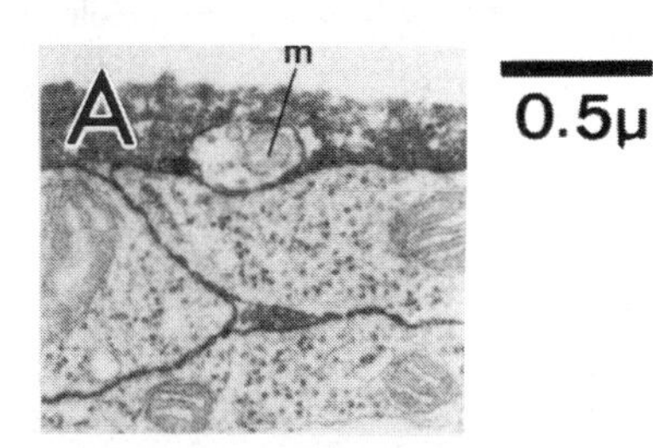

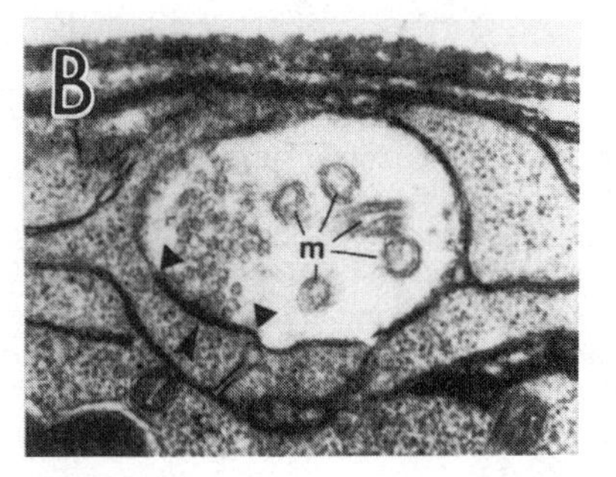

FIGURE 5. Electron micrograph from (**A**) a bottleneck and (**B**) a varicosity in a conditioned phasic terminal. As shown, the activity-induced formation of terminal varicosities was accompanied by the formation of synapses (bracketed by arrows), and the presence of large mitochondria (m). The many mitochondria profiles shown in the varicosity are all branches of the same mitochondrion as revealed by serial-section electron microscopy. (After reference 14.)

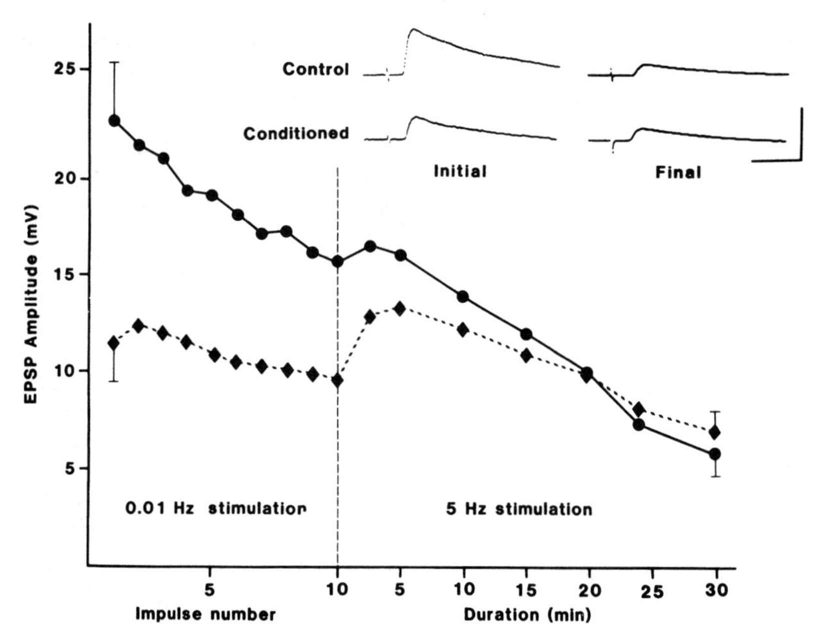

FIGURE 6. The effect of central stimulation of the phasic axon for three days on EPSPs recorded from closer muscle fibers. Impulse activity was increased exclusively in the central region of the motoneuron in one of the paired claws by electrical stimulation proximal to a reversible tetrodotoxin nerve block. Stimulation was applied *in vivo* for 2 hr/day for three consecutive days. One day after the third *in vivo* stimulation period, both conditioned and contralateral control claws were removed, and EPSPs were recorded from identified muscle fibers on the dorsal surface of the closer muscle during stimulation of the phasic axon. The data points represent the mean values from seven conditioned (◆) and seven contralateral control claws (●). The mean amplitude of the initial 10 EPSPs was significantly less in the conditioned claws (10.8 ± 1.7) than in the control claws (18.8 ± 2.0; $p < .01$). There was a trend toward less synaptic fatigue during repetitive stimulation; however, there was not a significant difference in the final EPSP amplitude. Thus, it appears that the conditioning produces a greater effect on initial transmitter release than on fatigability. It is unclear as to whether this is due to the short length of conditioning (three days), or the type of conditioning (central stimulation). *Inset:* representative EPSPs from the first impulse (initial) and after 30 min of stimulation at 5 Hz (final). Calibration: 10 msec; 20 mV. (Reproduced with permission from reference 68.)

the sprouting appears to be regulated by muscle activity,[57] the disuse-induced increase in transmitter release is not necessarily accompanied by terminal sprouting.[58]

Protein synthesis is necessary for this activity-dependent regulation of initial transmitter release. Removal of the motoneuron cell body[33] or the application of inhibitors of protein synthesis[57] blocks the stimulation-induced reduction seen at crayfish neuromuscular synapses. At Ia synapses in the spinal cord, the disuse-induced enhancement of initial EPSP amplitude is blocked by inhibitors of protein synthesis.[59] Activity may regulate the synthesis of macromolecules that are responsible for the effect, or the requisite macromolecules may be constitutively synthesized, but directly regulated by

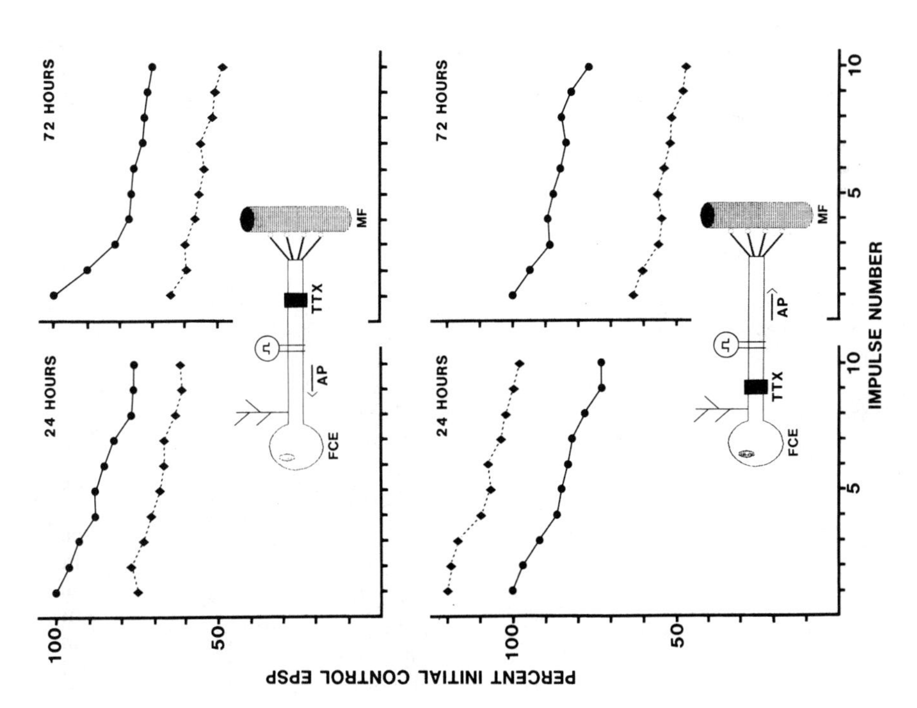

FIGURE 7. EPSP amplitudes produced by the phasic axon to the claw closer muscle 24 or 72 hr after a single *in vivo* central or peripheral stimulation period. The axon was stimulated *in vivo* proximal (central) or distal (peripheral) to a reversible tetrodotoxin block. Graphs show EPSP amplitudes measured during stimulation of the phasic axon at 0.01 Hz and normalized as a percentage of the initial control EPSP amplitude (first impulse) for each experiment. All statistical comparisons of EPSP amplitudes were performed by averaging the 10 EPSP amplitudes for each claw and comparing the mean values for paired claws from the same animal using a paired *t* test. *Upper graphs*: the *central* region of the motoneuron was stimulated *in vivo* for 2 hr in one of the paired phasic axons (*inset*), and the EPSP amplitudes produced by the conditioned (◆) and contralateral (●) phasic axons were examined 24 or 72 hr later. Twenty-four hours after conditioning, the mean EPSP amplitudes from 18 animals were significantly less in the conditioned claws (20.0 ± 4.0 mV; $p < .05$; $N = 10$ animals) than in the controls (25.2 ± 3.3 mV; $p < .05$). Seventy-two hours after a single *in vivo* stimulation, the EPSP amplitude in the conditioned claws (19.1 ± 2.6 mV) is also significantly less than in the control claws (26.0 ± 4.0 mV; $p < .05$; $N = 10$ animals). *Lower graphs*: the peripheral region of the motoneuron was stimulated *in vivo* for 2 hr (*inset*), and the EPSP amplitudes in the conditioned (◆) and control (●) claws were compared 24 or 72 hr later. Activation of the phasic axon neuromuscular synapses produced long-term facilitation, manifested as an increase in EPSP amplitude. Twenty-four hours after peripheral stimulation, long-term facilitation is still apparent, as shown by the significant increase in the EPSP amplitude in the conditioned (24.6 ± 4.4 mV) compared with the control claws (19.3 ± 3.1 mV; $p < .05$; $N = 8$ animals). Seventy-two hours after stimulation, long-term facilitation has decayed and the reduction in initial transmitter release is expressed. The EPSP amplitude is significantly less in the conditioned claws (16.3 ± 2.8 mV) than in the controls (26.3 ± 3.8 mV; $p < .02$; $N = 11$ animals). Thus selective stimulation of the central or peripheral region of the motoneuron is capable of inducing the reduction in initial EPSP amplitude. *Insets*: fast closer excitor (FCE), action potential (AP), tetrodotoxin nerve block (TTX), muscle fiber (MF). (Reproduced with permission from reference 68.)

activity in some manner. The latter possibility is suggested by the fact that inhibition of protein synthesis before stimulation blocks the effect, whereas the inhibition of protein synthesis during or after does not.[57]

Changes in Fatigability and the Formation of Synaptic Varicosities

The changes in initial transmitter release and fatigability appear to operate by different mechanisms because they can be temporally dissociated. It appears that the change in fatigability requires a longer period to develop than the changes in initial transmitter release. This has been recently demonstrated *in vitro* where a decrease in initial EPSP amplitude is seen 5 hr after stimulation,[52] without a change in fatigability (also see FIG. 6).

The activity-induced increase in fatigue resistance also appears to be dependent upon protein synthesis, because removing the cell body[33] or inhibiting protein synthesis[57] blocks the effect seen at crayfish neuromuscular synapses. The dependence upon protein synthesis is not surprising—the activity-induced increase in fatigue resistance is assumed to require the formation of varicosities and the growth of mitochondria.

One possible mechanism for the formation of synaptic varicosities is that they result from the activity-dependent movement of large mitochondria into the terminals. There is evidence that axonal[60] and dendritic[61] swellings can be produced by intracellular organelles. Indeed, the size of the mitochondria after conditioning would prohibit them from occupying the preexisting terminal in the absence of varicosities. According to this model, the formation of swellings would cause both synaptic and nonsynaptic membrane to move into the varicosities. This is supported by the finding that the ratio of synaptic membrane to total membrane is the same in varicosities and bottlenecks.[14] The greater synaptic membrane in varicosities is simply due to the greater amount of total membrane.

Mitochondrial mass has been shown to increase in a number of cell types because of increased energy demands,[62] and the increase in mitochondrial mass seen in the tonic terminals is presumably due to increased transport of mitochondria to the terminals. In both crustacean sensory neurons[63] and mammalian motoneurons,[23] there are more mitochondria in the cell bodies of tonically active than phasically active ones. Using differential interference contrast optics of fixed tissue, we have observed a greater number of mitochondria in the tonic motor axon to the claw closer muscle than in the phasic motor axon. However, the rates of mitochondrial transport in phasic and tonic axons or the possible effects of activity have not yet been explored.

Another important question is whether the increase in fatigue resistance requires synaptic activity for its induction. Both subthreshold stimulation and a central increase in impulse activity produced only a slight increase in fatigue resistance (see FIG. 6). The results at present are ambiguous, and more experiments are necessary.

Our findings at crayfish motor terminals generally support the hypothesis that the morphological changes are due to changes in the activity of the altered neuron and are not influenced by activity in, or competition from, other neurons. We also feel that postsynaptic activity is probably not essential for these changes, although this has not been directly tested. It is well established that the activity-dependent development of synaptic terminals in the visual cortex is influenced by competition and postsynaptic activity.[2] There have been few tests, however, of whether the synaptic terminal development is influenced by the neuron's own electrical activity. In one case, the normal growth and maturation of retinal terminals in the cat lateral geniculate were influenced directly by electrical activity.[48] It may be that the development of "normal" synaptic terminal morphology requires a basal level of electrical activity.

ADAPTATION OF SYNAPTIC PROPERTIES TO ACTIVITY LEVELS DURING DEVELOPMENT

What is the Adaptive Role of These Synaptic Changes?

The previously described matching of synaptic properties and impulse activity levels appears to be adaptive. Obviously, synapses that are normally active for prolonged periods require fatigue resistance, whereas synapses activated only briefly do not. In addition, synaptic fatigue may serve an adaptive function at phasic synapses because, as discussed previously, it has been shown to underlie behavioral habituation.

The activity-dependent changes in initial transmitter release also appear to serve an adaptive, although less obvious, function. The activity-induced reduction in initial transmitter release for the phasic terminals appears inconsistent with the increase in transmitter release normally seen after brief bursts of activity at a variety of synapses.[64] The phasic synapses studied here, however, do show long-term facilitation immediately after a single conditioning trial. The long-term adaptation of initial transmitter release is only seen after several conditioning trials or several days after a single conditioning trial (FIG. 7). Thus, the changes reported here represent longer-lasting synaptic changes resulting from altered activity.

Phasically active neural circuits often mediate defensive or aggressive behaviors. Thus, in spite of their low activity, they are important pathways, and it is essential that their synapses transmit reliably during their brief activation. A large release of transmitter during initial activation is appropriate for these synapses. Although the phasic synapses function mainly as a relay with a single action potential producing a maximal postsynaptic response, the tonic synapses must transmit the information encoded in the broad range of firing frequencies at which they normally operate. The low initial transmitter release at tonic synapses allows the frequency-dependent grading of postsynaptic depolarization through synaptic facilitation and temporal summation.

Does Activity Play a Role in Normal Development?

Electrical activity appears to play a role in the normal differentiation of the neuromuscular synapses because the phasic synapses gradually acquire their adult phenotype during postnatal development, the period in which they are most sensitive to experimental alternations in activity.[30] Additional evidence that the level of activity during development plays a role in the differentiation of these motor terminals is provided by studies in which animals were examined during different seasons. Young animals collected during the summer have more synaptic varicosities and greater fatigue resistance than those collected during the winter.[65] This is presumably due to the summer animals developing during a period of higher activity compared to winter animals.

The question remains as to the extent to which the differentiation of phasic and tonic synaptic terminals is regulated by activity. Evidence presented here demonstrates that the development of synaptic varicosities is dependent on activity levels. However, the incomplete morphological and physiological transformation of terminal type by imposed activity suggests that the cell may be partially predetermined to develop phasic or tonic properties. At this point, we cannot assess the relative contributions of nature and nurture; however, one might expect a greater effect of activity if the imposed activity levels were increased further, or animals were stimulated earlier in development. Indeed, the physiological changes appear to be age dependent: old animals show less physiological change after conditioning than young animals.[30]

ACKNOWLEDGMENTS

I would like to thank Robert Reuss and Christopher Case for assisting with the preparation of the manuscript, and Drs. Suzannah Tieman and John Schmidt for critical review of the manuscript.

REFERENCES

1. SCHMIDT, J. 1991. Long-term potentiation during the activity-dependent sharpening of the retinotopic map in goldfish. Ann. N.Y. Acad. Sci. This volume.
2. TIEMAN, S. 1991. Morphological changes in the geniculocortical pathway associated with monocular deprivation. Ann. N.Y. Acad. Sci. This volume.
3. TURNER, A. M. & W. T. GREENOUGH. 1985. Differential rearing effects on rat visual cortex synapses. I. Synaptic and neuronal density and synapses per neurone. Brain Res. **329:** 195-203.
4. MURPHEY, R. K. 1986. The myth of the inflexible invertebrate: Competition and synaptic remodelling in the development of invertebrate nervous systems. J. Neurobiol. **17:** 585-591.
5. HENNIG, R. & T. LOMO. 1984. Firing patterns of motor units in normal rats. Nature **34:** 164-166.
6. GERTLER, R. A. & N. ROBBINS. 1978. Differences in neuromuscular transmission in red and white muscle. Brain Res. **142:** 160-164.
7. CASTELLUCCI, V., H. PINSKER, I. KUPFERMANN & E. R. KANDEL. 1970. Neuronal mechanisms of habituation and dishabituation of the gill-withdrawal reflex in *Aplysia*. Science **167:** 1745-1748.
8. ZUCKER, R. S. 1972. Crayfish escape behavior and central synapses. II. Physiological mechanisms underlying behavioral habituation. J. Neurophysiol. **35:** 621-637.
9. DUDEL, J., I. PARNAS, I. COHEN & C. FRANKE. 1984. Excitability and depolarization-release characteristics of excitatory nerve terminals in a tail muscle of spiny lobster. Pfluegers Arch. **401:** 293-296.
10. DUDEL, J. 1982. Transmitter release by graded local depolarization of presynaptic nerve terminals at the crayfish neuromuscular junction. Neurosci. Lett. **32:** 181-186.
11. DUDEL, J. 1983. Graded or all-or-nothing release of transmitter quanta by local depolarizations of nerve terminals on crayfish muscle. Pfluegers Arch. **98:** 155-164.
12. BRIGANT, J. L. & A. MALLART. 1982. Presynaptic currents in mouse endings. J. Physiol. **333:** 619-636.

13. DUDEL, J. 1983. Transmitter release triggered by a local depolarization in motor nerve terminals of the frog: Role of calcium entry and of depolarization. Neurosci. Lett. **41:** 133-138.
14. LNENICKA, G. A., H. L. ATWOOD & L. MARIN. 1986. Morphological transformation of synaptic terminals of a phasic motoneuron by long-term tonic stimulation. J. Neurosci. **6**(8): 2252-2258.
15. HILL, R. J. & C. K. GOVIND. 1981. Comparison of fast and slow synaptic terminals in lobster. Cell Tissue Res. **221:** 303-310.
16. ATWOOD, H. L. & S. S. JAHROMI. 1978. Fast-axon synapses of a crab leg muscle. J. Comp. Physiol. **124:** 237-247.
17. ATWOOD, H. L. & H. S. JOHNSTON. 1968. Neuromuscular synapses of a crab motor axon. J. Exp. Zool. **167:** 457-470.
18. TITMUS, M. J. 1981. Ultrastructure of identified fast excitatory, slow excitatory and inhibitory neuromuscular junctions in the locust. J. Neurocytol. **10:** 363-385.
19. RHEUBEN, M. B. 1984. Quantitative comparison of the structural features of slow and fast neuromuscular junctions in manduca. J. Neurosci. **5:** 1704-1716.
20. DIAS, P. L. R. 1974. Surface area of motor end plates in fast and slow twitch muscles of the rabbit. J. Anat. **117:** 453-462.
21. FAHIM, M. A., J. A. HOLLEY & N. ROBBINS. 1984. Topographic comparison of neuromuscular junctions in mouse slow and fast twitch muscle. Neuroscience **13**(1): 227-235.
22. WAERHAUG, O. & H. KORNELIUSSEN. 1974. Morphological types of motor nerve terminals in rat hindlimb muscles, possibly innervating different muscle fiber types. Z. Anat. Entwicklungsgesch. **144:** 237-247.
23. SICKLES, D. W. & T. G. OBLAK. 1984. Metabolic variation among motoneurons innervating different muscle-fiber types. I. Oxidative enzyme activity. J. Neurophysiol. **51:** 529-537.
24. GOVIND, C. K., D. E. MEISS & J. PEARCE. 1982. Differentiation of identifiable lobster neuromuscular synapses during development. J. Neurocytol. **11:** 235-247.
25. ATWOOD, H. L. & L. MARIN. 1983. Ultrastructure of synapses with different transmitter-releasing characteristics on motor axon terminals of a crab, *Hyas areneas*. Cell Tissue Res. **231:** 103-115.
26. ATWOOD, H. L. & G. A. LNENICKA. 1986. Structure and function in synapses: Emerging correlations. Trends Neurosci. **9:** 248-250.
27. PUMPLIN, D. W., T. S. REESE & R. LLINAS. 1990. Are the presynaptic membrane particles the calcium channels? Proc. Natl. Acad. Sci. USA **78:** 7210-7213.
28. WALROND, J. P. & T. S. REESE. 1985. Structure of axon terminals and active zones at synapses on lizard twitch and tonic muscle fibers. J. Neurosci. **5:** 662-672.
29. PEARCE, J., C. K. GOVIND & R. R. SHIVERS. 1986. Intramembranous organization of lobster excitatory neuromuscular synapses. J. Neurocytol. **15:** 241-252.
30. LNENICKA, G. A. & H. L. ATWOOD. 1985. Age-dependent long-term adaptation of crayfish phasic motor axon synapses to altered activity. J. Neurosci. **5:** 459-467.
31. MERCIER, A. J. & H. L. ATWOOD. 1990. Long-term adaptation of a phasic extensor motoneurone in crayfish. J. Exp. Biol. **145:** 9-22.
32. BRADACS, H., A. J. MERCIER & H. L. ATWOOD. 1990. Long-term adaptation in lobster motor neurons and compensation of transmitter release by synergistic inputs. Neurosci. Lett. **108:** 110-115.
33. LNENICKA, G. A. & H. L. ATWOOD. 1985. Long-term facilitation and long-term adaptation at synapses of a crayfish phasic motoneuron. J. Neurobiol. **16:** 97-110.
34. PAHAPILL, P. A., G. A. LNENICKA & H. L. ATWOOD. 1985. Asymmetry of motor impulses and neuromuscular synapses produced in crayfish claws by unilateral immobilization. J. Comp. Physiol. **157:** 461-467.
35. ROBBINS, N. & G. D. FISCHBACH. 1971. Effect of chronic disuse of rat soleus neuromuscular junctions on presynaptic function. J. Neurophysiol. **34:** 570-578.
36. HINZ, I. & A. WERNIG. 1988. Prolonged nerve stimulation causes changes in transmitter release at the frog neuromuscular junction. J. Physiol. **401:** 557-565.
37. MURPHEY, R. K. & S. G. MATSUMOTO. 1976. Experience modifies the plastic properties of identified neurons. Science **191:** 564-566.

38. BLOOM, J. W. & H. L. ATWOOD. 1980. Effects of altered sensory experience on the responsiveness of the locust descending contralateral movement detector neuron. J. Comp. Physiol. **135:** 191-199.
39. GALLEGO, R. & E. GEIJO. 1987. Chronic block of the cervical trunk increases synaptic efficacy in the superior and stellate ganglia of the guinea-pig. J. Physiol. **382:** 449-462.
40. GALLEGO, R., M. KUNO, R. NUNEZ & W. D. SNIDER. 1979. Disuse enhances synaptic efficacy in spinal motoneurones. J. Physiol. **291:** 191-205.
41. MANABE, T., C. R. S. KANEKO & M. KUNO. 1990. Disuse-induced enhancement of Ia synaptic transmission in spinal motoneurons of the rat. J. Neurosci. **9**(7): 2455-2461.
42. LNENICKA, G. A., S. J. HONG, M. COMBATTI & S. LEPAGE. 1990. Activity-dependent development of synaptic varicosities at crayfish motor terminals. J. Neurosci. **11**(4): 1040-1048.
43. CHANG, F.-L. F. & W. T. GREENOUGH. 1984. Transient and enduring morphological correlates of synaptic activity and efficacy change in the rat hippocampus. Brain Res. **309:** 35-46.
44. LEE, K. S., F. SCHOTTLER, M. OLIVER & G. LYNCH. 1980. Brief bursts of high-frequency stimulation produce two types of structural change in rat hippocampus. J. Neurophysiol. **44**(2): 247-258.
45. WOJTOWICZ, J. M., L. MARIN & H. L. ATWOOD. 1989. Synaptic restructuring during long-term facilitation at the crayfish neuromuscular junction. Can. J. Physiol. Pharmacol. **67:** 167-171.
46. LOMO, T. & O. WAERHAUG. 1985. Motor endplates in fast and slow muscles of the rat: What determines their differences? J. Physiol. (Paris) **80:** 290-297.
47. BUDNIK, V., Y. ZHONG & C.-F. WU. 1989. Projection of motor axons in hyperexcitable *Drosophila* mutants. Soc. Neurosci. Abstr. **15:** 570.
48. KALIL, R. E., M. W. DUBIN, G. SCOTT & L. A. STARK. 1986. Elimination of action potentials blocks the structural development of retinogeniculate synapses. Nature **323:** 156-158.
49. GANZ, L., M. FITCH & J. A. SATTERBERG. 1968. The selective effect of visual deprivation on receptive field shape determined neurophysiologically. Exp. Neurol. **22:** 614-637.
50. WIESEL, T. N. & D. H. HUBEL. 1965. Comparison of the effects of unilateral and bilateral eye closure on cortical unit responses in kittens. J. Neurophysiol. **52:** 99-113.
51. LNENICKA, G. A. & H. L. ATWOOD. 1988. Long-term changes in neuromuscular synapses with altered sensory input to a crayfish motoneuron. Exp. Neurol. **100:** 437-447.
52. HONG, S. J. & G. L. LNENICKA. 1990. Long-term regulation of transmitter release from crayfish motor terminals by depolarization of the motoneuron cell body. Soc. Neurosci. Abstr. **16:** 1162.
53. NIWA, A. & K. NOBUFUMI. 1982. Tetrodotoxin-resistant propagating action potentials in presynaptic axons of the lobster. J. Neurophysiol. **47:** 353-361.
54. SUZUKI, J. 1976. Ca^{2+} activation in the giant axon of the crayfish. *In* Electrobiology of Nerve, Synapse and Muscle. J. P. Reuben, D. P. Purpura, M. V. L. Bennett & E. R. Kandel, Eds.: 27-36. Raven Press. New York, NY.
55. BROWN, M. C. & R. IRONTON. 1977. Motor neurone sprouting induced by prolonged tetrodotoxin block of nerve action potentials. Nature **265:** 459-461.
56. SNIDER, W. D. & G. L. HARRIS. 1979. A physiological correlate of disuse-induced sprouting at the neuromuscular junction. Nature **281:** 69-71.
57. NGUYEN, P. V. & H. L. ATWOOD. 1990. Expression of long-term adaptation of synaptic transmission requires a critical period of protein synthesis. J. Neurosci. **10**(4): 1099-1109.
58. TSUJIMOTO, T. & M. KUNO. 1988. Calcitonin gene-related peptide prevents disuse-induced sprouting of rat motor nerve terminals. J. Neurosci. **8**(10): 3951-3957.
59. MIYATA, Y. & H. YASUDA. 1988. Enhancement of Ia synaptic transmission following muscle nerve section: Dependence upon protein synthesis. Neurosci. Res. **5:** 338-346.
60. FRIEDE, R. L. & A. J. MARTINEZ. 1970. Analysis of the process of sheath expansion in swollen nerve fibers. Brain Res. **19:** 165-182.
61. SASAKI-SHERRINGTON, S. E., J. R. JACOBS & J. K. STEVENS. 1984. Intracellular control of axial shape in non-uniform neurites: A serial electron microscopic analysis of organelles and microtubules in AI and AII retinal amacrine neurites. J. Cell Biol. **98:** 1297-1290.

62. SMITH, R. A. & M. J. ORD. 1983. Mitochondrial form and function relationships *in vivo*: Their potential in toxicology and pathology. Int. Rev. Cytol. **83:** 63-133.
63. MAYES, J. I. & C. K. GOVIND. 1989. Higher mitochondrial density in slow versus fast lobster sensory neurons. Neurosci. Lett. **102:** 87-90.
64. ATWOOD, H. L. & J. M. WOJTOWICZ. 1986. Short-term and long-term plasticity and physiological differentiation of crustacean motor synapses. Int. Rev. Neurobiol. **28:** 275-362.
65. LNENICKA, G. A. & Y. ZHAO. 1991. Seasonal differences in the physiology and morphology of crayfish motor terminals. J. Neurobiol.: in press.
66. WIERSMA, C. A. G. 1961. The neuromuscular system. *In* The Physiology of Crustacea. Vol. 2: Sense Organs, Integration and Behavior. T. H. Waterman, Ed.: 191-240. Academic Press. New York, NY.
67. LNENICKA, G. A. 1988. The role of neuronal activity in the long-term regulation of synaptic performance at the crayfish neuromuscular junction. *In* Cellular Mechanisms of Conditioning and Behavioral Plasticity. C. D. Woody, D. L. Alkon & J. L. McGaugh, Eds.: 403-409. Plenum. New York, NY.
68. LNENICKA, G. A. & H. L. ATWOOD. 1989. Impulse activity of a crayfish motoneuron regulates its neuromuscular synaptic properties. J. Neurophysiol. **61:** 91-96.

Morphological Changes in the Geniculocortical Pathway Associated with Monocular Deprivation[a]

SUZANNAH BLISS TIEMAN

Neurobiology Research Center
State University of New York
Albany, New York 12222

BACKGROUND

As Hubel and Wiesel demonstrated in their Nobel Prize-winning research,[1,2] a kitten that is reared with the lids of one eye fused will appear blind when the eye is later opened and the animal is tested behaviorally for vision through that eye. Although there is some recovery with exposure, that recovery is limited; the animal's perceptual capabilities remain inferior when using the deprived eye.[3] For example, it will have severe difficulty learning visual discriminations with that eye.[4] Hubel and Wiesel first showed that this behavioral change is accompanied by morphological and physiological changes.[5,6] The effects of monocular deprivation have since been investigated in detail by both myself and others, and will be described below, after a brief review of the central visual pathways of the cat.[7,8]

The two eyes view overlapping regions of the visual world (FIG. 1). The area of overlap is called the binocular segment of the visual field. It is surrounded on either side by the monocular segments of the visual field, which can be seen by only one eye. The visual image is inverted by the eye so that the left half of each retina sees the right visual field, and vice versa. Fibers from the nasal hemiretina of each eye cross in the optic chiasm to innervate layer A of the contralateral lateral geniculate nucleus (LGN), whereas those from the temporal hemiretina remain uncrossed and innervate layer A1 of the ipsilateral LGN.[9] Thus, the left LGN views the right half of the visual world through both eyes, and the right LGN views the left half of the visual world. The retinal projections to the LGN are topographic, and although the inputs from the two eyes remain segregated, the topographic maps are in register.[10] The inputs from the two eyes are combined into a single topographic map in area 17[11,12] and a second topographic map in area 18.[13,14] Cells from corresponding regions of geniculate layers A and A1 project to overlapping ocular dominance patches in layer IV of areas 17 and 18.[15,16] Thus, visual cortex is the first place where the inputs from the two eyes are combined, and most cortical cells can be driven by stimulation of either eye, although they tend to be dominated by one eye or the other.[11] Cells in the monocular segments of visual cortex, of course, can be driven by stimulation of only the contralateral eye.

Monocular deprivation causes a shift in the distribution of ocular dominance such that most cells can be driven by only the experienced eye. Whereas in the cortex of normal cats, each eye has access to roughly 90% of the cells,[11] in monocularly deprived (MD) cats, the deprived eye can drive fewer than 10% of the cells,[6] at least a 10-fold

[a] This work was supported by grants from the National Institutes of Health (EY 02609) and the National Science Foundation (BNS 8217479 and BNS 8811039).

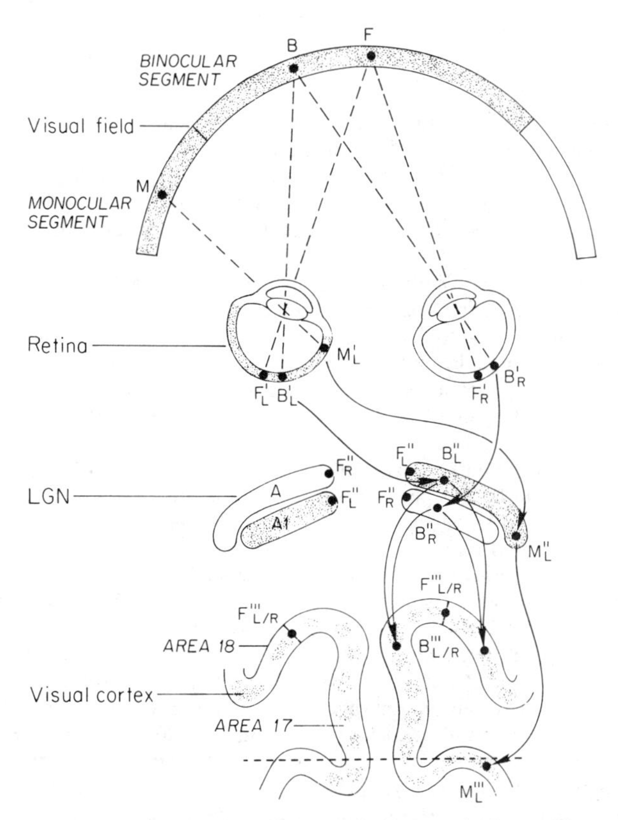

FIGURE 1. Diagram of the pathway from retina to visual cortex in the cat. The pathway for the left eye is shaded. The two eyes are converged on a fixation point F in the visual field, the image of which falls at F′ on each retina. The image of the visual field is inverted by the lens of each eye onto the retina such that a point M in the lateral, monocular visual field maps onto the medial, nasal retina of one eye (at M'_L), and a point B in the central, binocular field maps onto more central retina in both eyes (at B'_L and B'_R). The information is then relayed to the LGN, where it forms a topographic map of the contralateral visual field, so that points in the left visual field are mapped onto the right LGN, and vice versa. Information from F′ is relayed to both LGNs. (For simplicity, the projection lines for F′ and F″ have been omitted.) The inputs from the two eyes project to separate layers of the LGN. Layer A receives input from the nasal retina of the contralateral eye; layer A1 receives input from the temporal retina of the ipsilateral retina. From medial to lateral in the LGN corresponds to points in visual space from the midline (F″, at 0°) to 90° lateral in the contralateral hemifield. Thus, the monocular part of the visual field is represented at the lateral edge of layer A (at M″). Axons from corresponding points in the different layers of the LGN (B''_L and B''_R) then converge onto the same regions of both area 17 and area 18 ($B'''_{L/R}$), so that each hemisphere contains two binocular representations of the contralateral visual field, one in area 17 and one in area 18. The map in area 18 is more or less the reverse of the one in area 17. The two maps are joined at the representation of the midline ($F'''_{L/R}$), which is found at the 17/18 border. From dorsal to ventral in area 17 corresponds to points in visual space from medial to lateral, with the monocular segment represented in the splenial sulcus (at M'''_L). From medial to lateral in area 18 corresponds to points in visual space from medial to lateral, but there is probably no representation of the monocular segment in area 18.[13] The projection from one layer of the LGN (representing one eye) onto areas 17 and 18 is not continuous (except in the monocular segment), but ends in ocular dominance patches that partially overlap those from the adjacent layer of the LGN (representing the other eye). Although most cortical cells are binocular, those lying in a column extending above and below an ocular dominance patch for one eye (ocular dominance column) are dominated by that same eye. The dotted line at the bottom of the figure illustrates the approximate plane of section for FIGURE 2.

reduction in efficacy. In contrast, the effects of binocular deprivation are much less devastating; this led Wiesel and Hubel to postulate that the geniculocortical afferents representing the two eyes compete for cortical synaptic space during development.[17] The fact that, in MD cats, many cells in LGN but not cortex can be driven by the deprived eye[5] suggests that this competition disrupts the geniculocortical connections for the deprived pathway.

Although the gross morphology of the visual cortex of MD cats appears normal, there are striking changes in the LGN: the cell bodies of neurons in the deprived layers are shrunken in comparison to those of the experienced layer.[5] This shrinkage, however, is restricted to the binocular segment, and is not observed in the equally deprived monocular segment.[18] This suggests that the shrinkage is not due to the deprivation per se, but is a secondary effect of the competition between the afferents from the two eyes for access to cortical cells. Cells in the monocular segment would not shrink because there are no afferents from the other eye with which to compete. This suggestion has been confirmed in later experiments in which the prevention of competition or its consequences locally in small regions of the binocular segment of the geniculocortical pathway also prevented cell shrinkage in those same regions. Guillery[19] created an artificial monocular segment within the central visual field by making a small lesion in the retina of the open eye, resulting in the degeneration of the cells in the LGN to which this region of the retina had originally projected. Cells in the corresponding region of the deprived layer, having no cells with which to compete, did not shrink. More recently, Bear and Colman[20] prevented the shift in ocular dominance usually produced by monocular deprivation (and thus, presumably also prevented any loss of synaptic space by deprived geniculate neurons) by locally applying 2-amino-5-phosphonovalerate, an *N*-methyl-D-aspartate blocker, to visual cortex and showed that cells in the deprived geniculate layer projecting to this area of cortex (as demonstrated by the retrograde transport of horseradish peroxidase) again did not shrink. Thus it appears that the shrinkage of cells in the deprived geniculate layers of MD cats is due to their losing much of their synaptic input to visual cortex: preventing this loss prevents the shrinkage.

Thorpe and Blakemore[21] provided the first anatomical evidence for a loss of connections from the deprived pathway to visual cortex based on a decreased retrograde transport of horseradish peroxidase. However, because the uptake and retrograde transport of horseradish peroxidase is also affected by neural activity,[22] and because the deprived pathway is less active, this evidence was not conclusive. More convincing were the data of Shatz and Stryker,[23] who used transneuronal transport to label ocular dominance patches in MD cats. They showed that the patches for the deprived eye were shrunken and that those for the experienced eye were larger than normal. However, the shrinkage they observed (ocular dominance patches for the deprived eye occupied about 32% of layer IV[23,24]) is not sufficient to account for the dramatic shift in ocular dominance. This paper describes other changes in the geniculocortical pathways of MD cats that help to explain the dramatic physiological effects. These include changes in the cortical ocular activation columns, the geniculocortical synapses, and the amount of a possible geniculocortical transmitter.

OCULAR ACTIVATION COLUMNS

Nina Tumosa, David Tieman, and I studied ocular activation columns (regions of visual cortex activated by one eye) both in MD cats and in cats raised with equal and unequal alternating monocular exposure (AME).[25] Cats reared with equal AME receive equal, patterned exposure to the two eyes, but never simultaneously.[4,26] Cats reared with unequal AME receive alternating exposure to the two eyes, but one eye is systematically given more exposure than the other.[27] The advantage of this preparation is that one can manipulate the relative exposure of the two eyes without the confounding effects of pattern deprivation. We have typically studied AME 8/1 cats, in which one eye receives eight times as much stimulation as the other. We had previously shown that unequal AME has effects similar to those of monocular deprivation, but less severe. These effects included shifts in ocular dominance, shrinkage of cells in deprived layers of the LGN, and shrinkage of ocular dominance patches in visual cortex.[8]

To study the ocular activation columns, we silenced one eye, either by removing it or by injecting tetrodotoxin into the posterior chamber, and later administered [^{14}C]2-deoxyglucose through an intravenous catheter when the cats were active and alert. FIGURE 2 shows the pattern of ocular activation in an equal AME cat ipsilateral (top) and contralateral (bottom) to the stimulated eye. In the binocular segment, the pattern of activation is columnar, and is somewhat clearer ipsilateral than contralateral to the stimulated eye. In the monocular segment, the labeling is more uniform, and is much lighter ipsilateral to the stimulated eye. This pattern is similar to that seen in a normally reared cat, but is sharper and more clearly columnar.[25,28]

FIGURE 3 shows the pattern of ocular activation for (from top to bottom) the deprived eye of an MD cat, the 1-hr eye of an AME 8/1 cat, one eye of an equal AME cat, the 8-hr eye of an AME 8/1 cat, and the experienced eye of an MD cat. Note that the size of the columns increases with the relative competitive advantage that the eye received during rearing (that is, from top to bottom). The columns of the deprived eye are small, faint, difficult to see, and often fail to extend beyond layer IV, observations that are consistent with earlier reports.[29,30] They are somewhat clearer in area 18 than they are in area 17. The columns for the 1-hr eye are also small, but they are not as faint as those for the deprived eye, and they extend throughout the layers. This difference in the extent of the ocular activation columns into extragranular layers is consistent with earlier physiological reports that, in MD cats[23] but not unequal AME cats,[31] cells activated by the less-experienced eye are particularly difficult to record outside of layer IV.

The columns in the equal AME cat are somewhat larger than those for the 1-hr eye, with the labeled and the unlabeled columns being more nearly equal in size. Those for the 8-hr eye of an AME 8/1 cat are larger still. Largest of all are those for the experienced eye of an MD cat. These seem to have fused into much larger columns, which are visible mainly because of the occasional gaps in label (arrows), which probably represent the columns for the deprived eye. In fact, this was our most columnar example of the pattern of activation by the experienced eye. In our other two cats, the experienced eye seemed to have taken over virtually all of the visual cortex, and the pattern resembled that seen in animals with both eyes open during the administration of the 2-deoxyglucose, except that the monocular segment ipsilateral to the stimulated eye was not labeled.

We used image analysis to quantify the sizes of the ocular activation columns. We first smoothed the autoradiograms, and then set a threshold halfway between the peak and trough densities. The image processor then determined the percentage of visual cortex that was darker than threshold. Note that this measure takes into account the size of the ocular activation columns both in layer IV and in other layers. This percentage varied significantly with the competitive advantage of the stimulated eye.[25] When we correlated our anatomical measures of ocular activation with previously published physiological measures of ocular dominance from similarly reared cats,[17,31] we obtained a nearly perfect correlation of .98 (FIG. 4).

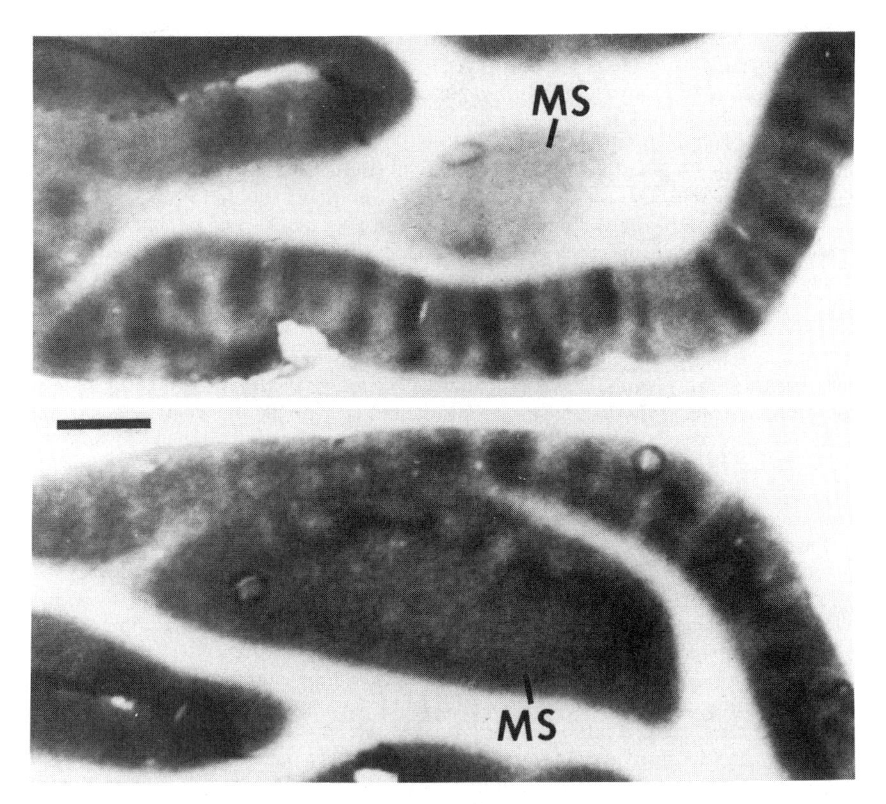

FIGURE 2. Autoradiograms of horizontal sections through visual cortex ipsilateral (above) and contralateral (below) to the stimulated eye in an equal AME cat. The approximate plane of section is indicated by the dotted horizontal line in FIGURE 1. The sections are oriented as if viewing from above an animal that is facing left. Thus, anterior is to the left, and the scale bar is near the animal's midline. The ipsilateral, unstimulated monocular segment (MS) was much lighter than the contralateral, stimulated monocular segment. Further, in this particular cat, there was a difference in the size of the ocular activation columns in the two hemispheres: the contralateral columns were larger than the ipsilateral columns. This was not a consistent finding. Scale bar: 2.0 mm. Reprinted with permission from Tumosa *et al.*[25]

FIGURE 3. Representative autoradiograms showing ocular activation columns ipsilateral to the stimulated eye in cats reared with equal or unequal exposure to the two eyes. Again, all sections are horizontal, with anterior to the left. In each case, medial is up, and lateral is down. **(A)** The deprived eye of an MD cat activated the smallest columns. These were widest in layer IV; in other layers they were narrower, sometimes disappearing altogether. The arrowheads point to two columns in area 17; the columns in area 18 were darker and easier to see. **(B)** The columns of the 1-hr eye of an AME 8/1 cat were smaller than the intervening unlabeled columns , and were of nearly uniform width throughout all the layers. **(C)** In an equal AME cat, the labeled columns were about the same size as the unlabeled columns. **(D)** The columns of the 8-hr eye of an AME 8/1 cat were wider than the intervening unlabeled columns. **(E)** The experienced eye of the MD cats activated the largest columns. The arrows point to gaps in the label, probably representing columns of the deprived eye. This section was chosen to illustrate columnar activity, but the experienced eye appeared to activate nearly all of both areas 17 and 18, and the columns were often difficult to see. Scale bars: 2.0 mm. Reprinted with permission from Tumosa *et al.*[25]

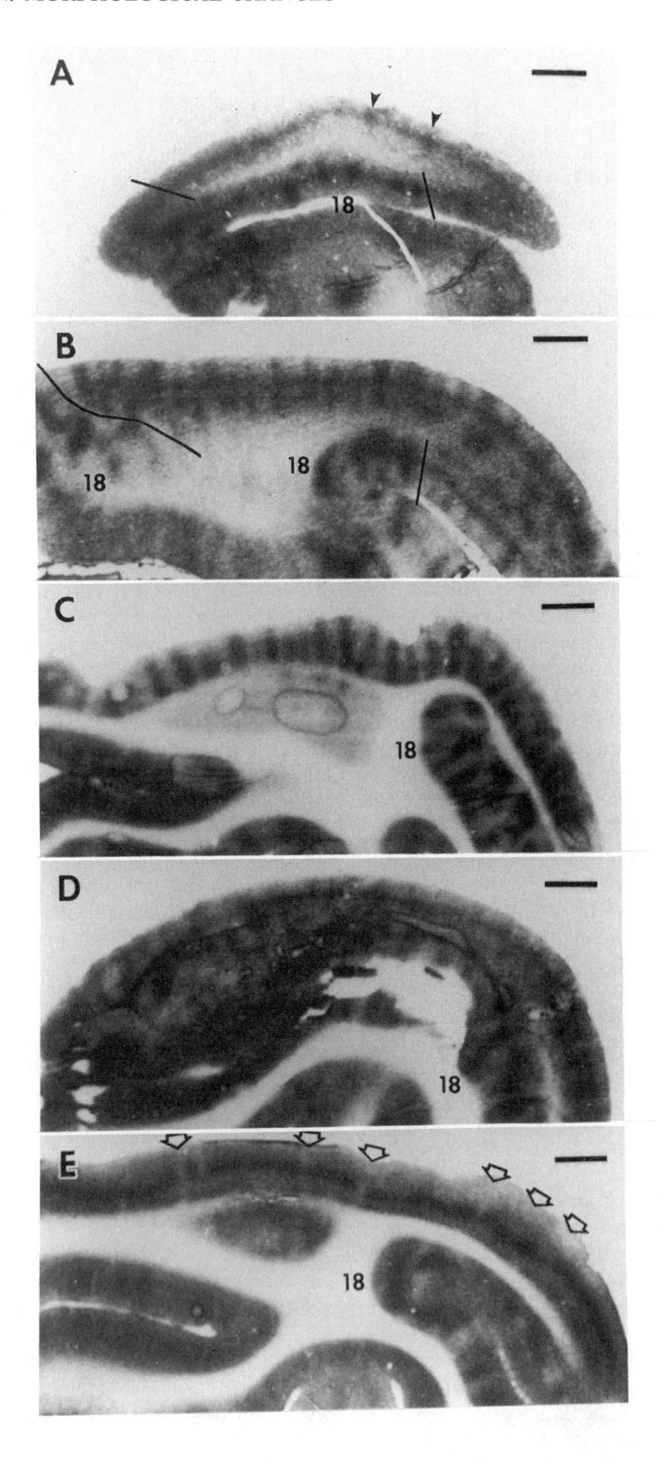
A
18
B
18
18
C
18
D
18
E
18

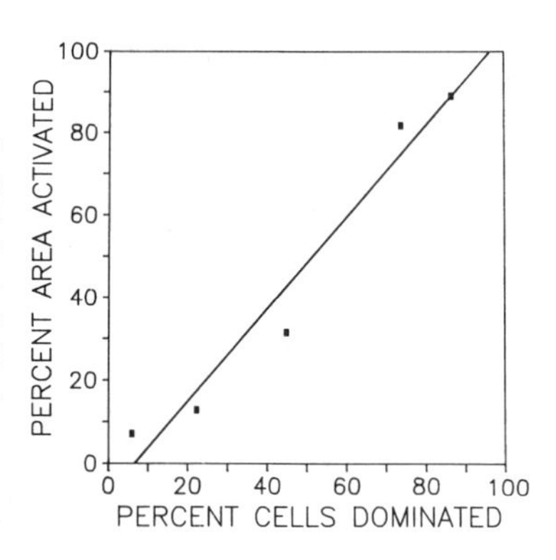

FIGURE 4. The percentage of cortex activated by the stimulated eye, as determined from 2-deoxyglucose studies,[25] is plotted against the percentage of cortical cells dominated by that eye, as recorded from other cats reared under similar conditions.[17,31] The correlation between the anatomical results and the physiological results is high ($r = .98$, $p < .01$). The best-fitting line is $Y = 1.1X - 7.5$. Modified with permission from Tumosa *et al.*[25]

As mentioned earlier, the columns for the deprived eye were not only small, but faint. This can be seen in FIGURE 5. On the upper left is an autoradiogram of a section through the binocular and monocular segments of area 17 in an MD cat whose deprived eye was stimulated during administration of the 2-deoxyglucose. Just below it is an image of the same section in which everything darker than a certain threshold has been set to black and everything lighter has been set to white. Note that we were able to choose a threshold that was darker than the columns in the binocular segment but lighter than the label in the monocular segment. In contrast, it was not possible to set such a threshold for the equal AME cats (upper right). A threshold that starts to lose the label in the columns also starts to lose the label in the monocular segment. Thus, for the deprived eye of MD cats, but not for either eye of equal AME cats, the label in the columns was lighter than the label in the monocular segment. As can be seen from the lower parts of the figure, again for the deprived eye of the MD cats, the label in the binocular segment of area 17 was also lighter than that in area 18.

We were able to quantify the density of label with our image analyzer. Because density varies with many variables, such as the state of the animal or length of film exposure, we normalized our data by taking ratios of the label in the activation columns of area 17 first, to the label in the stimulated layers of the LGN (FIG. 6A), and second, to the label in the monocular segment of area 17 (FIG. 6B). Note the similarity of results. It is clear that the deprived eye is very poor at activating cortex, even within its restricted activation columns. In contrast, the deprived eye is much better at activating the monocular segment, suggesting that the poor activation within the binocular segment is largely or entirely an effect of the binocular competition rather than of the deprivation. The faintness of the deprived eye columns was noted in an earlier study,[30] but was neither quantified nor compared to the label in LGN or the monocular segment of area 17.

It is clear from our data that the columns for the experienced eye of an MD cat are also not as dense as those for either eye of the equal AME cat. We feel that this decreased density reflects some kind of intrinsic limit on the cell's ability to sustain synaptic output. The arbor is spread over an area that is too large for it to cover at normal density; to make up for the greater area, it covers the area less densely (arbor conservation principle of DeVor and Schneider[32]). Note that this principle is not necessarily absolute: a neuron may well be able to increase its synaptic output above normal, as suggested by sprouting studies in spinal cord[33] and dentate gyrus[34,35] as well as by the work of Bailey and Chen[36,37] in *Aplysia,* but there will be some limit beyond which the neuron cannot go. This is illustrated in FIGURE 7, which shows schematic geniculocortical arbors for afferents representing either eye of an equal AME cat and an MD cat. Note that for the MD cat, the arbors representing each eye are less dense than those in the equal AME cat. In each case, the density of arbor depicted here correlates with the density of ocular activation columns that we observed with 2-deoxyglucose. Although this is the only evidence that the arbor for the experienced eye may be less dense than normal, additional evidence that the arbor for the deprived eye is less dense than that for the experienced eye is discussed below.

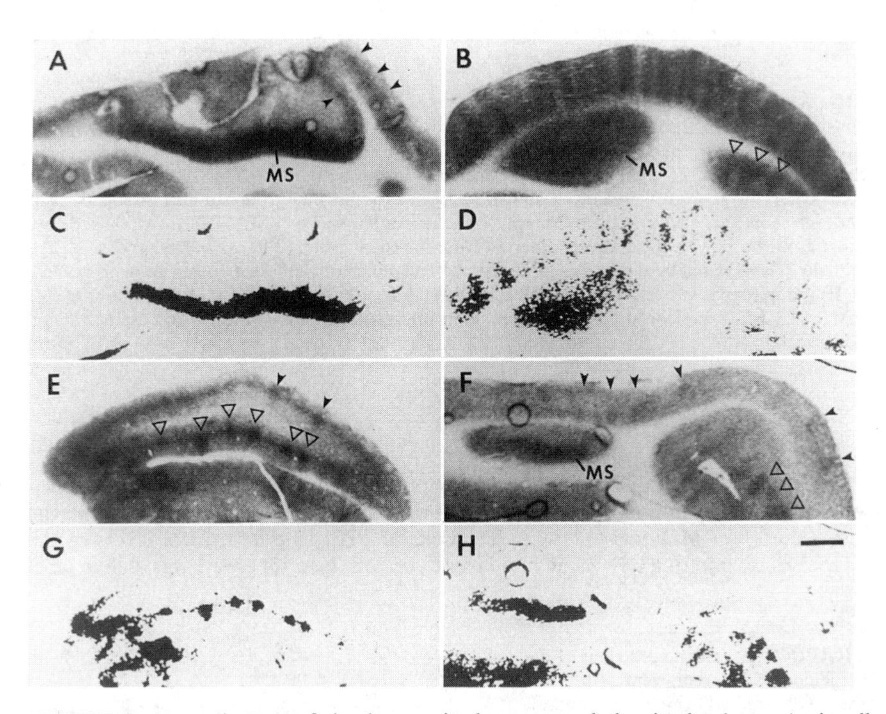

FIGURE 5. Autoradiograms of visual cortex in three monocularly stimulated cats. Again, all sections are horizontal, with anterior to the left and medial up. Each autoradiogram is presented twice: as a continuous tone print and with a threshold applied such that all pixels above threshold are shown as white and all pixels below threshold are shown as black. **(A & C)** Autoradiogram showing that the deprived eye columns in the binocular segment of cortex (arrowheads) were not as heavily labeled as the monocular segment of an MD cat. The threshold in **C**, while retaining all the label in the monocular segment, retains only artifact (air bubbles) in the binocular segment. **(B & D)** Autoradiogram from an equal AME cat showing that, in this animal, the ocular activation columns and the monocular segment were equally well labeled. The threshold in **D** causes a loss of label in the monocular segment and in the columns simultaneously. Note that the columns in area 18 (triangles) are disappearing at about the same rate. **(E & G)** Another autoradiogram from the same MD cat as in **A** and **C** showing that deprived eye columns were more heavily labeled in area 18 than in area 17. The threshold in **G** retains all of the columns in area 18, but very little of the label in area 17. **(F & H)** Autoradiogram and threshold from another MD cat showing that whereas the deprived eye columns in area 17 (arrowheads) were less heavily labeled than the monocular segment (MS), those in area 18 (triangles) were labeled as heavily as the monocular segment. Anterior to the left. Scale bar: 2.0 mm. Reprinted with permission from Tumosa *et al.*[25]

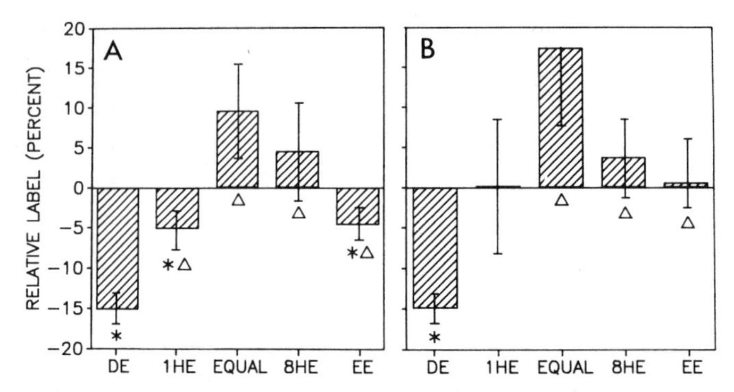

FIGURE 6. Binocular competition affects the density of ocular activation columns. (**A**) The density of label in the binocular segment of area 17 compared to the density of label in the stimulated layers in the LGN is shown for each experimental group. (**B**) The density of label in the binocular segment compared to the density of label of the monocular segment of area 17 contralateral to the stimulated eye is shown for each experimental group. In both **A** and **B,** each bar represents averaged data from three cats. Each error bar shows ±SEM. The asterisks indicate those groups that were significantly different from the equal group (8-hr eye of the AME 8/8 cat), and the triangles indicate those groups that were significantly different from the deprived eye (DE) group (pairwise t tests, $df = 4$, all $p < .05$). 1HE: 1-hr eye of AME 8/1; 8HE: 8-hr eye of AME 8/1; EE: experienced eye of MD cat. Reprinted with permission from Tumosa *et al.*[25]

SYNAPTIC NUMBER AND MORPHOLOGY

In order to study possible changes in synaptic structure following monocular deprivation, I used electron microscopic autoradiography to examine labeled profiles in the visual cortex of MD cats following injections of [^{3}H]lysine into single layers of the LGN.[38,39] By injecting [^{3}H]lysine into layer A of both hemispheres, I was able to label

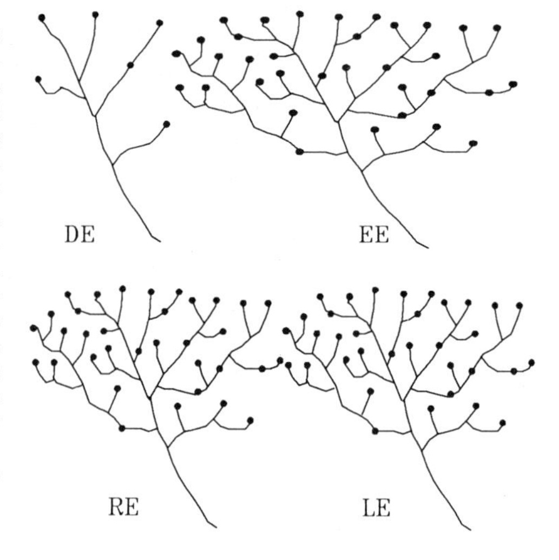

FIGURE 7. Schematic drawings of the geniculocortical arbors representing either eye of an MD cat (above) and an equal AME cat (below). In the equal AME cat, the arbors for the right eye (RE) and left eye (LE) are of equal density and size. In the MD cat, the arbor for the experienced eye (EE) is much larger than normal, and also less dense, because it cannot cover such a large area as well as it could cover its normal territory. The arbor for the deprived eye (DE) is much smaller than normal and is also less dense than the arbor for the experienced eye as well as those in the equal AME cat. This decreased density is clearly a result of the competition, because it does not occur in the monocular segment.

deprived afferents on one side and experienced afferents on the other, and thus to compare deprived and experienced afferents within a single animal simply by comparing labeled terminals in the left and right visual cortices.

Geniculocortical terminals could be labeled by injections into either the deprived or experienced layers of the LGN. In both cases, the terminals typically contained round synaptic vesicles and one or more mitochondrial profiles, and synapsed onto spines. Further, there were no striking morphological differences; that is, the deprived terminals were not obviously "sick" in some way. Quantitative comparisons, however, demonstrated that the geniculocortical terminals of the deprived pathway differed from those of the experienced pathway in several ways. These are summarized in FIGURE 8.

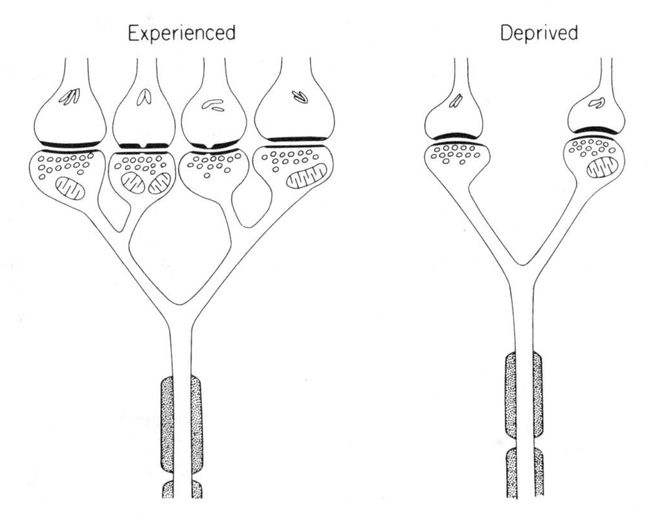

FIGURE 8. Schematic drawing of the changes in the synaptic connections from LGN to visual cortex produced by monocular deprivation. Deprived geniculocortical afferents branch less frequently and form fewer and smaller synaptic terminals, which contain fewer (or smaller) mitochondria, synapse onto smaller spines, and make more presynaptically convex and fewer perforated synapses than do experienced afferents. Reprinted with permission from Tieman.[39]

First, the terminals of deprived afferents were smaller than those of experienced afferents. This difference was consistent across animals; on average, the profiles of deprived terminals had 25% less area than did those of the experienced terminals. This may represent a stunted growth of the deprived terminals, because immature terminals are smaller than mature ones,[40,41] and seem to require activity for their normal growth.[42] Alternatively, because all cats studied were long-term deprived, the terminals may have been fully developed at one time and then atrophied. Larger terminals are presumably more effective than smaller ones; increased size is associated not only with maturation but also with kindling in rat cortical synapses,[43] with greater transmitter release at the frog neuromuscular junction,[44] and with decreased fatigability in the crayfish neuromuscular junction.[45,46]

Second, the deprived terminals contained 33% fewer mitochondrial profiles. This may simply be a function of their smaller size, since smaller terminals contained fewer mitochondrial profiles. This may reflect a reduction in mitochondrial number or a reduction in size and branching of mitochondria, as described by Lnenicka *et al.*[45] Certainly, however, it reflects a decrease in total mitochondrial volume, suggesting a decreased metabolic demand, and a decreased resistance to fatigue,[45] which is consistent with the report that cortical responses to stimulation of the deprived eye fatigue easily.[17,47] The decreased mitochondrial volume may be related to earlier reports of decreased activity of the mitochondrial enzyme cytochrome oxidase within the ocular dominance columns of the deprived eye of MD cats.[48] It is also likely, however, that individual mitochondria have decreased enzyme activity.[49] Further, because monocular deprivation also affects cytochrome oxidase levels in layers other than IV and VI,[48] the decreased mitochondrial content of geniculocortical terminals cannot be the entire explanation.

Third, although there was no difference in the length of the synaptic contact, there were differences in its shape. Experienced synapses were more likely to be presynaptically concave (FIG. 8). Petit[50,51] has called presynaptically concave synapses "frowning" and presynaptically convex synapses "smiling" because he illustrates them with the terminal above rather than below (imagine the synaptic vesicles as forming eyes and nose). He and others have shown that mature cortical[51,52] and spinal[53] synapses are more likely than immature ones to be frowning. Similarly, rearing in an enriched environment results in a preponderance of frowning synapses in rat cerebral cortex,[54] as does kindling,[43] and preventing the spontaneous electrical activity of developing cortical neurons in culture leads to an increase in smiling synapses,[55] as does deafferentation.[56] Thus, in spinal cord and cortex, frowning synapses are associated with maturation, training, and a history of activity. In contrast, potentiation in hippocampus is usually associated with smiling synapses.[57–59] This was first shown by Desmond and Levy,[57,58] who actually measured the curvature of the membrane apposition, rather than of the synaptic specialization itself. From the illustrations, it appears that in their studies the two variables are likely to be highly correlated. Further, Petit *et al.*[59] did measure the synaptic specialization, and found an increase in smiling synapses following repetitive stimulation due to systemic kainic acid. In contrast, Chang and Greenough[60] measured both variables; they found no change in the curvature of the specialization, and *fewer* smiling membrane appositions.

Thus, although there is some disagreement, in cortex frowning synapses are associated with conditions where one would expect more efficacious synapses, whereas in hippocampus the opposite is true. The reason for this apparent discrepancy is not clear. It seems unlikely that neocortex and hippocampus would use totally disparate mechanisms. Markus and Petit[51] have suggested that the different curvature responses may be due to different types of neurotransmitter or receptor. However, although the endogenous neurotransmitters have not been definitively established, glutamate receptors are known to be involved at both the potentiated hippocampal synapses[61,62] and the geniculocortical synapse,[63,64] suggesting that receptor type does not determine direction of curvature. One possible explanation may lie in the difference in timing: the hippocampal studies were done within hours of rather short-term stimulation, whereas the neocortical studies involve maturation, prolonged stimulation, or (in the case of the kindling study[43]) a 2-week poststimulation survival. Thus, enhancement in the short run may lead to smiling synapses and in the long run to frowning ones. A possible exception to this is the work of Pysh and Wiley,[65] who stimulated superior cervical ganglion during fixation (that is, poststimulation survival was 0) and examined the curvature of the membrane apposition. They found frowning synapses that they interpreted to reflect synaptic vesicle membrane that was fused with the presynaptic terminal membrane but that was not yet recycled. However, because these depleted synapses were *less* effective, not more, these data do not contradict the hypothesis that short-term enhancement is correlated with smiling synapses. A second possible exception is a kindling study in hippocampus in which the postseizure survival was 5 days.[66] The authors do not report a change in the proportion of smiling or frowning synapses, but do report that the enlargement of spine heads and stems normally associated with potentiation (see below) was seen only for the smiling synapses. To determine whether timing or synaptic type determines the curvature/efficacy correlation we need short- and long-survival kindling studies in hippocampus and cortex or a maturational study in hippocampus to determine whether hippocampal synapses change their curvature as they mature, and if so, in which direction.

The significance of change in curvature is also not clear, although it has been suggested that smiling synapses may be associated with more receptors or protein kinase[50,57–59,67] or larger postsynaptic spines[57,67] (see below), each of which could cause the synapse to be more effective. On the other hand, frowning synapses would be associated with larger presynaptic terminals[68] or might, by fanning out the presynaptic dense projections, facilitate access of the synaptic vesicles to fusion sites in the presynaptic membrane, and thereby enhance transmitter release. Additional discussion of this issue may be found in Jones[68] and Markus and Petit.[51]

Fourth, deprived synapses were less likely to have a gap in the postsynaptic density. Gaps in the postsynaptic density ("perforated" synapses) are associated with maturation,[52,69,70] rearing in an enriched environment[69] (although dark-rearing had no effect[70]), kindling in hippocampus,[71,72] and training.[73] The proportion of perforated synapses decreases in aged rats, and it has been suggested that this decrease may underlie the memory deficit seen in aged rats.[74] Again, the significance of the gap is not clear. Greenough *et al.*[69] originally examined this variable because they were intrigued by the suggestion of Peters and Kaiserman-Abramof[75] that the edge of the postsynaptic density might be the active site of the synapse, and thus, that gaps in the density might effectively strengthen the synapse. Thus, perforated synapses may represent a more efficacious type of synapse, either because the edge is the active site, as suggested by Peters and Kaiserman-Abramof,[75] or because it contains more attachment sites or calcium channels,[73] or perhaps the gap facilitates electrolyte or molecular movement across the membrane[69] or permits some specialized form of transmitter release.[76] Alternatively, perforated synapses may reflect a stage in the proliferation of synapses by division,[52,69,77] in their enlargement by accretion,[68] in synaptic renewal,[68] or in synaptic relocation (perhaps to a more efficacious site).[68] Whatever their significance in terms of synaptic mechanisms, however, their presence is clearly associated with conditions in which one would expect to find more efficacious synapses.

Fifth, deprived afferents synapsed onto smaller spines than did experienced afferents; the profiles of spines postsynaptic to deprived afferents were 28% smaller in area than those postsynaptic to experienced afferents. This finding is consistent with earlier observations that visual deprivation affects cortical spines[78–80] and with reports that long-term potentiation in the hippocampus,[57,81,82] training in honeybees,[83] and both maturation and experience in jewel fish[84,85] are associated with enlargement of dendritic spine heads. In many cases this is also associated with stem shortening and/or widening,[57,81,83–85] which would enhance the transmission of potentials from the spine head to the main dendrite.[86] In summary, then, monocular deprivation alters the geniculocortical synapse in a number of ways that may be correlated with its altered efficacy.

Does monocular deprivation also alter the number of geniculocortical synapses? To answer this, it was necessary to correct for variations in injection size and for a probable reduction in protein synthesis by cells in the deprived geniculate layers. Reasoning that larger injections or greater protein synthesis would label more axons as well as more terminals, for each hemisphere of each animal, I examined one or more strips 75 μm wide that extended through the depth of layer IV, counted all the labeled terminals and labeled myelinated axons, and then computed the ratio of labeled terminals to labeled myelinated axons. This ratio was 43% less for the deprived afferents than for the experienced afferents, suggesting that individual axons from deprived geniculate layers arborize less densely upon entering cortex and make fewer synapses. This suggestion is supported by the results of Friedlander and Vahle-Hinz,[87] who filled physiologically identified geniculocortical axons in MD cats with horseradish peroxidase and found that deprived axons arborized both over a small area and less densely within that area, and had fewer boutons.

Based on the faintness of the deprived eye activation columns relative to the labeling in the monocular segment, it is tempting to assume that this loss of synaptic terminals is a function of the competition rather than the deprivation per se. However, this is not entirely clear. First, I never examined the deprived geniculocortical synapses in the monocular segment. Second, binocular deprivation is reported to cause a selective loss in visual cortex of synapses from axons of subcortical origin.[88] Thus, the loss of deprived geniculocortical synapses may result at least in part from the deprivation itself.

Similarly, to the extent that they contributed to the faintness of the ocular activation columns, one might assume that the morphological changes were a function of the competition rather than the deprivation. It is also possible that deprivation and competition affect synaptic number and morphology differentially. Again, however, without a comparison of geniculocortical synapses within the deprived and experienced monocular segments, it is not possible to tell.

CANDIDATE NEUROTRANSMITTERS

One of the ways in which experience may modify synaptic efficacy is by regulating the amount of neurotransmitter available.[89–93] We have recently demonstrated that both retinal ganglion cells and LGN cells are immunoreactive for *N*-acetylaspartylglutamate (NAAG), a possible neurotransmitter.[94] In the LGN of a normal cat, both the neuropil and many cell bodies are heavily labeled. The neuropil labeling presumably represents the labeled terminals of retinal ganglion cells. To determine whether monocular deprivation affects the levels of NAAG in the geniculocortical pathway,[95] we reacted sections through the LGN of MD cats with antisera to NAAG[96] provided by Joe Neale of Georgetown University. Long-term monocular deprivation resulted in decreased labeling of the cell bodies, but not the neuropil, in the deprived layers of the LGN (FIG. 9). That is, the change in NAAG immunoreactivity was restricted to that part of the retinogeniculocortical pathway that is subject to binocular competition: the geniculocortical relay cells. There appeared to be no loss of label in the retinal ganglion cells of the deprived eye, just as there was no change in the labeling of their terminals in neuropil of the LGN.

In contrast to its effects on NAAG immunoreactivity, monocular deprivation had little or no effect on immunoreactivity for glutamate (FIG. 9b), the other obvious candidate for geniculocortical neurotransmitter.[97] Further, the changes in NAAG immunoreactivity of the LGN neurons in MD cats were much more striking than changes demonstrated with cytochrome oxidase histochemistry (FIG. 9c) or uptake and incorporation of [^{3}H]leucine (FIG. 9d). Thus, the changes are unlikely to be due to an overall decrease in metabolism or protein synthesis. These results suggest that long-term monocular deprivation selectively decreases the synthesis of NAAG in geniculate neurons. Whether this decrease simply reflects a decreased need for transmitter

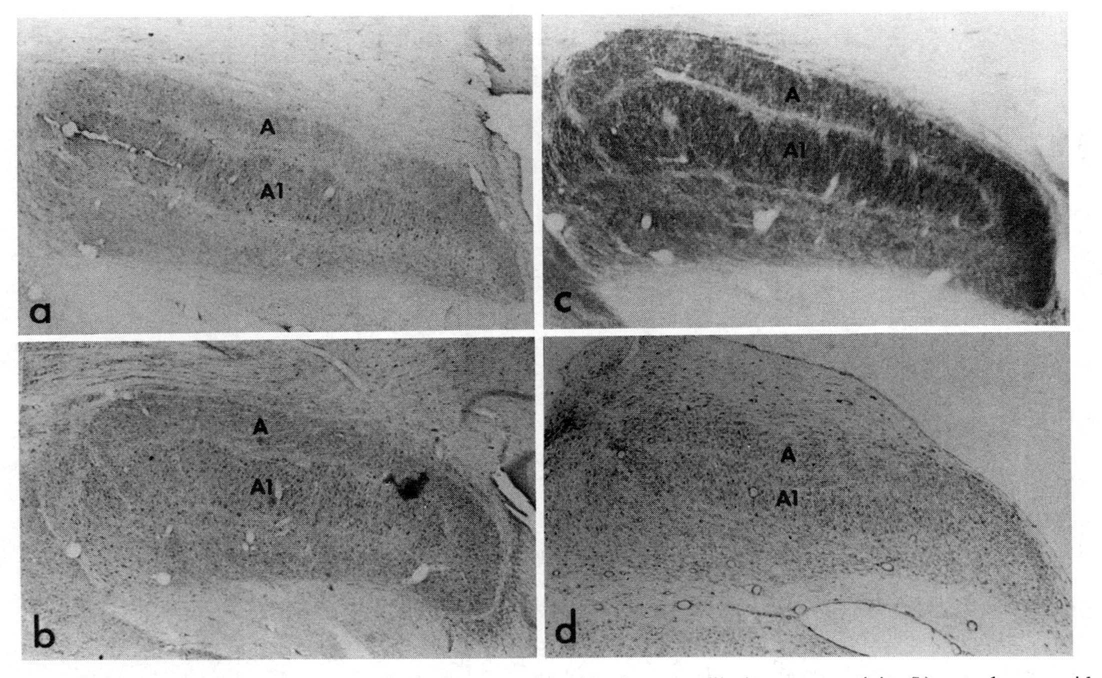

FIGURE 9. Effects of monocular deprivation on NAAG-like immunoreactivity (**a**), glutamate-like immunoreactivity (**b**), cytochrome oxidase histochemistry (**c**), and uptake and incorporation of [^{3}H]leucine into proteins (**d**). Parts **a**–**c** are from one MD cat, **d** from another. In each case, the left eye was deprived, and the right LGN is shown, so that A is the deprived layer and A1 is the experienced layer. In each case it is possible to discriminate the deprived and experienced layers, but the difference is clearest in **a,** where the cells in layer A show very little label compared to those in layer A1 (the neuropil label is the same). In **b** and **d,** A and A1 clearly differ in the sizes of cells, but the intensity of labeling is the same.

because the cell supports fewer synaptic terminals, or whether each terminal has less NAAG remains to be determined.

We have also examined the effect of monocular deprivation on immunoreactivity for GABA and its synthesizing enzyme, glutamic acid decarboxylase (GAD) (not shown). As has previously been reported for monocularly enucleated cats,[98] we observed no changes in immunoreactivity for GAD or GABA. In fact, it was very difficult to tell the deprived from the experienced layers, even on the basis of cell size. Because GAD and GABA are found not in relay cells but in interneurons,[99,100] which are not subject to binocular competition, these results perhaps are not surprising.

SUMMARY AND CONCLUSIONS

To summarize (FIG. 10), the structural consequences of monocular deprivation include the following changes: the relay cells in the binocular segments of the deprived geniculate layers shrink and contain less of the possible neurotransmitter NAAG. These changes appear to be secondary to a loss of terminal synaptic arbor. Certainly, deprived geniculocortical cells project to smaller ocular dominance patches in layer IV of visual cortex, where they make fewer and abnormal synapses. As a result, they activate ocular activation columns that, in addition to being small, are faint and usually fail to extend into extragranular layers. This failure to extend to other layers probably results from a failure of the poorly activated deprived-eye cells in layer IV to compete successfully with neighboring experienced-eye cells in layer IV, resulting in a loss of connections from layer IV to other layers (FIG. 11). Thus, the primary effect of monocular deprivation is probably the disruption of the geniculocortical synapse, with the other changes, such as cell size, and possibly the change in neurotransmitter content, being secondary. The disrupted synapse would result in poorly driven cortical cells and faint ocular activation columns, which in turn would bias a secondary competition for access to cells in extragranular layers.

There are certain general principles that unite the findings presented in this chapter with the others in this session.[37,46,101] First, there are similarities in the types of morphological changes observed, for example, changes in the number[37,101] and size[46] of synaptic

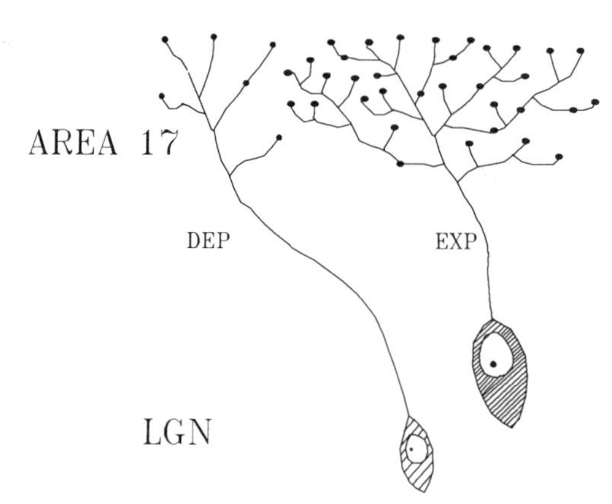

FIGURE 10. Summary of the effects of monocular deprivation on geniculocortical relay cells and their terminal arbors in area 17. On the left is shown a relay cell in deprived layer A1; on the right is a nearby relay cell in layer A. Note that the deprived cell is smaller, contains less NAAG (indicated by the lighter hatching), and terminates in a smaller, less dense arbor containing fewer and smaller boutons than does the experienced cell.

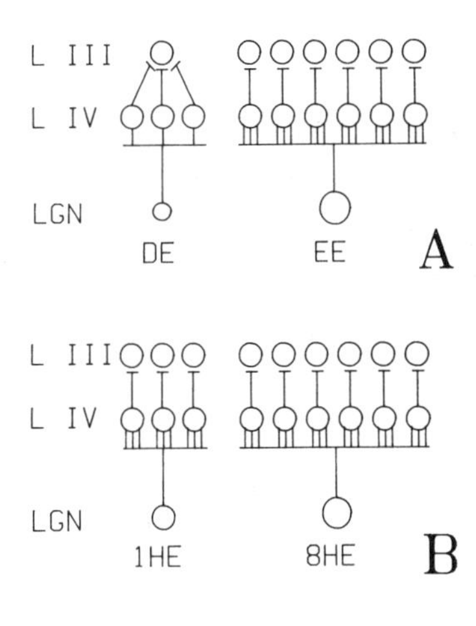

FIGURE 11. Schematic drawing illustrating the synaptic mechanisms thought to underlie the effects of binocular competition on size, shape, and density of ocular activation columns. (**A**) In MD cats, the geniculate afferents representing the deprived eye (DE) terminate over a smaller percentage of layer IV (L IV) than those representing the experienced eye (EE), and make fewer and abnormal synapses. As a result, the deprived eye drives the cells in layer IV so weakly that 1) they cannot effectively compete with the neighboring experienced eye cells for synaptic space on cells outside layer IV, and 2) it takes convergent inputs from many of these layer IV deprived eye cells to drive cells in the next layer. Therefore, the shape, as well as the size, of ocular activation columns in area 17 is altered. (**B**) In unequal AME cats, the 1-hr eye drives fewer cells in layer IV than the 8-hr eye, but drives them nearly as well. As a consequence, it competes effectively for its share of the cells outside layer IV, such as those in layer III (L III), and the columns are approximately the same width in all layers. Reprinted with permission from Tumosa *et al.*[25]

terminals, as well as mitochondrial changes.[46] This implies that there are similar changes during development and adult plasticity and also similar changes in vertebrates and invertebrates.

Second, it is not so much the amount of activity that determines these changes, but the pattern of activity.[46,101] In my results, the relative imbalance in activity is important, but not the absolute amount (for example, the columns activated by the 8-hr eye of an AME 8/1 are different from those activated by the 8-hr eye of an AME 8/8). Similarly, the binocular segment, where there was an imbalance and competition could occur, was affected, whereas the monocular segment, where there was no imbalance and competition could not occur, was not. Finally, the recent results of Reiter and Stryker[102] suggest that monocular deprivation produces changes only when the activity of the presynaptic cell and the postsynaptic cell are correlated. They showed that in the cortex of MD kittens where the postsynaptic cell was prevented from firing by the infusion of muscimol, a GABA agonist, most cells were dominated by the deprived eye, and not the experienced eye. Thus, both pre- and postsynaptic activity are important and must be correlated with each other. Correlated postsynaptic activity is apparently produced by simultaneously active presynaptic inputs and detected by activation of *N*-methyl-D-aspartate receptors as discussed by others in this volume.[103–106]

ACKNOWLEDGMENTS

I thank Dr. Joe Neale for his generous contribution of antibodies; Drs. Nina Tumosa and Dave Tieman for participating in some of the experiments described here; Drs. Greg Lnenicka, John Schmidt, and Dave Tieman for comments; and Mary Liu, Beth O'Neill, Margo Weldon, Denise Zimmerman, Susan LePage, and Susan Irtenkauf for excellent technical assistance.

REFERENCES

1. HUBEL, D. H. 1982. Nature **299:** 515-524.
2. WIESEL, T. N. 1982. Nature **299:** 583-591.
3. WIESEL, T. N. & D. H. HUBEL. 1965. J. Neurophysiol. **28:** 1060-1072.
4. GANZ, L., H. V. B. HIRSCH & S. B. TIEMAN. 1972. Brain Res. **44:** 547-568.
5. WIESEL, T. N. & D. H. HUBEL. 1963. J. Neurophysiol. **26:** 978-993.
6. WIESEL, T. N. & D. H. HUBEL. 1963. J. Neurophysiol. **26:** 1003-1017.
7. SHERMAN, S. M. & P. D. SPEAR. 1982. Physiol. Rev. **62:** 738-855.
8. HIRSCH, H. V. B., D. G. TIEMAN, S. B. TIEMAN & N. TUMOSA. 1987. *In* Imprinting and Cortical Plasticity. J. P. Rauschecker & P. Marler, Eds.: 286-320. Wiley. New York, NY.
9. HICKEY, T. L. & R. W. GUILLERY. 1974. J. Comp. Neurol. **156:** 239-254.
10. SANDERSON, K. J. 1971. J. Comp. Neurol. **143:** 101-118.
11. HUBEL, D. H. & T. N. WIESEL. 1962. J. Physiol. (London) **160:** 106-154.
12. TUSA, R. J., L. A. PALMER & A. C. ROSENQUIST. 1978. J. Comp. Neurol. **177:** 213-236.
13. TUSA, R. J., A. C. ROSENQUIST & L. A. PALMER. 1979. J. Comp. Neurol. **185:** 657-678.
14. HUBEL, D. H. & T. N. WIESEL. 1965. J. Neurophysiol. **28:** 229-289.
15. LEVAY, S. & C. D. GILBERT. 1976. Brain Res. **113:** 1-19.
16. SHATZ, C. J., S. LINDSTRÖM & T. N. WIESEL. 1977. Brain Res. **131:** 103-116.
17. WIESEL, T. N. & D. H. HUBEL. 1965. J. Neurophysiol. **28:** 1029-1040.
18. GUILLERY, R. W. & D. J. STELZNER. 1970. J. Comp. Neurol. **139:** 413-422.
19. GUILLERY, R. W. 1972. J. Comp. Neurol. **144:** 117-130.
20. BEAR, M. F. & H. COLMAN. 1990. Proc. Natl. Acad Sci. USA **87:** 9246-9249.
21. THORPE, P. A. & C. BLAKEMORE. 1975. Neurosci. Lett. **1:** 271-276.
22. SINGER, W., H. HOLLÄNDER & H. VANEGAS. 1977. Brain Res. **120:** 133-137.
23. SHATZ, C. J. & M. P. STRYKER. 1978. J. Physiol. (London) **281:** 267-283.
24. TIEMAN, S. B. 1990. Unpublished results.
25. TUMOSA, N., S. B. TIEMAN & D. G. TIEMAN. 1989. Vis. Neurosci. **2:** 391-407.
26. HUBEL, D. H. & T. N. WIESEL. 1965. J. Neurophysiol. **28:** 1041-1059.
27. TUMOSA, N., S. B. TIEMAN & H. V. B. HIRSCH. 1980. Science **208:** 421-423.
28. TIEMAN, S. B. & N. TUMOSA. 1983. Brain Res. **267:** 35-46.
29. BONDS, A. B., M. S. SILVERMAN, G. SCLAR & R. B. TOOTELL. 1980. Invest. Ophthalmol. Vis. Sci. **19**(Suppl.): 225.
30. KOSSUT, M., I. D. THOMPSON & C. BLAKEMORE. 1983. Acta Neurobiol. Exp. **43:** 273-282.
31. TIEMAN, D. G., M. A. MCCALL & H. V. B. HIRSCH. 1983. J. Neurophysiol. **49:** 804-818.
32. DEVOR, M. & G. E. SCHNEIDER. 1975. *In* Aspects of Neuronal Plasticity. F. Vital-Durand & M. Jeannerod, Eds.: 191-201. INSERM. Paris.
33. CHAMBERS, W. W., C. N. LIU & G. P. MCGOUCH. 1973. Brain Behav. Evol. **8:** 5-26.
34. STEWARD, O., C. W. COTMAN & G. S. LYNCH. 1976. Brain Res. **114:** 181-200.
35. COTMAN, C. W., C. GENTRY & O. STEWARD. 1977. J. Neurocytol. **6:** 455-464.
36. BAILEY, C. H. & M. CHEN. 1988. Proc. Natl. Acad. Sci. USA **85:** 2373-2377.
37. BAILEY, C. H. & M. CHEN. 1991. Ann. N.Y. Acad. Sci. This volume.
38. TIEMAN, S. B. 1984. J. Comp. Neurol. **222:** 166-176.
39. TIEMAN, S. B. 1985. Cell. Mol. Neurobiol. **5:** 35-45.
40. DYSON, S. W. & D. G. JONES. 1980. Brain Res. **183:** 43-59.
41. MASON, C. A. 1982. Neuroscience **7:** 541-559.
42. KALIL, R. E., M. W. DUBIN, G. SCOTT & L. A. STARK. 1986. Nature **323:** 156-158.
43. RACINE, R. & J. ZAIDE. 1978. *In* Limbic Mechanisms: The Continuing Evolution of the Limbic System Concept. K. E. Livingston & O. Hornykiewicz, Eds.: 457-493. Plenum. New York, NY.
44. KUNO, M., S. M. TURKANIS & J. N. WEAKLY. 1971. J. Physiol. (London) **213:** 545-556.
45. LNENICKA, G. A., H. L. ATWOOD & L. MARIN. 1986. J. Neurosci. **6:** 2252-2258.
46. LNENICKA, G. A. 1991. Ann. N.Y. Acad. Sci. This volume.
47. GANZ, L., M. FITCH & J. A. SATTERBERG. 1968. Exp. Neurol. **22:** 614-637.
48. WONG-RILEY, M. T. T. 1979. Brain Res. **171:** 11-28.

49. Wong-Riley, M. T. T. & C. Welt. 1980. Proc. Natl. Acad. Sci. USA **77:** 2333-2337.
50. Petit, T. L. 1988. *In* Neural Plasticity: A Lifespan Approach. T. L. Petit & G. O. Ivy, Eds.: 201-234. Alan R. Liss. New York, NY.
51. Markus, E. J. & T. L. Petit. 1989. Synapse **3:** 1-11.
52. Dyson, S. E. & D. G. Jones. 1984. Dev. Brain Res. **13:** 125-137.
53. Hayes, B. P. & A. Roberts. 1974. Cell Tissue Res. **153:** 227-244.
54. Wesa, J. M., F. F. Chang, W. T. Greenough & R. W. West. 1982. Dev. Brain Res. **4:** 253-257.
55. Van Huizen, F., H. J. Romijn & M. A. Corner. 1987. Dev. Brain Res. **31:** 1-6.
56. Benshalom, G. & E. L. White. 1988. Brain Res. **443:** 377-382.
57. Desmond, N. L. & W. B. Levy. 1983. Brain Res. **265:** 21-30.
58. Desmond, N. L. & W. B. Levy. 1986. J. Comp. Neurol. **253:** 466-475.
59. Petit, T. L., J. C. LeBoutillier, E. J. Markus & N. W. Milgram. 1989. Exp. Neurol. **105:** 72-79.
60. Chang, F. F. & W. T. Greenough. 1984. Brain Res. **309:** 35-46.
61. Collingridge, G. L. & T. V. P. Bliss. 1987. TINS **10:** 288-293.
62. Cotman, C. W., D. T. Monaghan, O. P. Ottersen & J. Storm-Mathisen. 1987. TINS **10:** 273-279.
63. Tsumoto, T., H. Masui & H. Sato. 1986. J. Neurophysiol. **55:** 469-483.
64. Hagihara, K., T. Tsumoto, H. Sato & Y. Hata. 1988. Exp. Brain Res. **69:** 407-416.
65. Pysh, J. J. & R. G. Wiley. 1972. Science **176:** 191-194.
66. Cronin, A. J., T. P. Sutula & N. L. Desmond. 1987. Soc. Neurosci. Abstr. **13:** 947.
67. Desmond, N. L. & W. B. Levy. 1988. *In* Long-Term Potentiation: From Biophysics to Behavior. P. W. Landfield & S. A. Deadwyler, Eds.: 265-305. Alan R. Liss. New York, NY.
68. Jones, D. G. 1988. *In* Neural Plasticity: A Lifespan Approach. T. L. Petit & G. O. Ivy, Eds.: 21-42. Alan R. Liss. New York, NY.
69. Greenough, W. T., R. W. West & T. J. DeVoogd. 1978. Science **202:** 1096-1098.
70. Müller, L., A. Pattiselanno & G. Vrensen. 1981. Brain Res. **205:** 39-48.
71. Geinisman, Y., F. Morrell & L. De Toledo-Morrell. 1990. Brain Res. **507:** 325-331.
72. Geinisman, Y., L. De Toledo-Morrell & F. Morrell. 1990. Brain Res. **513:** 175-179.
73. Vrensen, G. & J. Nunes-Cardozo. 1981. Brain Res. **218:** 79-97.
74. Geinisman, Y., L. De Toledo-Morrell & F. Morrell. 1986. Proc. Natl. Acad. Sci. USA **83:** 3027-3031.
75. Peters, A. & I. R. Kaiserman-Abramof. 1969. Z. Zellforsch. Mikrosk. Anat. **100:** 487-506.
76. Sirevaag, A. M. & W. T. Greenough. 1985. Dev. Brain Res. **190:** 215-226.
77. Carlin, R. K. & P. Siekevitz. 1983. Proc. Natl. Acad. Sci. USA **80:** 3517-3521.
78. Globus, A. & A. B. Scheibel. 1967. Exp. Neurol. **19:** 331-345.
79. Valverde, F. 1967. Exp. Brain Res. **3:** 337-352.
80. Freire, M. 1978. J. Anat. **126:** 193-201.
81. Fifková, E. & A. Van Harreveld. 1977. J. Neurocytol. **6:** 211-230.
82. Applegate, M. D., D. S. Kerr & P. W. Landfield. 1987. Brain Res. **401:** 401-406.
83. Brandon, J. G. & R. G. Coss. 1982. Brain Res. **252:** 51-61.
84. Coss, R. G. & A. Globus. 1978. Science **200:** 787-790.
85. Burgess, J. W. & R. G. Coss. 1983. Brain Res. **266:** 217-223.
86. Rall, W. 1974. *In* Cellular Mechanisms Subserving Changes in Neuronal Activity. C. D. Woody, K. D. Brown, T. J. Crow & J. D. Knipsel, Eds.: 13-21. Brain Research Institute. Los Angeles, CA.
87. Friedlander, M. J. & C. Vahle-Hinz. 1982. Soc. Neurosci. Abstr. **8:** 2.
88. Turlejski, K. & M. Kossut. 1985. Brain Res. **331:** 115-125.
89. Black, I. B., I. A. Hendry & L. L. Iversen. 1971. Brain Res. **34:** 229-240.
90. Kessler, J. A., J. E. Adler, M. C. Bohn & I. B. Black. 1981. Science **214:** 335-336.
91. Zigmond, R. E. & A. Chalazonitis. 1979. Brain Res. **164:** 137-152.
92. Chalazonitis, A. & R. E. Zigmond. 1980. J. Physiol. (London) **300:** 525-538.
93. Hendry, S. H. C. & E. G. Jones. 1986. Nature **320:** 750-753.
94. Tieman, S. B., C. B. Cangro & J. H. Neale. 1987. Brain Res. **420:** 188-193.

95. TIEMAN, S. B. & S. M. LEPAGE. 1988. Invest. Ophthalmol. Vis. Sci. **29**(Suppl): 32.
96. ANDERSON, K. J., D. T. MONAGHAN, C. B. CANGRO, M. A. A. NAMBOODIRI, J. H. NEALE & C. W. COTMAN. 1986. Neurosci. Lett. **72:** 14-20.
97. MONTERO, V. M. 1990. Vis. Neurosci. **4:** 437-443.
98. BEAR, M. F., D. E. SCHMECHEL & F. F. EBNER. 1985. J. Neurosci. **5:** 1262-1275.
99. FITZPATRICK, D., G. R. PENNY & D. E. SCHMECHEL. 1984. J. Neurosci. **4:** 1809-1829.
100. MONTERO, V. M. & J. ZEMPEL. 1985. Exp. Brain Res. **60:** 603-609.
101. GREENOUGH, W. T. & B. J. ANDERSON. 1991. Ann. N.Y. Acad. Sci. This volume.
102. REITER, H. O. & M. P. STRYKER. 1988. Proc. Natl. Acad. Sci. USA **85:** 3623-3627.
103. BEAR, M. F. & S. M. DUDEK. 1991. Ann. N.Y. Acad. Sci. This volume.
104. MEFFERT, M. K., K. D. PARFITT, V. A. DOZE, G. A. COHEN & D. V. MADISON. 1991. Ann. N.Y. Acad. Sci. This volume.
105. SCHMIDT, J. T. 1991. Ann. N.Y. Acad. Sci. This volume.
106. UDIN, S. B. & W. J. SCHERER. 1991. Ann. N.Y. Acad. Sci. This volume.

Cerebellar Synaptic Plasticity

Relation to Learning versus Neural Activity[a]

WILLIAM T. GREENOUGH AND
BRENDA J. ANDERSON

Departments of Psychology and Cell and Structural Biology
and
Neuroscience Program
Beckman Institute
University of Illinois at Urbana-Champaign
Urbana, Illinois 61801

INTRODUCTION: CEREBELLAR CORTICAL ORGANIZATION

The title of this volume, *Activity-Driven CNS Changes in Learning and Development,* has led us to emphasize what events are actually necessary and sufficient to initiate functional (and structural) change in the nervous system. We believe that there is substantial evidence that *activity,* per se, is not invariably the critical factor for the induction of a change in functional organization, although it appears to be sufficient to bring about changes in the metabolic support of highly active regions in some cases. Rather, there appears to be some very specific requirement for *learning,* or some other change in the pattern of neural activity, to drive the CNS to reorganize itself. This requirement must ultimately be expressed at the cellular level in a sequence of steps leading to changes in functional organization. Although the steps may well be understood ultimately in terms of the molecular events that underlie the cellular change, our goal is to understand this reorganization from the perspective of behavior. We seek to know whether the behavioral demands placed upon a subject can determine the neural response and if so how. As discussed later in this article, we have seen structural changes in the cerebellum following learning. As a result, we are curious as to how these changes come about, and how they might fit in the overall output of the cerebellar cortex.

Both experimental results and theory predict that the cerebellar cortex performs a plastic role in behavioral adaptation. As FIGURE 1 shows, there are two major excitatory informational inputs, both arising from brainstem and spinal loci, and a single output neuron, the Purkinje cell.[2] Climbing fibers, which originate largely or exclusively from the inferior olive, carry information from the somatosensory, vestibular, visual, and to a lesser extent auditory modalities. Mossy fibers arise from various locations in the brainstem and spinal cord and also carry sensory input, such that there is no simple, straightforward distinction with regard to the types of information the two afferent systems convey. On occasion the same Purkinje cell may be activated by both climbing fiber and parallel fiber input from the same peripheral location.[3,4] In both systems the input is modulated by descending projections from the forebrain. Climbing fibers terminate directly upon Purkinje cells, each Purkinje cell receiving multiple synaptic

[a] Preparation of this document and research not otherwise reported was supported by grants from the National Science Foundation (MH-43830, MH-40631, BNS 88-21219, and MH-18412).

contacts from a single climbing fiber. Climbing fibers also terminate on the various inhibitory interneurons discussed below. Mossy fibers synapse onto the dendrites of granule cells at a complicated glomerular synaptic junction that also receives input from inhibitory neurons. The granule cell axons project to the molecular layer above the Purkinje cell bodies where they become the parallel fibers that make contacts with large numbers of Purkinje and inhibitory interneuron dendrites. Whereas the parallel fibers synapse upon the distal *spiny branches* of Purkinje neurons, climbing fibers terminate on the *main branches,* which are closer to the soma. Parallel fibers synapse once or twice with a given Purkinje cell, whereas the climbing fiber makes many

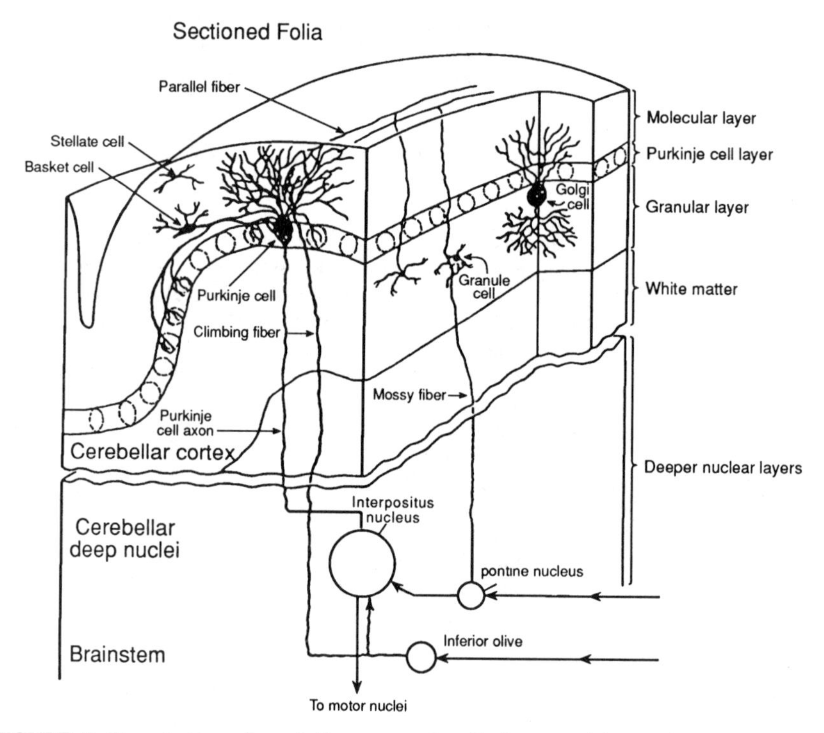

FIGURE 1. Organization of cerebellar cortex. Details in text. Adapted from Prosser and Greenough.[1]

contacts with each Purkinje cell it innervates. Because of this pattern, climbing fiber input gives rise to *complex spikes,* which are easily discriminable from the *simple spikes* that result from parallel fiber activity. The interaction between these two systems has been a primary topic of theories of cerebellar function and plasticity.

In parallel with these two excitatory systems, there exists an intrinsic network of inhibitory interneurons, consisting of Golgi, stellate, and basket neurons. These cells are also activated by parallel and climbing fibers and inhibit both Purkinje cells and other cells within the cerebellar cortex. Theorists have disagreed as to whether plasticity is confined to one or both excitatory systems or extends to the inhibitory neurons. The

Purkinje cell output from the cerebellar cortex is inhibitory and terminates on neurons in the deep nuclei of the cerebellum and the vestibular nuclei, modulating throughput to motor systems.

CEREBELLAR CORTICAL PLASTICITY

Theories of Cerebellar Plasticity

Many theorists have noted that the organization of cerebellar cortex lends itself easily to models that incorporate plastic changes of the sorts that might be involved in motor learning. Although there are theories that do not include plasticity in cerebellar function (see, for example, references 5-7), in this article we focus on those that do because they have provided hypotheses that can be tested. It should be noted that all theorists propose the strengthening and/or weakening of existing synapses, whereas the data we present below reflect the addition or loss of synapses. In principle there is no serious obstacle to considering these to be equivalent mechanisms. However, synapse formation can provide options potentially unavailable through existing connections, such as recruitment of new cells, depending upon the degree of redundancy in the network. The first thorough theory of cerebellar function was presented by Marr.[8] In Marr's model, "elemental movements" were initially instigated by the action of olivary climbing fiber input. In the learning process, parallel fibers, the activity of which reflect the temporal and physical context of the movement, become capable of initiating subsequent elemental movements in a movement sequence. The mechanism underlying this learning was facilitation, or strengthening, of parallel fiber synapses on Purkinje cells that were activated at the time the movement was performed under the governance of climbing fiber activity. Sequences of movement occurred based on peripheral feedback that, via parallel fibers, provided context for the next movement. Thus conjunctive activation of parallel fibers and climbing fibers was the critical element leading to synaptic change. Marr assumed that only parallel fiber synapses were modifiable, although he noted that modifiable inhibitory synapses could work in his theory. Thus, if synapses did form, Marr would predict an increase in the number of parallel fiber synapses on Purkinje neurons to underlie motor learning. Eccles,[9] like Marr, proposed increased parallel fiber to Purkinje cell synapse strength, but was explicit in specifying the selection from preexisting synapses of a population to be strengthened.

Albus[10] proposed a variation of Marr's theory compatible with the perceptron network models that were then popular and included an extensive quantitative analysis of cerebellar cortical organization that supported the application of this model. Albus noted that climbing fibers can transiently block Purkinje cell responsiveness to parallel fiber input (an "inactivation response" lasting for 40 msec or more) via a prolonged depolarization. Albus proposed that, through repeated conjunctive stimulation of climbing and parallel fibers, the parallel fibers to Purkinje synapses would weaken until they became insufficient to activate the Purkinje cell. Eventually, in behavioral conditions in which the weakened parallel fibers to Purkinje synapses are the only active synapses, the Purkinje cell will be shut down in a manner analogous to that resulting from the climbing fiber inactivation response. As the output of the Purkinje cell is inhibitory, a pattern of selective inactivation can "allow" an extrinsic movement to proceed. Albus, who also proposed that parallel fiber synapses on basket and stellate cells followed the same weakening rule, hypothesized that some inhibitory synapses, such as those on certain stellate neurons, might also be variable in strength. Thus, if synapse numbers change, Albus would predict a decrease in the number of parallel fiber synapses with learning, although he also speculated at one point that new parallel fiber synapses might "spontaneously and randomly grow and mature" to prevent total Purkinje cell deactivation and allow for new learning.

Gilbert[11] proposed that movement memories were stored in groups of Purkinje cells, with a collective pattern of output frequencies appropriate to each of the individual movements of a muscle. Thus the same collection of cells was reused for each of the multiple movements, and each memory constituted a separate pattern of frequencies. The strength of both parallel fiber synapses and basket cell synapses was adjusted to achieve a parallel fiber driven output frequency for each Purkinje cell that was specified by its climbing fiber input. Although increases in synapse strength were to occur initially, decreases could occur as well. For our purposes, the primary effect of learning should be an increase in parallel fiber synapses on Purkinje cells and in basket cell synapses on Purkinje cells.

Ito has proposed a model for the vestibular-ocular reflex (VOR). This reflex is responsible for adaptation to discrepancies between the vestibular-driven eye movement compensation for head rotation and the position of the retinal image (this paradigm is discussed more thoroughly in the paper by Peterson *et al.*[12]). Errors in retinal position are signaled via climbing fiber input, and parallel fiber-Purkinje cell synapses are adjusted to provide a phasic output that offsets the disparity in the brainstem circuitry. Fujita[13] has presented a formalization of this scheme. The form of plasticity proposed to underlie this parallel fiber change is long-term depression (LTD), a reduction in parallel fiber synaptic efficacy arising from conjunction of parallel fiber and climbing fiber activity. Although the mechanisms proposed to underlie LTD involve receptor desensitization at existing synapses,[14] an alternative, loss of synapses, seems compatible with most of the data.

Physiological Evidence for Cerebellar Cortical Plasticity

There have been a few tests of the associative interactions between parallel and climbing fibers as predicted by Marr and Albus. One effect of climbing fiber activity on Purkinje cell responses to parallel fiber activation is long-term depression. Rawson and Tilokskulchai[15] have reported a lasting depression of parallel fiber stimulation-evoked simple spike activity following low-frequency activation of climbing fiber inputs. This depression lasted for about the duration of the climbing fiber stimulation, up to as long as 8 min. When parallel fiber and climbing fiber stimulation were paired, a similar depression occurred, and when parallel fiber stimulation was combined with 4-Hz climbing fiber activation, the subsequent response to a normally effective parallel fiber stimulus was completely suppressed. Ito[16] reported a similar effect of low-frequency climbing fiber activation paired with a prolonged stimulus to the vestibular nerve. This depression was maximal after about 10 min and persisted to some degree for more than 1 hr. A stimulus-specific depression was reported by Ekerot and Kano[17] in response to 8 min of conjoint parallel fiber and climbing fiber activation. Subsequent responses to stimulation of the parallel fiber pathway were depressed over a period of 1-2 hr, with peak depression occurring at 10-20 min following the stimulation period. Responsiveness to stimulation of another set of parallel fibers was unaffected, indicating that the alteration was not a change in general Purkinje cell excitability but was specific to the pathway involved in the conjunctive stimulation. The effect was diminished as the interval between climbing fiber stimulation and parallel fiber stimulation was increased. The depression could be counteracted if inhibition was produced by stimulating parallel fibers "off-beam" from the Purkinje cell under study, a finding that was interpreted to support the involvement of climbing fiber-induced depolarization or "plateau potentials" in Purkinje cell dendrites in the depressed response. The modulatory effect of inhibition is not predicted by the Marr or Albus models. In general, long-term depression can be elicited with climbing fiber activation either preceding or following parallel fiber activation.[14]

A short-term bidirectional effect of pairing parallel fiber stimulation with spontaneous climbing fiber activity has also been reported.[18] If simple spike activity normally increases following climbing fiber spikes, that increase is accentuated when parallel fiber stimulation follows a climbing fiber spike. Similarly, Purkinje responses decrease further after climbing fiber spikes have originally depressed simple spike activity. These results suggest that climbing fibers change the gain of the Purkinje cell's response to parallel fiber input. It should be noted that the order of climbing fiber and parallel fiber activation in this study is precisely the reverse of that proposed to trigger plasticity in most theoretical models.

Parallel fiber stimulation (4 Hz) alone (or paired with subthreshold stimulation of climbing fibers in the white matter) has been found to result in potentiation of simple spikes in guinea pig cerebellar slices.[19] Significant potentiation lasted an average of 20 min, although potentiation in one cell lasted for 50 min. No changes in membrane potential or input resistance were seen, suggesting that the change is specific to parallel-Purkinje synapses. When parallel fibers were stimulated conjunctively with climbing fibers (4 Hz, 25 sec), long-term depression was seen, again for up to 50 min. Again, there were no changes in membrane potential or passive membrane properties of dendrites suggesting that the depression occurs at specific parallel-Purkinje cell synapses.

Taken together, the physiological results suggest that both increases and decreases in the response to parallel fiber activation may be elicited by climbing fiber activation. Despite this, most theories suggest only unidirectional efficacy change at this synapse. In addition, there has been little emphasis on truly long-lasting change, and there does not yet appear to be physiological evidence for effects that persist as long as do some forms of learning. In part, this lack reflects difficulty in maintaining extended, stable recordings.

Behavioral Evidence for Cerebellar Cortical Plasticity

In several motor learning paradigms cerebellar cortical activity changes as the animal learns. Whether or not these changes underlie learning or are a reflection of learning in an afferent structure is the subject of much controversy, as other articles in this volume attest. The data are more extensively covered in the articles by Bloedel *et al.*,[7] Peterson *et al.*,[12] and Yeo.[20] In VOR and associative eyeblink conditioning, both increases and decreases in simple spikes have been seen, whereas Gilbert and Thach report a decrease in simple spike activity during motor learning.[21] Adaptive modification of the VOR involves bidirectional changes in simple spike activity to synchronize eye movement rhythms with those of induced body rotation.[12,22] For associative eyeblink conditioning, bidirectional modification of simple spike activity is also reported. Foy and Thompson[23] reported that the predominant class of cells in conditioned animals decreased their simple spike firing in a manner that modeled the conditioned response (CR). Other cells showed simple spike increases prior to the CR. In contrast, Berthier and Moore[24] reported that most Purkinje neurons that responded to the conditioned stimulus increased their simple spike firing prior to the CR, with about 27% decreasing their activity.

MORPHOLOGICAL CHANGE AS A BASIS FOR CEREBELLAR PLASTICITY

Morphological Plasticity as a Widespread Feature of the Nervous System

Although the search for effects of experience upon cerebellar morphology constitutes a relatively small literature, an extensive backdrop of effects of experience upon the *morphology of neurons and supportive cells* in the cerebral cortex and elsewhere suggests the likelihood of such effects. Begun in the wake of the pioneering Berkeley studies showing that the housing environment altered cortical weight and thickness,[25] an extensive literature now indicates that the complexity of the rearing environment can affect dendritic branching of cortical neurons,[26–28] the number of synapses per neuron,[29,30] and the size and numbers of supportive structures such as glial cells[31–34] and capillaries.[35,36] Many similar effects have been reported for adult cerebral cortex[37–41] although the degree of responsiveness may diminish with advancing age (see, for example, reference 41). For the cerebral cortex, a detailed cell biology of morphological plasticity as it relates to synapses, neurons, glial cells, and vascular tissue is beginning to emerge, although it is far from complete, and the signaling mechanisms that give rise to these changes are completely unknown. Subcortical structures such as the hippocampal formation[43,44] and superior colliculus[45] also exhibit morphological plasticity of these sorts, although investigations have been much less detailed. A similarly detailed catalogue of effects of experience and other manipulations, such as long-term potentiation or other effects of electrical stimulation, on the *morphology of individual synapses* has been compiled, as indicated in some other articles in this volume. Changes have been described in the size and shape of various parts of synapses such as the postsynaptic density (PSD), for example, and in the structure of PSD densities (for reviews, see references 46 and 47). The probable relationship of at least some of these morphological phenomena to plastic functional processes has been emphasized by their occurrence in adult animals in learning situations.[48–51] One might almost expect similar plastic phenomena to be evident in cerebellar cortex, given proper conditions of looking for them.

Morphological Plasticity in the Cerebellar Cortex

Although most theorists refer to cerebellar plasticity as changes in synaptic strength, plasticity could be in the form of synaptic number changes. Several experiments have demonstrated the ability of the adult cerebellum to form new synapses. Hillman and Chen[52,53] found evidence for the formation of new synapses in adult rats within 6 to 8 hr following transections of parallel fibers. The short time interval between transection and the formation of new synapses has been suggested as evidence for a reserve pool of postsynaptic molecules ready to form new synapses. Climbing fibers also sprout and innervate new Purkinje cells following partial lesions of the inferior olive.[54] These data indicate that synapse numbers can change in adult animals and, consequently, such changes should be investigated as a potential mechanism underlying behavioral change.

Injections of harmaline (a monoamine oxidase inhibitor) have been shown to increase climbing fiber spikes from approximately 0.5 to 19 Hz.[55] There is a corresponding decrease in simple spikes despite a lack of any change in parallel fiber activity. Because these results correspond to Albus's theory, it is interesting to note that there were no changes in morphology of parallel and climbing fiber synapses onto Purkinje cells, examined as a group (measures: cross-sectional area of the postsynaptic density (PSD), diameter of apposition, thickness of PSD, vesicle numbers, curvature of PSD, and perforations in the PSD).[56] Synapse number changes were not investigated. The harmaline-induced depression in Purkinje cell simple spikes is a short-lasting response, a characteristic *not* consistent with Albus's theory. Biochemical and/or ionic processes seem more likely to underlie the short-lasting physiological changes, although changes in synaptic number cannot be ruled out.

EXPERIENCE AND MORPHOLOGICAL PLASTICITY IN THE CEREBELLAR CORTEX

Effects of Experience during Development

The first evidence that behavioral experience could influence the structure of neurons in the developing cerebellar cortex came from similar studies of mice and monkeys. Pysh and Weiss[57] reared mice in object-filled cages that provided various opportunities for motor activity. In addition, the mice were forced to swim and to climb wires and poles and were given other tasks involving exercise and presumably motor learning. In these mice, the nodulus, centralis, and pyramis regions had a thicker molecular layer and Purkinje cells with more extensive dendritic branching than in animals reared in a group in small cages.

Similarly Floeter and Greenough[58] studied monkeys that had been reared in large group *colony* environments (adult, infant, and juvenile monkeys) with a variety of structures that allowed for climbing, leaping, and so forth, as well as smaller objects that were often the focus of attentive exploration and play. After 6 months and a brief period of behavioral testing, various regions of their cerebella were compared with comparable regions from comparably aged monkeys that had been reared either *socially* in pairs in adjacent cages with a daily period of social interaction in one of the two cages or *individually* in closed environments that did not allow sight or sound of other monkeys. (The *Macaca fascicularis* monkeys in these experiments did not exhibit the postrearing "isolation syndrome" of the sort described by Harlow *et al.*[59] and others, which appears to be relatively unique to, or most pronounced in, the rhesus monkey, *Macaca mulatta.*[60]) Purkinje cell spiny branchlets, the sites of parallel fiber input, were larger in the ventral paraflocculus and nodulus of the colony animals relative to the other two groups, which did not differ, as shown in FIGURE 2A. In both the nodulus and the uvula, the mean diameter of Purkinje cell somata was greater in the colony animals than in the social and individual groups, which again did not differ. In contrast to the Purkinje cell measures, granule cell dendritic fields showed no significant or apparent effects of the different rearing conditions, as indicated for ventral paraflocculus in FIGURE 2B. Taken with the Pysh and Weiss report, these data make two major points: 1) the morphology of cerebellar Purkinje neurons is sensitive to the rearing environment, and 2) such plasticity may not be a feature of all cerebellar cortical cell types (for example, granule cells). In addition, the change in spiny branchlets, upon which the granule cell axons constitute the sole excitatory afferents, unaccompanied by a change in granule cell dendrites, argues against the simple hypothesis that mere *activity* drives morphological change.

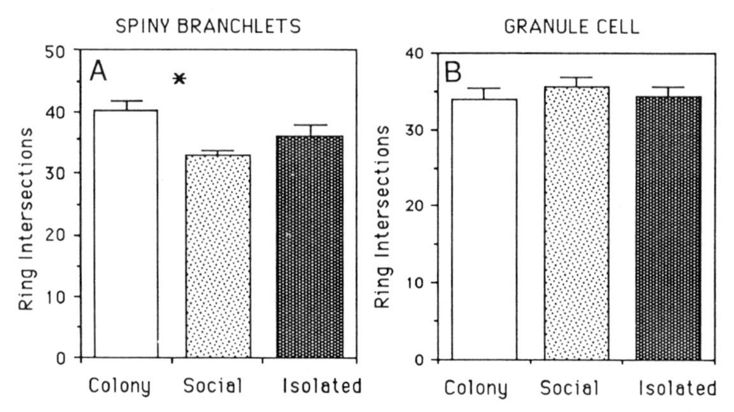

FIGURE 2. **(A)** Size of Purkinje cell spiny branchlets in ventral paraflocculus of monkeys reared from shortly after birth in a large, seminaturalistic colony, social cages, or individual chambers. **(B)** Size of granule dendritic tree in the same animals. "Ring intersections" refer to intersections of dendrites with a superimposed set of concentric spheres at 10-μm intervals from the origin of the branch from a main (smooth) branch. Data from Floeter *et al.*[58]

Effects of Experience during Adulthood

Morphological plasticity of the cerebellar cortex is not restricted to development. In a study of possible ameliorative effects of behavioral experience upon the well-documented cerebellar synaptic loss that occurs with aging,[61,62] Greenough *et al.*[63] discovered that Purkinje cells of rats housed from 24 to 26 months of age in a complex social and play-object-filled environment (EC), similar to that of the colony monkeys, added new spiny branchlets, compared to a baseline group sacrificed at the outset of the exposure period. (Sampling was broadly balanced across the culmen, simplex, declive, tuber, and pyramis of the vermis and the medial ansiform and paramedian lobules.) This experience effect occurred despite an overall tendency for Purkinje cell dendritic trees to degenerate during this period. Both of these effects are evident in FIGURE 3. FIGURE 3A shows the number of spiny branchlets per Purkinje neuron for the three groups in the study: a baseline group, the EC group, and a social condition (SC) group housed in pairs in standard laboratory cages. It is clear that the EC group has added new spiny branchlets relative to the baseline state whereas the SC group may have declined slightly (not significantly by this measure). FIGURE 3B shows

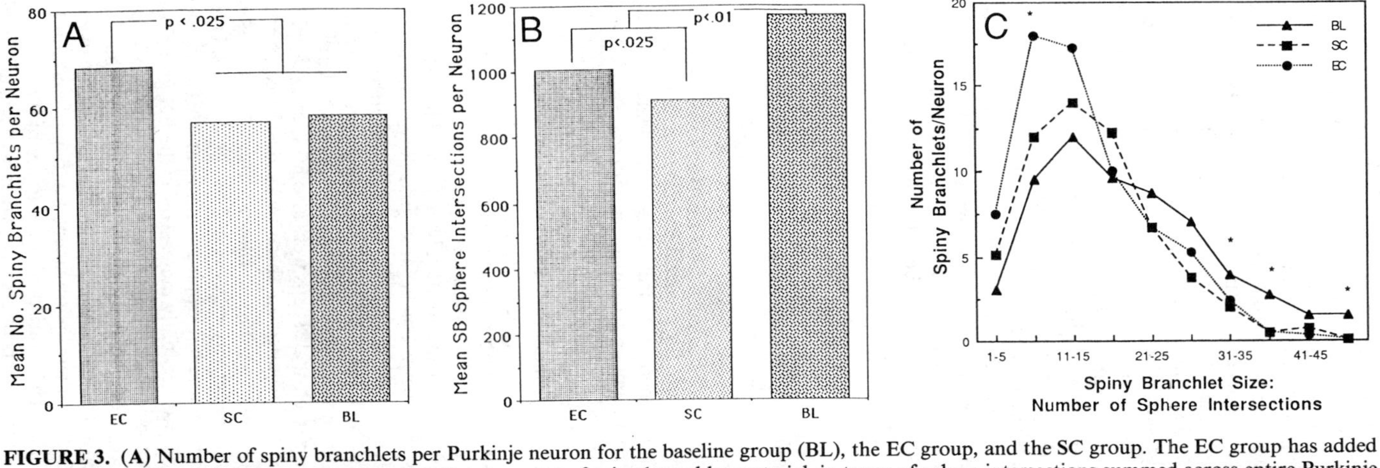

FIGURE 3. **(A)** Number of spiny branchlets per Purkinje neuron for the baseline group (BL), the EC group, and the SC group. The EC group has added new spiny branchlets relative to the baseline. **(B)** Total amount of spiny branchlet material, in terms of sphere intersections summed across entire Purkinje cells. BL have most, ECs less, and SCs still fewer. EC animals appear to be adding new spiny branchlets but to be losing spiny branchlet material as well. **(C)** Frequency distribution of spiny branchlets. Both ECs and SCs have fewer *large* spiny branchlets relative to BL, but the EC group has more *small* spiny branchlets than either of the other two groups. Data from Greenough *et al.*[63]

the total amount of spiny branchlet material per Purkinje cell—spiny branchlet ring intersection data, a measure of size, were cumulated for entire neurons. Here the baseline animals had the greatest amount, the EC group has less, and the SC group has still fewer. The EC animals appear to be adding new spiny branchlets but to be losing, through shrinkage, material from existing branchlets. This is confirmed in FIGURE 3C, in which the frequency of spiny branchlets of different sizes is presented. Both the EC and SC animals have fewer large spiny branchlets compared to the baseline group, but the EC group has more small spiny branchlets than either of the other two groups. We have examined some possible technical issues that would force a reexamination of these data (for example, whether large spiny branchlets could be breaking up into apparently smaller ones by selective spine loss on lower order branches) and conclude that the results indicate that elderly animals can still add to the dendritic field of Purkinje neurons (for example, the new, small spiny branchlets tend to be at *lower* main branch orders, whereas the interior spine loss scenario would put them at *higher* orders; spine loss is actually most evident near the tips of spiny branchlets). The net effect is analogous to a leaky finger in the dike—within the experience constraints of this study the tendency toward regression of Purkinje cell dendrites cannot be reversed; it can only be mitigated to some degree by the consequences of new functional demands.

Evidence that **Learning** *is Responsible for Adult Cerebellar Cortical Plasticity*

The preceding experiments confounded the opportunity to learn about the complex environment with the opportunity for physical exercise and associated neural activity, such that it was not clear whether neural activity or learning was the source of the morphological differences among treatment conditions. (Pysh and Weiss actually proposed that exercise was the critical element responsible for the morphological effects.) We recently conducted a study to assess the effects of these two potential sources of morphological change independently by varying the opportunity for learning and the opportunity for extensive physical exercise of a sort that might change neural morphology.[64] To maximize the opportunity for learning, we trained a group of mature adult female rats on the acrobatic condition (AC), a challenging series of eight successive pathways elevated above a white foam rubber floor. The pathways at the outset were 4-inch-wide boards. The pathways became gradually more demanding, and the AC rats, with physical encouragement from the experimenters, met the challenge. FIGURE 4 shows some of the tasks. Two- to 3-foot-long sections of rope, wire cloth, ¾-inch link chain, and ¼-inch wood dowels were all traversed. Barriers such as teeter-totters and high blocks on ⅝-inch-wide strips were also mastered. Despite the increasing task difficulty, the latency to solution of these tasks decreased dramatically across training sessions. Two of the other groups in this experiment served as controls for the physical and associated neural activity experienced by the AC rats. One group, forced exercise (FX), was forced by gentle tail tapping to move at a brisk walk along a moving treadmill. Session duration was gradually increased for the FX group, such that by the end of the 30-day-long experience they walked for a period of 30 min each day. A second physical exercise control group, voluntary exercise (VX), had access to a standard running wheel attached to their home cage. A final group, the inactivity condition (IC), remained undisturbed in their cages except for daily handling to simulate the treatment of the other groups. Thus statistical comparison of the exercise and inactivity condition groups could test for effects of exercise, whereas comparison of the acrobatic

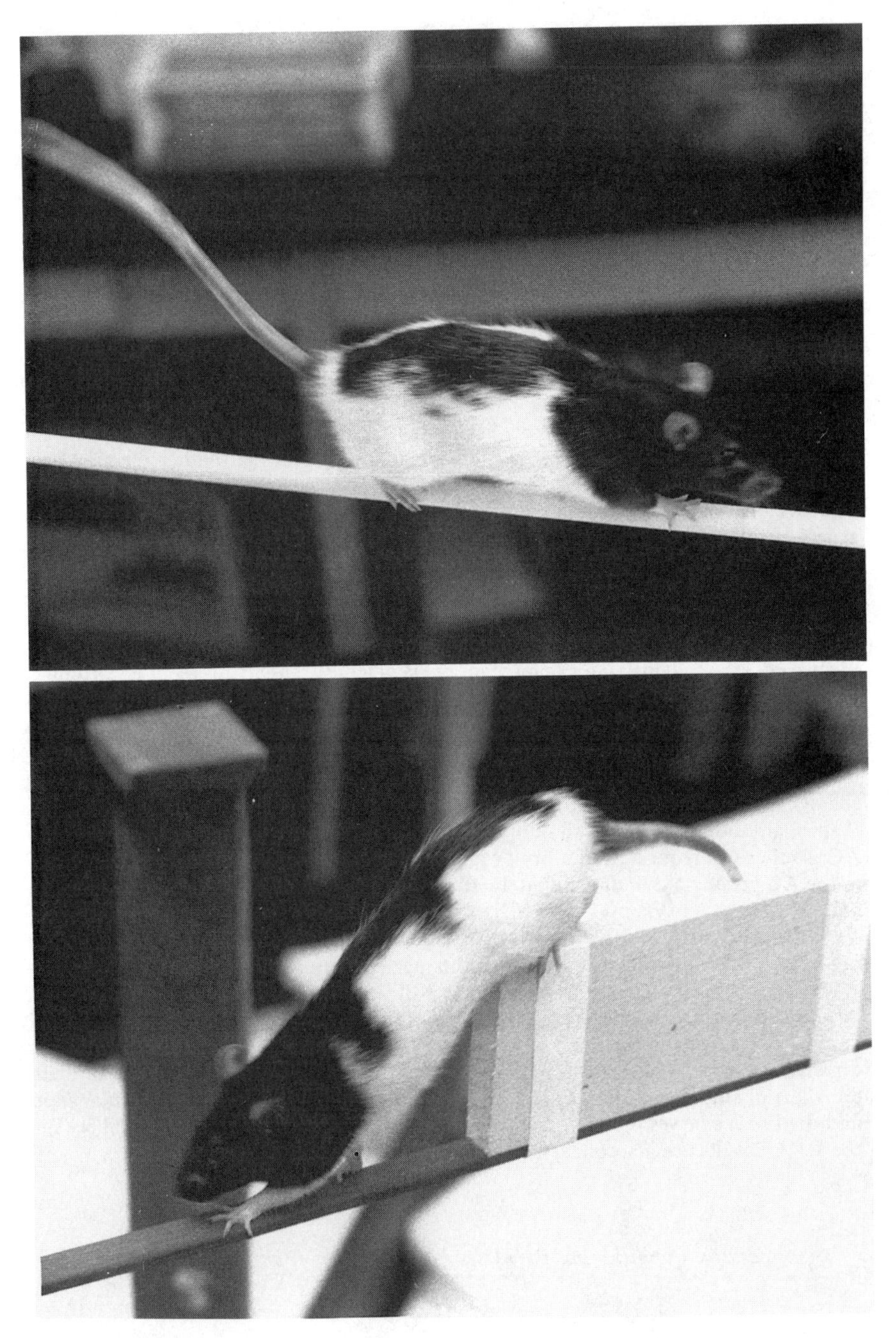

FIGURE 4. Typical acrobatic tasks used with AC rats. (Lower portion is reproduced with permission from Black *et al.*[64])

and inactivity conditions could test for effects due to activity and/or learning. We examined two variables, the density of capillaries as an indication of metabolic tissue support and the number of synapses per neuron as a measure of synapse formation or loss. Measurements were made in a stratified sample along the length of the paramedian lobule (PML). Capillaries were measured as the number of profiles per unit area of molecular layer. Neuronal density was estimated using the disector, an unbiased estimator of density that is unaffected by group differences in the shape or size of the objects whose density is being measured[65]; synaptic density was estimated according to Anker and Cragg[66]; and their ratio, synapses per Purkinje cell, a measure that removes possible differences in tissue volume due to addition of new components, was calculated. (This measure includes synapses on all types of cells and merely uses the Purkinje cell as a reference.)

The results appear in FIGURE 5. FIGURE 5D shows that capillary density was elevated by exercise and was unaffected by motor skill learning in the AC group. In contrast, the number of synapses per Purkinje cell was elevated by acrobatic motor learning and was unaffected by exercise (FIG. 5C). In short, activity, presumably the neuronal activity necessary to guide the behavior, was sufficient to alter the metabolic state of the tissue, but was insufficient to alter synaptogenesis. We take this as strong evidence that mere activity of cells need not trigger synaptogenesis and that some aspect of the activity that is related to skill acquisition is an essential feature necessary to cause synaptic reorganization.

As noted above, the number of synapses per Purkinje cell does not give any indication of either the sources or the targets of the synapses, many of which will terminate on stellate, basket, and Golgi cell components in the molecular layer. We are currently in the process of identifying synapses more specifically, classifying them according to literature-derived criteria, which has turned out to be a slow process. However, on the basis of the preliminary data we now have, we can confidently support the following conclusions: 1) there are significantly more parallel fiber synapses (with all cell types) in AC rats than in all other conditions (there were also more parallel fiber to Purkinje cell synapses in the AC group); 2) all excitatory synapses are similarly greater in AC than in any other group; 3) all nonparallel fiber synapses were significantly greater in AC; 4) there were more inhibitory synapses on dendritic shafts of inhibitory neurons in the AC group; and 5) unfortunately, the number of unclassifiable synapses (approximately 11% of all synapses) was higher in AC and VX than in IC (6%).

As FIGURES 5A & 5C make clear, the thickness of the molecular layer is a good predictor of the outcome of synapse counts per neuron, as was also true for previous studies of cerebral cortex.[25,29] To provide a quick assessment of another region, we have, hence, measured the average thickness of the molecular layer in a sample from vermis. Although the thickness of the vermis molecular layer (μm) was in the same direction as for the PML (AC = 238; others = 229 $\pm$ 4), the effect was weaker and fell short of statistical significance. As the vermis receives substantial vestibular input and the PML receives only indirect vestibular input, this result renders it unlikely that the PML results are a secondary consequence of vestibular activation.

Associative Conditioning Decreases Parallel Fiber Input to HVI Purkinje Cells

In collaboration with R. F. Thompson at the University of Southern California and J. Steinmetz at Indiana University, we have been examining the effects of associative eyeblink conditioning on the morphology of Purkinje cell dendritic fields in cerebellar lobule HVI, which has been implicated in the circuitry underlying this response,[23,24,67,68] particularly when, as in this case, electrical stimulation delivered to the dorsolateral pontine nucleus serves as the conditioned stimulus.[69] The unconditioned stimulus was an air puff to the cornea. In rabbits trained with standard delay pairing (350-msec stimulus, coterminating 100-msec air puff), we observed a significant *decrease* in the size of spiny branchlets in the trained hemisphere, relative to the contralateral side, indicating a reduction in the relative input from parallel fibers. This is consistent with one of the two types of neural responses observed in this area, a decrease in simple spike activity to the conditioned stimulus,[23,24,68] and it is also consistent with a plausible

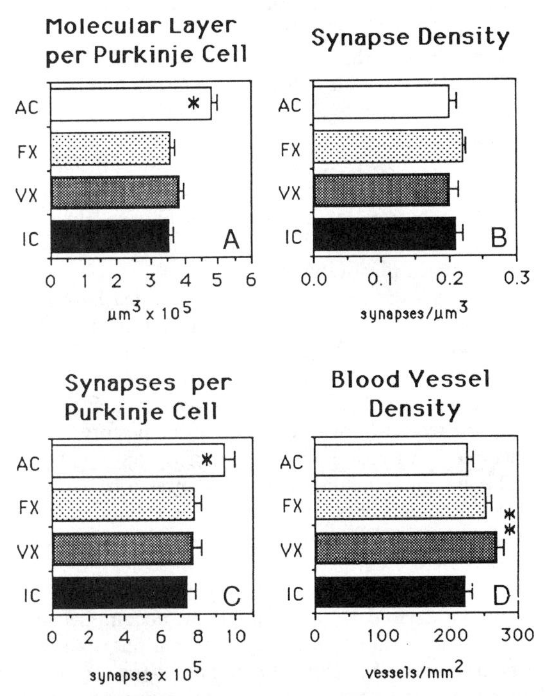

FIGURE 5. Molecular layer per Purkinje cell is increased in AC rats (**A**), while synapse density is similar across groups (**B**), such that the number of synapses per Purkinje cell is higher in AC rats (**C**). In contrast, capillary density is higher in physical exercise groups VX and FX (**D**). (Reproduced with permission from Black *et al.*[64])

role of the inhibitory output of the Purkinje cells—reduced inhibition could serve to "release" a blink response triggered by deep cerebellar nuclear excitatory afferents to brainstem motor nuclei. We have collected preliminary data from three rabbits with unilateral explicitly unpaired conditioned stimulus and unconditioned stimulus presentation and see no tendency toward asymmetry. With the acknowledgment that more control data would help to firm up this result, the ipsilateral branch *loss* in conditioned animals suggests that cerebellar Purkinje neurons may be capable of bidirectional modification of the amount of parallel fiber input. These results are compatible with physiological data suggesting that Purkinje cell simple spike responsiveness may vary bidirectionally[23,24]; that is, simple spike responses may be depressed or potentiated depending on the relationship of parallel fiber and climbing fiber input.

SPECULATIONS ON CEREBELLAR PLASTICITY

Fitting this result to a model is not straightforward, but a first approach might argue that in the associative task, conjunctive parallel fiber and climbing fiber input leads to long-term depression in simple spike responses, whereas in the acrobatic task, a preponderance of simple spike input results in potentiation of most parallel fiber synapses. Although most cerebellar models have emphasized unidirectional weight change at the parallel fiber-Purkinje cell synapse, either increasing or decreasing the weights (see, for example, references 8 and 10, respectively), some models have explicitly proposed the capacity for bidirectional change in the weights of these synapses (see, for example, reference 13). Likewise, the fact that both increases and decreases occur in Purkinje cell simple spike responses during associative eyeblink conditioning argues for the possibility of bidirectional modification of parallel fiber inputs. Additional physiological data supporting bidirectional parallel fiber change was described in the section on cerebellar cortical plasticity.[18,19]

Behaviorally, one might argue that the associative learning paradigm involves a focalized set of inputs, with most other inputs being relatively uninvolved in the behavior. Likewise, the motor task studied by Gilbert and Thach was a relatively focused forelimb tack. Acrobatic learning, by contrast, requires that a large amount of irrelevant or interfering ongoing behavior be suppressed, such that the appropriate set of focal behaviors is emitted without interference. The performance of behaviors thus occurs through selective focal reductions in inhibitory output. If one assumes that a broad range of focal behavior sets (and hence of suppression) occurs across the diverse circumstances faced by the acrobats, and that most of what the cerebellum does is to suppress irrelevant behaviors when complicated motor acts are called for, one might argue that a large part of the cerebellar role in motor learning is one of increasing broad inhibitory output that might be mediated by parallel fiber synapse increases. The focal disinhibition of the desired behavior might be accomplished both by reducing parallel fiber input (due to conjunctive climbing fiber input) and increasing inhibition of inhibitory interneurons (another type of synaptic change seen in the AC rats).

This general interpretation of cerebellar output is consistent with the comparatively unusual circumstance of a structure for which the entire output is inhibitory. A very crude representation of this concept appears in FIGURE 6. In summary, this view suggests that the cerebellar role in learned motor behavior is one of suppressing interfering components of behavior while permitting the desired behavior through focalized reductions in inhibitory output. When very little competing behavior needs to be suppressed, as might be the case for associative eyeblink conditioning, the predominant change may be disinhibition mediated by decreased parallel fiber input to Purkinje cells. When much competing behavior demands suppression, as in complex motor learning, inhibition, mediated by increased parallel fiber input, might predominate. Both changes in excitatory synapses and changes in inhibitory synapses could participate in this, and our current data suggest that both types of changes occur with motor learning.

In contrast to what we propose here, the theories both of Marr[8] and of Albus[10] share a common problem: theoretically, at some point *all* synapses would be strengthened (in

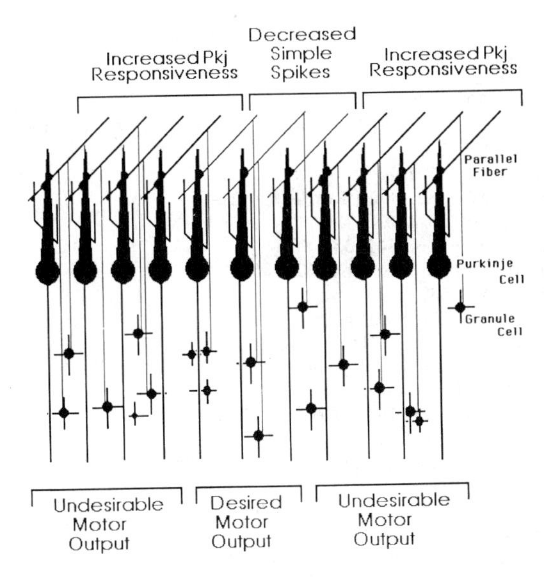

FIGURE 6. Speculative model of cerebellar cortical function in behavior. Parallel fibers to the left and right sides activate Purkinje cells, suppressing undesired behavior. Lighter center parallel fibers are relatively less effective, allowing desired focal behavior to be emitted.

the case of Marr) or weakened (Albus). Although Marr did not address this problem, Albus suggested that at some point new synapses must be added to provide a continuing population of synapses that can be decreased in strength. Albus's answer to the problem possesses two problems of its own. First, it is hard to imagine a system that has a *requirement* for a constant level of noise, which new, presumably random, connections would produce. Second, this solution seems to defeat the primary goal of refining the system. A system with bidirectional changes would not involve this constant background noise, while at the same time it could maintain a constant level of overall cerebellar excitation. Through learning, the Purkinje excitation would be redistributed over time and cerebellar (or Purkinje) space into a set of learned patterns of activity. We propose that while there may be an overall shift up or down in Purkinje excitation in specific cerebellar areas with specific tasks, the underlying changes most often include *both* increases and decreases in the parallel fiber input onto the Purkinje cell.

REFERENCES

1. PROSSER, C. L. & W. T. GREENOUGH. 1991. *In* Comparative Animal Physiology. 4th edit. C. L. Prosser, Ed. John Wiley & Sons. New York, NY.
2. ITO, M. 1989. *In* Neural Models of Plasticity: Experimental and Theoretical Approaches. J. H. Byrne & W. E. Berry, Eds.: 178-186. Academic Press. San Diego, CA.
3. MURPHY, J. T., W. A. MACKAY & F. JOHNSON. 1973. Brain Res. **55:** 263-289.
4. MURPHY, J. T., W. A. MACKAY & F. JOHNSON. 1973. J. Neurophysiol. **36:** 711-723.
5. LLINAS, R. R. & J. I. SIMPSON. 1981. *In* Handbook of Behavioral Neurobiology. A. L. Towe & E. S. Luschei, Eds.: 231-299. Plenum. New York, NY.
6. LISBERGER, S. G. 1988. TINS **11:** 147-152.
7. BLOEDEL, J. R., V. BRACHA, T. M. KELLY & J.-Z. WU. 1991. Ann. N.Y. Acad. Sci. This volume.

8. MARR, D. 1969. J. Physiol. **202:** 437-470.
9. ECCLES, J. C. 1977. Brain Res. **127:** 327-352.
10. ALBUS, J. S. 1971. Math. Biosci. **10:** 25-61.
11. GILBERT, P. F. C. 1974. Brain Res. **70:** 1-18.
12. PETERSON, B. W., J. F. BAKER & J. C. HOUK. 1991. Ann. N.Y. Acad. Sci. This volume.
13. FUJITA, M. 1982. Biol. Cybern. **45:** 195-206.
14. ITO, M. 1989. Annu. Rev. Neurosci. **12:** 85-102.
15. RAWSON, J. A. & K. TILOKSKULCHAI. 1982. Brain Res. **237:** 492-497.
16. ITO, M. 1982. Annu. Rev. Neurosci. **5:** 275-296.
17. EKEROT, C.-F. & M. KANO. 1985. Brain Res. **342:** 357-360.
18. EBNER, T. J. & J. R. BLOEDEL. 1984. Brain Res. **309:** 182-186.
19. SAKURAI, M. 1987. J. Physiol. **394:** 463-480.
20. YEO, C. H. 1991. Ann. N.Y. Acad. Sci. This volume.
21. GILBERT, P. F. C. & W. T. THACH. 1977. Brain Res. **128:** 309-328.
22. ITO, M. 1984. The Cerebellum and Neural Control. Raven Press. New York, NY.
23. FOY, M. R. & R. F. THOMPSON. 1986. Soc. Neurosci. Abstr. **12:** 518.
24. BERTHIER, N. E. & J. W. MOORE. 1986. Exp. Brain Res. **63:** 341-350.
25. ROSENZWEIG, M. R., E. L. BENNETT & M. C. DIAMOND. 1972. *In* Macromolecules and Behavior. J. Gaito, Ed.: 205-277. Appleton-Century-Crofts. New York, NY.
26. HOLLOWAY, R. L. 1966. Brain Res. **2:** 393-396.
27. GREENOUGH, W. T. & F. R. VOLKMAR. 1973. Exp. Neurol. **40:** 491-504.
28. JURASKA, J. M. 1984. Brain Res. **295:** 27-34.
29. TURNER, A. M. & W. T. GREENOUGH. 1985. Brain Res. **329:** 195-203.
30. BHIDE, P. G. & K. S. BEDI. 1984. J. Comp. Neurol. **227:** 305-310.
31. ALTMAN, J. & G. D. DAS. 1964. Nature (London) **204:** 1161-1163.
32. DIAMOND, M., F. LAW, H. RHODES, B. LINDER, M. ROSENZWEIG, D. KRECH & E. L. BENNET. 1966. J. Comp. Neurol. **128:** 117-125.
33. SIREVAAG, A. M. & W. T. GREENOUGH. 1987. Brain Res. **424:** 320-332.
34. SIREVAAG, A. M. & W. T. GREENOUGH. 1991. Brain Res. **540:** 273-278.
35. SIREVAAG, A. M., J. BLACK, D. SHAFRON & W. T. GREENOUGH. 1988. Dev. Brain Res. **43:** 299-304.
36. BLACK, J. E., A. M. SIREVAAG & W. T. GREENOUGH. 1987. Neurosci. Lett. **83:** 351-355.
37. UYLINGS, H. B. M., K. KUYPERS, M. C. DIAMOND & W. A. M. VELTMAN. 1978. Exp. Neurol. **62:** 658-677.
38. JURASKA, J. M., W. T. GREENOUGH, C. ELLIOTT, K. MACK & R. BERKOWITZ. 1980. Behav. Neural Biol. **29:** 157-167.
39. GREEN, E. J., W. T. GREENOUGH & B. E. SCHLUMPF. 1983. Brain Res. **264:** 233-240.
40. BLACK, J. E., A. M. SIREVAAG, C. S. WALLACE, M. H. SAVIN & W. T. GREENOUGH. 1989. Dev. Psychobiol. **22**(7): 727-752.
41. HWANG, H.-M. & W. T. GREENOUGH. 1986. Soc. Neurosci. Abstr. **12:** 1284.
42. BLACK, J. E., M. POLINSKY & W. T. GREENOUGH. 1989. Neurobiol. Aging **10:** 353-358.
43. JURASKA, J. M., J. M. FITCH, C. HENDERSON & N. RIVERS. 1985. Brain Res. **333:** 73-80.
44. JURASKA, J. M., J. M. FITCH & D. L. WASHBURNE. 1989. Brain Res. **479:** 115-119.
45. FUCHS, J. L., M. MONTEMAYOR & W. T. GREENOUGH. 1990. Behav. Neural Biol. **54:** 198-203.
46. GREENOUGH, W. T. & C. H. BAILEY. 1988. TINS **11:** 142-147.
47. GREENOUGH, W. T. & F.-L. F. CHANG. 1988. *In* Cerebral Cortex. Vol. 7. A. Peters & E. G. Jones, Eds.: 391-440. Plenum. New York, NY.
48. GREENOUGH, W. T., J. R. LARSON & G. S. WITHERS. 1985. Behav. Neural Biol. **44:** 301-314.
49. CHANG, F.-L. F. & W. T. GREENOUGH. 1982. Brain Res. **232:** 283-292.
50. VRENSEN, G. & J. N. CARDOZO. 1981. Brain Res. **218:** 79-97.
51. WENZEL, J., E. KAMMERER, M. FROTSCHER, W. KIRSCHE, H. MATTHIES & M. WENZEL. 1980. J. Hirnforschung. **21:** 647-654.
52. HILLMAN, D. E. & S. CHEN. 1984. Brain Res. **295:** 325-343.
53. CHEN, S. & D. E. HILLMAN. 1982. Brain Res. **240:** 205-220.
54. ROSSI, F., L. WIKLUND, J. J. L. VAN DER WANT & P. STRATA. 1989. Soc. Neurosci. Abstr. **15:** 89.

55. BERNARD, J. F., C. BUISSERET-DELMAS, C. COMPOINT & S. PLANTE. 1984. Exp. Brain Res. **57:** 128-137.
56. HAWDON, G., S. REES & J. A. RAWSON. 1988. Neurosci. Lett. **91:** 7-13.
57. PYSH, J. J. & G. M. WEISS. 1979. Science **206**(12): 230-231.
58. FLOETER, M. K. & W. T. GREENOUGH. 1979. Science **206:** 227-229.
59. HARLOW, H. F., R. O. DODSWORTH & M. K. HARLOW. 1965. Proc. Natl. Acad. Sci. USA **54:** 90-97.
60. SACKETT, G. P. 1966. Science **154:** 1468-1473.
61. ROGERS, J., M. A. SILVER, W. J. SHOEMAKER & F. E. BLOOM. 1980. Neurobiol. Aging **1:** 3-11.
62. ROGERS, J., S. F. ZORNETZER, F. E. BLOOM & R. E. MERVIS. 1984. Brain Res. **202:** 23-32.
63. GREENOUGH, W. T., J. W. MCDONALD, R. M. PARNISARI & J. E. CAMEL. 1986. Brain Res. **380:** 136-143.
64. BLACK, J., K. ISAACS, B. ANDERSON, A. ALCANTARA & W. GREENOUGH. 1990. Proc. Natl. Acad. Sci. USA **87:** 5568-5572.
65. STERIO, D. C. 1984. J. Microscopy **134**(2): 127-136.
66. ANKER, R. L. & B. G. CRAGG. 1974. J. Neurocytol. **3:** 725-735.
67. YEO, C. H., M. HARDIMAN & M. GLICKSTEIN. 1985. Exp. Brain Res. **60:** 99-113.
68. MCCORMICK, D. A. & R. F. THOMPSON. 1984. J. Neurosci. **4:** 2811-2822.
69. STEINMETZ, J. E., D. J. ROSEN, P. F. CHAPMAN, D. G. LAVOND & R. F. THOMPSON. 1986. Behav. Neurosci. **100:** 878-887.

PART V. ACTIVITY-DEPENDENT ALTERATIONS IN HIPPOCAMPAL NETWORKS

Introduction

At any given time, the operations of a neural network are dependent on the expression and the interactions of its constituent components on the molecular, cellular, and network levels. Persistent modification of any of these components with age or by experience can alter network function. This session focuses on the role that recurrent inhibitory synapses play in hippocampal network operations, particularly in limiting the spread of activity through local recurrent excitatory circuits. Repeated activation of these inhibitory synapses has been shown to produce short- and long-term depression of their inhibitory effects. At the same time, previously silent recurrent excitatory pathways become active. Thus, these activity-driven changes can generate small networks of cells that operate as functional units. Such units could serve as sites of memory storage.

Alger reviews the wide variety of neurophysiological mechanisms that contribute to alterations in GABA-mediated inhibitory postsynaptic potentials. One important issue centers on how transient depression of inhibitory postsynaptic potentials may serve an important role in permitting the gating of long-term potentiation by *N*-methyl-D-aspartate receptors.

Swann *et al.* report that CA3 networks are hyperexcitable during the second and third postnatal weeks. GABA inhibition is present and plays a central role in limiting further network excitability. An overabundance of recurrent excitatory synapses may exist during this critical period and markedly affect network operations.

Traub and Miles employ detailed information on hippocampal network connections and the intrinsic ionic conductances of single pyramidal cells to develop a new computer model of the CA3 subfield. Numerous network behaviors are mimicked by alterations in specific network components. One of these is the behavior seen when GABA synapses are suppressed and recurrent excitatory connections remain intact.

Together these papers emphasize the variety of behavior that hippocampal networks can produce. Activity-dependent alterations can be ascribed to changes in inhibitory as well as excitatory synaptic connections within the network.

Gating of GABAergic Inhibition in Hippocampal Pyramidal Cells[a]

BRADLEY E. ALGER

Department of Physiology
University of Maryland School of Medicine
Baltimore, Maryland 21201

INTRODUCTION

Neuronal excitability in the brain is regulated by a balance between excitatory and inhibitory influences, both intrinsic and synaptic. Synaptic control is exerted by excitatory and inhibitory postsynaptic potentials. Increases in excitability are associated with different phenomena that are important in learning and development, as well as with pathological phenomena. As discussed elsewhere in this volume, such diverse events as long-term potentiation (LTP), the shaping of retinotopic maps, critical periods in development, and epilepsy involve activation of *N*-methyl-D-aspartate (NMDA) receptors; hence, factors capable of regulating NMDA receptors are likely to be involved in the occurrence of these events. GABAergic inhibition is often ideally poised to exercise a restraining influence on the expression of NMDA receptor activity and other intrinsic excitatory processes; in so doing, it can alter the behavior of neuronal circuits. A great deal is known about inhibitory regulation of mammalian cortical neurons, different aspects of this topic having been reviewed previously.[1-8] This paper will emphasize recent intracellular results on cellular mechanisms of gating of GABA-ergic synaptic potentials obtained from the *in vitro* hippocampal slice of adult rodents.

In general terms, regulation of the GABAergic system can occur in at least four loci relative to the principal target, the pyramidal cell: at the postsynaptic site (that is, on the pyramidal cell itself) or at three presynaptic sites (that is, presynaptic with respect to the pyramidal cell). Modulation occurs at each of these sites. The following general inferences can be drawn: 1) inhibitory processes are very plastic and are modified by a variety of mechanisms, 2) the most common inhibitory modification identified to date is a transient depression of inhibition, although exceptions are found, and 3) a particularly important function of inhibitory plasticity is to "gate" NMDA-dependent synaptic transmission and other intrinsic voltage-dependent conductances.

BASIC GABAergic CIRCUITRY IN THE HIPPOCAMPAL CA1 REGION

In the hippocampus, GABAergic neurotransmission is largely carried out by local circuit interneurons of several different kinds. A simplified diagram of the basic GABAergic circuits in the CA1 pyramidal cell region is indicated in FIGURE 1. Hippocampal recurrent inhibition was first studied by Kandel *et al.*,[9] and the role of the

[a] This work was supported by a grant from the U.S. Public Health Service (NS22010).

basket cell interneuron in the feedback circuit was emphasized in classic experiments by Andersen and colleagues.[10,11] The morphological and physiological studies of Schwartzkroin's group[12–16] combined with the immunocytochemical work of Ribak and colleagues[17–19] and Somogyi and others[20–23] have added other elements and connections, and should be consulted for details. Nevertheless, FIGURE 1 illustrates the essential features of the circuits as currently understood. The pyramidal cell receives both feedback (interneuron 1) and feedforward (interneuron 3) GABAergic input. The recurrent circuit probably synapses on the pyramidal cell soma and mediates a simple monophasic inhibitory postsynaptic potential (IPSP). The feedforward GABAergic

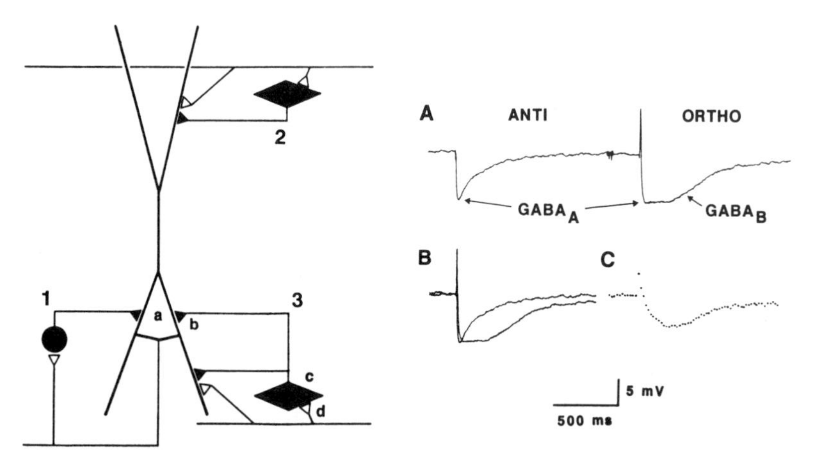

FIGURE 1. The diagram on the left represents a simplified organizational schematic of the inhibitory interneuron system in the hippocampal CA1 region. The pyramidal cell is contacted by three functional types of GABAergic interneurons. They form a recurrent loop (number 1) that provides somatic inhibition, a feedforward circuit restricted to the distal dendrites (number 2), and a feedforward circuit that contacts the cell soma as well as its dendrites (number 3). The recurrent circuit is activated by pyramidal cell axon collaterals (of this and neighboring cells). Feedforward circuits are activated by afferent fibers (open terminals) either intrinsic or extrinsic to the hippocampus. Traces on the right are intracellular membrane potential recordings from a CA1 pyramidal cell and illustrate responses to stimulation of the feedback loop (antidromic, "ANTI") and then of the feedforward circuits (orthodromic, "ORTHO"). The feedback IPSP is mediated by GABA$_A$ receptors as is the initial part of the orthodromic IPSP, the rates part of which is mediated by GABA$_B$ receptors (**A**). The upward deflection on the orthodromic response is an EPSP. Feedback and feedforward responses are overlapped in **B**; **C** shows the outline of the GABA$_A$ IPSP, obtained by subtracting the responses in **B**. This figure has been modified from references 25 and 46.

circuit involves at least one other interneuron activated by excitatory afferent fibers, as first postulated on the basis of physiological and pharmacological evidence.[24,25] Activation of this pathway produces a complex IPSP with both early and late phases. There may be another feedforward system (interneuron 2) in distal dendritic regions.

There are two general classes of GABA receptors, GABA$_A$ and GABA$_B$, that mediate bicuculline-sensitive and -insensitive responses, respectively (FIG. 1). GABA activates both kinds of receptors. Baclofen is a selective GABA$_B$ agonist; phaclofen and saclofen are selective GABA$_B$ antagonists. (Although there are several subtypes of $GABA_A$[26] and $GABA_B$[27] receptors, functional differences among them are not yet well established. Future work will undoubtedly reveal differences that may well pertain to the phenemona discussed below.) The $GABA_A$ response is Cl^- dependent.[9,28–30] $GABA_B$ receptors activate a postsynaptic K^+-dependent conductance[31–35] via a G protein-linked mechanism, and they hyperpolarize cells.[36–39] Activation of the $GABA_B$ receptors by synaptically released GABA causes the late, slow IPSP (FIG. 1). Support for this hypothesis comes from the similarities between the ionic mechanisms of $GABA_B$ receptor activation and the slow IPSP,[31–34] the antagonism of the slow IPSP by $GABA_B$ antagonists,[37,40] and the blockade of both by phorbol esters.[36,41,42]

Two functional sets of GABAergic interneurons were postulated to account for the observation that antidromic activation of pyramidal cells produces only a simple, $GABA_A$ IPSP, whereas orthodromic stimulation of afferent fibers produces a complex IPSP with both $GABA_A$ and $GABA_B$ components.[25,32,34,35,37,43–46] There is copious evidence for $GABA_A$ synaptic inhibition on neuronal somata. Increasing evidence indicates a preferential localization of $GABA_B$ receptors on pyramidal cell dendrites.[34] An intriguing question is whether or not a single GABA interneuron activates both postsynaptic $GABA_A$ and $GABA_B$ receptors. On the one hand, many treatments affect $GABA_A$ and $GABA_B$ IPSPs in common (for example, carbachol,[47] norepinephrine,[48] enkephalin,[32,49] and kainic acid[50,51]): a coincidence that is simple to explain if a single class of interneuron is involved. On the other hand, simultaneous recordings from interneuron-pyramidal cell pairs have detected slowly developing IPSPs following activation of some interneurons,[52] and rapidly rising IPSPs following activation of other interneurons,[53] in the coupled pyramidal cells. Correspondingly different electrical properties of the interneurons involved support the suggestion that different cells mediate $GABA_A$ and $GABA_B$ IPSPs. Drop application of 4-aminopyridine into the distal dendritic field of CA1 neurons can elicit a $GABA_B$-mediated response without an accompanying $GABA_A$ response.[54,55] A reasonable interpretation of these preliminary data is that there is a distinct class of GABA interneurons that is activated by the restricted 4-aminopyridine application that produces only dendritic $GABA_B$ IPSPs. If this interpretation is substantiated, the possibility for selective regulation of these two forms of inhibition is opened up. This would have important implications for understanding hippocampal circuit behavior.

Hippocampal pyramidal cell dendrites receive excitatory synaptic inputs and are electrically excitable[56–59]; they can conduct both Na^+- and Ca^{2+}-dependent action potentials. When GABAergic inhibition is depressed excitatory postsynaptic potentials (EPSPs) become larger[57,60,61] and an NMDA component of the EPSP becomes evident.[51,62,63] The dendritic location of much of the GABAergic inhibition is, therefore, a particularly important feature of the circuit because it helps control these excitatory drives. The feedforward connection ensures optimal temporal conjunction between IPSPs and EPSPs. Localized application of agonists and antagonists to pyramidal cell dendrites,[25,64–67] intradendritic recording,[57,68] anatomical studies cited above, and work discussed below all strongly support the hypothesis of feedforward dendritic GABAergic inhibition.

PRESYNAPTIC REGULATION OF GABA IPSPs

The evidence of presynaptic regulation of GABAergic IPSPs has been developed using the $GABA_A$ IPSP as the subject response. If $GABA_A$ and $GABA_B$ IPSPs are produced by GABA liberation from the same interneuron, then the conclusions concerning presynaptic regulation will apply to both types of IPSPs. If different interneurons are involved, the present conclusions may not apply to the $GABA_B$ system. Regulation of GABAergic transmission can occur at three different sites relative to the interneuron: 1) the terminals, 2) the soma, and 3) the excitatory inputs to the interneuron (FIG. 1). All three sites are subject to extrinsic control.

In the CNS, indirect methods are most often used to discriminate presynaptic from postsynaptic sites of action of the GABA system. A common test is to compare the effects of a treatment on the evoked IPSP and on the response to GABA applied directly by iontophoresis or pressure injection. If the former is influenced and the latter is not, it is tentatively concluded that the site of action is presynaptic, although the universal caveat for this experiment is that the population of receptors activated by direct application is not the same as the population activated by synaptically released GABA. Direct application probably affects a group of receptors that is larger than, or different from (or both), the synaptically activated group of receptors. Conceivably, the GABA application method would be insufficiently discriminating to detect relevant changes. A more straightforward approach is to impale an inhibitory interneuron with a microelectrode and record its activity directly. With traditional intracellular recording methods in the slice preparation this is tedious at best, however, and confirmation of the identity of the presumed inhibitory interneuron is not always possible. This technique has rarely been used to assess regulation of interneuron activity (see references 69 and 70 for exceptions). The new "patch-slice" technologies[71,72] offer great promise for simplifying recordings from interneurons. A convenient indirect method for assessing GABAergic function is to record "spontaneous" IPSPs from pyramidal cells with KCl-filled intracellular electrodes.[73] The large majority of these events are blocked by tetrodotoxin, indicating that they depend on voltage-dependent Na^+ channel activity for their occurrence. They are evidently elicited by action potential firing in the presynaptic terminals,[53] unlike true quantal events at frog neuromuscular junction. The IPSPs are spontaneous in the sense of not being explicitly elicited by electrical stimulation, and, although certain experimental treatments affect their occurrence by affecting the interneurons, the "spontaneous" nomenclature remains useful. Tetrodotoxin-insensitive, quantal GABAergic events do occur.[74] Interestingly, quantal $GABA_A$ events seem to be caused by activation of only 10-30 $GABA_A$ channels,[74a,74b] an observation with important functional implications (see below).

PRESYNAPTIC INHIBITION OF GABA RELEASE (GABA, ENKEPHALIN, GLUTAMATE, NOREPINEPHRINE, ACh)

Once released from its nerve terminals, GABA can inhibit its own further release. The earliest distinction between $GABA_A$ and $GABA_B$ receptors was the observation that GABA could inhibit neurotransmitter release from peripheral tissues in a bicuculline-insensitive way, suggesting the existence of a novel receptor type localized on presynaptic nerve terminals of autonomic neurons.[27]

GABA may regulate its own release via activation of $GABA_B$ receptors on interneuron terminals in the hippocampus.[75–78] In the dentate gyrus, baclofen can block $GABA_A$ IPSPs[79] and, indeed, act as a convulsant, apparently by depressing GABA release.[80] Presynaptic inhibition by $GABA_B$ receptors in CA1 is also likely as baclofen profoundly depresses CA1 field potentials. Paired-pulse depression of the monosynaptic $GABA_A$ IPSP is caused by $GABA_B$-mediated reduction of GABA release.[80a] Presynaptic $GABA_B$ receptors may be of a different subtype than postsynaptic $GABA_B$ receptors.

The latter are readily blocked by pertussis toxin,[36,39] which blocks activation of some G proteins, whereas the former are pertussis insensitive. Phaclofen is reportedly more effective in blocking postsynaptic than presynaptic $GABA_B$ receptors.[27] In neonatal rats the $GABA_A$ agonist, muscimol, also has convulsant effects in the CA1 pyramidal cell region[81] possibly related to a presynaptic inhibitory role. $GABA_A$ antagonists increase $GABA_B$-mediated slow IPSPs,[34,46] suggesting that $GABA_A$ receptors normally inhibit the slow IPSP interneurons presynaptically.[34] The evidence that these presynaptic inhibitory effects occur on the GABAergic terminals is not conclusive. An alternative possibility in some cases is that GABA neurons inhibit one another via somatically located synapses. There is good electron microscopical-immunocytochemical evidence for the occurrence of glutamic acid decarboxylase (GAD; the synthetic enzyme for GABA)-containing terminals on the somata of GAD-positive cells,[4] strongly suggesting that at least some of the GABAergic interneurons are subject to somatic inhibition. Extrinsic GABAergic afferents project to the hippocampus[82] where many of them terminate on inhibitory interneurons.[83] Paired-cell recordings indicate that interneuron-to-interneuron inhibitory connections exist.[14]

OTHER NEUROTRANSMITTERS AFFECT GABAergic INTERNEURONS

The endogenous opiate receptor agonist, enkephalin, blocks evoked and spontaneous IPSPs with no effect on pyramidal cell membrane properties or iontophoretically applied GABA, suggesting a presynaptic site of action.[32,49,84,85] Enkephalin-induced hyperpolarization of presumed inhibitory interneurons apparently explains its disinhibitory effects.[70] Bath application of kainic acid to hippocampal neurons *in vitro* also decreases evoked GABAergic IPSPs.[50] Slices obtained from kainic acid-lesioned hippocampi show substantial decreases in IPSPs.[51,86,87] GABA interneurons in the slice are initially excited by kainic acid, as indicated by an increased frequency of spontaneous IPSPs recorded in the postsynaptic cell, but are later depressed.[50] Both spontaneous and evoked IPSPs decline with no change in the response to iontophoretic GABA, suggesting that the depressant effects of kainic acid receptor activation are presynaptic, although the exact mechanism has not been determined. Interestingly, folic acid, a vitamin that contains glutamic acid as a constituent and binds to kainic acid receptors, also blocks $GABA_A$ IPSPs.[88] The effects in this case, however, are probably postsynaptic rather than presynaptic because folic acid blocks iontophoretic GABA responses. Finally, at least some interneurons seem to "follow" afferent stimulation only poorly[52,89]; some undefined failure of excitatory transmission onto the interneurons or of cell responsiveness occurs.

EVIDENCE FOR PRESYNAPTIC INHIBITION OF INPUT TO GABAergic INTERNEURONS

Acetylcholine has complex actions on $GABA_A$-mediated inhibition. Activation of muscarinic receptors (mAChRs) depresses excitatory transmission via presynaptic inhibition.[90–93] Depression of evoked $GABA_A$ IPSPs by cholinergic agonists has also been reported,[47,92,94] and a presynaptic inhibition of GABA release was the suggested mechanism, although presynaptic inhibitory effects of synaptically released ACh have evidently not been found. We have recently found, however, that mAChR activation with carbachol increases the frequency of spontaneous IPSP input to pyramidal cells; spontaneous IPSP size does not decrease even when evoked IPSPs are reduced[95] (Pitler and Alger, unpublished observations). This effect is persistent and independent of the membrane depolarization caused by carbachol. The increased frequency of spontaneous IPSPs is best explained by an increase in the excitability of the interneurons. Spontaneous IPSP persistence implies that evoked IPSP depression was not due to block of GABA release per se, or to any postsynaptic factors. Perhaps presynaptic inhibition of excitatory synapses onto the inhibitory interneurons accounts for evoked IPSP depression. Increase in spontaneous IPSP frequency indicates that mAChRs directly excite GABA interneurons. This hypothesis is in agreement with other observations that ACh transiently hyperpolarizes neocortical[96] and hippocampal[97] neurons by exciting GABA interneurons prior to depolarizing the principal cells. In dentate gyrus muscarinic activation of interneurons may induce a $GABA_B$-mediated inhibitory effect,[98] although previous pharmacological data had suggested the induced inhibition might be $GABA_A$ mediated.[99] Activation of GABA interneurons in CA1 by muscarinic agonists is consistent with a great deal of anatomical work indicating that many muscarinic fibers terminate in the zone populated by the interneurons.[100] This circuit appears to be activatable physiologically because stimulation of cholinergic axons causes an increase in spontaneous IPSP frequency during the resultant muscarinic slow EPSP (Pitler and Alger, unpublished observations). The results suggest a novel mode of regulation of $GABA_A$ inhibition by muscarinic agonists, namely, switching of control by one afferent system to control by another.

Norepinephrine, acting via an α-receptor, has effects on the GABA system that are strikingly similar to those of carbachol in the CA1 region.[48] Both $GABA_A$- and $GABA_B$-mediated evoked IPSPs are reduced with no charge in iontophoretic GABA responses; spontaneous IPSP frequency is enhanced. Of the explanations suggested by Madison and Nicoll,[48] one was that norepinephrine has a direct excitatory effect on the inhibitory interneurons and a presynaptic inhibitory effect on the afferent excitatory fibers to the interneurons. Presynaptic inhibition mediated by α-receptors in the hippocampus is not as well established as in mAChR-induced presynaptic inhibition, and more is needed to determine the site of action of norepinephrine. It will also be necessary to learn if synaptically released norepinephrine affects GABA interneurons.

USE-DEPENDENT IPSP DEPRESSION

Repetitive stimulation depresses hippocampal evoked IPSPs, in some cases both the $GABA_A$- and $GABA_B$-mediated components.[67,101–105] Although, as discussed below, much of the depression is due to postsynaptic factors, some of it is probably attributable to the presynaptic phenomena discussed above. $GABA_A$ and $GABA_B$ responses differ greatly in their receptors, ionic mechanisms, second messenger involvement, and postsynaptic sites of susceptibility to modification. Most work on postsynaptic mechanisms of IPSP depression thus far has focused on use-dependent depression of the $GABA_A$ IPSP, and results from this work are emphasized.

GABA RECEPTOR DESENSITIZATION

The first widely recognized postsynaptic factor was "desensitization": a decrease in neurotransmitter-activated conductance produced by a given dose of transmitter. Prolonged GABA applications result in a decrease in the $GABA_A$-activated conductance in hippocampus.[67,101,106,107] There is no evidence for desensitization of the $GABA_B$ conductance. In view of recent demonstrations of decreased $GABA_A$ receptor responsiveness resulting, for example, from interactions of second messengers, increases in intracellular calcium, and phosphorylation factors, it is possible that what has been called $GABA_A$ "desensitization" will not be explained by a simple unitary mechanism. Here the word refers to a decreased $GABA_A$ conductance attributable solely to the prior application of GABA. Some data suggest that desensitization can be influenced by the permeant anions in the recording electrode. In acutely isolated hippocampal cells using the whole-cell voltage-clamp technique, desensitization is greater when the intracellular Cl^- concentration ($[Cl^-]_i$) is high than when it is low.[107] These results suggest that the degree of desensitization might vary with the physiological conditions. Apart from one early report,[101] however, there has been little evidence that repetitively activated IPSPs decline because of desensitization. Either most stimulus paradigms used to study use-dependent IPSP depression are insufficient to induce substantial desensitization of synaptically released GABA, or the direct GABA application method of testing for desensitization is inadequate to detect it.

SHIFTS IN $GABA_A$ REVERSAL POTENTIAL

There is other evidence that desensitization is not a major factor in $GABA_A$ IPSP depression with repetitive stimulation. Wong and Watkins[67] observed that following a train of synaptic stimuli an intradendritically recorded response to a pressure pulse of GABA changed polarity from a hyperpolarization to a depolarization; a depolarizing shift in the GABA reversal potential (E_{GABA}) apparently occurred. Experimentally induced increases in extracellular potassium ($[K^+]_o$) facilitated this process. McCarren and Alger[102] found that repetitive synaptic activation of IPSPs in CA1 cells caused a clear shift in the reversal potential of the IPSP and the iontophoretic $GABA_A$ response and suggested that accumulation of Cl^-_i was responsible. Concomitant stimulation-induced changes in $[K^+]_o$ implied that the shifts in E_{Cl} might have been brought about by the dependence of Cl^-_i on the transmembrane K^+ gradient under the quasi-Donnan equilibrium conditions that prevail in neurons. The magnitude of the inhibitory GABA current is directly related to the driving force on Cl^-, so a depolarizing shift in E_{Cl}, which decreases the driving force, will decrease inhibitory current. Because the balance between excitation and inhibition is often quite delicate, this may represent an important mechanism of IPSP regulation during repetitive stimulation.

Increases in $[K^+]_o$ induce epileptiform burst activity in hippocampal slices,[108–110] evidently via depression of GABAergic inhibition.[111] Increases in $[K^+]_o$ subsequent to block of the Na^+/K^+ pump inhibition with high concentrations of ouabain also increase epileptiform activity, probably associated with a decrease in IPSPs.[112] Ouabain does cause shifts in the reversal potentials for both the $GABA_A$ and $GABA_B$ IPSP.[46] McCarren and Alger[113] found that dihydroouabain, a less potent cardiac glycoside, also resulted in epileptiform bursting and depression of IPSPs. Negligible increases in $[K^+]_o$ were produced by the comparatively low doses of dihydroouabain used, however, implying the involvement of factors apart from increased $[K^+]_o$ in both induction of hyperexcitability and IPSP depression.

Besides hyperpolarizing pyramidal cells, GABA can depolarize them when applied in the dendritic region.[24,25,64–67,114] The depolarization is most readily demonstrated with direct GABA application, but under certain conditions synaptically released GABA also depolarizes the cells.[24,66,115,116] Depolarizations seem to require higher concentrations of GABA for induction and may be caused by activation of an unusual, perhaps "extrasynaptic," $GABA_A$ receptor. The ionic mechanism of this effect is not understood; Cl^- probably participates and, in dentate gyrus at least, may fully account for the depolarization.[117] The functional role of the depolarization is similarly unclear. Depolarization could offset the inhibitory effectiveness of a hyperpolarizing response, and increases in GABAergic depolarizations could be involved in the depression of IPSPs.

In acutely isolated neurons, contributions of both desensitization and shifts in $[K^+]_o$ can be eliminated or controlled. Nevertheless, Huguenard and Alger[107] found that marked shifts in E_{GABA} continued to occur in response to repetitive application of brief pulses of GABA. Because the $GABA_A$ response could be fully accounted for by Cl^- permeability, alterations in E_{GABA} were equivalent to alterations in E_{Cl}. Changes in transmembrane ionic gradients under these conditions were remarkable for several reasons, one of which being that the cell was impaled with a large-bore pipette, and it might be assumed that the mobile intracellular constituents are identical to those in the pipette. These results emphasize, however, that the entire cell is not continuously controlled by pipette constituents. Apparently, in the very near neighborhood of the membrane, tiny surpluses or deficits of ions can cause shifts in measured equilibrium potentials. In any event, these results confirmed that the Cl^- gradient is very labile. One criticism of our results[107] is that they were obtained at 22-24 °C, temperatures at which ion transport mechanisms operate more sluggishly than they do at physiological temperatures.[118] Nevertheless, the experiments were important in demonstrating that Cl^- accumulation can contribute to the shifts in E_{IPSP} measured in slice experiments. Thompson and Gahwiler, working at 35 °C, recently reported shifts in the reversal potential of the $GABA_A$-induced current in voltage-clamp experiments in cells in the hippocampal organotypic slice culture.[105,119] Their results supported and extended the earlier observations. Thompson and Gahwiler also directly demonstrated a role for $[K^+]_o$ by showing that reducing $[K^+]_o$ in the bathing solution reduced the change in IPSP driving force, without affecting the decrease in IPSP conductance brought about by repetitive stimulation.

ROLE OF Ca^{2+} AND "PHOSPHORYLATION FACTORS" IN GABA RESPONSE

Stelzer, Wong, and colleagues have observed that increases in intracellular Ca^{2+} concentration ($[Ca^{2+}]_i$) depress $GABA_A$ conductance. Increases in Ca^{2+} influx suppress the $GABA_A$-activated Cl^- conductance in frog sensory neurons by decreasing the apparent affinity of the $GABA_A$ receptor for GABA.[120] Intracellular injection of EGTA, however, did not prevent the reduction in GABA response due to $Ca^{2+}{}_i$. Direct evidence for the regulation of $GABA_A$ conductance by Ca^{2+} has been provided by Chen *et al.*[121] using an intracellular perfusion technique in acutely isolated cells that permitted study of GABA responses as a function of $[Ca^{2+}]_i$. Increased $[Ca^{2+}]_i$ did, in fact, decrease the $GABA_A$-mediated conductance. Stelzer *et al.*[122] and Chen *et al.*[121] showed that "phosphorylation factors" were responsible for down-regulation of the GABA response caused by $Ca^{2+}{}_i$. Maintenance of the $GABA_A$ receptor function required Mg-ATP, and, conversely, depression of the $GABA_A$ response was speeded by phosphatase perfusion. The conclusion of these studies was that the $GABA_A$ receptor function is jointly controlled by the interplay between $[Ca^{2+}]_i$ and a phosphorylation-dephosphorylation mechanism. Although extracellular cAMP can reduce IPSPs through a kinase-independent mechanism,[122a] phosphorylation mediated by the catalytic subunit of protein kinase A reduces $GABA_A$ IPSPs in cultured spinal neurons.[122b] The beta subunit of the $GABA_A$ receptor is a substrate for both protein kinase A and C.[122c]

Increases in $[Ca^{2+}]_i$, induced by physiological activity, may reduce the effects of synaptically released GABA. IPSPs were recorded from CA1 pyramidal cells with KCl-filled microelectrodes. A train of action potentials enhances Ca^{2+} influx and activation of a Ca^{2+}-dependent K^+ afterhyperpolarization (AHP).[123–127] Spontaneous IPSPs were blocked during the AHP. The AHP was not responsible for the IPSP block, however, because IPSPs were still blocked by direct stimulation in the presence of carbachol, which blocks the AHP.[97,128,129] The IPSP decrease by action potential firing occurs in the presence of D-2-amino-5-phosphonovalerate (to block NMDA receptors) and phaclofen (to block $GABA_B$ receptors), and so is not secondary to the release of other neurotransmitters. IPSPs were not decreased by trains of action potentials following intracellular injection of Ca^{2+} chelators, implying that Ca^{2+} that entered through voltage-dependent channels was the active agent.

LONG-TERM CHANGES IN IPSPs

Stimulus paradigms that produce LTP of excitatory transmission have sometimes been found to cause lasting decreases in inhibitory transmission,[130,131] but often do not decrease IPSPs,[132,133] GABA responsiveness,[134] or the response of GABA interneurons to synaptic stimulation.[69] Long-term depression of GABA transmission is not necessary to induce or maintain LTP; induction proceeds quite well in the absence of IPSPs[135] and is, indeed, abetted by pharmacological blockade of inhibition.[136] As discussed above, transient IPSP depression accompanies trains of stimuli used to induce LTP. Feedforward IPSPs overlap and truncate EPSPs.[61] When IPSPs are pharmacologically blocked, EPSPs are markedly enhanced in size.[60,137] $GABA_B$-mediated IPSP depression facilitates the development of LTP,[137a, 137b] and, indeed, Davies *et al.* report that activation of presynaptic $GABA_B$ receptors by released GABA may be a crucial factor in LTP induction.[137c] Perhaps most importantly, the NMDA component of glutamatergic response is released when GABAergic IPSPs are blocked.[51,63] Because NMDA receptor activation is an essential step in the induction of LTP,[138,139] the transient depression of IPSPs induced by the stimulus train evidently serves an important permissive role in "gating" LTP processes (see reference 140). Long-term depression of IPSPs would not be required. Two features of this idea deserve emphasis: 1) the importance of feedforward inhibition, which, because it acts concurrently on the EPSP, prevents the crucial

LTP induction step from occurring, and 2) the transient nature of the IPSP reduction. If depression of IPSPs serves the function of a gate, opening to allow LTP to take place, then it might be important that, under most circumstances, the gate closes after it has briefly opened. Specificity, both temporal and spatial, may be best achieved in this way. The finding that quantal GABA events result from activation of only a few ion channels[74a, 74b] may be especially important, since so few channels could easily be turned on or off by regulatory factors.

Persistent IPSP depression can occur, however, usually under circumstances that model pathological change. Repeated bouts of stimulation of afferent pathways in the slice can cause "kindling"—a model form of epileptic discharge—and a gradual reduction of spontaneous IPSPs in CA1 cells.[104] Blockade of this effect by D-2-amino-5-phosphonovalerate implicates NMDA receptor activation as a necessary element. The reason for the persistent IPSP depression by NMDA receptor activation is not known, but a role for $Ca^{2+}{}_i$ is suggested by the data reviewed above. Connor *et al.*[141] showed that NMDA receptor activation leads to a prolonged Ca^{2+} entry into CA1 cells in the acutely isolated neuron preparation. An increase in $Ca^{2+}{}_i$ alone may not be sufficient for the prolonged effect, however, because Ca^{2+}-dependent IPSP depression following vigorous action potential firing in rat pyramidal cells (Pitler and Alger, unpublished observations) produced only a transient IPSP reduction. Apparently additional features of the NMDA-induced depression account for its more pronounced effects.

In some cases[130,131] LTP production in CA3 has been accompanied by a lasting IPSP depression. In addition, Miles and Wong[103] showed, using paired-cell recording techniques, that latent excitatory synaptic connections between two hippocampal CA3 cells could be revealed by repeated tetanic stimulation of afferent fibers. Initially stimulation of one of the cells in a pair did not produce a synaptic response in the second cell; they were not demonstrably coupled. Following repetitive activation of afferent fibers in the slice, a synaptic connection between the cells was revealed when direct stimulation of one elicited a synaptic response in the other. The pair had evidently possessed the anatomical capability of interacting physiologically, but were not able to do so because they were functionally disconnected by GABAergic inhibition. Associated with the induced physiological detectability of the latent synapses was a depression of IPSPs. The hypothesized role of IPSP reduction in revealing the connections was strengthened by the demonstration that pharmacological blockade of $GABA_A$ receptors also led to a higher incidence of excitatory connections between nearby, but randomly selected, pairs of cells than was seen in control conditions.[103,142]

The reason for the disparate observations regarding long-term IPSP depression is a mystery. Most often, however, persistent IPSP depression has been found in guinea pig CA3 neurons, whereas only transient IPSP depression is seen in studies of rat CA1 cells, suggesting possible species or cell-type differences. It may also be relevant in this context that, of ten studies of the effects of repetitive stimulation of afferent pathways, the five reporting some evidence for persistent IPSP reduction[103,104,130,131,142] used $[Mg^{2+}]_o$ from 1.0 to 1.5 mM, whereas the five that found no lasting IPSP depression[69,102,105,132,133] used $[Mg^{2+}]_o$ from 2.0 to 4.0 mM. Magnesium ions block NMDA channels,[143] and hence the lower concentrations of Mg^{2+} may have led to prolonged IPSP block via NMDA receptor activation.

ENDOGENOUS FACTORS CAN INCREASE $GABA_A$ RECEPTOR ACTIVATION

Most is known about GABA system modulation by suppression. Certain endogenous factors can enhance GABA responses as well, however, although there is no good evidence for a physiological role for any of these factors yet. Majewska *et al.*[144] found that metabolites of the steroid hormones progesterone and deoxycorticosterone act on the GABA receptor, enhancing and prolonging GABA actions as barbiturates[145] and steroid anesthetics[146] do. The metabolites are present in the CNS and hence could act as physiological regulators. Carlen *et al.*[147] reported that arachidonic acid and its hepoxilin metabolites enhance both $GABA_A$ and $GABA_B$ IPSPs in the hippocampus. Arachidonic acid is generated by a number of hippocampal neurotransmitters, including acetylcholine[148] and glutamate,[149] and so a physiological role for this action is conceivable, although some of these neurotransmitters do not increase, or in fact decrease, GABA actions. Thus the link between these neurotransmitters and the arachidonic acid actions is tenuous at present.

In acutely isolated hippocampal neurons $GABA_A$ responses can be potentiated by low concentrations of glutamate, several glutamate analogues, and even the NMDA receptor antagonist D-2-amino-5-phosphonovalerate.[150] Because endogenous glutamate in the hippocampal slice achieves levels of 50 μM,[151] the potentiating effect of low concentrations of glutamate would appear to be maximal in resting conditions. It is hard to envisage a regulatory role for glutamate unless resting levels are lowered.

POSTSYNAPTIC REGULATION OF $GABA_B$ POTENTIALS

The postsynaptic actions of G protein-linked neurotransmitter receptors such as the $GABA_B$ and adenosine receptors are subject to regulation by protein kinase C[36,41] and receptor systems that activate the phosphatidylinositol-protein kinase C system.[42] Baraban *et al.*[41] reported that, among other effects, phorbol esters block the $GABA_B$-mediated slow IPSP in hippocampus. Subsequent work indicated that phorbol esters and muscarinic agonists that cause phosphatidylinositol breakdown block the effects of iontophoretically applied adenosine and baclofen.[42] Muscarinic agonists that are weakly coupled to phosphoinositide breakdown[152] were commensurately less potent in inhibiting the G protein-linked neurotransmitter effects. In a recent twist, Muller and Misgeld[47] have confirmed that muscarinic agonists block baclofen, but not the $GABA_B$ conductance activated by GABA itself, and raise the question as to whether the same $GABA_B$ receptor is involved in all cases. (A similar suspicion is raised by Segal's recent data.[55]) A possible physiological role for these muscarinic actions was proposed based on evidence that the depressive effects of adenosine on hippocampal field potentials could be relieved by repetitive stimulation that activated the muscarinic slow EPSP.[42] Given the numerous possibilities for regulation offered by G protein-linked systems, discovery of other postsynaptic mechanisms of $GABA_B$ regulation will not be surprising.

CONCLUSIONS

For many years Roberts,[4] Krnjevic,[153] and others have championed the idea of GABAergic disinhibition as a major regulatory principle in the brain. The work reviewed here supports this hypothesis by emphasizing the wide variety of mechanisms available to effect disinhibition. The diversity of disinhibitory mechanisms leads to the speculation that subtle distinctions exist among the physiological consequences of each, and that a variety of functions is served by disinhibition. Although the importance of gross disruption of GABAergic inhibition in the etiology of epileptiform activity has been evident for many years (see references 1 and 5-7 for reviews), the functional implications of more modest alterations in inhibition are just emerging. Principles developed from work on invertebrate systems seem likely to apply to mammalian CNS.

In reviewing the invertebrate work, Getting[154] has emphasized that "operation of a neural network depends on interactions among multiple nonlinear processes at the cellular, synaptic, and network levels," and that "modulation of these underlying processes can alter network operation." Anatomical connectivity provides the substrate; functional connectivity reflects the moment-to-moment pattern of network interactions. A functional circuit can involve a subset of the neurons within an anatomical circuit or a pattern of functional connections that can be modified. Transient IPSP depression can act as a gate that temporarily allows stronger or more direct connections to exist between neurons and facilitates LTP induction, thus stabilizing the strengthened connections. Persistent IPSP depression establishes lasting realignments. Increased inhibition can disconnect cells intimately connected. Selection of discrete functional neuronal circuits from among the myriad of anatomical circuits is likely to be a major consequence of fine-scale intervention in the workings of the GABAergic system. A crucial task for the future will be the elucidation of the roles played by GABAergic inhibition in shaping the functional neuroanatomy of the brain.

REFERENCES

1. ALGER, B. E. 1984. *In* Brain Slices. R. Dingledine, Ed.: 155-199. Plenum. New York, NY.
2. ALGER, B. E. 1985. *In* Neurotransmitter Actions in the Vertebrate Nervous System. M. A. Rogawski & J. L. Barker, Eds.: 33-69. Plenum. New York, NY.
3. NICOLL, R. A. 1988. Science **241:** 545-551.
4. ROBERTS, E. 1986. *In* Advances in Neurology. Vol. 44. A. V. Delgado-Escueta, A. A. Ward, Jr., D. M. Woodbury & R. J. Porter, Eds.: 319-341. Raven Press. New York, NY.
5. SCHWARTZKROIN, P. A. 1986. *In* The Hippocampus. Vol. 3. R. L. Isaacson & K. H. Pribram, Eds.: 113-136. Plenum. New York, NY.
6. PRINCE, D. A. & B. W. CONNORS. 1986. *In* Advances in Neurology. Vol. 44. A. V. Delgado-Escueta, A. A. Ward, Jr., D. M. Woodbury & R. J. Porter, Eds.: 275-299. Raven Press. New York, NY.
7. TRAUB, R. D., R. MILES & R. K. S. WONG. 1989. Science **243:** 1319-1325.
8. NICOLL, R. A., R. C. MALENKA & J. A. KAUER. 1990. Physiol. Rev. **70:** 513-565.
9. KANDEL, E. R., W. A. SPENCER & F. J. BRINLEY, JR. 1961. J. Neurophysiol. **24:** 225-242.
10. ANDERSEN, P., J. C. ECCLES & Y. LOYNING. 1964. J. Neurophysiol. **27:** 592-607.
11. ANDERSEN, P., J. C. ECCLES & Y. LOYNING. 1964. J. Neurophysiol. **27:** 608-619.
12. SCHWARTZKROIN, P. A. & L. H. MATHERS. 1978. Brain Res. **157:** 1-10.
13. TAUBE, J. S. & P. A. SCHWARTZKROIN. 1987. Neurosci. Lett. **78:** 85-90.
14. LACAILLE, J.-C., A. L. MUELLER, D. D. KUNKEL & P. A. SCHWARTZKROIN. 1987. J. Neurosci. **7:** 1979-1993.
15. LACAILLE, J.-C. & P. A. SCHWARTZKROIN. 1988. J. Neurosci. **8:** 1411-1424.
16. LACAILLE, J.-C. & P. A. SCHWARTZKROIN. 1988. J. Neurosci. **8:** 1400-1410.
17. RIBAK, C. E., J. E. VAUGHN & K. SAITO. 1978. Brain Res. **140:** 315-332.
18. RIBAK, C. E. & L. SERESS. 1983. J. Neurocytol. **12:** 577-597.
19. RIBAK, C. E. 1985. Brain Res. **326:** 251-260.
20. SOMOGYI, P., M. G. NUNZI, A. GORIO & A. D. SMITH. 1983. Brain Res. **259:** 137-142.
21. SOMOGYI, P., A. J. HODGSON, A. D. SMITH, M. G. NUNZI, A. GORIO & J.-Y. WU. 1984. J. Neurosci. **4:** 2590-2603.
22. SOMOGYI, P., T. F. FREUND, A. J. HODGSON, J. SOMOGYI, D. BEROUKAS & I. W. CHUBB. 1985. Brain Res. **332:** 143-149.
23. WOODSON, W., L. NITECKA & Y. BEN-ARI. 1989. J. Comp. Neurol. **280:** 254-271.
24. ALGER, B. E. & R. A. NICOLL. 1979. Nature **281:** 315-317.
25. ALGER, B. E. & R. A. NICOLL. 1982. J. Physiol. (London) **328:** 105-123.
26. SIEGHART, W. 1989. Trends Pharmacol. Sci. **10:** 407-411.

27. Bowery, N. 1989. Trends Pharmacol. Sci. **10:** 401-407.
28. Eccles, J., R. A. Nicoll, T. Oshima & F. J. Rubia. 1977. Proc. R. Soc. London Ser. B **198:** 345-361.
29. Segal, M. & J. L. Barker. 1984. J. Neurophysiol. **51:** 500-515.
30. Gray, R. & D. Johnston. 1985. J. Neurophysiol. **54:** 134-142.
31. Newberry, N. R. & R. A. Nicoll. 1984. Nature **308:** 450-452.
32. Newberry, N. R. & R. A. Nicoll. 1984. J. Physiol. (London) **348:** 239-254.
33. Gahwiler, B. H. & D. A. Brown. 1985. Proc. Natl. Acad. Sci. USA **82:** 1558-1562.
34. Newberry, N. R. & R. A. Nicoll. 1985. J. Physiol. (London) **360:** 161-185.
35. Hablitz, J. J. & R. H. Thalmann. 1987. J. Neurophysiol. **58:** 160-179.
36. Andrade, R., R. C. Malenka & R. A. Nicoll. 1986. Science **234:** 1261-1265.
37. Dutar, P. & R. A. Nicoll. 1988. Nature **332:** 156-158.
38. Thalmann, R. H. 1984. Neurosci. Lett. **46:** 103-108.
39. Thalmann, R. H. 1988. J. Neurosci. **8:** 4589-4602.
40. Lambert, N. A., N. L. Harrison, D. I. B. Kerr, J. Ong, R. H. Prager & T. J. Teyler. 1989. Neurosci. Lett. **107:** 125-128.
41. Baraban, J. M., S. H. Snyder & B. E. Alger. 1985. Proc. Natl. Acad. Sci. USA **82:** 2538-2542.
42. Worley, P. F., J. M. Baraban, M. McCarren, S. H. Snyder & B. E. Alger. 1987. Proc. Natl. Acad. Sci. USA **84:** 3467-3471.
43. Fujita, Y. 1979. Brain Res. **175:** 59-69.
44. Thalmann, R. H. & G. F. Ayala. 1982. Neurosci. Lett. **29:** 243-248.
45. Knowles, W. D., J. H. Schneiderman, H. V. Wheal, C. E. Stafstrom & P. A. Schwartzkroin. 1984. Cell. Mol. Neurobiol. **4:** 207-230.
46. Alger, B. E. 1984. J. Neurophysiol. **52:** 892-910.
47. Muller, W. & U. Misgeld. 1989. Neurosci. Lett. **102:** 229-234.
48. Madison, D. V. & R. A. Nicoll. 1988. Brain Res. **442:** 131-138.
49. Masukawa, L. M. & D. A. Prince. 1982. Brain Res. **249:** 271-280.
50. Fisher, R. S. & B. E. Alger. 1984. J. Neurosci. **4:** 1312-1323.
51. Ashwood, T. J. & H. V. Wheal. 1987. Br. J. Pharmacol. **91:** 815-822.
52. Knowles, W. D. & P. A. Schwartzkroin. 1981. J. Neurosci. **1:** 318-322.
53. Miles, R. & R. K. S. Wong. 1984. J. Physiol. (London) **356:** 97-113.
54. Segal, M. 1987. Brain Res. **414:** 285-293.
55. Segal, M. 1990. Brain Res. **511:** 163-164.
56. Schwartzkroin, P. A. & M. Slawsky. 1977. Brain Res. **135:** 157-161.
57. Wong, R. K. S. & D. A. Prince. 1979. Science **204:** 1228-1231.
58. Wong, R. K. S., D. A. Prince & A. I. Basbaum. 1979. Proc. Natl. Acad. Sci. USA **76:** 986-990.
59. Benardo, L. S., L. M. Masukawa & D. A. Prince. 1982. J. Neurosci. **2:** 1614-1622.
60. Dingledine, R. & L. Gjerstad. 1980. J. Physiol. (London) **305:** 297-313.
61. Turner, D. A. 1990. J. Physiol. (London) **422:** 333-350.
62. Herron, C. E., R. Williamson & G. L. Collingridge. 1985. Neurosci. Lett. **61:** 255-260.
63. Dingledine, R., M. A. Hynes & G. L. King. 1986. J. Physiol. (London) **380:** 175-189.
64. Andersen, P., R. Dingledine, L. Gjerstad, I. A. Langmoen & A. M. Laursen. 1980. J. Physiol. (London) **305:** 279-296.
65. Andersen, P., B. Bie & T. Ganes. 1982. Exp. Brain Res. **45:** 357-363.
66. Alger, B. E. & R. A. Nicoll. 1982. J. Physiol. (London) **328:** 125-141.
67. Wong, R. K. S. & D. J. Watkins. 1982. J. Neurophysiol. **48:** 938-951.
68. Masukawa, L. M. & D. A. Prince. 1984. J. Neurosci. **4:** 217-227.
69. Taube, J. S. & P. A. Schwartzkroin. 1987. Brain Res. **419:** 32-38.
70. Madison, D. V. & R. A. Nicoll. 1988. J. Physiol. (London) **398:** 123-130.
71. Blanton, M. G., J. J. Lo Turco & A. R. Kriegstein. 1989. J. Neurosci. Meth. **30:** 203-210.
72. Edwards, F. A., A. Konnerth, B. Sakmann & T. Takahashi. 1989. Pfluegers Arch. **414:** 600-612.
73. Alger, B. E. & R. A. Nicoll. 1980. Brain Res. **200:** 195-200.
74. Rutecki, P. A., F. J. Lebeda & D. Johnston. 1987. J. Neurophysiol. **57:** 1911-1924.

74a. EDWARDS, F. A. *et al.* 1990. J. Physiol. (London) **430:** 213-249.
74b. ROPERT, N. *et al.* 1990. J. Physiol. (London) **428:** 707-722.
75. DEISZ, R. A. & D. A. PRINCE. 1989. J. Physiol. (London) **412:** 513-541.
76. THOMPSON, S. M. & B. H. GAHWILER. 1989. J. Neurophysiol. **61:** 524-533.
77. HARRISON, N. L. 1990. J. Physiol. (London) **422:** 433-446.
78. MOTT, D. D., A. C. BRAGDON & D. V. LEWIS. 1990. Neurosci. Lett. **110:** 131-136.
79. MISGELD, U., W. MULLER & H. BRUNNER. 1989. Pfluegers Arch. **414:** 139-144.
80. MOTT, D. D., A. C. BRAGDON, D. V. LEWIS & W. A. WILSON. 1989. J. Pharmacol. Exp. Ther. **249:** 721-725.
80a. DAVIES, C. H. *et al.* 1990. J. Physiol. (London) **424:** 513-531.
81. CHESNUT, T. J. & J. W. SWANN. 1989. Brain Res. **502:** 365-374.
82. SHINODA, K., M. TOHYAMA & Y. SHIOTANI. 1987. Brain Res. **409:** 181-186.
83. FREUND, T. F. & M. ANTAL. 1988. Nature **336:** 170-173.
84. ZIEGLGANSBERGER, W., E. D. FRENCH, G. R. SIGGINS & F. E. BLOOM. 1979. Science **205:** 415-417.
85. NICOLL, R. A., B. E. ALGER & C. E. JAHR. 1980. Nature **287:** 22-25.
86. ASHWOOD, T. J. & H. V. WHEAL. 1986. Brain Res. **367:** 390-394.
87. FRANCK, J. E., D. D. KUNKEL, D. G. BASKIN & P. A. SCHWARTZKROIN. 1988. J. Neurosci. **8:** 1991-2002.
88. OTIS, L. C., D. V. MADISON & R. A. NICOLL. 1985. Brain Res. **346:** 281-286.
89. FINCH, D. M. & T. L. BABB. 1977. Brain Res. **130:** 354-359.
90. YAMAMOTO, C. & N. KAWAI. 1967. Exp. Neurol. **19:** 176-187.
91. HOUNSGAARD, J. 1978. Exp. Neurol. **62:** 787-797.
92. VALENTINO, R. J. & R. DINGLEDINE. 1981. J. Neurosci. **1:** 784-792.
93. DUTAR, P. & R. A. NICOLL. 1988. J. Neurosci. **8:** 4214-4224.
94. HAAS, H. L. 1982. Brain Res. **233:** 200-204.
95. PITLER, T. A. & B. E. ALGER. 1990. Soc. Neurosci. Abstr. **16:** 467.
96. MCCORMICK, D. A. & D. A. PRINCE. 1985. Proc. Natl. Acad. Sci. USA **82:** 6344-6348.
97. BENARDO, L. S. & D. A. PRINCE. 1982. Brain Res. **249:** 315-331.
98. BRUNNER, H. & U. MISGELD. 1988. Neurosci. Lett. **88:** 63-68.
99. BILKEY, D. K. & G. V. GODDARD. 1985. Brain Res. **361:** 99-106.
100. LYNCH, G., G. ROSE & C. GALL. 1978. *In* Functions of the Septo-Hippocampal System (Ciba Foundation Symposium 58): 5-24. Elsevier/Excerpta Medical/North-Holland. Amsterdam.
101. BEN-ARI, Y., K. KRNJEVIC & W. REINHARDT. 1979. Can. J. Physiol. Pharmacol. **57:** 1462-1466.
102. MCCARREN, M. & B. E. ALGER. 1985. J. Neurophysiol. **53:** 557-571.
103. MILES, R. & R. K. S. WONG. 1987. Nature **329:** 724-726.
104. STELZER, A., N. T. SLATER & G. TEN BRUGGENCATE. 1987. Nature **326:** 698-701.
105. THOMPSON, S. M. & B. H. GAHWILER. 1989. J. Neurophysiol. **61:** 501-511.
106. NUMANN, R. E. & R. K. S. WONG. 1984. Neurosci. Lett. **47:** 289-294.
107. HUGUENARD, J. R. & B. E. ALGER. 1986. J. Neurophysiol. **56:** 1-18.
108. SCHWARTZKROIN, P. A. & D. A. PRINCE. 1980. Brain Res. **185:** 169-181.
109. RUTECKI, P. A., F. J. LEBEDA & D. JOHNSTON. 1985. J. Neurophysiol. **54:** 1363-1374.
110. TRAYNELIS, S. F. & R. DINGLEDINE. 1988. J. Neurophysiol. **59:** 259-276.
111. KORN, S. J., J. L. GIACCHINO, N. L. CHAMBERLIN & R. DINGLEDINE. 1987. J. Neurophysiol. **57:** 325-340.
112. HAGLUND, M. M. & P. A. SCHWARTZKROIN. 1990. J. Neurophysiol. **63:** 225-239.
113. MCCARREN, M. & B. E. ALGER. 1987. J. Neurophysiol. **57:** 496-509.
114. THALMANN, R. H., E. J. PECK & G. F. AYALA. 1981. Neurosci. Lett. **21:** 319-324.
115. THALMANN, R. H. 1988. Neurosci. Lett. **95:** 155-160.
116. PERREAULT, P. & M. AVOLI. 1988. Can. J. Physiol. Pharmacol. **66:** 1100-1102.
117. MISGELD, U., R. A. DEISZ, H. U. DODT & H. D. LUX. 1986. Science **232:** 1413-1416.
118. THOMPSON, S. M. & D. A. PRINCE. 1986. J. Neurophysiol. **56:** 507-522.
119. THOMPSON, S. M. & B. H. GAHWILER. 1989. J. Neurophysiol. **61:** 512-523.
120. INOUE, M., Y. OOMURA, T. YAKUSHIJI & N. AKAIKE. 1986. Nature **324:** 156-158.

121. CHEN, Q. X., A. STELZER, A. R. KAY & R. K. S. WONG. 1990. J. Physiol. (London) **420:** 207-221.
122. STELZER, A., A. R. KAY & R. K. S. WONG. 1988. Science **241:** 339-341.
122a. LAMBERT, N. A. & N. L. HARRISON. 1990. J. Pharmacol. Exp. Ther. **255:** 90-94.
122b. PORTER, N. M. *et al.* 1990. Neuron **5:** 789-796.
122c. BROWNING, M. D. *et al.* 1990. Proc. Natl. Acad. Sci. USA **87:** 1315-1318.
123. ALGER, B. E. & R. A. NICOLL. 1980. Science **210:** 1122-1124.
124. SCHWARTZKROIN, P. A. & C. E. STAFSTROM. 1980. Science **210:** 1125-1126.
125. HOTSON, J. R. & D. A. PRINCE. 1980. J. Neurophysiol. **43:** 409-419.
126. WONG, R. K. S. & D. A. PRINCE. 1981. J. Neurophysiol. **45:** 86-97.
127. NICOLL, R. A. & B. E. ALGER. 1981. Science **212:** 957-959.
128. COLE, A. E. & R. A. NICOLL. 1983. Science **221:** 1299-1301.
129. GAHWILER, B. H. 1984. Neuroscience **11:** 381-388.
130. YAMAMOTO, C. & T. CHUJO. 1978. Exp. Neurol. **58:** 242-250.
131. MISGELD, U., J. M. SARVEY & M. R. KLEE. 1979. Exp. Brain Res. **37:** 217-229.
132. HAAS, H. L. & G. ROSE. 1982. J. Physiol. (London) **329:** 541-552.
133. GRIFFITH, W. H., T. H. BROWN & D. JOHNSTON. 1986. J. Neurophysiol. **55:** 767-775.
134. SCHARFMAN, H. E. & J. M. SARVEY. 1985. Neuroscience **15:** 695-702.
135. YAMAMOTO, C., K. MATSUMOTO & M. TAKAGI. 1980. Exp. Brain Res. **38:** 469-477.
136. WIGSTROM, H. & B. GUSTAFSSON. 1983. Nature **301:** 603-604.
137. JOHNSTON, D. & T. H. BROWN. 1981. Science **211:** 294-297.
137a. OLPE, H.-R. & G. KARLSSON. 1990. Naunyn-Schmied. Arch. Pharmacol. **342:** 194-197.
137b. MOTT, D. D. *et al.* 1990. Neurosci. Lett. **113:** 222-226.
137c. DAVIES, C. H. *et al.* 1990. Nature **349:** 609-611.
138. MULLER, D., M. JOLY & G. LYNCH. 1988. Science **242:** 1694-1697.
139. NICOLL, R. A., J. A. KAUER & R. C. MALENKA. 1988. Neuron **1:** 97-103.
140. DINGLEDINE, R. 1986. Trends Neurosci. **9:** 47-49.
141. CONNOR, J. A., W. J. WADMAN, P. E. HOCKBERGER & R. K. S. WONG. 1988. Science **240:** 649-653.
142. MILES, R. & R. K. S. WONG. 1987. J. Physiol. (London) **388:** 611-629.
143. MAYER, M. L., G. L. WESTBROOK & P. B. GUTHRIE. 1984. Nature **309:** 261-263.
144. MAJEWSKA, M. D., N. L. HARRISON, R. D. SCHWARTZ, J. L. BARKER & S. M. PAUL. 1986. Science **232:** 1004-1007.
145. NICOLL, R. A., J. C. ECCLES, T. OSHIMA & F. RUBIA. 1975. Nature **258:** 625-627.
146. HARRISON, N. L., S. VICINI & J. L. BARKER. 1987. J. Neurosci. **7:** 604-609.
147. CARLEN, P. L., N. GUREVICH, P. H. WU, W.-G. SU, E. J. COREY & C. R. PACE-ASCIAK. 1989. Brain Res. **497:** 171-176.
148. CONKLIN, B. R., M. R. BRANN, N. J. BUCKLEY, A. L. MA, T. I. BONNER & J. AXELROD. 1988. Proc. Natl. Acad. Sci. USA **85:** 8698-8702.
149. DUMUIS, A., M. SEBBEN, L. HAYNES, J.-P. PIN & J. BOCKAERT. 1988. Nature **336:** 68-70.
150. STELZER, A. & R. K. S. WONG. 1989. Nature **337:** 170-173.
151. SAH, P., S. HESTRIN & R. A. NICOLL. 1989. Science **246:** 815-818.
152. FISHER, S. K., J. C. FIGUEIREDO & R. T. BARTUS. 1984. J. Neurochem. **43:** 1171-1179.
153. KRNJEVIC, K. 1981. *In* GABA and Benzodiazepine Receptors. E. Costa, G. Di Chiara & G. L. Gessa, Eds.: 111-120. Raven Press. New York, NY.
154. GETTING, P. A. 1989. Annu. Rev. Neurosci. **12:** 185-204.

Age-Dependent Alterations in the Operations of Hippocampal Neural Networks[a]

JOHN W. SWANN,[b,c,d] KAREN L. SMITH,[b] AND ROBERT J. BRADY [b,c]

[b]*Wadsworth Center for Laboratories and Research*
New York State Department of Health
and
[c]*Department of Biomedical Sciences*
State University of New York at Albany
Albany, New York 12201

The operations of neural networks, including those in the central nervous system, are generally agreed to be the product of 1) the patterns of interconnections among neurons within a given network, 2) the types of interactions that take place at the points of synaptic contact, and 3) the intrinsic physiological properties of the neuronal elements.[1] Alterations in cellular processes within each of these categories would be expected to modify network behavior and ultimately animal behavior. Sprouting or withdrawal of axon collaterals, increases or decreases in the efficacy of already existing synapses, or alterations of the firing properties of neurons could each result in a different output for a given fixed network input. In recent years significant advances have been made in understanding the cellular and molecular events that underlie different forms of plasticity in both vertebrate and invertebrate neuronal systems. Studies reviewed in this volume attest to the progress made in understanding a variety of activity-dependent alterations in neurons. How such changes can modify behavior have been described in some simple systems. However, our understanding of use-dependent alterations of network functioning in the mammalian brain is far from resolution.

Even less is known about the ontogeny of neural networks. With the advent of *in vitro* slice procedures, local circuits within brain regions became available for detailed neurophysiological investigations. However, only recently have laboratories explored such preparations to understand alterations in the behavior of local circuits during CNS maturation.[2,3] Results from studies of the hippocampal slice suggest that, during a critical period early in postnatal life, the CA3 hippocampal neuronal network has a marked capacity to undergo synchronized discharges.[4] This period of hyperexcitability appears to coincide with the time when CA3 pyramidal cells possess an overabundance of recurrent excitatory axon collaterals.

[a]This work was supported by grants from the National Institutes of Health (NS18309 and NS23071).

[d]Address for correspondence: Department of Biomedical Sciences, State University of New York at Albany, P. O. Box 509, Empire State Plaza, Albany, New York 12201.

INTRACELLULAR RECORDINGS OF LOCAL CIRCUIT SYNAPTIC INTERACTIONS IN IMMATURE HIPPOCAMPUS

When intracellular recordings are obtained from slices taken from rats 1-2 weeks of age, responses to electrical stimulation of afferent fiber tracts are quite different from those reported in mature hippocampus. In individual mature pyramidal cells a single orthodromic stimulus usually generates a brief excitatory postsynaptic potential (EPSP) followed by a prolonged hyperpolarization.[5] This hyperpolarization has been shown to consist of two inhibitory postsynaptic potentials (IPSPs) mediated by $GABA_A$ and $GABA_B$ receptors.[6] With repetitive stimulation the cells often undergo a tonic hyperpolarization, which is likely due to the temporal summation of these IPSPs.[7] During a prolonged train of stimuli the membrane potential can slowly return to resting potential; in some instances, it will overshoot and the cell will depolarize. Immediately following a train, a response to a single orthodromic stimulus is quite different from that of the pretetanus control period. IPSPs are often depressed, and pronounced depolarizations can be recorded.[7,8]

FIGURE 1 shows recordings from the CA3 subfield of a hippocampus taken from a rat on postnatal day 16. These are markedly different from their counterparts in adults. It has been previously demonstrated that responses to single orthodromic or antidromic stimulation at this time in life can be similar to those recorded in mature hippocampus.[9] However, prolonged depolarizing responses can occur. FIGURE 1 shows an example of such events. The response labeled "a" consists of a slow depolarization that produced several action potentials. When a relatively low-frequency train of nine orthodromic stimuli was applied, the depolarization temporally summated. Immediately after the train unusually intense and prolonged depolarizing responses were recorded. Responses gradually returned to pretetanus control values over the next few minutes. Panel B shows that as few as eleven electrical stimuli produced self-sustained discharges that persisted for more than 30 sec. This record differs dramatically from the results reported in mature hippocampus. Typically only brief bursts of action potentials occur.[7] Only after repeated and prolonged trains of stimuli do slices from older animals produce prolonged discharges like those shown in Panel B.[10] Thus the sensitivity of the developing hippocampus to repetitive stimulation is unusual at this stage of postnatal life. Consequently, studies have been undertaken to determine the physiological processes underlying this form of hyperexcitability.

One clue to the underlying processes was obtained during the course of recordings such as those shown in FIGURE 1. Usually large and prolonged spontaneous EPSP- and IPSP-like potentials were often observed in baseline recordings. FIGURE 2 shows an example of such spontaneous events. In this instance the events were primarily depolarizing. It seemed very unlikely that such events were simply unitary synaptic potentials like those obtained in intracellular recordings in mature hippocampus. Instead, they appeared to be more complex, network-driven bursts of synaptic potentials.[2] In support of this contention are the following observations. In addition to the potentials being large (5-10 mV) and prolonged (100-500 msec), they often had very complex waveforms. Bursts of such potentials were routinely recorded simultaneously in cell pairs during dual intracellular recordings. The frequency of these events does not change with alterations in membrane potential. One further observation that suggested that these events were network driven was that, simultaneous with the intracellular potentials, negative field potentials were recorded in the basilar dendritic layer (Traces B in FIG. 3). The amplitude and time course of the field potentials covaried with those of the simultaneously recorded intracellular events. Thus, the field potentials likely

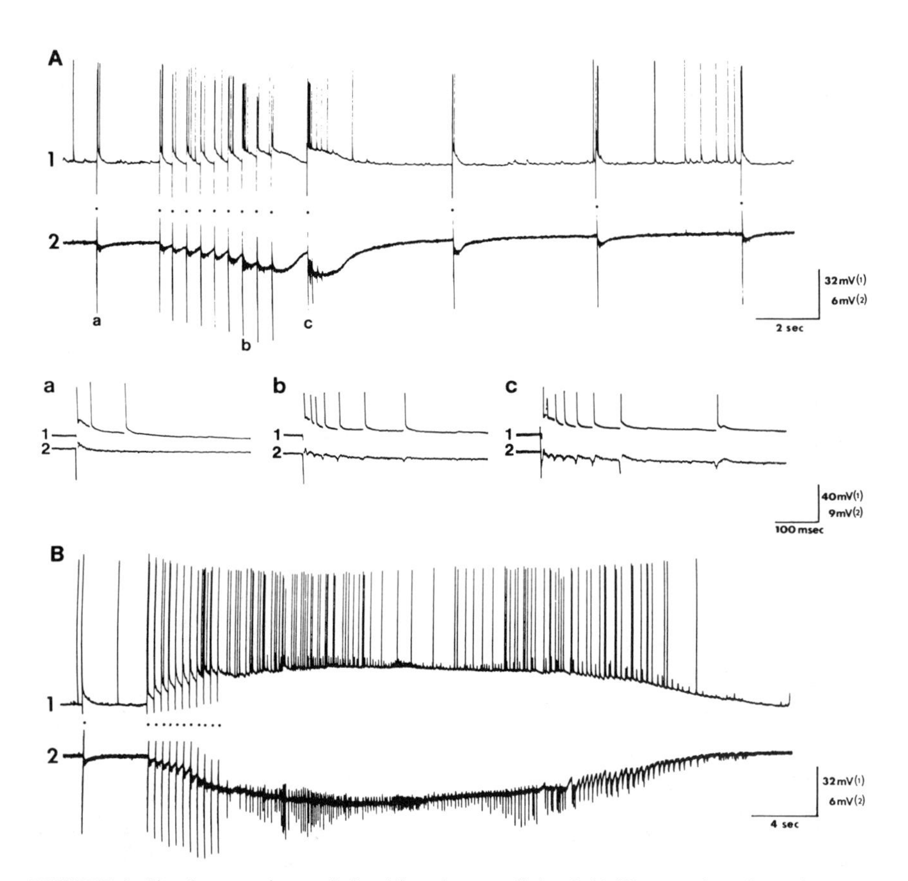

FIGURE 1. Simultaneous intracellular (**1**) and extracellular field (**2**) recordings from the CA3 subfield of hippocampus taken from a postnatal day 16 rat. Recordings were made in normal medium. The intracellular microelectrode was positioned in the cell body layer while the field electrode was placed extracellularly in the infrapyramidal zone (proximal portion of basilar dendritic layer). During recording in Panel A, a train of orthodromic stimuli (2 Hz) was given. Selected responses a, b, and c are shown at a faster time base below. A slightly longer train of stimuli was given in Panel B. This resulted in a self-sustained discharge. In this panel the neuron was hyperpolarized with DC current. The resting potential of the cell was -61 mV. A dot between traces indicates the application of an electrical stimulus (100-μsec square-wave constant current pulse).

reflect the generation of synaptic events simultaneously in a number of hippocampal pyramidal cells. The synaptic potential bursts can be recorded in isolated segments of the CA3 subfield. Thus they are not produced by networks of dentate granular cells or CA1 hippocampal neurons.

Further support for the idea that the intracellular bursts are synaptically generated is that their amplitude was a monotonic function of the membrane potential.[2] Reversal potentials could differ from cell to cell, however. This latter property would be explained if the events were mixtures of local circuit synaptic potentials and if different cells received different complements of inhibitory and excitatory synaptic inputs. This notion was supported by pharmacological studies in which $GABA_A$ receptor antagonists were applied locally to regions in the slice where bursts of synaptic events were recorded. FIGURE 3 shows that, when microdroplets of bicuculline were applied to slices near the intracellular recording pipette, both the synaptic potential bursts and simultaneously

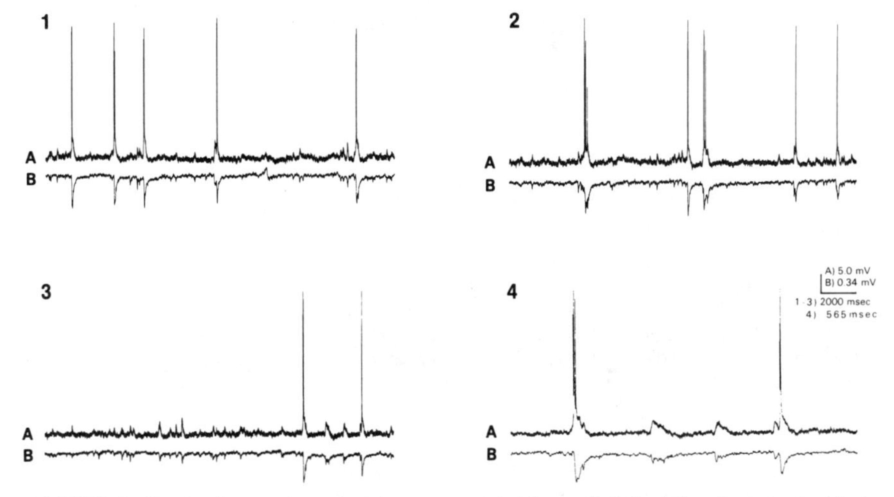

FIGURE 2. Bursts of synaptic potentials were recorded intracellularly (**A**) and were coincident with the occurrence of small negative field potentials (**B**) in the infrapyramidal zone. Panels 1, 2, and 3 show spontaneous activity in three 20-sec epochs taken from an immature rat hippocampal slice. The latter portion of traces in Panel 3 are shown in Panel 4 at a faster time base.

recorded field potentials were dramatically increased in amplitude. This effect is likely explained by elimination of GABA IPSPs that would tend to hold the membrane potential near resting. Furthermore, current shunting by the IPSPs would attenuate depolarizations generated by simultaneous EPSPs. It is also possible that by blocking IPSPs the number of pyramidal cells that excite each other via recurrent excitatory pathways would increase since the ability of inhibitory interneurons to prevent the spread of excitation through local networks of mutually excitatory pyramidal cells would be suppressed (Miles and Wong,[11] see discussion below).

Because the bursts occur spontaneously, it seemed likely that they would also be entrained by electrical stimulation. Thus they might contribute to responses such as those shown in FIGURE 1. Indeed, the bursts of synaptic events were easily elicited by orthodromic stimulation. Panel 1 of FIGURE 4 shows an example of a spontaneous

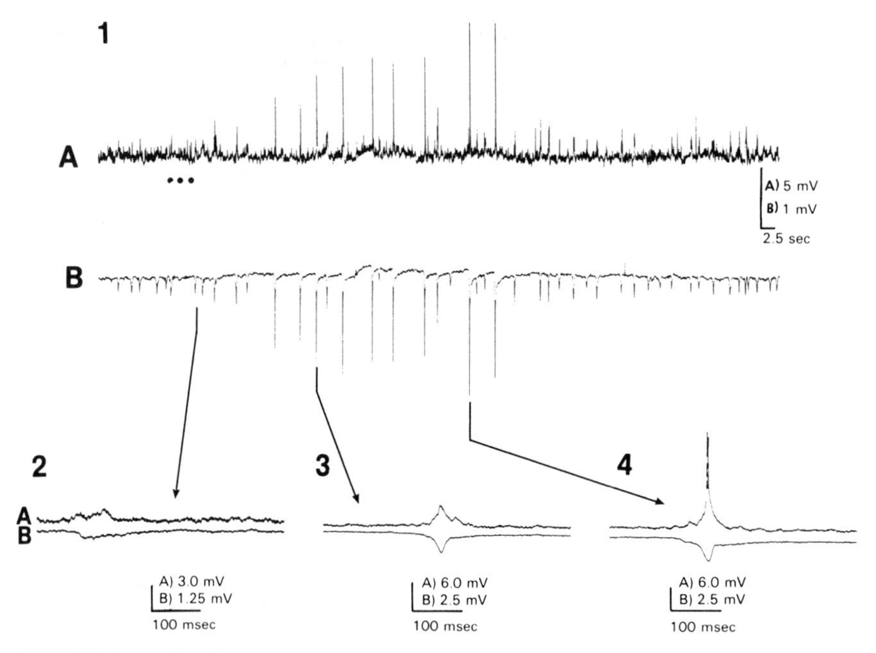

FIGURE 3. Local application of bicuculline dramatically increases the amplitude of depolarizing components of synaptic potential bursts. Panel 1 is a slow time base recording of intracellularly recorded bursts of synaptic potential **(A)** and simultaneous extracellular field potentials **(B)**. At the dots bicuculline (100 μM) was pressure injected into the cell body layer near the recording electrodes. Panels 2, 3, and 4 show events indicated by the arrows at a faster time base. In panel 2, bicuculline had yet to exert its effect. Notice changes in calibrations between Panel 2 and those in Panels 3 and 4.

event and another event evoked in the same cell. At threshold stimulation strength the bursts often occurred in an all-or-none manner and after long delays, suggesting a network origin. With increasing stimulus intensity their latency decreased. Panel 3 shows that with further increases in stimulus intensity the evoked events increased in amplitude. Thus, it seems likely that the orthodromically elicited prolonged depolarizations in Panel 1 of FIGURE 1 are produced at least in part by synaptic events resulting from the discharge of local neural networks.

RECURRENT INHIBITION AND NETWORK FUNCTION IN IMMATURE HIPPOCAMPUS

Synchronous bursts of synaptic potentials such as those shown in FIGURES 2, 3 & 4 have not been reported in mature hippocampus under normal physiological conditions. However, Schneiderman[12] and Miles and Wong[11,13] have recorded such events in CA3 hippocampal neurons once GABAergic synaptic transmission was partially suppressed, either by application of $GABA_A$ receptor antagonists or tetanic stimulation. Thus the existence of these events under normal conditions in developing hippocampus might suggest that recurrent inhibition is not fully developed early in postnatal life. However, results of intracellular recordings from a number of laboratories have shown that synaptic inhibition is well developed early in postnatal life in the CA3 subfield.[9,14] In the rat, measures of antidromically elicited IPSPs made on postnatal days 4-6 were similar to those obtained in month-old rats. During the second postnatal week IPSP conductances were actually larger than those recorded in older animals. Thus it appears that the marked propensity of the CA3 neurons to generate network-driven bursts of synaptic potentials early in postnatal life cannot be explained simply by a deficit in local circuit synaptic inhibition.

Support for the notion that GABAergic synaptic transmission plays an important role in local-circuit functioning early in postnatal life comes from studies in which $GABA_A$ receptor antagonists have been added to slices as microdroplets (see FIG. 3) or by bath application. When $GABA_A$ antagonists, such as penicillin, are added to slice perfusate, rhythmic synchronized discharging of the developing CA3 hippocampal subfield occurs.[4] Such discharges are very reminiscent of, if not identical to, those that occur during electrographic seizures recorded *in vivo*. Traces B in FIGURE 5 show examples of such events. The discharges occurred spontaneously in the presence of penicillin and cycle between brief and prolonged electrographic events. In contrast, slices from adult rats generate only brief discharges as shown in Traces A. While the differences between these recordings in immature and mature hippocampus are dra-

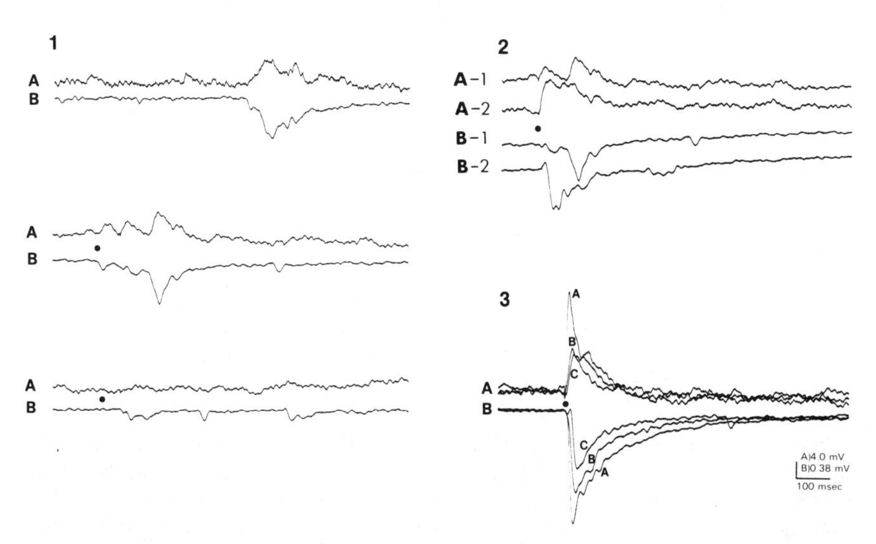

FIGURE 4. Intracellularly recorded synaptic potential bursts (**A**) and coincident extracellular field potentials (**B**) can be recruited by low-amplitude orthodromic stimulation. Upper traces in Panel 1 show a spontaneous burst of synaptic potentials and a coincident field potential recorded in subfield CA3a. The middle traces show that at stimulus threshold the events can be elicited but at a long latency. In the lower traces the same stimulus failed to produce a response. Panel 2 shows that increasing the stimulus intensity decreases the latency of the events. A higher stimulus strength was used in events A2-B2. Panel 3 shows that further increases in stimulus intensity produce increasingly larger responses. The stimulating electrode was positioned in stratum lucidum of CA3c. Dots between traces indicate time of stimulus (100-μsec square-wave pulse) application.

matic, they share two common features. First, at either age such events are never recorded under normal conditions. Second, they are both produced by the suppression of synaptic inhibition. Thus synaptic inhibition appears to play a central role in network functioning throughout life, but evidently it has an even more pronounced role in regulating network behavior in the immature brain.

Miles and Wong[11] have examined the mechanism by which GABA receptor antagonists produce their effect on network operations in mature hippocampus. Their results suggest that GABAergic interneurons normally limit the spread of excitation through networks of synaptically interconnected CA3 hippocampal pyramidal cells. Dual intracellular recordings[15] as well as anatomical studies of single pyramidal cells filled with horseradish peroxidase[16] have shown that each CA3 pyramidal cell activates neighboring cells through a network of local recurrent excitatory axon collaterals. It has been suggested that each pyramidal cell makes contact with as many as 20 other pyramidal cells (see, for example, Traub and Miles[17]). However, since thousands of

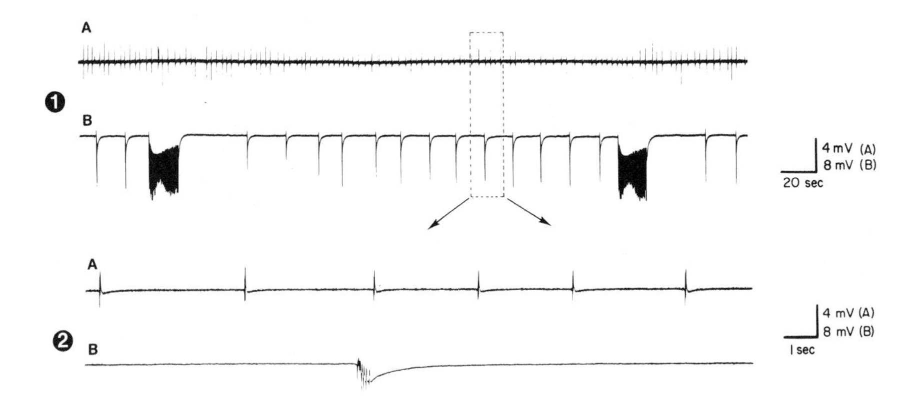

FIGURE 5. Comparison of spontaneous field potentials recorded in a hippocampal slice from a mature rat (**A**) and an 11-day-old rat pup (**B**). The synchronous discharging of the CA3 population was produced by bath application of 1.7 mM penicillin. Extracellular field recordings were made from the cell body layer. The events outlined in the frames of Panel 1 are shown in Panel 2 at a faster time base. (From Swann *et al.*,[47] reproduced by permission of Alan R. Liss, Inc.).

pyramidal neurons are present in the CA3 subfield of a hippocampal slice, the likelihood, based on chance alone, of impaling two cells that are monosynaptically coupled is remote. Indeed, under normal conditions action potentials in a single pyramidal cell rarely (2% of pairs) produce EPSPs in a second simultaneously impaled neuron.[15] However, in the presence of GABA antagonists, the frequency of excitatory synaptic interaction between pyramidal cell pairs increases dramatically.[13] These synaptic interactions have been shown to be polysynaptically mediated.

The most parsimonious explanation for this latter observation is as follows. When a pyramidal cell discharges it activates not only other nearby pyramidal cells but also inhibitory interneurons that use GABA as their neurotransmitter. The basket cell of the hippocampal formation would be one such cell type. Discharging of basket cells in response to pyramidal cell input would result in disynaptically mediated IPSPs in hundreds of nearby pyramidal cells. These IPSPs would tend to prevent neuronal discharging. Thus, the likelihood that second-order pyramidal cells will fire in response to discharging of the first pyramidal cell would be small when inhibitory interneurons are functional. When synaptic inhibition is blocked by GABA antagonists, IPSPs will not occur, and the discharging of the second-order pyramidal cells would now be much more likely. These cells in turn would excite many other cells through their own local recurrent excitatory collaterals, and a cascade of excitation through the local network would occur. Thus, if two pyramidal cells are impaled when GABA IPSPs are blocked, one would be more likely to record EPSPs, although usually these would be polysynaptically mediated. As discussed above, this is exactly what is observed experimentally. Indeed, often pairs of pyramidal cells that are not synaptically coupled in normal medium became polysynaptically coupled after application of $GABA_A$ receptor antagonists.[13]

On the basis of this concept of network functioning and the role that GABA synaptic inhibition seems to play in immature hippocampal networks, one may attempt to explain why repetitive activation of hippocampal neural networks early in life generates prolonged depolarizations and synchronized discharges (see FIG. 1). One explanation is that with repeated activation the effectiveness of GABA synapses is diminished. Actually, use-dependent depression of GABA IPSPs is well documented in the hippocampal formation[7,8] and the physiologic mechanisms underlying this depression have been studied extensively.[18] By contrast EPSPs are usually not depressed by repeated activation but instead are most often potentiated by such patterns of activity. As suggested earlier, orthodromic stimuli recruit small networks of neurons in immature hippocampus. These produce complex synaptic bursts that are mixtures of both EPSPs and IPSPs. If during repetitive stimulation the IPSP components of the bursts become selectively depressed, then one would expect that trains of such stimuli would result in large prolonged depolarizations like those recorded when $GABA_A$ antagonists are applied (FIG. 3). Indeed, that is what is observed upon repeated stimulations such as those employed in FIGURE 1, Panel 1. Furthermore, if repetitive stimulation is excessive, GABA IPSPs may be suppressed to such a degree that prolonged discharges such as shown in FIGURE 5 would be recorded. This also can occur as illustrated in FIGURE 1, Panel 2. Thus results suggest that one use-dependent feature of developing hippocampal networks might be ascribed to activity-dependent suppression of local circuit synaptic inhibition. However, one fundamental observation still remains unexplained. Why is network behavior so different in immature hippocampus once GABA inhibition is suppressed?

RECURRENT EXCITATION AND NETWORK FUNCTIONING IN IMMATURE HIPPOCAMPUS

Recently, Traub and colleagues[19] used computer simulations to study properties of CA3 hippocampal neural networks. Their model predicts that, when synaptic inhibition is suppressed, bursts of synaptic potentials are likely to occur. As discussed above, this has been observed experimentally.[11–13] The model also demonstrates that the presence of these synchronized synaptic events is highly dependent upon recurrent excitation. When recurrent excitation is not included in simulations, synaptic potential bursts do not occur. Furthermore, if the strength of the recurrent excitation is increased in the model, the amplitude of these events also increases. Because synaptic inhibition is present in the developing CA3 subfield, the possibility exists that enhanced recurrent excitation may be present early in postnatal life and that this is the underlying process responsible for the occurrence of synaptic potential bursts. Furthermore, the synchronized discharging of larger populations of neurons that occurs when GABA inhibition is abolished would also be explained by enhanced recurrent excitation.

One result that strongly suggests that recurrent excitation plays an important role in the generation of the synchronized discharging is that excitatory amino acid receptor antagonists abolish penicillin-induced discharging in immature hippocampus.[20,21] Since this is in keeping with the notion that enhanced recurrent excitation may exist early in postnatal life, recent studies have attempted to test this hypothesis more directly. One observation consistent with this idea is that, in disinhibited developing hippocampus, action potentials in an unusually high percentage of individual pyramidal cells are able to entrain the entire CA3 population. Miles and Wong[22] were the first to demonstrate that, upon blockade of local circuit synaptic inhibition, action potentials in single CA3 pyramidal cells were able to trigger a synchronous discharge of the CA3 subfield. This finding is fully consistent with the notion of neural network behavior reviewed above. That is, given that discharging of a single CA3 neuron produces monosynaptic EPSPs in 20 other pyramidal cells, and that each of these cells would in turn activate 20 other cells, a cascade of excitation through the network could be produced from a single cell, and a population discharge could ensue. In their study, Miles and Wong[22] found that approximately 30% of cells could entrain the population. However, in hippocampus from one- and two-week-old rats, we have found that over 60% of cells have this capacity.

Dual intracellular recordings performed in one-week-old rats also suggest that an overabundance of recurrent excitatory synapses may be present at this time. The frequency of synaptic interaction recorded in pairs of CA3 pyramidal cells in disinhibited slices taken from 10-16-day-old rats is unusually high. In 51% of 150 paired intracellular recordings, action potentials in one neuron of the pair produced EPSPs in the other neuron. Thirty-five percent of these EPSPs were polysynaptically mediated while 4% and 7% were produced mono- and disynaptically, respectively. Six percent of pairs were mutually excitatory. These numbers are uniformly higher than those reported under similar conditions in mature hippocampus.[11,15] Indeed, mutually excitatory pairs of cells have not been reported in hippocampus from adult animals, suggesting that the extent of local recurrent excitatory collateralization may be extensive early in postnatal life.

Preliminary anatomical examination of axonal arbors from CA3 hippocampal pyramidal cells on postnatal days 4-6, 10-16, and 40-50 support these physiological observations.[23] Individual CA3 cells have been filled intracellularly with biocytin.[24] While during the first postnatal week few axons from single neurons have been observed in the CA3 subfield, these axons often terminate in clearly identifiable growth cones. This would suggest that axon addition occurs during this period. By day 10 extensive collateralization of pyramidal cell axons is observed throughout the CA3 subfield. Markedly fewer collaterals have been observed in mature rats. Axon collaterals observed during the second postnatal week have been found to be studded with hundreds and often thousands of "fiber swellings" or varicosities. Ultrastructural studies have shown that these are synaptic specializations.[25] Thus both physiological and anatomical results support the hypothesis that early in postnatal life an overabundance of recurrent excitatory collaterals exist in the CA3 subfield and that this form of local circuit synaptic transmission plays an important role in age-dependent changes in hippocampal network functioning.

AXON PRUNING: A REGRESSIVE EVENT IN HIPPOCAMPAL NETWORK FORMATION

Over the last ten years numerous laboratories have demonstrated that early in postnatal life a transient overproduction of axonal projections occurs in various neural systems both in the periphery and brain (for reviews, see references 26-29). With maturation, long axon collaterals as well as terminal arbors are pruned. The elimination of axons and associated synaptic contacts is thought to be an important process that underlies age-dependent rearrangements of neural networks. One of the best studied and most dramatic examples of "exuberant" projections of axons in neonatal brain is the interhemispheric axons of the corpus callosum.[30] At birth the corpus callosum of rhesus monkey contains 188 million axons. The adult has a complement of only 56 million axons. Axons are thought to be lost in two phases during the first three months of life. During the first stage of regression, LaMantia and Rakic[31] estimate that 4.4 million axons are lost each day. This translates into the withdrawal of 50 axons/sec.

Separate ultrastructural studies by Huttenlocher and colleagues[32,33] in humans and Rakic *et al.*[34] in primates have also shown that the density of synapses in the cortex is higher early in life than in the adult. Moreover, age-dependent elimination of functional synapses has been demonstrated at the neuromuscular junction, autonomic ganglion,[35] and climbing fiber innervation of cerebellar Purkinje cells.[36]

In the rodent CNS, several groups have shown that there is a transient overproduction of markers for excitatory amino acid synapses early in postnatal life (for review, see McDonald and Johnston[37]). In hippocampal studies, transient overshoots in binding sites for excitatory amino acid neurotransmitters have been demonstrated. In the rat both glutamate binding and *N*-methyl-D-aspartate (NMDA)-sensitive glutamate binding exceed adult levels during the second postnatal week. In the CA3 subfield, NMDA receptor binding reaches a maximal value by day 10. Thus neurochemical results are consistent with the notion that, at a time when networks within the CA3 subfield demonstrate marked excitability, a transient overproduction of excitatory synapses may exist.

EXCITATORY AMINO ACID RECEPTORS IN IMMATURE HIPPOCAMPUS

Numerous studies of the developing visual pathways have suggested that NMDA receptors play an important role in the consolidation of synapses and thus in the rearrangements in network circuitry that takes place during normal brain development (for review, see Constantine-Paton *et al.*,[38] Schmidt,[39] and Udin and Scherer[40]). Chronic treatments with NMDA receptor antagonists have been shown to prevent ocular dominance shifts in monocularly deprived kittens and the segregation of inputs from normal and supernumerary eyes in tadpoles.[41,42]

In hippocampus NMDA receptors are present at times when the formation of recurrent excitatory synapses on CA3 hippocampal pyramidal cells appears to be taking place. One promising avenue for future study will be examination of the role that NMDA receptors play in synapse consolidation in developing hippocampus. In this regard it is interesting that several laboratories have reported that NMDA receptors in developing hippocampus have unique properties that distinguish them from those in mature hippocampus. Ben-Ari *et al.*[43] have suggested that, during the first postnatal week, NMDA receptors lack voltage dependency. Nadler *et al.*[44] have reported that NMDA receptors on developing CA1 neurons are less sensitive to changes in extracellular Mg^{2+} than cells from older animals. Based on age-dependent differences in expression of binding sites from NMDA, TCP, and glycine (strychnine insensitive), McDonald and Johnston[37] have suggested that a different isoform of the NMDA receptor may exist in the immature hippocampus.

In the CA3 subfield of one-week-old rat hippocampus, NMDA receptors have a voltage dependency resembling that of receptors in hippocampus from mature rats. However, the voltage dependency appears to arise in a unique way. Extracellular Ca^{2+} appears to impart a large measure of the voltage dependency to the channel associated with the NMDA receptor at this stage in neuronal development.[45] In mature cells Mg^{2+} has this role. The impact of an NMDA Ca^{2+}-mediated voltage dependency on synaptic transmission and plasticity in the developing brain could be substantial. During repetitive activation of synapses, extracellular Ca^{2+} transiently decreases as it enters neurons through channels that increase Ca^{2+} permeability (see, for example, Heinemann *et al.*[46]). Among these would be the NMDA receptor channel itself. Thus, upon synapse activation, and especially during repetitive stimulation, NMDA-mediated processes might be enhanced not only by membrane depolarization but also by a reduction in the concentration of Ca^{2+} extracellularly. The presence of such a unique synaptic process could have implications for both age-dependent differences in network functioning and the establishment of synaptic connectivity, which ultimately contribute importantly to network behavior.

SUMMARY

Results from numerous studies suggest that the functioning of rat hippocampal neural networks during the second postnatal week of life differs distinctly from that in the mature brain. During this critical period, network behavior might be considered hyperexcitable. Spontaneous network-driven bursts of synaptic potentials, which have not been reported in mature hippocampus, are commonly observed. While these events could be attributable to a late onset of GABAergic synaptic transmission, results suggest that this is not the case. In the immature hippocampus orthodromic stimulation leads to prolonged depolarizations and often repetitive synchronized discharging of the entire CA3 population. These events are in many ways reproduced by application of drugs that suppress GABAergic synaptic transmission. The synchronized discharging of the CA3 population is blocked by excitatory amino acid antagonists. This finding, coupled with our growing understanding of the role that recurrent excitation plays in CA3 network functioning, has led to the hypothesis that the differences in network behavior early in life may be largely attributable to an overabundance of local-circuit recurrent excitatory synapses. With maturation, axon collaterals and attendant synapses would regress to achieve an adult complement. Results from dual intracellular recordings as well as anatomical studies of individual CA3 pyramidal cells support this hypothesis. Unique properties of the NMDA receptor at these recurrent excitatory synapses early in life may also promote network excitability. The participation of extracellular Ca^{2+} in the voltage dependency of the NMDA receptor-linked iontophore could also contribute to synapse consolidation during maturation and thus in the establishment of network connectivity.

REFERENCES

1. Getting, P. A. 1989. Emerging principles governing the operation of neural networks. Annu. Rev. Neurosci. **12:** 185-204.
2. Swann, J. W., K. L. Smith & R. J. Brady. 1990. Neuronal networks and synaptic transmission in immature hippocampus. *In* Excitatory Amino Acids and Neuronal Plasticity: Advances in Experimental Medicine and Biology. Y. Ben Ari, Ed.: 161-171. Plenum. New York, NY.
3. Ben-Ari, Y., E. Cherubini, R. Corradetti & J.-L. Gaiarsa. 1989. Giant synaptic potentials in immature rat CA3 hippocampal neurones. J. Physiol. (London) **416:** 303-325.
4. Swann, J. W. & R. J. Brady. 1984. Penicillin-induced epileptogenesis in immature rat CA3 hippocampal pyramidal cells. Dev. Brain Res. **12:** 243-254.
5. Alger, B. E. & R. A. Nicoll. 1990. Feed forward dendritic inhibition in rat hippocampal pyramidal cells in vitro. J. Neurophysiol. **328:** 105-123.
6. Newberry, N. R. & R. A. Nicoll. 1984. A bicuculline-resistant postsynaptic potential in rat hippocampal pyramidal cells in vitro. J. Physiol. **348:** 239-254.
7. McCarren, M. & B. E. Alger. 1985. Use-dependent depression of IPSPs in rat hippocampal pyramidal cells in vitro. J. Neurophysiol. **53:** 557-571.
8. Ben-Ari, Y., K. Krnjevic & W. Reinhardt. 1979. Hippocampal seizures and failure of inhibition. Can. J. Physiol. Pharmacol. **57:** 1462-1466.
9. Swann, J. W., R. J. Brady & D. L. Martin. 1989. Postnatal development of GABA-mediated synaptic inhibition in rat hippocampus. Neuroscience **28:** 551-561.
10. Stasheff, S. F., W. W. Anderson, S. Clark & W. A. Wilson. 1989. NMDA antagonists differentiate epileptogenesis from seizure expression in an in vitro model. Science **245:** 648-651.
11. Miles, R. & R. K. S. Wong. 1987. Inhibitory control of local excitatory circuits in the guinea pig hippocampus. J. Physiol. (London) **388:** 611-629.
12. Schneiderman, J. H. 1986. Low concentrations of penicillin reveal rhythmic, synchronous synaptic potentials in hippocampal slice. Brain Res. **398:** 231-241.
13. Miles, R. & R. K. S. Wong. 1987. Latent synaptic pathways revealed after tetanic stimulation in the hippocampus. Nature **329:** 724-726.
14. Schwartzkroin, P. A. 1982. Development of rabbit hippocampus: Physiology. Dev. Brain Res. **2:** 469-486.
15. Miles, R. & R. K. S. Wong. 1986. Excitatory synaptic interactions between CA3 neurons in the guinea pig hippocampus. J. Physiol. (London) **373:** 397-418.
16. Ishizuka, N., J. Weber & D. G. Amaral. 1990. Organization of intrahippocampal projections originating for CA3 pyramidal cells in the rat. J. Comp. Neurol. **295:** 580-623.
17. Traub, R. D. & R. Miles. 1991. Multiple modes of neuronal population activity emerge after modifying specific synapses in a model of the CA3 region of the hippocampus. Ann. N.Y. Acad. Sci. This volume.
18. Alger, B. E. 1991. Gating of GABAergic inhibition in hippocampal pyramidal cells. Ann. N.Y. Acad. Sci. This volume.
19. Traub, R. D., R. Miles & R. K. S. Wong. 1989. Model of the origin of rhythmic population oscillations in the hippocampal slice. Science **243:** 1319-1325.
20. Brady, R. J. & J. W. Swann. 1988. Suppression of ictal-like activity by kynurenic acid does not correlate with its efficacy as an NMDA receptor antagonist. Epilepsy Res. **2:** 232-238.
21. Swann, J. W., K. L. Smith & R. J. Brady. 1991. Localized synaptic interactions mediate the sustained depolarization of electrographic seizures. J. Neurosci. Submitted.
22. Miles, R. & R. K. S. Wong. 1983. Single neurones can initiate synchronized population discharge in the hippocampus. Nature **306:** 371-373.
23. Gomez, C. M., F. L. Rice, K. L. Smith & J. W. Swann. 1990. Morphological studies of CA3 hippocampal neurons in the developing rat. Neurosci. Abstr. **16:** 1290.
24. Horikawa, K. & W. E. Armstrong. 1988. A versatile means of intracellular labeling: Injection of biocytin and its detection with avidin conjugates. J. Neurosci. Methods **25:** 1-11.

25. DEITCH, J. S., K. L. SMITH, J. W. SWANN & J. N. TURNER. 1991. Ultrastructural investigation of neurons identified and localized using the confocal scanning laser microscope. J. Electron Microsc. Tech.: in press.
26. COWAN, W. M., J. W. FAWCETT, D. D. M. O'LEARY & B. B. STANFIELD. 1984. Regressive events in neurogenesis. Science **225:** 1258-1265.
27. EASTER, S. S., JR., D. PURVES, P. RAKIC & N. C. SPITZER. 1985. The changing view of neural specificity. Science **230:** 507-511.
28. STANFIELD, B. B. 1984. Postnatal reorganization of cortical projections: The role of collateral elimination. Trends Neurosci. **7:** 37-41.
29. PURVES, D. 1988. Regulation of developing neural connections. *In* Body and Brain: A Trophic Theory of Neural Connections: 75-91. Harvard University Press. Cambridge, MA.
30. KOPPEL, H. & G. M. INNOCENTI. 1983. Is there a genuine exuberancy of callosal projections in development? A quantitative electron microscopic study of the cat. Neurosci. Lett. **41:** 33-40.
31. LAMANTIA, A.-S. & P. RAKIC. 1990. Axon overproduction and elimination in the corpus callosum of the developing rhesus monkey. J. Neurosci. **10:** 2156-2175.
32. HUTTENLOCHER, P. R., C. DECOURTEN, L. J. GAREY & H. VAN DER LOOS. 1982. Synaptogenesis in human visual cortex—evidence for synapse elimination during normal development. Neurosci. Lett. **33:** 247-252.
33. HUTTENLOCHER, P. R. 1984. Synapse elimination and plasticity in developing human cerebral cortex. Am. J. Mental Defic. **88:** 488-496.
34. RAKIC, P., J.-P. BOURGEOIS, M. F. ECKENHOFF, N. ZECEVIC & P. S. GOLDMAN-RAKIC. 1986. Concurrent overproduction of synapses in diverse regions of the primate cerebral cortex. Science **232:** 232-235.
35. PURVES, D. & J. W. LICHTMAN. 1980. Elimination of synapses in the developing nervous system. Science **210:** 153-157.
36. CREPEL, F. 1982. Regression of functional synapses in the immature mammalian cerebellum. Trends Neurosci. **5:** 266-269.
37. MCDONALD, J. W. & M. V. JOHNSTON. 1990. Physiological and pathophysiological roles of excitatory amino acids during central nervous system development. Brain Res. Rev. **15:** 41-70.
38. CONSTANTINE-PATON, M., H. T. CLINE & E. DEBSKI. 1990. Patterned activity, synaptic convergence, and the NMDA receptor in developing visual pathways. Annu. Rev. Neurosci. **13:** 129-154.
39. SCHMIDT, J. T. 1991. Long-term potentiation during the activity-dependent sharpening of the retinotopic map in goldfish. Ann. N.Y. Acad. Sci. This volume.
40. UDIN, S. B. & W. J. SCHERER. 1991. Experience-dependent formation of binocular maps in frogs: Possible involvement of *N*-methyl-D-aspartate receptors. Ann. N.Y. Acad. Sci. This volume.
41. KLEINSCHMIDT, A., M. F. BEAR & W. SINGER. 1987. Blockade of NMDA receptors disrupts experience-dependent plasticity of the kitten striate cortex. Science **238:** 355-358.
42. CLINE, H. T., E. DEBSKI & M. CONSTANTINE-PATON. 1987. NMDA receptor antagonist desegregates eye-specific stripes. Proc. Natl. Acad. Sci. USA **84:** 4342-4345.
43. BEN-ARI, Y., E. CHERUBINI & K. KRNJEVIC. 1988. Changes in voltage dependence of NMDA currents during development. Neurosci. Lett. **94:** 88-92.
44. NADLER, J. V., D. MARTIN, M. A. BOWE, R. A. MORRISETT & J. O. MCNAMARA. 1990. Kindling, prenatal exposure to ethanol and postnatal development selectively alter responses of hippocampal pyramidal cells to NMDA. *In* Excitatory Amino Acids in Neuronal Plasticity. Plenum. New York, NY.
45. BRADY, R. J., K. L. SMITH & J. W. SWANN. 1990. Calcium modulation of the NMDA response and electrographic seizures in immature hippocampus. Neurosci. Lett.: in press.
46. HEINEMANN, U., A. KONNERTH & H. D. LUX. 1981. Stimulation induced changes in extracellular free calcium in normal cortex and chronic alumina cream foci of cats. Brain Res. **213:** 246-250.
47. SWANN, J. W., R. J. BRADY, K. L. SMITH & M. G. PIERSON. 1988. Synaptic mechanisms of focal epileptogenesis in the immature nervous system. *In* Disorders of the Developing Nervous System. J. W. Swann & A. Messer, Eds.: 19-50. Alan R. Liss. New York, NY.

Multiple Modes of Neuronal Population Activity Emerge after Modifying Specific Synapses in a Model of the CA3 Region of the Hippocampus

ROGER D. TRAUB

IBM Research Division
IBM T. J. Watson Research Center
Yorktown Heights, New York 10598

RICHARD MILES

Pasteur Institute
Laboratory of Cellular Neurobiology
75724 Paris Cedex 15, France

Single neurons have complex morphologies and electrical properties. The surface membrane is branched and probably contains different densities of channels in different places. Many ionic voltage-dependent conductances exist with dynamics spanning a wide range of time constants, from perhaps 0.1 msec for activation of the sodium channel to more than 1000 msec for relaxation of a slow Ca^{2+}-activated g_K. As a result, individual neurons fire in several distinct modes depending upon how they are activated.[1] In the hippocampus, inhibitory cells and pyramidal cells may have different firing patterns.

Further complexities arise at the level of neuronal populations. Examples of such complexities include 1) the topology of the interconnections; 2) the number of cell types, particularly inhibitory cells (basket cells, chandelier cells, stratum lacunosum-moleculare cells, and so on); 3) the existence of synaptic interactions that take place on different time scales: that is, there exist different kinetics within both excitatory and inhibitory types of synaptic interaction. For instance, we may note the rapid (voltage-independent) actions of glutamate mediated by quisqualate receptors, as compared with the slower (and voltage-dependent) actions of glutamate mediated by *N*-methyl-D-aspartate (NMDA) receptors.[2] Again, there is the rapid action of γ-aminobutyric acid (GABA) on A-receptors,[3] as compared with the slower (in latency and in decay time course) action of GABA on B-receptors.[4]

One expects, therefore, that the electrophysiological behaviors generated by even a small population of neurons would be incomprehensible. Nevertheless, it does appear possible to classify several distinctive types or "modes" of population activity that occur with variations in strength of particular synapses, specifically the unitary conductances of $GABA_A$- and $GABA_B$-activated synapses. Our model describes the *global* properties of a network of CA3 neurons, and does not predict precisely which neurons will fire at which times, at least not in a manner insensitive to the presence of noise. Indeed, our results suggest that such precise predictions may be impossible.[5]

Our methods involve a synthesis of physiological experimentation and computer modeling. This note concentrates on the latter: the means by which a model of *in vitro* CA3 circuitry can be created, and the types of behavior it generates. We draw correspondences with experimental observations wherever possible.

REVIEW OF PHYSIOLOGY OF CA3 CELLS AND CIRCUITRY

Construction of a faithful model of the CA3 region requires attention to a wealth of experimental detail on membrane channel kinetics and synaptic physiology. Yet we must also extract underlying functional principles of single cell mechanisms, synaptic organization, and emergent population behaviors. In this note, we attempt to give some feel for both details and principles.

Pyramidal Cell Models

We have used two different models for pyramidal cells. Both employ a single or branching cable to simulate passive electrotonic properties,[6] adding nonlinear voltage- and/or calcium-dependent conductances to simulate active membrane currents. In the original model, four types of membrane conductances were incorporated.[7] The conductances were concentrated in only one or two membrane compartments, and the kinetics were reconstructed as best we could from current-clamp records. The original model, when incorporated into network simulations, was quite adequate for the study of fully and partially synchronized bursts and rhythmical population waves,[8–10] but it was not able—even in networks—to generate synchronized multiple bursts[11] without strong *ad hoc* assumptions.[12]

Recently, we have developed a more realistic model that uses six types of ionic conductance, each distributed across a large part of the cell membrane.[13] The specific conductances are 1) a fast g_{Na}; 2) a high-voltage-activated g_{Ca}; 3) the delayed rectifier type of g_K; 4) the A-type of transient g_K; 5) the C-type of voltage- and calcium-dependent g_K; 6) the prolonged afterhyperpolarization (AHP) calcium-activated g_K. The kinetics of these conductances have been at least partially characterized from whole-cell patch experiments performed on isolated hippocampal cells[14–17] or in other neuronal preparations.[18–21] In using these data to model CA3 cells, we must contend with certain difficulties: i) Most of the experiments were done on CA1 cells. We assume the channel kinetics to be similar in CA3 cells. ii) Isolated cells lose most of their dendrites,[22] yet intradendritic recordings indicate the presence of significant g_{Na} and g_{Ca} in dendrites.[23,24] We assume that somatic and dendritic channel kinetics are similar, although channel densities need not be similar. While it is possible to estimate the total maximum conductance of, say, g_{Na} on an isolated cell, the total conductance on different regions of dendritic membrane is not known.

In spite of these difficulties, it have proven possible to find a set of conductance distributions (for example, $\overline{g_{Na}}, \overline{g_{Ca}}$) as functions of membrane location that are consistent with a variety of voltage-clamp and current-clamp experiments.[13] Our single-cell model accounts for two major forms of pyramidal cell firing: intrinsic bursting (at resting potentials not too depolarized, or after a brief excitatory stimulus) and repetitive firing (at relatively depolarized holding potentials)[1]; it also accounts for transitional forms with grouped action potentials. The single-cell model further predicts the occurrence of a third distinct form of firing: repetitive brief bursts at about 10 Hz in response to a steady dendritic depolarizing stimulus.[13] This revised model of the single pyramidal cell, when used in network simulations, produces similar results to the original model on synchronized and partially synchronized bursts; but it reproduces other phenomena as well (synchronized multiple bursts and tonic seizures—see the following). The figures used in this note were generated using the revised model.

For simulating *inhibitory* cells (i-cells), we have used three types of model. Again, all types start with a cable representation of soma-dendritic membrane, with voltage-dependent conductances then superimposed. These latter conductances are confined to the soma. The i-cells may be modeled i) as pyramidal cells, since some i-cells fire in patterns similar to pyramidal cells[9,25]; ii) as pyramidal cells, but without g_{Ca} and $g_{K(Ca)}$—such model cells are repetitively firing rather than bursting (see FIG. 2 of Traub *et al.*[9]); iii) as for ii, but with rate functions modified to allow for the narrower spikes, rapid spike AHPs, and high firing rates of many i-cells. In the present note, we use model iii, with $R_{input} = 56$ MΩ, $R_m = 10{,}000$ Ω − cm², $\tau_m = 10$ msec. The lack of kinetic data on i-cell membrane and the variability in i-cell physiology make this aspect of circuity modeling somewhat problematic.

Two types of inhibition ("fast" and "slow") occur in the hippocampus, mediated by different types of GABA receptor (A and B, respectively). We make the assumption that fast and slow IPSPs are generated by distinct subpopulations of inhibitory neurons.[26,5]

We have devoted considerable attention to modeling chemical synaptic interactions; interactions mediated via field effects or gap junctions will not be considered here. Synaptic interactions have both *functional* and *topological* (or connectivity) aspects. From a *functional* point of view, we have tried to establish experimentally the effect of a single pyramidal cell or inhibitory cell action potential on any connected cell (be the postsynaptic cell either pyramidal or inhibitory).[3,27,28] It must then be established what a train of action potentials from one or more cells (of a given type) does to a given postsynaptic neuron. We consider four types of chemical synaptic interaction: fast and slow excitation (corresponding to glutamate actions at quisqualate (QUIS) and NMDA receptors, respectively), and fast and slow inhibition (corresponding to GABA actions at A and B receptors, respectively). Excitation and inhibition are organized differently in our model: we assume that any excitatory synapse onto a pyramidal cell is capable of eliciting both types of excitation, while excitatory synapses onto inhibitory cells in our model produce QUIS effects only. In contrast, an inhibitory synapse is either fast or slow, depending upon the presynaptic cell; fast and slow inhibitory synapses are also located on distinct portions of the postsynaptic cell membrane. All synapses work through a conductance g(t,V), so that the synaptic current is $g(t,V)(V - V_{reversal})$. (See TABLES 1 & 2). Postsynaptic conductances induced by multiple presynaptic spikes add linearly.

Topological aspects of network structure must also be considered. For example: How many pyramidal cells (or inhibitory cells) does a given pyramidal cell (or inhibitory cell) contact? Does the probability of synaptic contact depend on the distance between the respective cells, and, if so, what is the dependence? Can a reasonable network structure be generated by a series of independent random choices—subject to connection probability constraints? Or, to the contrary, does a nonrandom structure—that is, statistical correlations—exist in the connections?

We have used several approaches to estimate synaptic connectivity. First, there are anatomical data.[29,30] Second, we have attempted to estimate connectivity by performing large numbers of dual intracellular recordings, searching for cases where cell 1 influences cell 2 monosynaptically; cell 1 can be characterized as excitatory or inhibitory based upon its action on cell 2, while cell 2 itself must be characterized by its firing pattern and other intrinsic properties.[3,27,31] This experiment is usually done with the

TABLE 1. Properties of Simulated Synaptic Interactions

Synapse	Reversal	V-dependent?	Decay τ	Location	Time to Peak
QUIS (e→e)	+60 mV	No	3 msec	dendrites	3 msec
QUIS (e→i)	+60 mV	No	2 msec	soma	2 msec
NMDA (e→e)	+60 mV	Yes	85 msec	dendrites	3 msec
$GABA_A$	−15 mV	No	7 msec	perisomatic	0.5 msec
$GABA_B$	−15 mV	No	100 msec	dendrites	150 msec

NOTES: 1) e→e refers to a synapse from one excitatory cell to another; e→i refers to a synapse from an excitatory cell onto an inhibitory cell.
2) Reversal potentials are relative to resting potential.
3) The time constant for the NMDA conductance decay is taken from Forsythe and Westbrook.[2]
4) For the NMDA-activated conductance, $G(t,V) = g(t)h(V)$, where (using C. E. Jahr and C. F. Stevens, personal communication, and assuming $[Mg^{2+}]_o = 2$ mM in slice experiments) $h(V) = 1/[1 + (2/3)(e^{-.07(V-60)})]$, V relative to an assumed resting potential of −60 mV.
5) The time to peak conductance of the $GABA_B$ is close to that reported by Hablitz and Thalmann.[4]

TABLE 2. Unitary Synaptic Conductances in Network Model

Synapse Type	Unitary Conductance Maximum	Saturating Conductance
QUIS (e→e)	$4 \times 3/e$ nS (apical)	400 nS
	$4 \times 3/e$ nS (basilar)	400 nS
QUIS (e→i)	10 nS	50 nS
NMDA (e→e)	$0.4 \times 3/e$ nS (apical)	40 nS
	$0.4 \times 3/e$ nS (basilar)	40 nS
$GABA_A$ (i→e)	5-10 nS	—
$GABA_A$ (i→i)	1-2 nS	—
$GABA_B$ (i→e)	1 nS	30 nS
$GABA_B$ (i→i)	0.2 nS	30 nS

NOTES: 1) 3/e factor arises for QUIS synapses because the conductance is $c_e t e^{-t/3}$. The maximum conductance occurs when t = 3 msec and is $c_e \times 3/e$. We use $c_e = 4$ nS.
2) Excitatory and slow inhibitory conductances are *dendritic.* The conductance measured at the soma for a unitary excitatory synaptic input will be smaller. Note that Miles and Wong[27] estimate about 1 nS for a unitary EPSP.
3) The saturating conductance is the maximum conductance any cell can receive at a given time. It is intended to represent the finite number of synaptic receptors existing on the cell. Hablitz and Thalmann[4] estimated 15.3 nS as the maximum *somatic* conductance for slow inhibition evoked by mossy fiber stimulation.
4) The 10:1 ratio for QUIS/NMDA conductances derives from Forsythe and Westbrook.[2]
5) The $GABA_A$ conductance is in the range observed by Miles and Wong.[3] More recent data (R. Miles, unpublished data) suggest an 8- to 10-fold spread of unitary $GABA_A$ conductances in CA3; however, in any one simulation we use a constant value.
6) A unitary $GABA_B$ input produces a conductance $.01 \times g(t)$ $[0 < t < 150$ msec], where $g(t) = 100 - e^{-t/100}$. The maximum conductance occurs when t = 150 msec and is $.01 \times (100 - e^{-150/100}) = 1 - .01 \times e^{-150/100} \sim 1$ nS.

cells within about 200 μ of each other; one assumes that connection probability is constant on such a spatial scale. Data of this sort indicate that the probability of an e→e connection locally is about 0.02, of an e→i connection about 0.1, and of an i→e connection perhaps as high as 0.6 (R. Miles, unpublished data; see also Traub & Miles[5]). Physiological experiments suggest that excitatory connection probabilities fall off with distance,[32] but in systems with 1000 cells, such as illustrated in this paper, the falloff is not likely to be significant. A third means of estimating connectivity is to compare, say, a unitary inhibitory postsynaptic potential (IPSP) with a maximal IPSP evoked by an afferent shock. This suggests that a pyramidal cell has on the order of 20 (presumably $GABA_A$) inhibitory inputs (R. Miles, unpublished data). It is difficult to apply this method to excitatory postsynaptic potentials (EPSPs), comparing a unitary EPSP with the conductance developing during a synchronized burst (when all excitatory inputs are active). The reason is that the inputs are discharging repetitively during a synchronized burst (see FIG. 4). The connection probabilities used in the 1000-pyramidal-cell-model described here are similar to (but not identical to) the experimental estimates (TABLE 3).

TABLE 3. Average Connection Parameters Used in Model with 1000 Pyramidal Cells and 100 Inhibitory Cells

Connection Type	Mean Connection Probability
e→e	0.02
e→i	0.04
i→e	0.4
i→i	0.1

NOTES: 1) The 100 inhibitory cells consist of 50 cells that activate $GABA_A$ synapses and 50 cells that activate $GABA_B$ synapses.
2) Data are not available on the probability of i→i interactions.
3) The e→i and i→e connectivities used here may be too small.

On the Possibly Nonrandom Nature of Excitatory Connectivity

We assume, for purposes of network simulations, that synaptic connectivity is either random or locally random. Data suggest, however, that recurrent excitatory connectivity within CA3 is not random. These data derive from pair recordings of polysynaptic (probably usually disynaptic) excitatory connections. Given a pair of nearby e-cells (call them cell 1 and cell 2), the probability that bursting in cell 1 can be shown to elicit an EPSP in cell 2 is about 0.1.[5,31] Yet, we have not seen an interaction wherein both cell 1 polysynaptically excites cell 2 *and* cell 2 polysynaptically excites cell 1 (R. Miles, unpublished data). Such observations suggest that the CA3 excitatory network may not contain short cycles, that is, cycles of 3 or 4 cells. Nevertheless, *longer* cycles must exist. For when a population burst is elicited in picrotoxin by stimulating a single cell, the stimulated cell itself bursts in phase with the population, implying the existence of paths from cells in the population *back* to the initiating cell.[31,33] It is intriguing to speculate that immature hippocampus is randomly connected, but that the peculiar dynamics of the system, together with appropriate synaptic modification "rules," act to eliminate the short cycles. Why, and if, the resulting topology might be useful is not clear.

Functional Aspects of Simple Synaptic Circuits

Our knowledge of the properties of single synapses allows some predictions on the behavior of simple synaptic circuits. We shall give a few examples: i) Excitatory synapses are usually not powerful enough for a presynaptic spike to evoke a postsynaptic spike; however, presynaptic bursts cause both facilitation and temporal summation of EPSPs at recurrent synapses between pyramidal cells. Thus, a presynaptic burst may evoke a postsynaptic burst with mean probability 0.4 to 0.5 and latency of 10 to 30 msec. ii) Excitatory connections onto i-cells do allow spike-spike transmission with high probability (often greater than 0.5) and with very brief latency (2.5 to 5 msec). iii) Disynaptic inhibition ($e \rightarrow i \rightarrow e$) is observed[3] with (functional) connection probability 0.3. This probability appears to be less than what would be expected from the $e \rightarrow i$ and $i \rightarrow e$ probabilities (0.1 and 0.6, respectively). For example, consider a population of 1000 pyramidal cells with 50 i-cells producing $GABA_A$ inhibition, and 2 pyramidal cells (called cell 1 and cell 2). For a fixed inhibitory cell, the probability of cell 1 $\rightarrow$ fixed i-cell $\rightarrow$ cell 2 is .06. The probability that at least one i-cell exists so that cell 1 $\rightarrow$ i-cell $\rightarrow$ cell 2 is $1 - .94^{50} = .955$, a number much larger than 0.3. This discrepancy could arise from occasional refractoriness of intercalated i-cells or synaptic failures, from nonrandomness in the connections, or from the possibility that measured connection probabilities apply only to a subclass of i-cells rather than to i-cells as a whole. Thus, matching functional connectivity to anatomical connectivity is not always straightforward. iv) Disynaptic excitatory connections are rarely observed in the presence of functional $GABA_A$-mediated inhibition, even though morphological arguments suggest that such paths must exist. Disynaptic inhibitory pathways from cell 1 to cell 2 can prevent burst propagation even if a monosynaptic excitatory connection exists from cell 1 to cell 2. It is notable that the *latency* of disynaptic inhibition (3.5 to 6 msec) is less than the 10 to 30 msec required for burst propagation across a single excitatory synapse. Similarly, functional polysynaptic $e \rightarrow e \rightarrow e$ connectivity occurs with probability 0.1 (in picrotoxin). Again, this number is smaller than would be expected from the monosynaptic probabilities: in a 1000-cell population with a connection probability of .02, the disynaptic connection probability should be $1 - (1 - .02 \times .02)^{1000} = 0.33$. Once more, we must consider the possibilities of refractoriness of the intercalated pyramidal cell or of nonrandomness in the excitatory connections.

What Aspects of Dynamical Behavior Might be Relevant for Synaptic Plasticity?

Long-term potentiation (LTP) requires activation of NMDA receptors, and NMDA-activated currents have two salient features: a long time after agonist-receptor binding (perhaps hundreds of milliseconds) during which synaptic current might be able to flow,[2] and a requirement for dendritic membrane depolarization for the synaptic current actually to flow.[34,35] This suggests that for an excitatory connection from cell 1 to cell 2 to become potentiated, a portion of cell 2 dendrite must become sufficiently depolarized (perhaps by a dendritic calcium spike?) within a characteristic time, say 100 or 200 msec, of cell 2 firing or bursting. This notion has received some experimental support.[36,37] While LTP has not, to our knowledge, been clearly demonstrated in CA3 recurrent excitatory synapses, we shall direct our attention to conditions of population behavior where potentiation by the above mechanism might occur.

DYNAMICAL MODES OF POPULATION BEHAVIOR IN THE CA3 MODEL

Having outlined the structure and functional aspects of the cells and synapses, we can now present some of the global behaviors exhibited by our CA3 model. (Some of these behaviors are discussed in much more detail in Traub and Miles.[5] The model there has more cells, but each cell has a simpler structure than in the present paper.) The conditions of our simulations are admittedly simplified with respect to the actual slice: Each pyramidal cell is biased with a small current (0 to 0.1 nA) chosen randomly at the beginning of the simulation but then held fixed: were the cells to be isolated from each other, a cell biased with 0.0 nA would burst at about 0.3 Hz, whereas a cell biased at 0.1 nA would burst at about 0.5 Hz. In the full network, therefore, spontaneous bursting occurs, but the total population activity is strongly modified by the synaptic interactions. These interactions lead to synaptically induced firing in pyramidal cells and to widespread IPSPs under conditions when inhibition is not blocked.

We describe population activities that result from postsynaptic alterations in the strength of inhibitory synapses operated by either $GABA_A$ receptors, $GABA_B$ receptors, or both.

Mode 1: $GABA_A$ Synapses Powerful, $GABA_B$ Synapses Blocked

In this case (FIG. 1) there is continuous, but low-level, firing throughout the population, without apparent rhythmicity. Synaptic potentials in pairs of cells are uncorrelated. To our knowledge, this form of collective behavior has not yet been demonstrated in the slice. The pattern is reminiscent of a desynchronized EEG.

Mode 2: Both Types of Inhibition Present

In this case (FIG. 2) firing occurs rhythmically in a series of waves, but each wave represents a small fraction of the population. The period in the case illustrated is about 500 msec. Randomly chosen pairs of cells exhibit correlated synaptic potentials during each wave (lower traces). This type of activity has been demonstrated in slices with apical dendritic field potential recordings[38] or with dual intracellular recordings.[10,31,39] It is more likely to occur after partial blockade of inhibition.[31] The pattern of activity is reminiscent of rhythmical EEG waves.

The structure of an individual wave of firings is revealing (FIG. 3). Here, we have made a raster plot of the firings of those 31 cells active during a particular wave, and have drawn in (FIG. 3, right) the synaptic relations between connected pairs of active cells, where onset of firing in one cell precedes the termination of firing in the other. Such synapses are "causal" in that they contribute to the excitatory spread of firing and they are precisely those synapses that one might expect to become potentiated. The wave is initiated by spontaneous firing in one cell (★). Because of slow IPSPs produced by the previous wave, spontaneous firing is suppressed for some interval after the previous wave, and thus several cells may begin to fire at about the same time.

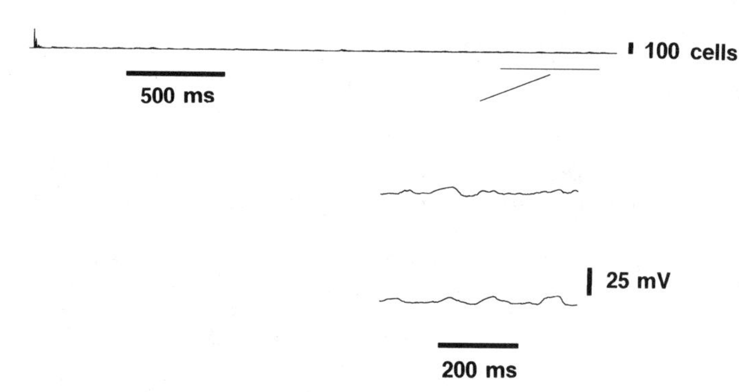

FIGURE 1. Mode 1: Population behavior when $GABA_A$ is functional (unitary IPSP conductance = 10 nS) *and* $GABA_B$ synapses are blocked. *Upper trace*: number of cells "firing." After an initial transient, there is a low amplitude of background firing without rhythmical waves. *Bottom two traces*: simultaneous somatic potentials from a pair of pyramidal cells. Synaptic potentials in the two cells are uncorrelated. This type of population behavior is a prediction of the model, not yet (to our knowledge) experimentally verified.

Firing spreads along recurrent excitatory collaterals: Recall that these are powerful enough to allow burst propagation, and that disynaptic inhibition is not widespread enough to suppress all spread of bursting. Nevertheless, firing spreads along only a few of the available collaterals. For while each cell excites an average of 20 others, any one cell evokes firing in at most half of these. This is so because of refractoriness caused by previous waves, as well as by recurrent inhibition. As the wave proceeds, more inhibitory cells are recruited, making spread even more difficult. Thus, the initiating cell excites more followers than any later cell; these followers can be recruited by firing in the initiating cell alone, whereas cells that begin firing toward the end of the wave

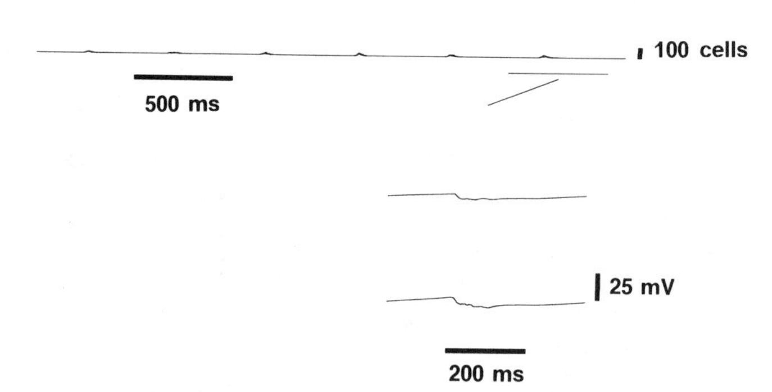

FIGURE 2. Mode 2: Population behavior when $GABA_A$ synapses are functional (unitary IPSP conductance = 5 nS) *and* $GABA_B$ are also functional (with maximal slow IPSP conductance = 30 nS in any neuron). Low-amplitude rhythmical waves of firing occur throughout the population (upper trace); during any such wave, only a small fraction of pyramidal cells fire. These cells recruit inhibitory cells having divergent axons that generate synchronized synaptic potentials in pairs of neurons (lower two traces). The synaptic potentials in the present case are predominantly inhibitory. This type of activity has been observed in hippocampal slices, particularly after partial blockade of inhibition or after tetanic stimulation.

are only recruited by firing in several of their synaptic precursors. Finally, note that activity tends not to spread cyclically (that is, cell 1 → cell 2 → cell 3 → cell 4 → cell 1), because if cell 1 fires early, its AHP renders it refractory and unable to fire again. This situation is in contrast to a fully synchronized burst, wherein firing in *all* of a cell's precursors can overcome intrinsic refractoriness.[8] Synaptic potentiation would be expected to occur during the Mode 2 type of activity, and potentiation along "straight" pathways would be more favored than potentiation along "cyclic" pathways.

We find both experimentally and in simulations that as $GABA_A$ synapses are progressively blocked, then more cells fire on average during each wave, the variability in wave amplitude increases, and the mean interval between waves increases.

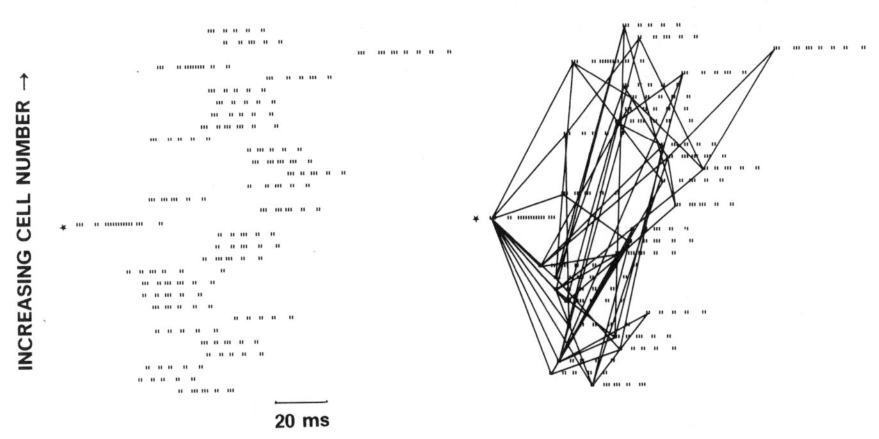

FIGURE 3. Structure of population activity during a single wave, Mode 2 behavior (as in FIG. 2). During a particular single wave, lasting about 150 msec, 31 pyramidal cells fire, out of a total of 1000 pyramidal cells in the model population. On the left of the Figure, these cells are arranged in numerical order along the vertical axis. A row of dots signifies depolarization of a cell (above 20 mV) at the corresponding times. The wave is initiated by a burst in a single cell, marked by the star (★). The right part of the figure is the same as the left, but we have drawn in lines from the dots of cell A to the dots of cell B, whenever A is connected monosynaptically to B *and* cell A begins firing before B finishes. The lines indicate the flow of activity during the wave, and also indicate synaptic connections that might become potentiated. Note that *no loops* occur in this structure.

Mode 3: $GABA_A$ Synapses Blocked Nearly Completely, $GABA_B$ Synapses Functional

In this case (FIG. 4) single synchronized bursts occur with intervals greater than 1 sec. Such events are reminiscent of interictal spikes *in vivo.* Initiation of each event occurs from a small number of cells and spreads along excitatory collaterals, relatively unimpeded by $GABA_A$-mediated synaptic inhibition. Delayed inhibition develops slowly enough that excitatory spread can "outrun" the slow IPSP.[9] Because bursts in different cells overlap each other in time (FIG. 4), there is opportunity, in principle, for *all* recurrent excitatory synapses to become potentiated.

Mode 4: $GABA_A$ Synapses Blocked Completely, $GABA_B$ Synapses Functional

In this case (FIG. 5), population-synchronized *multiple* bursts occur at intervals of more than 1.5 sec. Within a synchronized event, the interburst interval is about 75 msec. This type of activity occurs *in vitro* in picrotoxin[11,40] and is reminiscent of the EEG pattern called polyspike-and-wave recorded in some patients with primary generalized epilepsy. The mechanism of this type of event in the model is as follows: The initial burst proceeds as in Mode 3 by excitatory spread from one or a few cells. Once all the cells are firing, each pyramidal cell will receive a sustained excitatory input, particularly with fast inhibition completely blocked. Under such conditions, each pyramidal cell will develop a series of brief dendritic bursts, as described in the section on single cell properties. Since the dendritic bursts in turn evoke somatic bursts, and since the pyramidal cells are synaptically coupled, cellular bursts will tend to be

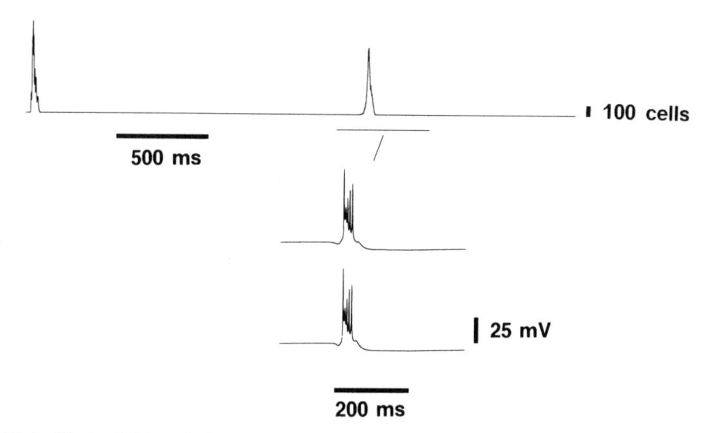

FIGURE 4. Mode 3: Population activity after near-complete blockade of $GABA_A$ synapses (unitary conductance now 1.0 nS) and $GABA_B$ synapses functional. Synchronized population bursts occur at intervals of 1 to 2 sec. Pairs of neurons (traces below) generate single bursts that are nearly simultaneous. This type of behavior is recorded in slices from mature guinea pigs that are bathed in penicillin. It corresponds to interictal spikes *in vivo,* although synchrony *in vivo* may not be this extreme.

synchronized between cells. Synchrony will be favored if the intrinsic time constants determining interburst interval (for example, for calcium channel kinetics and intracellular Ca^{2+} ion removal) are similar for all of the pyramidal cells. An event is terminated by intrinsic outward currents and slow IPSPs. During one of these synchronized events, there is repeated opportunity, in principle, for potentiation of *all* of the recurrent synapses.

Mode 5: $GABA_A$ and $GABA_B$ Both Blocked

An initial population-synchronized multiple burst merges into high-frequency sustained repetitive firing in all of the cells (FIG. 6). This activity is reminiscent of a tonic

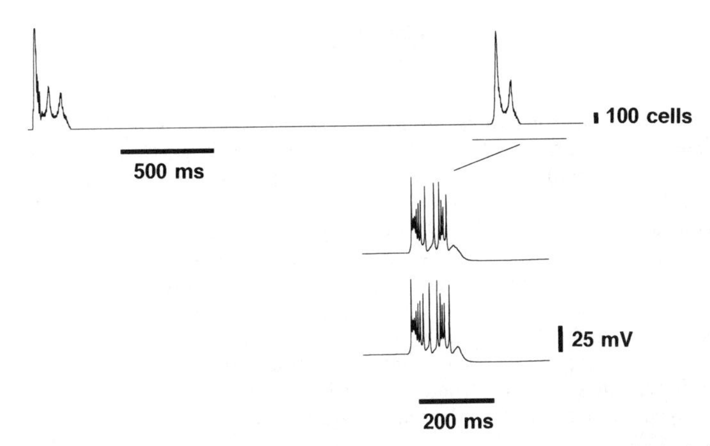

FIGURE 5. Mode 4: Population activity with $GABA_A$ synapses completely blocked, but with $GABA_B$ synapses functional. Synchronized multiple bursts occur at intervals of several seconds. This type of activity occurs in mature slices bathed in picrotoxin. It may correspond to so-called polyspike-and-wave in human EEG (a pattern sometimes recorded in patients with primary generalized epilepsy who have grand mal seizures).

seizure. Similar activity can occur in CA1 in slices bathed in high $[K^+]_o$. While high $[K^+]_o$ can diminish the effectiveness of fast inhibition by a depolarizing shift of the Cl^- reversal potential, it remains unclear if loss of slow inhibition plays a role in the transition from single synchronized bursts to ictal behavior[41–43] (R. Dingledine, personal communication). Such collective activity would provide sustained activation of all synapses, unless extracellular $[Ca^{2+}]$ fell to levels too low to sustain transmitter release.[44]

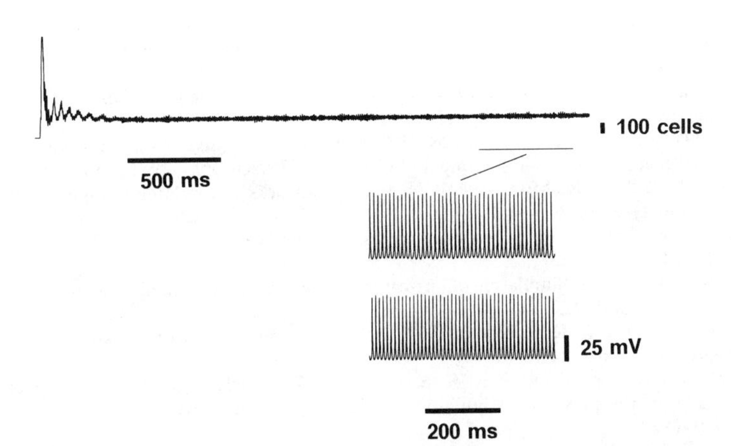

FIGURE 6. Mode 5: Population activity with *both* $GABA_A$ and $GABA_B$ synapses blocked. High-frequency repetitive firing occurs in all of the cells. We speculate that this activity may correspond to ictal events sometimes generated in CA1 in high $[K^+]_o$ and to tonic seizures *in vivo*.

CONCLUSIONS

Modeling a population of neurons in the style we have chosen requires consideration of channel kinetics, the membrane distribution of ionic conductances, and synaptic function and topology. It is remarkable, in view of the many uncertainties, that certain population behaviors emerge that appear to agree well with experiments: Modes 2, 3, and 4—low-amplitude population waves and synchronized bursts and multiple bursts. We would like to suggest the existence of other types of behavior when slow ($GABA_B$) inhibition is blocked. Analysis of the different population behaviors is a prerequisite for understanding the computational possibilities of real neuronal networks. It may be useful for better understanding of EEG waves and epileptic transients, particularly interictal spikes, and possibly the interictal-ictal transition. Finally, we hope that such an analysis will suggest how structure, that is, nonrandomness, might emerge from circuitry dynamics combined with synaptic plasticity.

ACKNOWLEDGMENTS

Roger D. Traub would like to acknowledge helpful discussions with G. Grinstein, E. Pytte, and G. Tesauro.

REFERENCES

1. Wong, R. K. S. & D. A. Prince. 1981. Afterpotential generation in hippocampal pyramidal cells. J. Neurophysiol. **45:** 86-97.
2. Forsythe, I. D. & G. L. Westbrook. 1988. Slow excitatory postsynaptic currents mediated by *N*-methyl-D-aspartate receptors on cultured mouse central neurones. J. Physiol. **396:** 515-533.
3. Miles, R. & R. K. S. Wong. 1984. Unitary inhibitory synaptic potentials in the guinea-pig hippocampus *in vitro.* J. Physiol. **356:** 97-113.
4. Hablitz, J. J. & R. H. Thalmann. 1987. Conductance changes underlying a late synaptic hyperpolarization in hippocampal CA3 neurons. J. Neurophysiol. **58:** 160-179.
5. Traub, R. D. & R. Miles. 1991. Neuronal Networks of the Hippocampus. Cambridge University Press. New York, NY.
6. Rall, W. 1962. Theory of physiological properties of dendrites. Ann. N.Y. Acad. Sci. **96:** 1071-1092.
7. Traub, R. D. 1982. Simulation of intrinsic bursting in CA3 hippocampal neurons. Neuroscience **7:** 1233-1242.
8. Traub, R. D. & R. K. S. Wong. 1982. Cellular mechanism of neuronal synchronization in epilepsy. Science **216:** 745-747.
9. Traub, R. D., R. Miles & R. K. S. Wong. 1987. Models of synchronized hippocampal bursts in the presence of inhibition. I. Single population events. J. Neurophysiol. **58:** 739-751.
10. Traub, R. D., R. Miles & R. K. S. Wong. 1989. Model of the origin of rhythmic population oscillations in the hippocampal slice. Science **243:** 1319-1325.
11. Miles, R., R. K. S. Wong & R. D. Traub. 1984. Synchronized afterdischarges in the hippocampus: Contribution of local synaptic interactions. Neuroscience **12:** 1179-1189.
12. Traub, R. D., W. D. Knowles, R. Miles & R. K. S. Wong. 1984. Synchronized afterdis-

charges in the hippocampus: Simulation studies of the cellular mechanism. Neuroscience **12:** 1191-1200.

13. TRAUB, R. D., R. K. S. WONG & R. MILES. 1990. A model of the CA3 hippocampal pyramidal cell based on voltage-clamp data. Soc. Neurosci. Abstr. **16:** 1297.
14. SAH, P., A. J. GIBB & P. W. GAGE. 1988. The sodium current underlying action potentials in guinea pig hippocampal CA1 neurons. J. Gen. Physiol. **91:** 373-398.
15. SAH, P., A. J. GIBB & P. W. GAGE. 1988. Potassium current activated by depolarization of dissociated neurons from adult guinea pig hippocampus. J. Gen. Physiol. **92:** 263-278.
16. KAY, A. R. & R. K. S. WONG. 1987. Calcium current activation kinetics in isolated pyramidal neurones of the CA1 region of the mature guinea-pig hippocampus. J. Physiol. **392:** 603-616.
17. NUMANN, R. E., W. J. WADMAN & R. K. S. WONG. 1987. Outward currents of single hippocampal cells obtained from the adult guinea-pig. J. Physiol. **393:** 331-353.
18. ADAMS, P. R., A. CONSTANTI, D. A. BROWN & R. B. CLARK. 1982. Intracellular Ca^{2+} activates a fast voltage-sensitive K^+ current in vertebrate sympathetic neurones. Nature **296:** 746-749.
19. LANCASTER, B. & P. R. ADAMS. 1986. Calcium-dependent current generating the afterhyperpolarization of hippocampal neurons. J. Neurophysiol. **55:** 1268-1282.
20. LANCASTER, B. & R. A. NICOLL. 1987. Properties of two calcium-activated hyperpolarizations in rat hippocampal neurones. J. Physiol. **389:** 187-203.
21. ROSS, W. N. & R. WERMAN. 1987. Mapping calcium transients in the dendrites of Purkinje cells from the guinea-pig cerebellum *in vitro.* J. Physiol. **389:** 319-336.
22. KAY, A. R. & R. K. S. WONG. 1986. Isolation of neurons suitable for patch-clamping from adult mammalian central nervous systems. J. Neurosci. Methods **16:** 227-238.
23. WONG, R. K. S., D. A. PRINCE & A. I. BASBAUM. 1979. Intradendritic recordings from hippocampal neurons. Proc. Natl. Acad. Sci. USA **76:** 986-990.
24. BENARDO, L. S., L. M. MASUKAWA & D. A. PRINCE. 1982. Electrophysiology of isolated hippocampal pyramidal dendrites. J. Neurosci. **2:** 1614-1622.
25. KAWAGUCHI, Y. & K. HAMA. 1988. Physiological heterogeneity of nonpyramidal cells in rat hippocampal CA1 region. Exp. Brain Res. **72:** 494-502.
26. LACAILLE, J.-C. & P. A. SCHWARTZKROIN. 1988. Stratum lacunosum-moleculare interneurons of hippocampal CA1 region. II. Intrasomatic and intradendritic recordings of local circuit synaptic interactions. J. Neurosci. **8:** 1411-1424.
27. MILES, R. & R. K. S. WONG. 1986. Excitatory synaptic interactions between CA3 neurones in the guinea-pig hippocampus. J. Physiol. **373:** 397-418.
28. MILES, R. 1990. Synaptic excitation of inhibitory cells by single CA3 pyramidal cells in the guinea pig hippocampus. J. Physiol. **428:** 61-77.
29. ISHIZUKA, N., J. WEBER & D. G. AMARAL. 1990. Organization of intrahippocampal projections originating from CA3 pyramidal cells in the rat. J. Comp. Neurol. **295:** 580-623.
30. SOMOGYI, P., M. G. NUNZI, A. GORIO & A. D. SMITH. 1983. A new type of specific interneuron in the monkey hippocampus forming synapses exclusively with the axon initial segments of pyramidal cells. Brain Res. **259:** 137-142.
31. MILES, R. & R. K. S. WONG. 1987. Inhibitory control of local excitatory circuits in the guinea-pig hippocampus. J. Physiol. **388:** 611-629.
32. MILES, R., R. D. TRAUB & R. K. S. WONG. 1988. Spread of synchronous firing in longitudinal slices from the CA3 region of the hippocampus. J. Neurophysiol. **60:** 1481-1496.
33. MILES, R. & R. K. S. WONG. 1983. Single neurones can initiate synchronized population discharge in the hippocampus. Nature **306:** 371-373.
34. JAHR, C. E. & C. F. STEVENS. 1990. A quantitative description of NMDA receptor-channel kinetic behavior. J. Neurosci. **10:** 1830-1837.
35. BEKKERS, J. M. & C. F. STEVENS. 1989. Dual modes of excitatory synaptic transmission in the brain. *In* Molecular Neurobiology. Proceedings of the First NIMH Conference. S. Zalcman & R. Schiller, Eds.: 39-50.
36. LARSON, J. & G. LYNCH. 1986. Induction of synaptic potentiation in hippocampus by patterned stimulation involves two events. Science **232:** 985-988.
37. GUSTAFSSON, B., H. WIGSTRÖM, W. C. ABRAHAM & Y.-Y. HUANG. 1987. Long-term potentiation in the hippocampus using depolarizing current pulses as the conditioning stimulus to single volley synaptic potentials. J. Neurosci. **7:** 774-780.

38. SCHNEIDERMAN, J. H. 1986. Low concentrations of penicillin reveal rhythmic, synchronous synaptic potentials in hippocampal slice. Brain Res. **398:** 231-241.
39. SCHWARTZKROIN, P. A. & M. M. HAGLUND. 1986. Spontaneous rhythmic synchronous activity in epileptic human and normal monkey temporal lobe. Epilepsia **27:** 523-533.
40. HABLITZ, J. J. 1984. Picrotoxin-induced epileptiform activity in hippocampus: Role of endogenous versus synaptic factors. J. Neurophysiol. **51:** 1011-1027.
41. KORN, S. J., J. L. GIACCHINO, N. L. CHAMBERLIN & R. DINGLEDINE. 1987. Epileptiform burst activity induced by potassium in the hippocampus and its regulation by GABA-mediated inhibition. J. Neurophysiol. **57:** 325-340.
42. TRAYNELIS, S. F. & R. DINGLEDINE. 1988. Potassium-induced spontaneous electrographic seizures in the rat hippocampal slice. J. Neurophysiol. **59:** 259-276.
43. JENSEN, M. S. & Y. YAARI. 1988. The relationship between interictal and ictal paroxysms in an *in vitro* model of focal hippocampal epilepsy. Ann. Neurol. **24:** 591-598.
44. PUMAIN, R., C. MENINI, U. HEINEMANN, J. LOUVEL & C. SILVA-BARRAT. 1985. Chemical synaptic transmission is not necessary for epileptic seizures to persist in the baboon *Papio papio.* Exp. Neurol. **89:** 250-258.

Introduction

This sixth section focuses on those changes that are initiated by activity in sensory pathways and that affect activity in motor pathways, thereby altering behavior. The studies presented begin with intact animals and seek to learn how changes in behavior are explained by activity-driven neuronal and synaptic plasticity. The central purpose is to move from specific behavioral changes to the CNS modifications responsible for them, and thence to the activity-driven processes that produce the modifications. The prerequisite for delineating the CNS modifications underlying behavioral change is knowledge of their locations. Thus, these studies focus on simple behaviors subserved by pathways that are reasonably well-defined and accessible to study. At present, in each model system, the search for the site of change focuses on a small number of strong possibilities.

Both Yeo and Bloedel *et al.* discuss current evidence concerning the role of the cerebellar cortex and deep nuclei in eyeblink conditioning. Whether cerebellar lesions destroy the sites of plasticity and/or simply impair performance remains unresolved. New data concerning the differing effects of cerebellar cortical lesions on conditioned and unconditioned responses and concerning the effects of conditioning in the decerebrate/decerebellate animal shed new light on crucial issues.

Peterson *et al.* describe recent studies of vestibuloocular reflex conditioning that focus on cross-axis adaptation. This approach, which allows better analysis of adaptive changes, has led to formulation of a new model of vestibuloocular adaptive control in which plasticity occurs in a number of cerebellum-brainstem circuits controlling eye position and velocity.

Wolpaw *et al.* discuss operant conditioning of the monosynaptic pathway of the spinal stretch reflex. This conditioning changes the spinal cord. Current studies focus on the most probable sites of plasticity, the primary afferent-motoneuron synapse and the motoneuron itself.

A theme common to each of these model systems is that activity-driven changes appear to occur at multiple sites. This finding emphasizes both the complexity of the plasticity underlying even simple changes in behavior and the extreme importance of simplicity in the model systems used to investigate mechanisms of behavioral change.

Cerebellum and Classical Conditioning of Motor Responses

CHRISTOPHER H. YEO

Department of Anatomy and Developmental Biology
University College London
London WC1E 6BT, England

Nowhere is the crucial importance of learning and memory more evident than in severely amnesic patients who are unable to form new memories. Damage to limbic structures, particularly hippocampus, produces amnesic symptoms, but these lesions leave some memory systems intact. Since motor skills can be acquired and retained, different forms of learning must depend upon different circuits within the brain.

In order to simplify the neural analysis of learning, Pavlov introduced reflex conditioning procedures that provide a closely controlled learned response isolated from other, background behaviors. He rang a bell immediately before presenting food to his dogs; after several such pairings, they began to salivate as the bell was sounded and before presentation of the food.[1] Using Pavlov's terminology, the food is an unconditioned stimulus (US) because it unconditionally elicits salivation and the bell is a conditioned stimulus (CS) because it elicits salivation only conditionally upon it having been paired with food. The reflex response to food is the unconditioned reflex (UR) and the new, learned response to the CS is the conditioned response (CR). This basic, delay conditioning design (so named because there is a short delay between the onsets of the CS and US) is the simplest of several conditioning procedures that Pavlov developed, but much of our understanding of associative learning processes derives from it.

Pavlov began his conditioning studies with the salivation response, but the procedure was extended to skeletal muscle reflexes such as limb flexion. Conditioned motor responses are now widely used in the analysis of learning mechanisms. And yet such conditioning is very little affected in many instances of human amnesia. A patient with Korsakoff's syndrome and exhibiting severe amnesia was shown to have normal levels of eyeblink conditioning even though he had no recollection of the experimental procedures to which he was exposed daily.[2] Should this lead us to abandon Pavlovian conditioning as a model for our study of learning mechanisms? Probably not. Evidence is increasing from studies in animals and in normal and amnesic humans that, though there may be different forms of learning and memory and these may depend critically on different brain areas, the cellular mechanisms that underlie them may be rather similar. The advantage of Pavlov's precisely defined unit of behavioral change for analysis of the neural mechanisms of learning remains unchallenged.

EYEBLINK CONDITIONING IN RABBITS

The classically conditioned nictitating membrane response (NMR) of the rabbit[3] has several advantages for analyzing the neural mechanisms of conditioning.[4] Normally driven by a touch close to the eye, under experimental conditions the NMR may be elicited by a puff of air to the cornea or by brief electrical stimulation of the periocular area. In contrast to the external eyelids, which also blink to these stimuli, there are very few spontaneous NM blinks.

The NMR may be conditioned by pairing a sound or light CS with an airpuff or periocular shock US. In a typical delay conditioning procedure, a tone CS is sounded for 350 msec with the final 50 msec accompanied by electrical periocular stimulation as the US. Initially, NM movement is driven only by the US. After approximately 200 paired presentations of the CS and US, NM movement is driven by the CS. On paired trials, this conditioned response (CR) is seen as movement of the NM before US onset. Occasional presentations of the CS alone elicit the full form of the CR with no contribution from the unconditioned response (UR) component which accompanies it on paired trials. It is on such unpaired trials that the precise timing of the CR can be fully appreciated. The peak of the CR is timed to coincide with the onset of the US. This has presumed adaptive value for circumstances where lid closure would ameliorate the effects of the US, though this is not allowed to happen in the true Pavlovian conditioning procedure.

The motor output pathways for the external eyelid blink and the NMR are distinct and different. The orbicularis oculi muscle closes the external eyelids and is innervated exclusively by the seventh nerve from motoneurons located in a dorsal/dorsomedial subgroup of the facial nucleus. In contrast, the NM is swept horizontally across the cornea in a nasal to temporal direction as a consequence of eyeball retraction into the orbit. The eyeball retraction is primarily accomplished by the retractor bulbi muscle innervated by the sixth nerve from neurons mainly located in the accessory abducens nucleus,[5,6] which lies ventrolateral to the main abducens nucleus. The other extraocular muscles may contribute to small NM excursions.[7]

WHAT IS ESSENTIAL FOR EYEBLINK CONDITIONING?

Lesions of the cerebral neocortex,[8,9] hippocampus,[10] or all of the forebrain to the level of the thalamus[11] do not abolish NMR conditioning. NMR conditioning is even spared in the decerebrate.[12] As in all lesion experiments we should be cautious in the interpretation of these findings; conditioning can be supported by brainstem structures *but may involve forebrain structures in the normal animal.*

In 1981, Thompson and his colleagues reported an intriguing finding.[13] Large unilateral lesions of the cerebellum abolished NMR conditioning ipsilateral to the lesion and prevented reacquisition. If the US was applied contralateral to the lesion, conditioning proceeded normally, and the unconditioned NM response was normal on both sides. The unilateral cerebellar lesion had abolished only the ipsilateral conditioned response; in all other respects the subject appeared to be normal. This important finding is subject to three alternative, but not mutually exclusive, explanations.

1. The cerebellum is the site of neural plasticity for NMR conditioning.
2. The cerebellum is a necessary conduit through which the conditioned response is expressed.
3. The cerebellum provides a necessary input upon a remote center critical for conditioning.

Given the crucial role of the cerebellum in the normal execution of movements, explanations 2 and 3 are *a priori* equally plausible as the exciting suggestion 1, that the critical associative processes are within the cerebellum. What additional evidence is there for a cerebellar role in NMR conditioning?

CEREBELLUM AND NMR CONDITIONING—AGREEMENTS

In the original investigation by Thompson and his colleagues the cerebellar lesions were large. Because the cerebellum has a relatively simple anatomical organization with well-described input and output pathways, additional studies using discrete lesions were used to establish where within the cerebellum is the circuitry essential for NMR conditioning. These studies demonstrated that the interpositus nucleus[14,15] or dentate/interpositus border[16,17] are essential. Given the difficulties in identifying the dentate-interpositus boundary in rabbits, the location of the lesions in our studies and those of Thompson and his colleagues are essentially the same. Welsh and Harvey[18] have also seen impairment or abolition of CRs following lesions of the interpositus nucleus.

Lesions of cerebellar outflow, in the superior cerebellar peduncle[19] and in the magnocellular red nucleus[20]/rubro-bulbar tract,[21] are equally effective in abolishing conditioned NMR responses.

CEREBELLUM AND NMR CONDITIONING—ARGUMENTS

Cerebellar Cortex

The importance of the cerebellar cortex for NMR conditioning is not entirely agreed. We observed that lesions of cerebellar cortex that invaded the hemispheral part of lobule VI - HVI of Larsell's classification,[22] or lobulus simplex of Bolk,[23] impaired conditioning.[24] When we made lesions that removed HVI entirely, there was a complete loss of conditioned NM responses, and retraining over five days did not reinstate them. The conditioning deficit was complete and as permanent as nuclear lesions had been shown to be. In sharp contrast, Thompson and his colleagues found no loss of conditioned responses following any cerebellar cortical lesion, but only disruptions of the topography of the CR.[17] More recently there has been some resolution of these differences. Here we report that, with extended postoperative training, there can be substantial recovery of conditioning with lesions restricted to HVI. If the lesion is extended to include the adjacent medial parts of ansiform lobe (HVII) the conditioning deficits are permanent, though there may be continued low levels of responding. Thompson and his colleagues have now reported substantial initial deficits following HVI lesions but with subsequent recovery.[25]

At present, the different findings following cerebellar cortical lesions cannot be fully explained. One important factor appears to be the performance levels before the lesion. Our subjects were not at asymptotic levels, perhaps rendering them more vulnerable to the effects of cerebellar lesion. The attractive suggestion that the cerebellar cortex is particularly involved in the acquisition of conditioning might be supported by our findings but is not borne out in a recent study by Lavond and Steinmetz[26] who reported acquisition following HVI lesions. It should be noted, however, that the acquisition was impaired and CR amplitudes were small.

One recent interpretation of these findings is that "the essential neural plasticity . . . resides in the interpositus nucleus or its afferents: i.e., brainstem structures and/or multiple cerebellar cortical sites not completely lesioned to date."[25] Though this interpretation has the advantage of being all-embracing, it is unlikely that the reportedly small critical region of the interpositus nucleus receives input from widely separate cortical areas.

Anatomical studies revealed that the critical region of anterior interpositus nucleus does receive a major input from HVI.[27] Additionally these studies indicated that the major olivary input to HVI is from the medial part of the rostral dorsal accessory olive (DAO), which is the face somatosensory region of the olive. We also saw a variety of mossy fiber inputs and noted, in particular, strong inputs from the dorsolateral and lateral divisions of the basilar pontine nuclei. These are known to receive auditory and visual inputs.[28,29] It cannot escape notice that the inputs to lobule HVI include auditory and visual mossy fiber projections and a face somatosensory climbing fiber input. We suggested that these inputs satisfied an extremely simple implementation of the Marr/Albus model[30,31] of motor learning in the cerebellar cortex. A visual or auditory CS-related mossy fiber/parallel fiber input to the Purkinje cell of the cerebellar cortex may be modified by climbing fiber activity driven by face somatosensory input to the inferior olive. Such an associative mechanism is, of course, similar to the cerebellar learning model suggested by Ito and his colleagues in their investigations of modification of the vestibuloocular reflex.[32]

Interpositus Nucleus Lesions—Learning or Performance Deficits

If a lesion damages any part of the olivo-cerebellar system that influences the face musculature, then impairments of the CR will be seen. So, lesions in the face area of the inferior olive, lobule HVI of cerebellar cortex, lateral parts of the anterior interpositus, and rostral/dorsal parts of the red nucleus all impair eyeblink conditioning. All of these lesions will result in changes in tonic activity of the premotor and motoneuron populations. In our olivary lesion study[33] we had recognized that the inevitable consequence of the lesion was to deregulate that part of the cerebellar cortex that we knew to be critical. We could conclude only that the face somatosensory area of the olive is essential for conditioning; no conclusions may be drawn on its specific role, if any, in conditioning.

Welsh and Harvey[34] suggested an alternative interpretation of the effects of interpositus nucleus lesions upon NMR conditioning. Interpositus nucleus lesions cause changes in tonic activity of the motor system and thereby impair conditioning through a nonlearning, or performance, deficit. Using lidocaine to temporarily inactivate the interpositus nucleus of a conditioned subject, Welsh and Harvey obtained clear abolition of conditioned responses and observed that the UR to the standard intensity of corneal airpuff was normal. But they made an important, additional, observation. When lower intensities of the corneal airpuff were used URs were smaller. But these smaller URs were further reduced in amplitude during interpositus nucleus inactivation. Deep nuclear inactivation produced impairments of the conditioned *and* the unconditioned response. But why should this matter if, at training levels of the US, there were no deficits? The reasons have been made clear in earlier studies of the effects of drugs on conditioning.[35] Even when a subject is strongly conditioned, the CS is a weaker excitor of the response than is the US. If there is a response performance deficit, then this may go unnoticed with presentations of the US well above threshold. With the weaker response activation of the CS, these deficits may be severe enough to prevent expression of the CR. The appropriate test for reflexivity of the preparation is to use much lower levels of the US, which give response amplitudes comparable with those elicited by the CS.

The UR amplitude deficit was seen immediately after interpositus nucleus inactivation using lidocaine. What of the cerebellar lesion studies, where testing was carried out some days after the lesion? Would such UR deficits also be seen? Welsh and Harvey demonstrated that there were no UR amplitude deficits following interpositus nucleus lesions, but there was a lower frequency of emission of the UR and longer latencies to onset and peak when low intensities of US were used.[18] So, Welsh and Harvey demonstrated an impairment of the UR, and this allowed their conclusion that, rather than learning deficits, cerebellar lesions produce performance deficits.

If interpositus nucleus lesions cause longer latency URs at weak US intensities, should they not cause longer latency CRs? Welsh and Harvey found that 15 subjects, classified as CR abolished in paired CS-US trials, did show CRs but at latencies longer than the 285-msec CS-US interval. Using CS-alone trials and extending the analysis window to 800 msec, small numbers of CRs of small amplitude were seen in some subjects. They give a straightforward interpretation of these results. Cerebellar lesions produce performance deficits within which there may or may not be learning deficits. But there are problems even with this simplest of interpretations. In the 15 subjects categorized as CR abolished in the CS-US interval, the CRs that were detected within the 800-msec window were impaired in amplitude. Indeed, in four of these subjects there were no CRs at all. So, although interpositus lesions extend the onset latencies of both the CR and the UR, there are also measured differences in the lesion effects upon the CR and the UR. The UR does not show amplitude decreases while the CR may have decreased amplitude or even be abolished. Interpositus nucleus lesions are seen to have differential effects upon CR and UR.

Lesion studies serve as a useful guide to identifying circuitry that is importantly involved in a particular behavior. Rarely can lesion studies directly throw light on the underlying neural mechanism, they must serve as guides for electrophysiological or pharmacological investigation. But some of the charges leveled at lesion studies of the cerebellum must be challenged. It seems clear that interpositus nucleus lesions can cause performance deficits due to a loss of tonic excitatory outflow from the cerebellum. But no evidence exists to suggest that they do not cause learning deficits too. If the mechanisms critical for conditioning are within the cerebellum, then we can be sure that they are contained within those areas that when lesioned, cause loss of conditioning. That this conditioning loss is accompanied by a performance deficit is the inevitable consequence of the lesion method.

Cerebellar Lesions in Decerebrates

Lesions of the olivo-cerebellar system produce deficits in conditioning. There has been debate concerning the magnitude of these deficits and whether they are secondary to general decrements in performance. In all studies, cerebellar lesions always cause CR impairment. Recently, work has appeared that challenges this conclusion. Bloedel and his colleagues reported that eyeball retraction and external eyelid conditioning are not impaired following complete cerebellar ablation if the conditioning is established in a decerebrate preparation.[36,37] Decerebrate rabbits were conditioned using a 350-msec CS-US interval and an unusually short intertrial interval of 8-10 sec. Asymptotic levels of performance were slightly unstable at around 60% to 70%. The surprising finding was that a large cerebellar lesion was without effect upon this performance. Although the lesions were not complete cerebellectomies, they included most of the cerebellar cortex and deep nuclei ipsilateral to the conditioning. Why was conditioning spared? Bloedel and his colleagues have suggested that cerebellum is not necessary for condi-Tioning even in normal animals. In the decerebrate, the usual performance deficit following cerebellar lesions is not seen and learning is, therefore, not disrupted.

CEREBELLAR LESIONS AND NMR CONDITIONING—A REEXAMINATION

Recently, my colleagues and I reevaluated the effects of cerebellar lesions upon NMR conditioning in the light of the studies reviewed above. We made lesions of cerebellar cortex lobule HVI and adjacent cortical areas. In addition to testing for impairments of the CR, we have critically evaluated possible changes in the UR following the lesion. In a second study, we have examined the effects of cerebellar lesions upon NMR conditioning established in the decerebrate.

Cerebellar Cortical Lesions and NMR Conditioning

Following cerebellar interpositus nucleus lesions, CRs may be abolished or made smaller and delayed.[18] We showed that CRs are abolished in the CS-US interval following cerebellar cortical lesions,[24] but we did not know if the CRs are truly lost or whether their latencies are extended into the post-US period. We have, therefore, conditioned rabbits using a procedure in which every tenth trial is a CS-alone presentation. This enabled us to test for the presence of a CR up to 1000 msec after CS onset. We also continued to condition the subjects for very much longer than is usual following the cerebellar lesion, in order to demonstrate further the permanence of the lesion effects. Because interpositus nucleus lesions alter the UR as well as the CR,[18] we tested the reflexivity of the subjects at the conclusion of training. URs for a variety of US intensities were measured.

Rabbits were trained using methods identical to those of our earlier experiments (a mixture of light and white noise CSs each of 550-msec duration coterminous with a 50-msec periocular electrical US), but each daily session consisted of 90 paired CS-US trials and 10 unpaired, CS-alone trials for each modality. Hence, there were 200 trials each day. NM movement was measured for 350 msec before, and for 1000 msec after CS onset. The criterion for a CR was 0.5-mm movement at any time after CS onset within the 1000-msec window. Preoperative training was for 5 daily sessions to the right side. Aspiration lesions of the right cerebellar cortex were then made and there was a postoperative period of one week. Fifteen postoperative conditioning sessions to the right side were given, followed by 5 sessions to the left, unoperated, side, and then 15 more sessions to the right. After these conditioning sessions, the UR to a range of US intensities was measured on the right and left sides. Some of these data have been reported briefly.[38]

The cerebellar cortical lesions of three representative cases, CTX1, CTX2, and CTX3 are shown in FIGURE 1 and their NMR conditioning data in FIGURE 2. Lesions

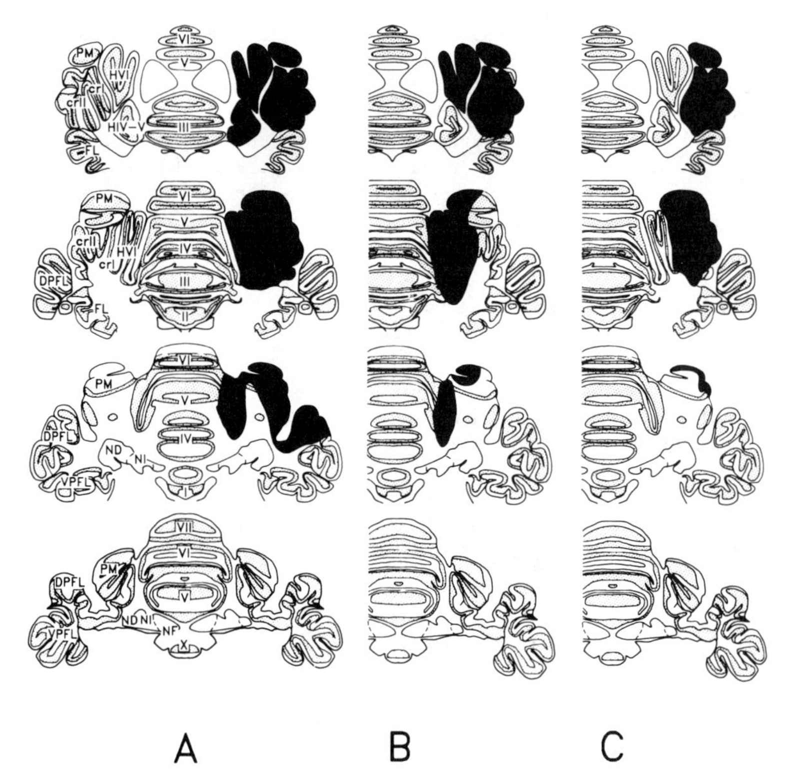

FIGURE 1. Reconstruction of cerebellar cortical lesions of subjects CTX1 (**A**), CTX2 (**B**), and CTX3 (**C**). For each case, the area of primary tissue loss and degeneration is shown in black on four transverse sections through rostral to middle levels of the cerebellum. Abbreviations: crI and crII, crus 1 and 2 (of ansiform lobe); DPFL, dorsal paraflocculus; FL, flocculus; HIV-V, HVI, hemispheral lobules 4-5 and 6 (of Larsell); ND, dentate nucleus; NF, fastigial nucleus; NI, interpositus nucleus; PM, paramedian lobe; VPFL, ventral paraflocculus; II-X vermis lobules 2-10 (of Larsell).

mainly restricted to lobule HVI, as in CTX2, abolished conditioned NM responses for 3 to 10 postoperative sessions. Recovery was to normal frequencies of CR for the white noise CS, but the CR latencies were widely variable and longer than those measured preoperatively (FIGS. 2C & 2D).

Lesions such as CTX1, which removed all of lobule HVI in addition to ansiform and rostral paramedian lobes (FIG. 1A), caused more profound and permanent deficits. The frequency of CRs was greatly reduced. Over 6000 postoperative conditioning trials to the lesioned side, mean CR frequency on unpaired CS trials was 1.3% and 8% for light and white noise, respectively (FIG. 2A). The latencies of these very few CRs were even more variable than those after lesion of HVI only, and included both abnormally short and long latency responses (FIG. 2B).

Control lesions of ansiform lobe and rostral paramedian lobe without damage to HVI, as in case CTX3 (FIG. 1C), caused no significant changes to CR frequency or latency (FIGS. 2E & 2F).

If the loss of CRs in the HVI and ansiform lobe lesioned subjects had been accompanied by a depression of UR amplitude or velocity, then the loss of CRs might have been due to a performance, rather than a learning, deficit. But these cerebellar cortex lesions that abolish CRs actually enhance the amplitude of URs to periocular shock. This enhancement may be seen at all US intensities—it is not restricted to low intensities close to threshold. FIGURES 3A & 3B show UR amplitudes for a range of US intensities from 0.1 to 2.5 mA on the right (lesioned) side and left (nonlesioned) side of subjects CTX1 and CTX2 with permanently impaired or partially recovered

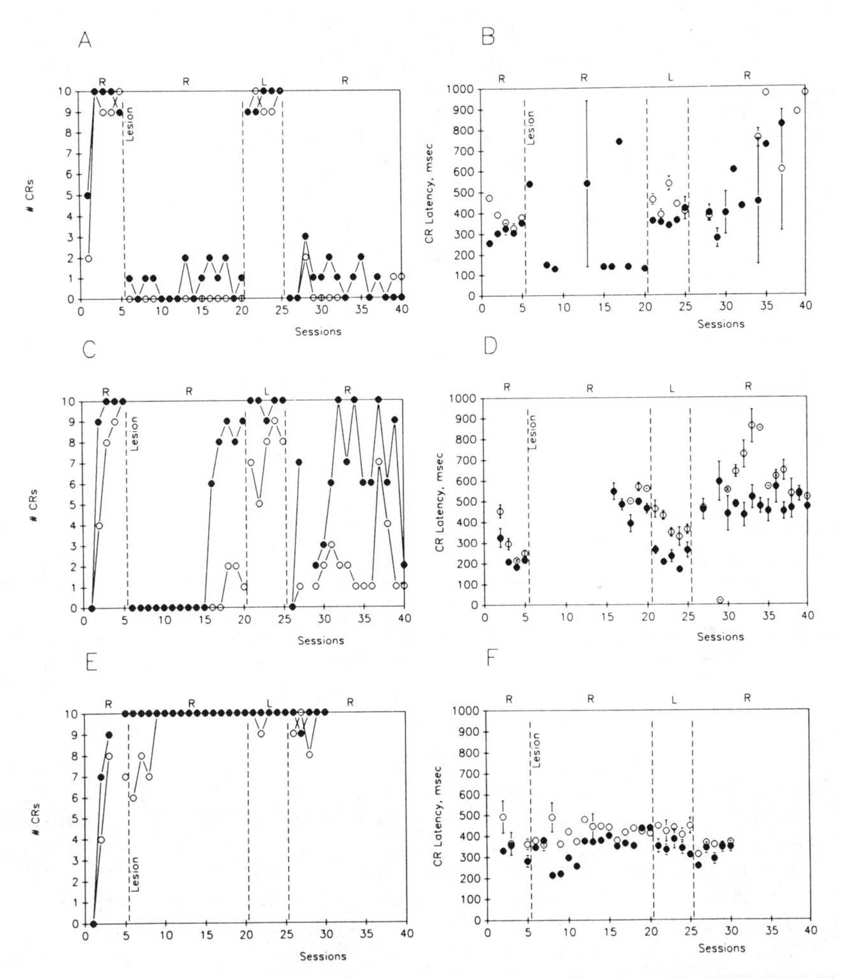

FIGURE 2. CR frequency and CR onset latency for subjects CTX1 (**A & B**), CTX2 (**C & D**), and CTX3 (**E & F**). Sessions 1-5 were preoperative training to the right side. Sessions 6-20 were postoperative training to the right (lesioned) side. Sessions 21-25 were to the left side and 26-40 again the right. Responses to the white noise are indicated by black circles; responses to the light are indicated by open circles. Panels **A**, **B**, and **C** show the number of CRs (NM movement larger than 0.5 mm) for the 20 CS-alone (10 white noise and 10 light) trials of each session. Panels **B**, **D**, and **F** show the mean and standard error of CR latencies for these CS-alone trials. Missing points indicate no CRs in that session. Recovered CRs are seen to have grossly abnormal and variable latencies in sessions 6-20 and 26-40 of subject CTX1 and long and variable latencies for subject CTX2. In contrast CTX3 has normal and stable latencies.

conditioning, respectively. In case CTX1, at every tested level of US, the UR was significantly larger on the lesioned than on the nonlesioned side. In case CTX2, the right UR was only slightly greater at intermediate levels of US intensity. All subjects with conditioning impaired at the conclusion of training showed UR amplitudes larger on the lesioned side. Subjects without significant impairment of conditioning displayed no left-right differences.

Our reexamination of cerebellar cortical lesions has revealed that HVI is critical for conditioning. Though lesions of HVI alone may allow some eventual recovery of conditioning, when accompanied by lesions of ansiform and rostral paramedian lobe, conditioning deficits are permanent. It seems that ansiform and paramedian lobes are subsidiary in their importance, since lesions restricted to them have very little effect upon NMR conditioning. Anatomical and physiological investigation will reveal whether there are face somatosensory climbing fiber inputs to these additional cortical areas in the rabbit. That cerebellar cortical lesions enhance unconditioned response amplitudes is entirely consistent with reports that interpositus nucleus lesions depress them. The removal of cerebellar cortical inhibitory drive upon the deep nuclei provides

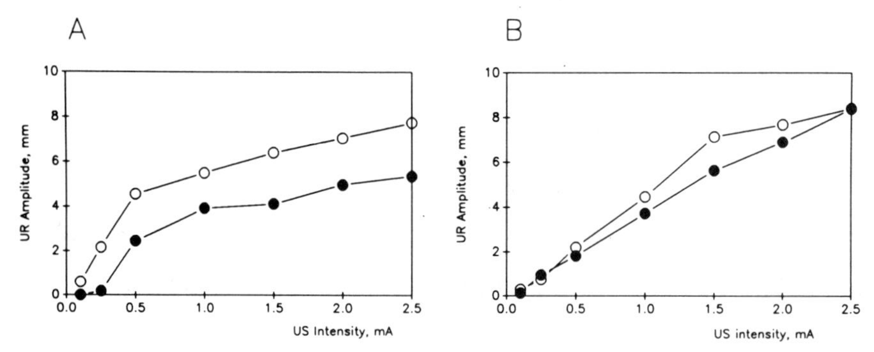

FIGURE 3. UR amplitudes for US intensities 0.1-2.5 mA for subjects CTX1 (**A**) and CTX2 (**B**). Each point is a mean value for 10 US presentations. Responses on the right (lesioned) side are indicated by open circles, responses on the left are shown as black circles.

tonic levels of excitatory outflow; interpositus nucleus lesions deprive the system of excitatory drive. But the cerebellar cortical lesions have sharply contrasting effects upon conditioned responses. Despite significantly increased UR amplitudes, conditioned responses are profoundly depressed. That we see this clear differentiation between the effects of cortical lesions upon the CR and the UR points clearly to a crucial cerebellar function in generating the CR. But there are other interpretations. Is the level of tonic, excitatory drive to the motor apparatus now too high to allow modulation by the CR signal? This question awaits resolution. The cortical lesion findings provide new insight into the performance versus learning debate.

Cerebellum and NMR Conditioning in Decerebrates

We have recently investigated the effects of cerebellar ablation upon NMR conditioning established in decerebrates. We used procedures similar to those of Bloedel and his colleagues[37] but with certain important differences. In all of the experiments we measured movement of the NM with a position transducer as usual. Additionally, electromyograms from the orbicularis oculi of the upper eyelid were recorded via implanted microwires or with percutaneous clips.

Decerebrations were performed by aspiration under halothane anesthesia. The level of decerebration was critical. Transections of the neuraxis immediately rostral to the superior colliculus inevitably resulted in poor levels of conditioning. Histological checks revealed that such decerebrations damage rostral red nucleus. When transections were made approximately 1.5 mm rostral to the superior colliculus, conditioning was possible within a normal number of trials and to asymptotic levels. These findings are essentially similar to those of Mauk and Thompson.[12]

Conditioning was established using a white noise CS. The CS-US interval was 250 msec and the CS was coterminous with a 60-msec, 50-Hz biphasic square-wave pulse of 2-mA intensity. In contrast to Kelly *et al.*,[36] who used an intertrial interval (ITI) of 8-10 sec, we used an ITI of 25-35 sec (mean: 30 sec) to insure against sensitization which can occur at shorter ITIs. Conditioning proceeded at rates similar to the rate seen in normal subjects. In most subjects, conditioning was well established by 500 trials. FIGURE 4 shows NMR and orbicularis oculi EMG data from a representative case, DC1. In FIGURE 4A is seen the NMR and EMG response to the white noise CS before conditioning. A small, short-latency inhibition of background EMG response is followed by an excitation at a latency around 50 msec. In one trial a test corneal airpuff was applied to demonstrate the NM and EMG components of a full-sized blink. In FIGURE 4B are shown, on CS-alone trials, the conditioned NM and EMG responses after 500 conditioning trials.

The subjects were reanesthetised and the cerebellum was completely removed by aspiration. After 30-60 min of recovery, conditioning recommenced. In all subjects, there were no conditioned NM or EMG responses following cerebellectomy. FIGURE 4C shows NM and EMG responses after 20 conditioning trials following cerebellectomy. The baseline EMG is depressed and there are no detectable CRs. But the airpuff-evoked reflex blink is vigorous. Between 200 and 500 conditioning trials were given. Further conditioning was limited by the health of the preparation. With 500 trials before, and 200 trials after cerebellectomy, 13 hr had elapsed since decerebration. But, in the conditioning session after cerebellectomy, some CRs were seen to develop in some subjects. These were usually seen when the baseline orbicularis oculi EMG began to return to precerebellectomy levels. CRs were seen both as NM movement and orbicularis oculi EMG. These CRs were of small amplitude and long latency (see Trace 3 in FIG. 4D). To date, we have not maintained the preparation long enough to determine if these few, small CRs would continue to develop to precerebellectomy levels.

Some similarities and important differences exist between our findings and those of Kelly *et al.*[36] We have established conditioning in the decerebrate, and, using a standard ITI, this conditioning appears very similar to that of normal subjects. In general, levels of conditioning in the other study were lower than normal. Our cerebellar lesions always caused a profound loss of conditioning; histological verification confirms that all of our effective lesions completely and bilaterally removed the deep cerebellar nuclei. The mainly unilateral cerebellar lesions of Kelly *et al.*[36] caused no change in the conditioned eyeblink. They argue that their decerebration frees the cerebellectomy from its normally attendant loss of excitatory drive, and hence conditioning, established in some putative brainstem site, is normal.

If this interpretation is correct, then there must be some critical input between the levels of decerebration in the two studies which alone mediates this effect. Equally plausible is the suggestion that the lower level of decerebration has partially damaged

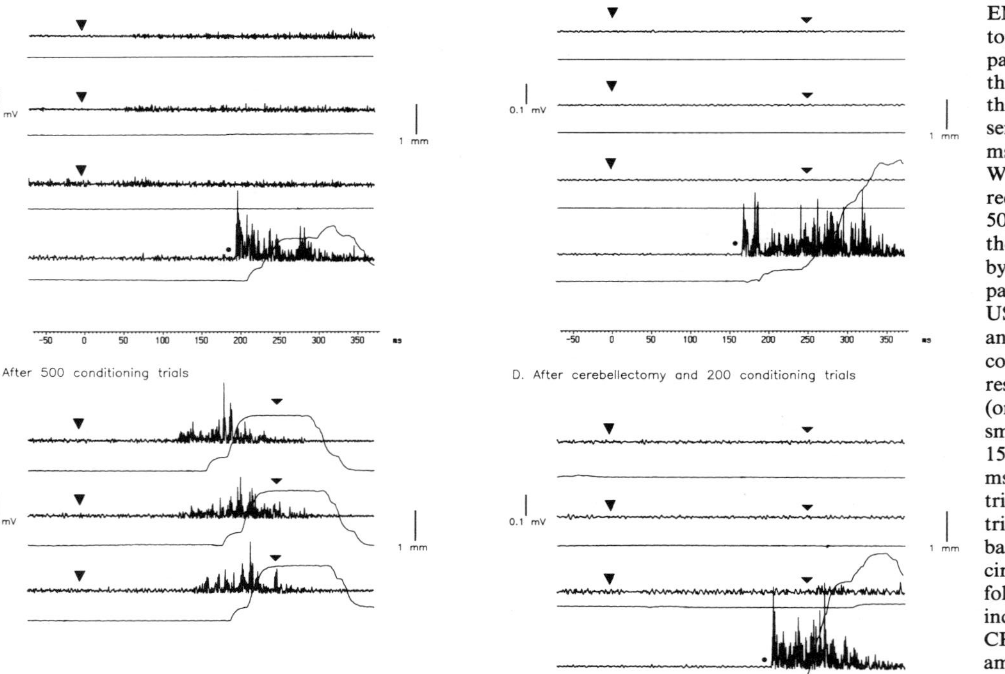

FIGURE 4. NM and orbicularis oculi (o.o.) EMG responses in a decerebrate and cerebellectomized subject, DC1. All panels show data as paired traces, with the EMG record above and the related NM movement record below. **(A)** In the upper three pairs of traces, responses to presentation of the white noise CS. Onset was at 0 msec (larger pointer) and offset at 310 msec. Where baseline EMG activity is low (upper two records) the CS elicits a mild EMG response at 50 msec. Where baseline EMG activity is higher, the CS elicits a short latency inhibition followed by excitation at 50 msec. In the fourth record pair, a corneal airpuff test stimulus (not the shock US) was introduced (black circle). The EMG and NM responses are vigorous. **(B)** After 500 conditioning trials, NM and EMG conditioned responses on CS-alone trials. Time of US onset (on the CS-US paired trials) is indicated by the smaller pointer. NMR onset latency is around 150 msec, o.o. EMG activity onset around 110 msec. **(C)** NM and EMG responses on CS-alone trials after cerebellectomy and 20 conditioning trials. There are no CS-elicited responses. The baseline EMG is flat, but corneal airpuff (black circle) elicits a full blink. **(D)** After 200 trials following cerebellectomy, the baseline EMG has increased. There is a small NM and o.o. EMG CR on the third record, but it is late and of small amplitude. The corneal airpuff test stimulus (black circle) continues to elicit a full blink.

the cerebello-rubro-bulbar pathway where the face area of the red nucleus is relatively rostral and dorsal. This, then, would account for the absence of cerebellar lesion effects, since conditioning would have been established in an essentially acerebellar system.

If this latter suggestion is correct, then it must follow that conditioning can be established without cerebellum, albeit in a decerebrate preparation. This suggestion is consistent with observed low levels of eyeblink conditioning in chronic, decerebrate cats[39] in which the level of decerebration was caudal to the red nucleus. We too, of course, observed emerging low levels of conditioning following the cerebellar ablation in our decerebrates.

The main issue to resolve is whether these data implicate the cerebellum in NMR conditioning. It seems that, in our decerebrates, cerebellum is importantly involved because profound loss of conditioning follows its removal. That conditioning apparently occurs in the absence of cerebellum in decerebrates is indicated by the final phase of our own experiment, by Norman *et al.*[39] and by Kelly *et al.*[36] What is not resolved is whether the CS-elicited responses seen in these studies are true CRs. Are they formed associatively? Why are they seen only at low frequencies? Why are their amplitudes lower than in intact subjects? Is their timing appropriate? Only when we have answers to all of these questions can we rule out the cerebellum as a generator of the conditioned responses that we observe in the intact animal.

CONCLUSIONS

Lesion experiments continue to fuel the debate over the role of cerebellum in motor learning. Cerebellar lesions always cause conditioning deficits in normal animals. That they do so entirely due to performance deficits is challenged by our finding that cerebellar cortical lesions enhance unconditioned responses yet they depress conditioned responses. That conditioning can be established in the decerebrate/cerebellectomized preparation awaits verification that the CS-driven responses are produced associatively. If so, then we will still need to know whether their obvious deficiencies in amplitude and timing relate to the loss of a performance- or learning-related cerebellar mechanism.

REFERENCES

1. Pavlov, I. P. 1927. Conditioned Reflexes. Oxford University Press. Oxford.
2. Weiskrantz, L. & E. K. Warrington. 1979. Neurophysiology **17:** 187-194.
3. Gormezano, I., N. Schneiderman, E. Deaux & I. Fuentes. 1962. Science **138:** 93-106.
4. Thompson, R. F. 1976. Am. Psychol. **31:** 209-227.
5. Cegavske, C. J., R. F. Thompson, M. M. Patterson & I. Gormezano. 1976. J. Comp. Physiol. Psychol. **90:** 411-423.
6. Gray, T. S., S. E. McMaster, J. A. Harvey & I. Gormezano. 1981. Brain Res. **226:** 93-106.
7. Berthier, N. E. & J. W. Moore. 1980. Physiol. Behav. **24:** 931-937.
8. Oakley, D. A. & I. S. Russell. 1972. Physiol. Behav. **8:** 915-926.
9. Oakley, D. A. & I. S. Russell. 1977. Physiol. Behav. **18:** 931-937.
10. Schmaltz, L. W. & J. Theios. 1972. J. Comp. Physiol. Psychol. **79:** 328-333.

11. Enser, L. D. 1976. A Study of Classical Nictitating Membrane Conditioning in Neodecorticate, Hemidecorticate and Thalamic Rabbits. Ph.D. thesis. University of Iowa. Iowa City, IA.
12. Mauk, M. D. & R. F. Thompson. 1987. Brain Res. **403:** 89-95.
13. McCormick, D. A., D. G. Lavond, G. A. Clark, R. E. Kettner, C. E. Rising & R. F. Thompson. 1981. Bull. Psychon. Soc. **18:** 103-105.
14. Yeo, C. H., M. J. Hardiman, M. Glickstein & I. S. Russell. 1982. Soc. Neurosci. Abstr. **8:** 22.
15. Yeo, C. H., M. J. Hardiman & M. Glickstein. 1985. Exp. Brain Res. **60:** 87-98.
16. McCormick, D. A., G. A. Clark, D. G. Lavond & R. F. Thompson. 1982. Proc. Natl. Acad. Sci. USA **79:** 2731-2735.
17. McCormick, D. A. & R. F. Thomson. 1984. Science **223:** 296-299.
18. Welsh, J. P. & J. A. Harvey. 1989. J. Neurosci. **9:** 299-311.
19. McCormick, D. A., P. E. Guyer & R. F. Thompson. 1982. Brain Res. **244:** 347-350.
20. Rosenfield, M. E. & J. W. Moore. 1983. Behav. Brain Res. **10:** 393-398.
21. Rosenfield, M. E., A. Dovydaitis & J. W. Moore. 1985. Physiol. Behav. **34:** 751-759.
22. Larsell, O. 1953. Anat. Rec. **115:** 341.
23. Bolk, L. 1906. Das Cerebellum der Säugetiere. De Erven F. Bohn, Haarlem. G. Fischer, Jena.
24. Yeo, C. H., M. J. Hardiman & M. Glickstein. 1985. Exp. Brain Res. **60:** 99-113.
25. Lavond, D. G., J. E. Steinmetz, M. H. Yokaitis & R. F. Thompson. 1987. Exp. Brain Res. **67:** 569-593.
26. Lavond, D. G. & J. E. Steinmetz. 1989. Behav. Brain Res. **33:** 113-164.
27. Yeo, C. H., M. J. Hardiman & M. Glickstein. 1985. Exp. Brain Res. **60:** 114-126.
28. Kawamura, K. 1975. Brain Res. **9:** 309-322.
29. Wells, G. R., M. J. Hardiman & C. H. Yeo. 1989. J. Comp. Neurol. **279:** 629-652.
30. Marr, D. 1969. J. Physiol. (London) **202:** 437-470.
31. Albus, J. S. 1971. Math. Biosci. **10:** 25-61.
32. Ito, M. 1984. The Cerebellum and Neural Control. Raven Press. New York, NY.
33. Yeo, C. H., M. J. Hardiman & M. Glickstein. 1986. Exp. Brain Res. **63:** 81-92.
34. Welsh, J. P. & J. A. Harvey. 1989. *In* The Olivocerebellar System in Motor Control. P. Strata, Ed.: 374-379. Springer. Berlin.
35. Harvey, J. A. 1987. *In* Psychopharmacology, The Third Generation of Progress. H. Meltzer, Ed.: 1485-1491. Raven Press. New York, NY.
36. Kelly, T. M., C.-C. Zuo & J. R. Bloedel. 1990. Behav. Brain Res. **38:** 7-18.
37. Bloedel, J. R., V. Bracha, T. M. Kelly & J.-Z. Wu. 1991. Ann. N.Y. Acad. Sci. This volume.
38. Yeo, C. H. & M. J. Hardiman. 1988. Am. Soc. Neurosci. Abstr. **14:** 3.
39. Norman, R. J., J. S. Buchwald & J. R. Villablanca. 1977. Science **196:** 551-553.

Substrates for Motor Learning

Does the Cerebellum Do It All?[a]

JAMES R. BLOEDEL,[b] VLASTISLAV BRACHA,
THOMAS M. KELLY, AND JIN-ZI WU

Division of Neurobiology
Barrow Neurological Institute
Phoenix, Arizona 85013

Revealing the location of the plastic changes underlying the acquisition of learned behavior has been an illusive experimental objective for several decades. Various behaviors falling within the loosely defined category of "motor learning" intuitively provided the scientific community with certain advantages in pursuing this goal: simplicity of behaviors, relatively well-known anatomical substrates for the involved reflexes, and quantitative well-accepted methods for measuring changes in the behavior. Despite the hope that the characteristics of the systems involved in motor learning would rapidly lead to a definition of the anatomical pathways by which the modifications were produced, in general the data have been consistently controversial.

Many of the controversies revolve around the supposition that one structure, the cerebellum, may serve as the site of plastic changes required for a broad range of conditioning and adaptive phenomena classified as motor learning. This proposal is based on studies that have investigated the basis for modifications in the vestibulo-ocular reflex,[1–5] the nictitating membrane reflex of the rabbit,[6,7] the withdrawal reflex,[8–10] as well as conditioned autonomic reflexes.[11] Although this proposal had a theoretical basis in the papers of Marr[12] and Albus,[13] the greatest initial experimental impetus was provided by studies examining the adaptation of the vestibuloocular reflex (VOR). Early studies illustrated that ablation of critical cerebellar regions resulted in an inability to acquire this type of adaptive behavior.[14] In addition, some single-unit studies indirectly supported this view.[3,15] On the basis of these observations, specific types of neuronal interactions, most notably involving the climbing fiber system, were proposed as the mechanisms for inducing these plastic changes.[1]

Major support for the argument that the cerebellum serves as the critical site for the plastic processes required for motor learning was further provided by a series of experiments from at least two laboratories indicating that lesions in specific regions of the cerebellar cortex and cerebellar nuclei were capable of eliminating the conditioned nictitating membrane reflex without generating a parallel change in the unconditioned response.[4–7] These lesions selectively abolished conditioned responses in animals that had already acquired the behavior and, in addition, prevented the acquisition of the behavior in naive animals. This view also was supported by single-unit recording data.[7,16] Despite this support, several studies have presented evidence that the plastic changes underlying conditioned and/or adaptive behavior do not occur within the cerebellum. Single-unit recordings in monkeys revealed that neural elements in the cerebellar cortex were not modulated in a manner consistent with their involvement in the plastic processes required for VOR adaptation.[17,18] More recent studies revealed

[a] This work was supported by a grant from the National Institutes of Health (NS21958).

[b] Address for correspondence: James R. Bloedel, Division of Neurobiology, Barrow Neurological Institute, 350 W. Thomas Road, Phoenix, Arizona 85013.

that the putative site for the plastic changes underlying VOR adaptation may be occurring at the level of the vestibular nuclei,[19] although information processing in the cerebellum certainly may contribute to the time course and extent of the modifications. In addition, lidocaine blockade of conduction in the climbing fiber projection, the afferents felt to be essential for establishing the memory trace in the cerebellum, immediately modified the VOR gain, an observation inconsistent with the proposed climbing fiber induction of the plastic changes required for VOR adaptation.[20]

Data from other experiments, including those from our laboratory, suggest that the cerebellum may not be required for the acquisition and performance of the conditioned nictitating membrane reflex in the rabbit. Two sets of completed experiments now support this alternate view. Experiments performed in our laboratory using decerebrate-decerebellate rabbits revealed that decerebrate animals that were previously conditioned could execute the conditioned reflex behavior following complete removal of the cerebellum.[21] Furthermore, naive decerebrate rabbits could acquire the conditioned behavior following cerebellar removal. In studies in intact cats Welsh and Harvey[22] demonstrated that the effects produced by cerebellar lesions may be due to modifications in the performance of the conditioned reflex rather the elimination of the critical storage site for the underlying memory trace. More recently the same laboratory[23] reported that rabbits were able to acquire the conditioned nictitating membrane reflex even though a local anesthetic had been injected into the critical regions of the cerebellar nuclei during training.

This paper will present an overview of two lines of experimentation in our laboratory that address this controversy by examining the substrates required for the acquisition and performance of two very different types of conditioned behavior. The first series is based on a new conditioning paradigm in which modifications of the locomotor cycle are produced in decerebrate ambulating ferrets and cats. The second series of studies addresses the effects of cerebellar removal as well as the microinjection of lidocaine in specific brainstem sites on the conditioned and unconditioned nictitating reflex in the rabbit.

MODIFICATION OF THE LOCOMOTOR CYCLE IN DECEREBRATE AND DECEREBRATE-DECEREBELLATE ANIMALS

These experiments have been performed in ferrets and cats in an effort to examine the modifiability of the locomotor cycle in ambulating, acutely decerebrate animals. All animals were decerebrated at a supracollicular level using a series of electrolytic lesions across the midbrain at a prerubral level appropriate for maintaining locomotion with a treadmill moving underfoot. The animal's head was secured to a stereotaxic head holder in order to appropriately mount stimulating and recording electrodes required for some aspects of these experiments. The animal was held over the treadmill with an abdominal sling. In several animals bipolar EMG electrodes were inserted into the biceps and triceps muscles of each forelimb. In others the electrodes were implanted in the brachioradialis as well. The forelimb to be perturbed was attached by a string tied just above the wrist joint to the lever arm of a low-friction potentiometer in order to measure the forward and backward movement of the animal's extremity. The signal from this potentiometer also was used to trigger the timing of a perturbation applied with an air-driven rod that could be interjected into the step cycle at specifiable times during swing phase. Modifications in the step cycle were also induced by changing the speed of the treadmill both in the presence and in the absence of continuous unilateral forelimb perturbation. The kinematics of the forelimb locomotor cycle were assessed using a standard VCR camera together with a Watscope/Watsmart System that was time coupled to the other components of the data acquisition system. The latter was employed to monitor the position of the wrist, elbow, and shoulder joints and a specific point on the scapula using infrared emitting diodes. This method made it possible to quantify alterations in the limb's position in three-dimensional space and in addition to calculate modifications in joint angle occurring during the experiment.

The principal observation from these initial studies[24] is summarized in FIGURE 1. When the bar was first activated, the trajectory of the forelimb was interrupted by the limb's contact with the bar (B). In the example shown, the foot stumbled over the bar after contact occurred. After 5-10 steps, the height of the trajectory was elevated, and the bar no longer produced a modification in swing phase as the limb was brought forward (C). Once this behavior was acquired, discontinuation of the perturbation resulted in a maintained elevation of swing phase for approximately five steps followed by an extinction period during which the height of the wrist in swing phase progressively decreased to prestimulus values (E). Across all of the animals in this study, the behavior was acquired in approximately 5 to 25 trials and could be extinguished over a comparable time course. In some animals the extinction phase had two components: an initial rapid component lasting 3 to 15 step cycles and a later component that often lasted as long as 30 sec during which the amplitude of stepping progressively decreased to control values.

In later experiments performed on both ferrets and cats, a very sensitive touch sensor was placed on the rod to determine if the foot completely missed the bar during swing phase once the conditioned behavior was acquired. Although the bar was missed in some steps, in most instances the foot was found to contact the bar lightly in the majority of step cycles even when contact was so minimal that the trajectory of the extremity was not modified. Additional analysis of this behavior further demonstrated that the increased height of stepping was associated with an increased duration of the step cycle, consistent with what has been reported in responses to perturbations applied in single steps.[25] Because the rate of ambulation remained the same, the increase in step cycle duration resulted in a decrease in the number of steps per unit time.

In another series of animals the capacity to perform this type of behavior was examined following chronic removal of the cerebellum. Ferrets used in this study were decerebellated two months to one year before an acute decerebration identical to the procedure described above. Their capacity to acquire the modification in gait observed in the nondecerebellated animals is shown in FIGURE 2. This plot shows the time course of the changes in the height of the step cycle in one of the ferrets in the study. Clearly the animal was capable of the acquisition and extinction of the behavior with a time course comparable to what was observed in ferrets with an intact cerebellum.

Even though the behavior could be acquired, performance deficits were definitely present. In general, successive step cycles differed slightly due to an apparent disorganization of movement. This feature was accentuated when the animal was required to step over the bar, resulting in more failures to clear the bar after the acquisition phase of the paradigm than were observed in animals before cerebellectomy.

Additional studies in our laboratory examining neuronal interactions in the cerebellum during the acquisition of this type of conditioned behavior also are consistent with the proposal that this structure plays a role in ongoing regulation of motor behavior rather than the plastic processes underlying its acquisition. In a preliminary report of studies in the decerebrate, ambulating ferret, in which the activity of up to five sagittally oriented Purkinje cells were recorded simultaneously within a single folium,[26] the relationship of the simple and complex spike responses were found to be different than predicted from the learning theories of climbing fiber function. According to most of these views, the climbing fiber system would be expected to induce changes in simple

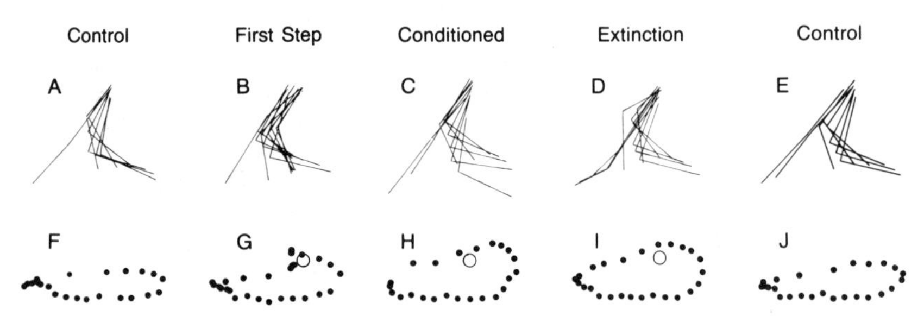

FIGURE 1. Characteristics of the modifications induced by perturbations of the step cycle in decerebrate ambulating ferrets. (**A-E**): Reconstructed stick figures of swing phase under the five indicated conditions. Stick figures are reconstructed from the position of the diodes located over the shoulder, elbow, and wrist. (**F-J**): Dots representing the position of the wrist joint over the same five conditions. The open circles in **G**, **H**, and **I** indicate the position of the perturbation rod relative to the step cycle. Because extinction was demonstrated after cessation of perturbation, the open circle in **I** indicates only the previous position of the bar when perturbations were applied during conditioning.[23]

spike responses that would persist following a decrease in the climbing fiber activity after the behavior was acquired.[1] Rather than this type of interaction, a statistically significant *parallel* relationship was found between the simple and complex spike activity during the acquisition as well as the performance phase of this conditioned behavior. The occurrence of complex spikes was associated with an enhanced modulation of simple spikes.

These experiments have recently been continued in cats in which five Purkinje cells in each of two sagittal zones were simultaneously recorded. The overall objective of these studies is to compare simultaneously the information processing in identified sagittal zones during the acquisition and performance of modifications in the step cycle. Zones were identified electrophysiologically using the criteria of Oscarsson.[27] The data in FIGURE 3 illustrate this type of experiment. These data were obtained from five cells in zone B and five cells from zone C2, all recorded simultaneously. The records for

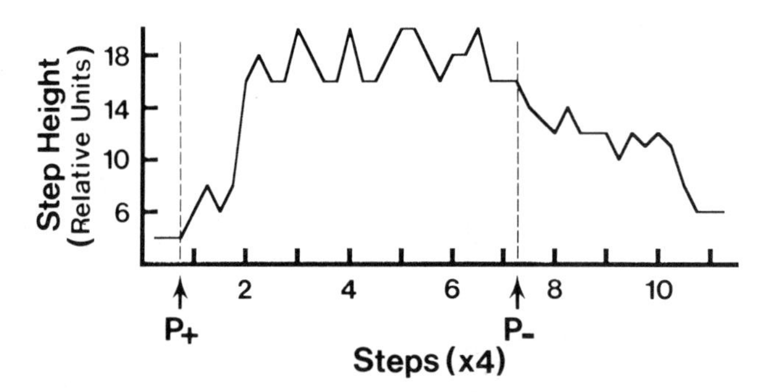

FIGURE 2. Plot of the change in step height as a function of number of steps during modification of the locomotor cycle in a decerebrate-decerebellate ferret. The step cycle at which the perturbation was initiated and concluded is indicated by P+ and P−, respectively. In this example, the animal increased its stepping height approximately three times during control locomotion. The step number can be obtained by multiplying the number on the abscissa times 4.

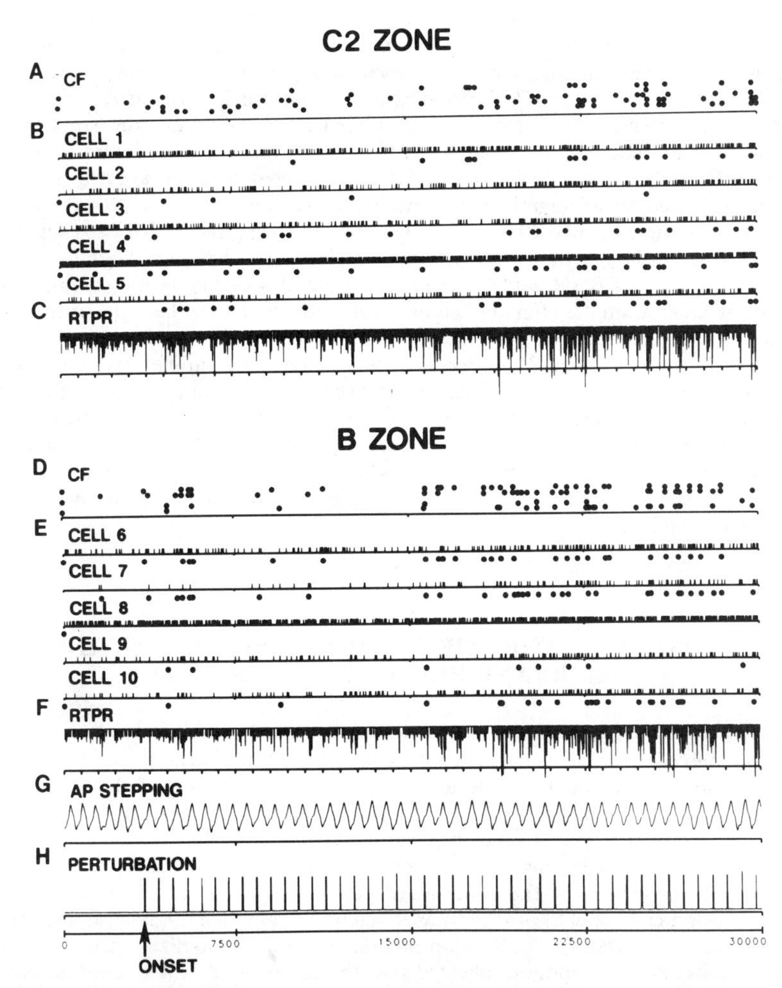

FIGURE 3. Change of Purkinje cell activity associated with modification of the step cycle in an acutely decerebrate ambulating cat. This figure illustrates the simultaneously recorded unitary activity of ten Purkinje cells, five in zone C2 (**A**-**C**) and five in zone B (**D**-**F**). The digitized spike trains of each simultaneously recorded neuron are shown in separate traces. The times of occurrence of each cell's complex spikes are shown below each digitized spike train as dots. The RTPRs, calculated from the unitary activity in each zone, are shown in **C** and **F**, respectively. The arrow indicates the time at which the perturbations were initiated. The anterior/posterior excursion of the animal's forelimb during locomotion as monitored using the potentiometer as described in the text is shown in **G**. The times of occurrence of the perturbations are indicated in **H**. The composite records of the climbing fiber activity for all the cells recorded in each zone are shown in **A** and **D**, respectively.

each cell in B and E illustrate their digitized simple spike activity, with the time of occurrence of their complex spikes shown by dots below each record. The combined complex spike activity for all cells on each array is shown in A and D, respectively; the dots corresponding to each cell's complex spikes are on separate lines. The combined response of the cells in each zone are presented as real-time postsynaptic responses (RTPRs) in C and F. The RTPRs represent the summed hyperpolarizations evoked by all isolated neurons in a zone on a simulated postsynaptic neuron (see Lou and Bloedel[28] for more details on the method). This analytical approach makes it possible to quantify the population's response relative to a behavior or an event in real time without the need for averaging or histogram construction.

In this example the perturbation bar was applied during successive steps beginning at the arrow. Although these data are quite preliminary, the initial findings indicate that at least in zones B, C1, and C2, the modulation of both simple and complex spike activity is most dramatic after the animal begins to acquire the behavior rather than at the very initial stages of the acquisition process. This feature is apparent in the example shown in FIGURE 3. Notice that the complex and simple spike modulation becomes more dramatic as the animal acquires the behavior, not at the initiation of the perturbation. As in our initial study on this question,[26] there continues to be a parallel relationship between simple and complex activity rather than the type of relationship predicted from learning theories[2,12,13] in which the climbing fiber input acts as a "teacher," disappearing after the induction of a pattern of modulation in the Purkinje cell's simple spike responses.

SUBSTRATES FOR THE CONDITIONED NICTITATING MEMBRANE REFLEX IN THE RABBIT

Two lines of experiments have been done in our laboratory to address the neuroanatomical substrate required for the acquisition and performance of the conditioned nictitating membrane reflex (NMR). The first study focused on whether the cerebellum was required for this type of reflex behavior.[21] As reviewed above, previous studies[4–7] indicated that discrete lesions in the cerebellar cortex and cerebellar nuclei resulted in a selective suppression of the conditioned reflex without a parallel change in the amplitude or incidence of the unconditioned reflex. To approach this issue as definitively as possible, a preparation was developed utilizing cerebellectomized, acutely decerebrate rabbits. Previous studies indicated that the conditioned NMR could be evoked in decerebrate animals.[29,30]

Decerebration was performed electrolytically across the brainstem just rostral to the red nucleus. The red nucleus was not damaged by the lesion in any animal included in the study. The cerebellectomy was performed by aspirating the cerebellum through a craniectomy. In all of the reported animals all of the cerebellar nuclei ipsilateral to the eye receiving the airpuff were either aspirated or aspirated substantially enough to be completely disconnected from the brainstem. Animals were conditioned using a traditional paired-stimulus delay paradigm with a 1-kHz, 85-db tone superimposed on 75-db white noise serving as the conditioned stimulus and a 75-msec duration airpuff as the unconditioned stimulus. The interstimulus interval ranged from 250-400 msec.

As previously reported by Kelly *et al.*,[21] all of the decerebrate rabbits capable of acquiring the conditioned behavior before the cerebellum was removed also demonstrated or were capable of reacquiring the conditioned reflex following cerebellectomy.

An example of the conditioned behavior observed in these animals is shown in FIGURE 4. This animal acquired the conditioned behavior before cerebellectomy. However, following the surgical removal of the cerebellum, the incidence of CR responding was near control levels. Upon resuming paired stimuli the animal clearly was capable of reacquiring the conditioned behavior. Furthermore the behavior could be repeatedly extinguished and reacquired. It should be emphasized that the published data of Kelly *et al.*[21] illustrate that the conditioned responses obtained in most of these experiments were quite robust. All responses classified as conditioned responses were no less than 10% of the amplitude of the unconditioned response. These same experiments illustrated that naive decerebrate rabbits could acquire the conditioned behavior after cerebellar removal.

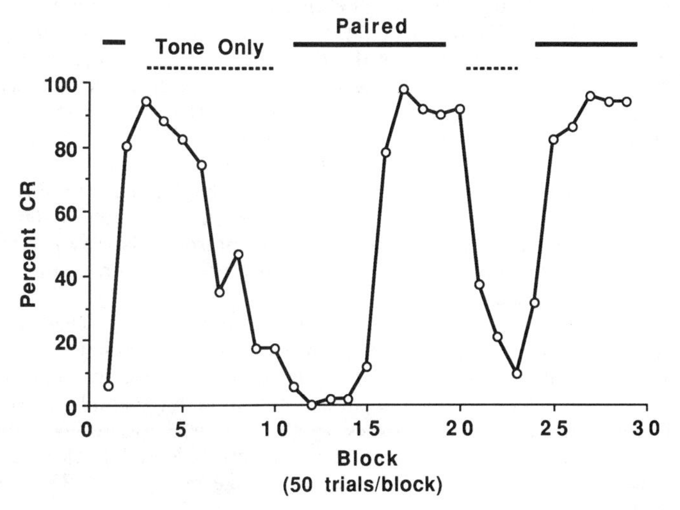

FIGURE 4. Time course of the acquisition of the conditioned nictitating membrane reflex (CR) in a decerebrate-decerebellate rabbit. This plot illustrates three successive epochs during which the conditioned reflex behavior was acquired during the use of the paired stimulus paradigm described in the text. (From Kelly *et al.*[21]; used with permission.)

After demonstrating that decerebrate-decerebellate rabbits can acquire as well as execute the classically conditioned NMR and following the demonstration by Welsh and Harvey[22] that the effects of cerebellar ablation in the intact rabbits were likely the consequence of a performance deficit, the focus in our laboratory turned to the brainstem as the potential substrate for the plastic changes required for the acquisition of the conditioned NMR.

The possible involvement of brainstem nuclei was investigated by determining the effects of lidocaine microinjection into specific brainstem sites on the properties of the conditioned and unconditioned NMRs. Initial experiments were performed using injections in the red nucleus, the principal target nucleus of the cerebellar nuclear regions proposed by some to be required for the plastic changes underlying the acquisition of the conditioned reflex.[7] In addition, inferences made in other studies implied that the red nucleus itself may be involved in the acquisition process.[31–33]

All of the lidocaine microinjection experiments were performed in intact albino New Zealand rabbits instrumented with guide tubes for injection cannulae as well as for the devices required for applying the airpuff stimuli, for applying the light stimuli (when required), and for measuring the movement of the nictitating membrane. Conditioned stimuli consisted of a 1-kHz, 85-db tone superimposed on the 75-db white noise. The unconditioned stimulus, applied 350 msec after the onset of the conditioned stimulus, consisted of a 75- to 100-msec duration airpuff that was 30 psi at the source. The conditioned and unconditioned stimuli were coterminal.

All animals were conditioned using a delayed paired-stimulus paradigm for approximately 10 days until a baseline performance above the 95% criterion was consistently achieved by the animal. Once the six- to seven-day training period was completed, the paradigm was altered in order to assess the effects of lidocaine injection. Stimuli were applied in three-trial sequences: airpuff alone, a paired tone-airpuff, and a light stimulus alone adequate for evoking an NMR. The unconditioned light stimulus was used to provide a reflexively evoked movement of the nictitating membrane that was dependent on the same efferent projection but not the same afferent projection as the response to airpuff. This made it possible to determine, at least to some degree, whether any changes in conditioned and unconditioned reflex behavior may be the consequence of the spread of lidocaine to critical output projections of the motor neurons.

After one day of applying the three-trial sequences to assess baseline responding, the effects of injecting 4% buffered lidocaine in the rubral and perirubral region were determined. The injecting needles were localized through a guide tube system installed at the beginning of the experiment. After adequate control trials were performed, 0.3 to 1 μl of the lidocaine solution was slowly injected over a 5- to 6-min period as the three-trial sequence paradigm was continued.

Data from one of these experiments is shown in FIGURE 5. These bar graphs illustrate the effects of the lidocaine injection on both the conditioned and unconditioned responses. Note that the effect of this injection, localized to the lower medial region of the red nucleus, resulted in a parallel reduction in the amplitude of both the conditioned and unconditioned responses. In all cases ($n = 9$) parallel reductions in the conditioned response and unconditioned response (CR and UR) were observed when the injection site was in or just adjacent to the red nucleus. These effects of

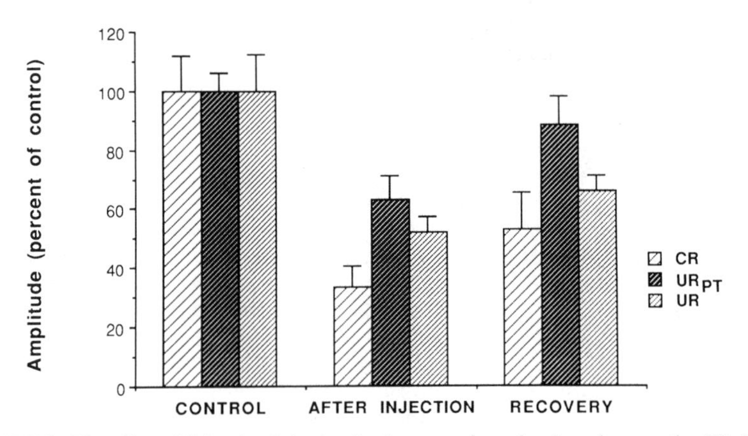

FIGURE 5. The effect of lidocaine injection in the contralateral red nucleus on the CR, UR_{PT} (unconditioned response, paired trials), and UR in an intact, awake rabbit.

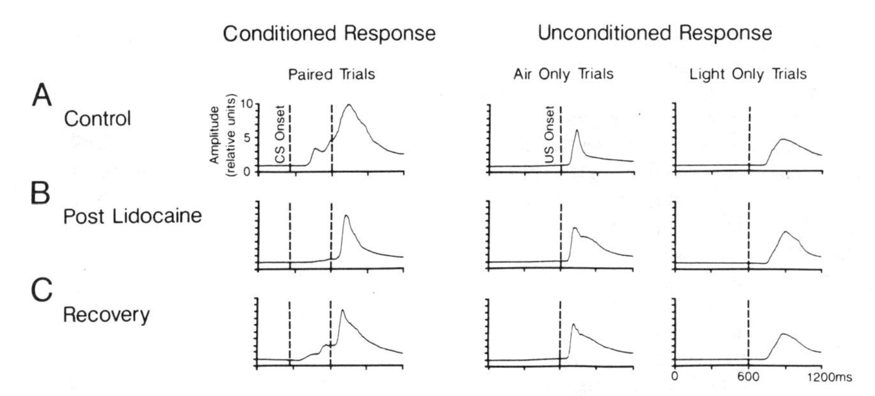

FIGURE 6. The effect of lidocaine injection in the region of the medial spinal trigeminal nucleus on the CR and UR in an intact, awake rabbit. These responses were selected to show the paired CR-UR records and the unpaired UR records before (**A**), immediately after (**B**), and near recovery (**C**) from the injection.

lidocaine injection were well localized. When a comparable amount of lidocaine was injected approximately 1 mm above or below the original injection site, the same changes in the CR and UR were not observed. In this experiment all of the responses had not yet returned to control values at the time (15 min postinjection) the recovery measurements were obtained. Furthermore, the injection of saline at the original site had no effect.

In contrast to the effects observed in the red nucleus study, microinjection of lidocaine in the vicinity of specific regions of the pars oralis of the spinal trigeminal nucleus produced selective changes in the conditioned reflex. The effect of injecting 1 μl of lidocaine at the medial border of this structure on the conditioned and unconditioned NMR in one animal is shown in FIGURE 6. Following the injection there was a clear reduction in the amplitude of the conditioned reflex with no parallel reduction in the amplitude of the unconditioned reflex. In fact the amplitude of the latter was increased. Examples of the conditioned and unconditioned responses approximately 20 min following the injection are shown. In addition to the reduction in the amplitude of the conditioned response, its incidence also was reduced substantially, whereas the incidence of the unconditioned response remained at 100%.

In contrast to these observations, parallel effects on the conditioned and unconditioned responses were often observed at injection sites in other regions of the spinal trigeminal nucleus. In fact parallel changes were observed in two animals following an injection located at the same AP level but on the lateral edge of this structure in the absence of any statistically significant change in the amplitude of the unconditioned responses evoked by light.

CONCLUSIONS AND INFERENCES

The data presented here summarize the findings in our laboratory supporting the argument that extracerebellar sites are responsible for the plastic changes required for motor learning in two types of conditioned behaviors. Furthermore the findings obtained from decerebrate locomoting ferrets and cats reveal that the modified behavior in these preparations can be quite complicated, involving the appropriate sequential activation of muscles as well as the bilateral coordination of limb movements required for the asymmetrical responses to the perturbations of the step cycle.

Substantial discrepancies remain between our findings and those of at least two other laboratories regarding the role of the cerebellum in this type of motor learning. These studies all support the argument that the cerebellum is not a required storage site of the plastic changes underlying VOR adaptation,[34] the classical conditioning of the NMR,[23] or the modifications of the locomotor cycle described above. Nevertheless, there continue to be reports that discrete lesions of the cerebellar cortex and nuclei can produce selective decreases in the acquisition as well as the amplitude of the conditioned response, including the report by Yeo[35] in this volume. The basis for the dramatic difference in these findings remains unclear.

An hypothesis regarding this issue has recently been presented by Bloedel and Kelly.[36] They proposed that the primary effect of cerebellar lesions on the conditioned response results from a dramatic change in the excitability (and consequently the functional properties) of neurons in the brainstem circuits responsible for mediating the CR due to the loss of tonically active inputs from the cerebellar nuclei. In other aspects of motor function the effects of cerebellar lesions on the excitability of pathways *outside* the cerebellum are well known.[37] These effects are usually most profound at times ranging from one to three weeks following the lesion, depending on the species.[38] Although some exceptions do exist in the literature, those studies advocating a permanent learning deficit[4–7] in general acquired observations sooner after the cerebellar lesions than those suggesting an effect due to a performance deficit.[22] Consequently it is possible that the failure to demonstrate conditioned responses and their acquisition in the studies supporting the cerebellar learning hypothesis is related to this mechanism. Based on this same argument, our findings in decerebrate preparations may have resulted from a fortuitous modification in the tonic excitability of brainstem neurons following decerebration that offset the change induced by cerebellar lesions. This effect could also explain why the conditioned responses observed in our studies were more robust (see FIG. 1 of Kelly *et al.*[21]) than those observed in the study of Welsh and Harvey.[22]

Last, this mechanism could also be the basis for the difference between our studies and those reported by Yeo in this volume.[35] The difference in the level of decerebration employed in these two experiments may be at least 2-3 mm. It is well known that in the cat differences as small as 0.5-1.0 mm in the level of decerebration can produce dramatic changes in muscle tone as well as locomotion. Consequently it is possible that decerebration at the two levels represented in the experiments on classical NMR conditioning do not result in the same changes in the tonic activity of brainstem neurons responsible for the CR.

It should be emphasized that the results from our laboratory do not represent an isolated finding or a single series of experiments. Rather they represent a consistent series of findings ranging from data obtained in decerebellate-decerebrate rabbits to data acquired during temporary lesions of selected brainstem nuclei. Furthermore all of our studies on the cerebellar cortical interactions involving the climbing fiber system,[39] the system originally proposed as the one responsible for establishing the engram in the cerebellum,[12,13] support the notion that this afferent system is involved in operational, short-term interactions rather than those underlying prolonged modifications in the excitability of cerebellar neurons. Last, it must be emphasized that to date there is no evidence that any long-term neuronal interaction is produced by the action of the climbing fibers in the cerebellar nuclei, another location suggested as the storage site required for the acquisition and execution of the CR.[7]

One final point must be stressed regarding the experiments supporting an essential role of the cerebellum as a storage site required for the acquisition and performance of conditioned reflexes. Causal relations between structure and function are very difficult to establish definitively using ablation techniques, the approach used by most investigators favoring a cerebellar storage site for at least the conditioning of the NMR. For example, ablating the cerebellum has been shown to produce profound deficits in types of quadruped behavior for which this structure is not essential, particularly during the first one to three weeks following the lesion.[37] These include locomotion, standing, and reaching. Just as the cerebellum is not solely essential for these functions, it may not be essential for the acquisition and execution of the conditioned NMR. Lesions of the cerebellum may dramatically affect this reflex for the reasons discussed above rather than because a storage site required for the plastic changes had been removed. The recent studies of Weisz and LoTurco[40] support this contention by showing that UR facilitation, a phenomenon also resulting from the conditioning procedure, is not abolished by a cerebellar lesion that eliminates the performance of the CR. This finding might be expected if cerebellar lesions affect the excitability of extracerebellar circuits that are required for the plastic processes rather than eliminate the critical storage site.

The same caution must be exercised in interpreting the findings reported here showing that lidocaine injection at the medial edge of the spinal trigeminal nucleus selectively abolishes the CR. Although these data offer an alternative to the cerebellum as a site for the plastic changes, clearly the results also could be due to interrupting the afferent limb of a circuit in which the plastic changes are occurring elsewhere along the pathway. Additional experiments are required to resolve this issue. Nevertheless, the findings do demonstrate that the CR pathway can be separated from the UR pathway even at the afferent-most side of these reflex circuits.

Despite the differences in the observations reviewed above, the fact remains that the execution of conditioned NMRs has been reported in both intact and decerebrate rabbits following substantial cerebellar lesions. Furthermore in a recent abstract Welsh and Harvey[23] reported that it was possible for an animal to actually acquire the conditioned behavior during the application of an anesthetizing dose of lidocaine in the critical regions of the cerebellar nuclei. These data clearly represent positive findings supporting the argument that the cerebellum is not an *essential* site for the storage of engrams required for this type of motor learning.

Ongoing studies utilizing the locomotor paradigm to examine changes in the behavior of the small populations of Purkinje cells during conditioning (see FIG. 3), although preliminary, are consistent with the role of the cerebellum in regulating the ongoing performance of coordinated movements, a role that could contribute to the acquisition or learning of a motor task without this structure serving as the storage site for the learned behavior. The contribution of the cerebellum to the process of motor learning by enhancing the quality of motor performance is likely to apply to the acquisition of many types of skilled movements.

More generally, in our view the combined data regarding substrates for plastic changes required for the modifications of conditioned motor behaviors imply that the plastic changes likely occur, not within a common site serving as a shared memory bank among all of the subsystems involved in different types of motor learning, but within the anatomical pathways required for the execution of the specific behavior undergoing the modification. FIGURE 7 illustrates this concept. According to this view engrams established during conditioning or reflex adaptation are *substrate specific.* This hypothesis further implies that substrates for motor learning are distributed and utilize the properties of pathways specifically related to the behavior undergoing modification. Although additional studies are required to fully test this view, it should be emphasized that it is not only supported by the work reviewed in this manuscript but also by studies examining response modifications in spinal reflex circuits that clearly do not employ a common structure such as the cerebellum.[41] Finally this concept leaves open the possibility that plastic changes in the cerebellum could be essential for the acquisition of other behaviors that actually require this structure as a component of the circuitry required for their execution.

In summary, in our view the cerebellum can no longer be considered to be an essential substrate for the plastic changes required for NMR conditioning and very likely for most conditioned behavior. Rather, it likely plays its primary role in most (but possibly not all) types of motor learning by contributing to the optimal performance of the task to be learned. This argument does not exclude the possibility that plastic changes may occur in this structure 1) early in the development of motor behavior, 2) when there are extreme and prolonged modifications in the activity level of the animal, and 3) when the cerebellum is an essential component of the circuit required for the execution of the task. All three of these features characterize the bases for plastic

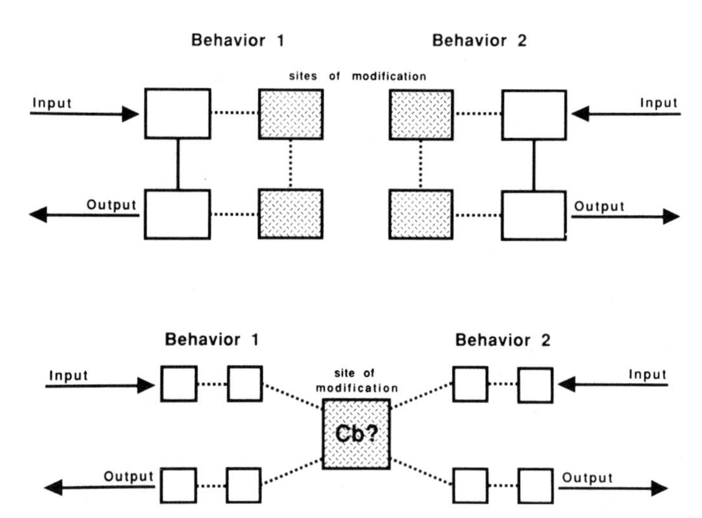

FIGURE 7. Diagrams illustrating the difference between two concepts pertaining to the substrate for motor learning. *Top diagram:* Concept of substrate-specific plastic changes. *Bottom diagram:* Concept of a shared cerebellar network utilized as the site of plastic changes by many central systems involved in motor learning.

changes in other central structures (see other papers in this volume). Consequently plastic changes that may occur in the cerebellum are not likely to be due to its unique role in motor learning. Rather it is likely to be the result of processes that govern plastic changes elsewhere in the brain that are either developmentally or activity dependent.

ACKNOWLEDGMENTS

The authors wish to thank Jan Carey for her assistance in preparation of the manuscript and Joel McAlduff and Shelley Cartwright for their assistance during the experiments.

REFERENCES

1. ITO, M. 1972. Neural design of the cerebellar motor control system. Brain Res. **40:** 81-84.
2. ITO, M. 1982. The cerebellar control of the vestibulo-ocular reflex, around the flocculus hypothesis. Annu. Rev. Neurosci. **5:** 275-296.
3. ITO, M. 1982. The role of the cerebellum during motor learning in the vestibular reflex: Different mechanisms and different species. Trends Neurosci. **5:** 416.
4. YEO, C. H., M. J. HARDIMAN & M. GLICKSTEIN. 1985. Classical conditioning of the nictitating membrane response of the rabbit. I. Lesions of the cerebellar nuclei. Exp. Brain Res. **60**(1): 87-98.
5. YEO, C. H., M. J. HARDIMAN & M. GLICKSTEIN. 1985. Classical conditioning of the nictitating membrane of the rabbit. II. Lesions of the cerebellar cortex. Exp. Brain Res. **60**(1): 99-113.
6. MCCORMICK, D. A. & R. F. THOMPSON. 1984. Cerebellum: Essential involvement in the classically conditioned eyelid response. Science **223:** 296-299.
7. MCCORMICK, D. A. & R. F. THOMPSON. 1984. Neuronal responses of the rabbit cerebellum during acquisition and performance of a classically conditioned nictitating membrane-eyelid response. J. Neurosci. **4:** 2811-2822.
8. DONEGAN, N. H., R. W. LOWERY & R. F. THOMPSON. 1983. Effects of lesioning cerebellar nuclei on conditioned leg-flexion responses. Soc. Neurosci. Abstr. **9:** 331.
9. KARAMIAN, A. I., V. V. FANARDJIAN & A. A. KOSAREVA. 1969. The functional and morphological evolution of the cerebellum and its role in behavior. *In* Neurobiology of Cerebellar Evolution and Development. First International Symposium. R. Llinas, Ed.: 639-673. American Medical Association. Chicago, IL.
10. POPOV, I. P. 1929. The role of the cerebellum in elaborating the motor conditioned reflexes. *In* Higher Nervous Activity. D. S. Fursikov, M. O. Gurevich & A. N. Zalmanzon, Eds. Vol. 1: 140-148. Com. Acad. Press. Moscow.
11. KRASUSKY, V. K. 1957. General nature of changes of food conditioned reflexes in dogs following a surgical lesion of the cerebellum. Zh. Vyssh. Nerv. Deiat. **7:** 733-740.
12. MARR, D. 1969. A theory of cerebellar cortex. J. Physiol. (London) **202:** 437-470.
13. ALBUS, J. S. 1971. A theory of cerebellar function. Math. Biosci. **10:** 25-61.
14. ROBINSON, D. A. 1976. Adaptive gain control of vestibuloocular reflex by the cerebellum. J. Neurophysiol. **39:** 954-969.
15. WATANABE, E. 1984. Neuronal events correlated with long-term adaptation of the horizontal vestibulo-ocular reflex in the primate flocculus. Brain Res. **297:** 169-174.
16. BERTHIER, N. E. & J. W. MOORE. 1986. Cerebellar Purkinje cell activity related to the classically conditioned nictitating membrane response. Exp. Brain Res. **63:** 341-350.
17. MILES, F. A., D. J. BRAITMAN & B. M. DOW. 1980. Long-term adaptive changes in primate vestibuloocular reflex. IV. Electrophysiological observations in flocculus of adapted monkeys. J. Neurophysiol. **43:** 1477-1493.
18. MILES, F. A. & S. G. LISBERGER. 1981. Plasticity in the vestibuloocular reflex: A new hypothesis. Annu. Rev. Neurosci. **4:** 273-299.
19. LISBERGER, S. G. 1988. The neural basis for motor learning in the vestibuloocular reflex in monkeys. Trends Neurosci. **11:** 147-152.
20. DEMER, J. L. & D. A. ROBINSON. 1982. Effects of reversible lesions and stimulation of olivocerebellar system on vestibuloocular reflex plasticity. J. Neurophysiol. **47:** 1084-1107.
21. KELLY, T. M., C.-C. ZUO & J. R. BLOEDEL. 1990. Classical conditioning of the eyeblink reflex in the decerebrate-decerebellate rabbit. Behav. Brain Res. **38:** 7-18.
22. WELSH, J. P. & J. A. HARVEY. 1989. Cerebellar lesions and the nictitating membrane reflex: Performance deficits of the conditioned and unconditioned response. J. Neurosci. **9**(1): 299-311.
23. WELSH, J. P. & J. A. HARVEY. 1989. Intra-cerebellar lidocaine: Dissociation of learning from performance. Soc. Neurosci. Abstr. **15:** 639.
24. LOU, J.-S. & J. R. BLOEDEL. 1988. A new conditioning paradigm: Conditioned limb movements in locomoting decerebrate ferrets. Neurosci. Lett. **84:** 185-190.
25. MATSUKAWA, K., H. KAMEI, K. MINODA & M. UDO. 1982. Interlimb coordination in cat locomotion investigated with perturbation. A behavioral and electromyographic study on symmetric limbs of decerebrate and awake walking cats. Exp. Brain Res. **46:** 425-437.

26. LOU, J.-S. & J. R. BLOEDEL. 1988. A study of cerebellar cortical involvement in motor learning using a new avoidance conditioning paradigm involving limb movement. Brain Res. **445:** 171-174.
27. OSCARSSON, O. 1979. Functional units of the cerebellum-sagittal zones and microzones. Trends Neurosci. **2:** 143-145.
28. LOU, J.-S. & J. R. BLOEDEL. 1986. The responses of simultaneously recorded Purkinje cells to the perturbations of the step cycle in the walking ferret: A study using a new analytical method—the real-time post-synaptic response (RTPR). Brain Res. **365:** 340-344.
29. MAUK, M. D. & R. F. THOMPSON. 1987. Retention of classically conditioned eyelid responses following acute decerebration. Brain Res. **403:** 89-95.
30. NORMAN, R. J., J. S. BUCHWALD & J. R. VILLABLANCA. 1977. Classical conditioning with auditory discrimination of the eyeblick responses in decerebrate cats. Science **196:** 551-553.
31. ROSENFIELD, M. E. & J. W. MOORE. 1983. Red nucleus lesions disrupt the classically conditioned nictitating membrane response in rabbits. Behav. Brain Res. **10:** 393-398.
32. ROSENFIELD, M., A. DOVYDAITIS & J. W. MOORE. 1985. Brachium conjunctivum and rubrobulbar tract: Brain stem projections of red nucleus essential for the conditioned nictitating membrane response. Physiol Behav. **34:** 751-759.
33. ROSENFIELD, M. E. & J. W. MOORE. 1985. Red nucleus lesions impair acquisition of the classically conditioned nictitating membrane response but not eye-to-eye savings or unconditioned response amplitude. Behav. Brain Res. **17:** 77-81.
34. BAKER, R., M. WEISER & J. G. MCELLIGOTT. 1988. Adaptive gain control of the vestibulo-ocular reflex in goldfish. I. Hemicerebellectomy. Soc. Neurosci. Abstr. **14:** 169.
35. YEO, C. H. 1991. Cerebellum and classical conditioning of motor responses. Ann. N.Y. Acad. Sci. This volume.
36. BLOEDEL, J. R. & T. M. KELLY. In press. The dynamic selection hypothesis: A proposed function for cerebellar sagittal zones. *In* Neurobiology of Cerebellar Systems: A Centenary of Ramon y Cajal's Description of Cerebellar Circuits. R. Llinas & C. Sotelo, Eds. Oxford University Press. New York, NY.
37. GILMAN, S., J. R. BLOEDEL & R. LECHTENBERG. 1981. Disorders of the Cerebellum. F. A. Davis. Philadelphia, PA.
38. DOW, R. S. & G. MORUZZI. 1958. The Physiology and Pathology of the Cerebellum University of Minnesota Press. Minneapolis, MN.
39. BLOEDEL, J. R. & T. J. EBNER. 1985. Climbing fiber function: Regulation of Purkinje cell responsiveness. *In* Cerebellar Functions. J. R. Bloedel, J. Dichgans & W. Precht, Eds.: 247-259. Springer-Verlag. Berlin.
40. WEISZ, D. J. & J. J. LOTURCO. 1988. Reflex facilitation of the nictitating membrane response remains after cerebellar lesions. Behav. Neurosci. **102:** 203-209.
41. WOLPAW, J. R., C. L. LEE & J. S. CARP. 1991. Operantly conditioned plasticity in spinal cord. Ann. N.Y. Acad. Sci. This volume.

A Model of Adaptive Control of Vestibuloocular Reflex Based on Properties of Cross-Axis Adaptation[a]

BARRY W. PETERSON,[b] JAMES F. BAKER, AND JAMES C. HOUK

Department of Physiology
Northwestern University Medical School
Chicago, Illinois 60611

Since its description by Gonshor and Melvill-Jones,[1] plastic adaptive control of the vestibuloocular reflex (VOR) has become an accepted model of motor learning. Its role is to adjust the size, direction, and timing of eye movements produced by the VOR so that they precisely compensate for any motion of the head. Adjustment is crucial since the VOR is an "open loop" system in that the eye movements it produces are not sensed by the labyrinthine receptors that generate those movements. Errors are only sensed by visual receptors, and the eye movements that they elicit are too slow to modify the VOR during an individual head movement. Instead, these errors, which take the form of slip of images on the retina, activate adaptive mechanisms, which modify VOR circuits in a way that will reduce the error during future head movements.

Initial studies of VOR plasticity were one-dimensional. They concentrated on changes in VOR gain (ratio of amplitudes of VOR eye movement and head movement) that resulted when subjects wore magnifying, reducing, or reversing lenses that altered the image slip produced by a head movement.[1-3] Studies of VOR plasticity in three dimensions began in 1981. Berthoz *et al.*[4] reported that prism-induced VOR gain changes are specific to planes in which image reversal occurs. Schultheis and Robinson[5] showed that pairing pitch rotation with horizontal optokinetic rotation alters the direction of VOR induced by pitch in the dark by adding an adaptive horizontal component. This chapter concentrates on cross-axis plastic changes in VOR direction produced either by pairing pitch rotation with horizontal optokinetic rotation[6] or by pairing horizontal rotation with pitch optokinetic rotation.[7] Because these plastic responses occur in a direction orthogonal to the normal VOR, we have been able to measure their properties with high fidelity, obtaining data that have led us to new models of the plastic process. In this chapter we will first discuss the three-dimensional transformations that take place in the normal VOR, which must be understood before considering VOR plasticity. We will then present our current model of VOR plasticity followed by the data that led to its generation.

[a] Work in the authors' laboratories is supported by grants from the National Institutes of Health (EY 05289, EY 06485, EY 07342, NS 17489, and NS 21015) and by a contract from the Office of Naval Research (N00014-90-J-1822).

[b] Address for correspondence: Department of Physiology, Northwestern University Medical School, 303 East Chicago Avenue, Ward Building 5-311, Chicago, Illinois 60611.

THREE-DIMENSIONAL TRANSFORMATIONS IN VOR

The most direct pathways mediating the VOR consist of three neurons: a semicircular canal afferent fiber, a vestibuloocular relay neuron (VORN), and an extraocular motoneuron. Dashed lines in FIGURE 3 indicate the three-neuron arc that connects the right horizontal canal to left lateral rectus motoneurons. Less direct pathways involving neurons in cerebellum, prepositus hypoglossi nucleus, reticular formation, and other structures are in parallel with the three-neuron arcs. FIGURE 1 illustrates the technique we and others use to quantify the spatial transformations that occur in these pathways. Activity of a neuron, or in this case of an extraocular muscle, is recorded during sinusoidal rotation in many different planes in space. As illustrated by the data traces, when the rotation frequency is ≥ 0.5 Hz, VORNs[8] and eye muscles[9,10] show sinusoidal modulation, the amplitude of which varies as the cosine of the angle between the axis of the rotation and a best axis, which we refer to as the "maximum activation direction" or MAD. Responses of semicircular canal afferents exhibit similar cosine tuning.[11] Because of the approximately linear behavior of the system, its spatial properties can be completely described by finding MADs of its neural elements. We then seek to understand how VOR circuitry transforms the MADs of canal afferents into MADs of extraocular muscles that are appropriate to generate a compensatory VOR.

FIGURE 2A shows the MADs of the elements of the three-neuron VOR arc in the cat. Comparing the MADs of horizontal canals and horizontal rectus eye muscles reveals that they are nearly perfectly aligned. Thus connecting one to the other will suffice to generate the observed responses.[9] This is not the case for vertical canals and eye muscles. Relative to the canal MADs, those of oblique muscles are shifted toward the roll axis, requiring additive convergence of signals from anterior and posterior canals on the same side to produce the required spatial transformation. A number of the VORNs in FIGURE 2A have MADs aligned with those of oblique muscles, which indicates that they receive the convergent input required to mediate this transformation.[8,10] However, none of the VORNs had MADs aligned with those of vertical rectus muscles, which require convergent input from bilateral posterior or anterior canals. We believe that this convergence may arise from the bilateral projections of VORNs to vertical rectus motoneurons.[12,13] Thus the three-neuron arc circuitry is sufficient to explain the spatial transformations underlying the VOR. Part of the process involves convergence of canal signals upon VORNs, part the divergent projections of VORNs to motoneurons.

Can the responses and projections of VORNs also account for cross-axis VOR plasticity? Perhaps so, as shown by FIGURE 2B. In this case the MAD of an anterior canal-type vertical VORN was measured and then the animal was exposed to 40 min of horizontal rotation paired with vertical optokinetic rotation as described by Harrison *et al.*[7] Initially the VORN had little response to horizontal rotation (that is, its MAD had little upward or downward component in the front view). After adaptation, the MAD had a downward direction indicating that the neuron was now excited by rightward horizontal rotation.[14] Activation of anterior canal VORNs generates upward eye movements. Thus the VORNs response was appropriate to contribute to the adapted VOR in which rightward rotation was coupled to an upward eye movement.

MODEL OF VOR PLASTICITY

FIGURE 3 is a model of the neuronal substrates of the cross-axis adaptation that occurs when downward pitch rotation, which activates the anterior canals, is coupled

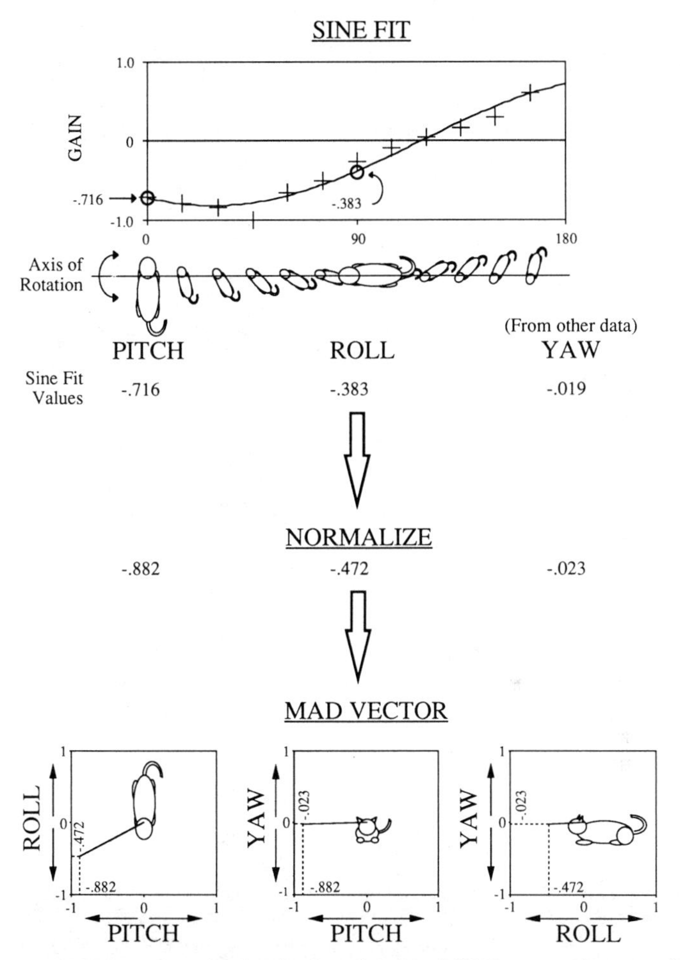

FIGURE 1. Calculation of maximal activation direction (MAD) vector. At top, a sinusoid is fitted to gains of left inferior rectus muscle EMG responses obtained by rotating the cat about 12 earth horizontal axes whose angle with the cat's longitudinal axis is plotted on abscissa. Cartoons show the orientation for each rotation. The sinusoid yields estimates of the muscle's response to pitch and roll rotations. An estimate of response to yaw is derived from other data series. The vector formed by these three values (pitch, roll, yaw) is normalized to a length of 1.0 to obtain the MAD vector, which is plotted in three views of the cat at the bottom. According to the "right hand rule," if the right thumb is placed along the vector, the curl of the right fingers indicates the direction of rotation (nose up and slightly to the right in this case) that maximally activates the neuron or muscle.

with leftward motion of a visual pattern. The model has been simplified by showing only the pathways that connect three receptors of the right labyrinth with lateral, but not medial, rectus motoneurons. It contains three categories of pathways. The first are the three-neuron arcs that connect the right horizontal and anterior semicircular canals (RHC and RAC) to horizontal and vertical extraocular motoneurons. Only the first stage of the vertical three-neuron arc is shown; solid lines indicate that this pathway is activated by the stimulus. Dashed lines indicate that the RHC and the contralaterally projecting horizontal VORN, which does not receive inhibition from the flocculus,[15,16]

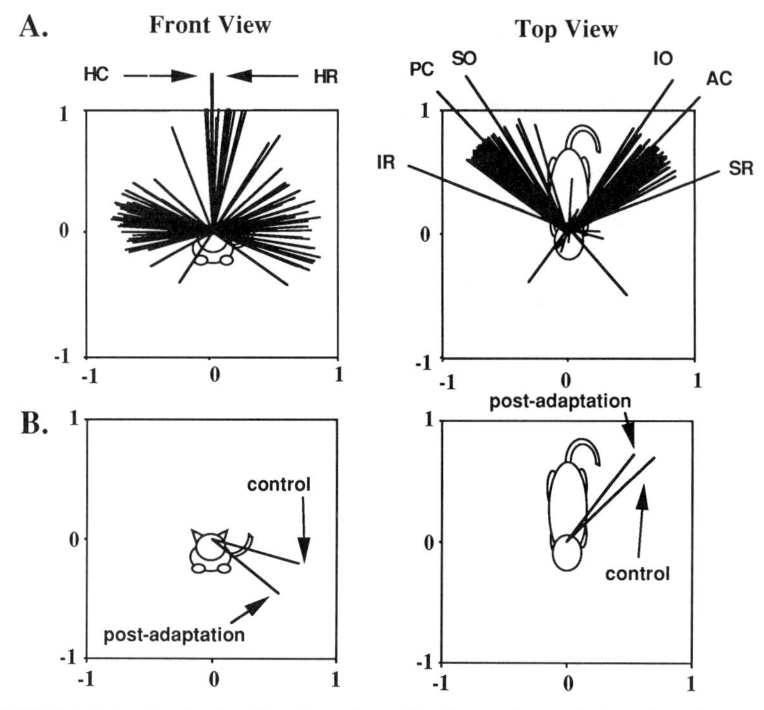

FIGURE 2. Maximal activation directions of semicircular canals, vestibuloocular relay neurons, and extraocular muscles. (**A**) MADs of VORNs compared with those of ipsilateral canals (HC, PC, AC) and eye muscles (HR, SR, IR, SO, IO) receiving primary excitatory connections from these canals. Note the alignment of some neuron MADs with MADs of oblique eye muscle, but the lack of appropriate signals to drive the vertical recti. (**B**) AC-type VORN that developed a large change in horizontal sensitivity during 40 min of coupling yaw rotation with vertical optokinetic rotation.

are not activated. The diamond-shaped boxes in the pathways connecting RHC to vertical VORNs and RAC to horizontal VORNs indicate that these pathways may involve additional relays in vestibular or fastigial nuclei. Brainstem connections between otolith organs and lateral rectus motoneurons are not shown because these would not be expected to produce any net modulation during pitch rotations.

The second category of pathway involves relays in the prepositus hypoglossi nucleus and adjacent marginal zone of the medial vestibular nucleus. These regions contain neurons that carry a horizontal eye position signal,[17,18] which we postulate is generated

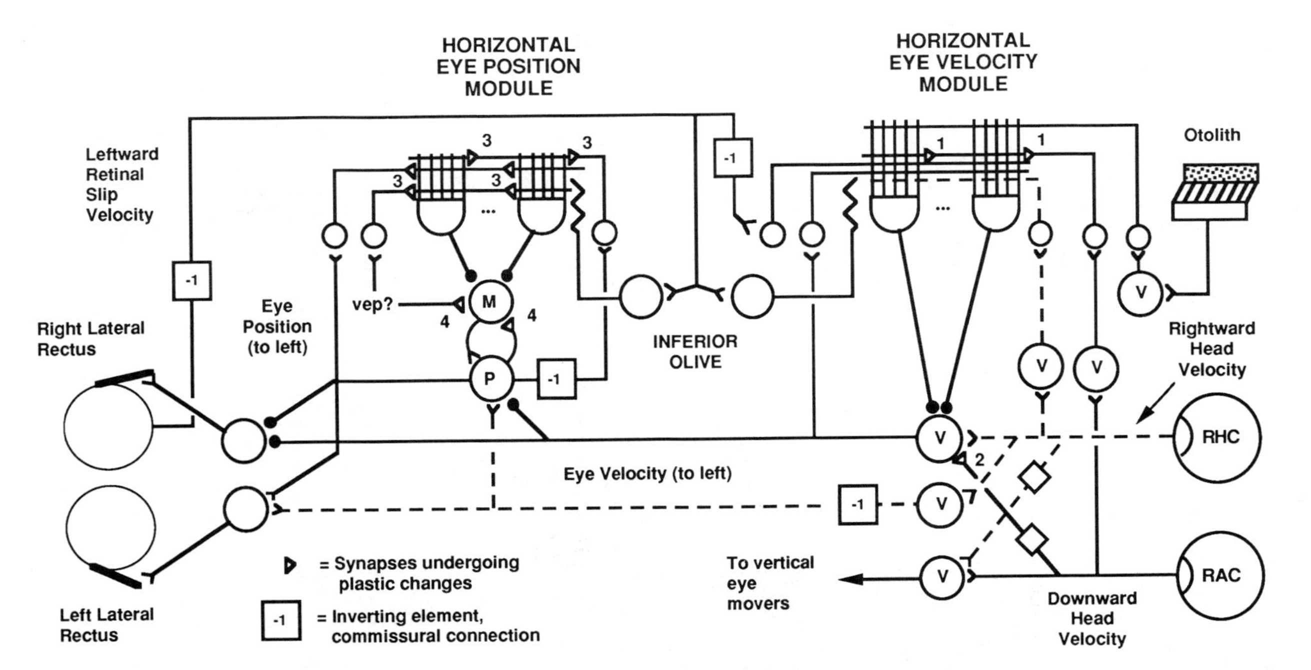

FIGURE 3. Circuit-level model of VOR plasticity. Diagram shows circuits involved in coupling pitch rotation to horizontal eye movement so that eyes move to right during downward pitch. For simplicity, model shows only pathways controlling lateral rectus muscles. At bottom are three neuron arc pathways containing right horizontal and anterior canals (RHC, RAC), vestibular neurons (V), and motoneurons. Coupling of RAC signals elicited by pitch rotation to horizontal VORNs is mediated by a horizontal eye velocity module, involving flocculus granule and Purkinje cells and their target neurons in vestibular nuclei. Adaptive control of the phase of this cross-coupled response is produced by altering output of the horizontal eye position module, which changes the proportion of eye velocity and position signals reaching horizontal rectus motoneurons. This module includes flocculus granule and Purkinje cells and their target neurons in the marginal zone of the medial vestibular nucleus (M) and prepositus hypoglossi nucleus (PH). Also shown is a possible convergent input of vertical eye position or velocity signals (vep?) to neurons in this module. Synapses at which adaptive changes occur are shown as open triangles. Small circles are flocculus granule cells. Diamond-shaped boxes in afferent pathways coupling RHC to vertical VORNs and RAC to horizontal VORNs indicate that these connections may involve additional relay neurons in vestibular or fastigial nuclei.

by positive feedback interconnections between them. As shown by Skavenski and Robinson,[19] extraocular motoneurons must receive a combination of signals related to velocity and position to match their discharge to the dynamic properties of the oculomotor plant and thereby generate vestibuloocular eye movements that are exactly in phase with (and opposite in direction to) head rotation. The medial vestibular and prepositus nuclei are thought to generate the horizontal eye position signal, some of which is routed back to VORNs (not shown), some of which is sent directly to lateral rectus motoneurons and internuclear neurons in the abducens nucleus.[20] As shown here, the horizontal eye position module consists of coupled medial vestibular and prepositus nucleus neurons that receive input from horizontal VORNs and send excitatory projections to ipsilateral and inhibitory projections to contralateral lateral rectus motoneurons. Also shown is a possible input of convergent vertical eye position or velocity signals.

The final category consists of pathways involving the inferior olive and cerebellar flocculus that modulate the activity of the two brainstem circuits. Within the horizontal eye velocity module are Purkinje cells of the middle zone of the cerebellar flocculus that receive climbing fiber signals related to leftward retinal slip velocity and mossy fiber signals related to a variety of vestibular, retinal slip, and eye velocity (efference copy) signals.[16,21,22] Purkinje cells process these mossy fiber inputs to generate a motor command signal, related in this case to the correction necessary to obtain the desired horizontal eye velocity. Analogous processing takes place in the horizontal eye position module where Purkinje cells process mossy fiber inputs related to activity of medial and prepositus nucleus circuits to generate an inhibitory output appropriate to regulate the eye position signal generated by those circuits. The actual manner in which Purkinje cells process mossy fiber input is probably highly nonlinear. In a recent model of limb movement control,[23] Purkinje cells were assumed to be turned on and off in a bistable fashion, inspired by the plateau potentials seen in dendritic recordings.[24,25] As pointed out in the cited paper, a more realistic model of Purkinje cells should treat them as multistable elements, assuming that local dendritic regions each behave in a bistable manner and sum to produce a net somatic depolarization. Recent calcium imaging results lend support to the concept that local dendritic regions exhibit plateau potentials that can be controlled independently.[26] From a functional standpoint, multistability would give Purkinje cells the capacity to operate more like function generators than conventional feedback elements.

In our model, as in that of Galiana,[27] plasticity occurs both at parallel fiber synapses in the cerebellar cortex and at synapses along the VOR pathway. The training rules proposed for these two sites are based on current hypotheses regarding the mechanisms that mediate plasticity at the cellular level.

The training scheme proposed for parallel fiber synapses is summarized in the top half of FIGURE 4. It is based on the scheme discussed in Houk *et al.*,[23] which should be consulted for additional references. Activity in a parallel fiber releases the neurotransmitter glutamate, which binds to quisqualate (Q) receptors located in the adjacent dendritic spine.[28] This reaction leads to excitatory current that causes local depolarization of a dendrite. When other inputs are present to make the local dendritic depolarization sufficiently strong, calcium channels are opened leading to the sustained plateau potentials discussed above. The local dendritic depolarization then spreads to the soma where it combines with depolarization from other dendrites to control somatic depolarization and Purkinje cell discharge. This chain of events is assumed to control the immediate response of the neuron. A slower process controlled by calcium levels in the dendrite and by local chemical signals (such as diacylglycerol) regulates the strength of parallel fiber synapses.[28,29] According to our scheme, the local chemical signals serve to render a synapse eligible for modification. When both the synaptic

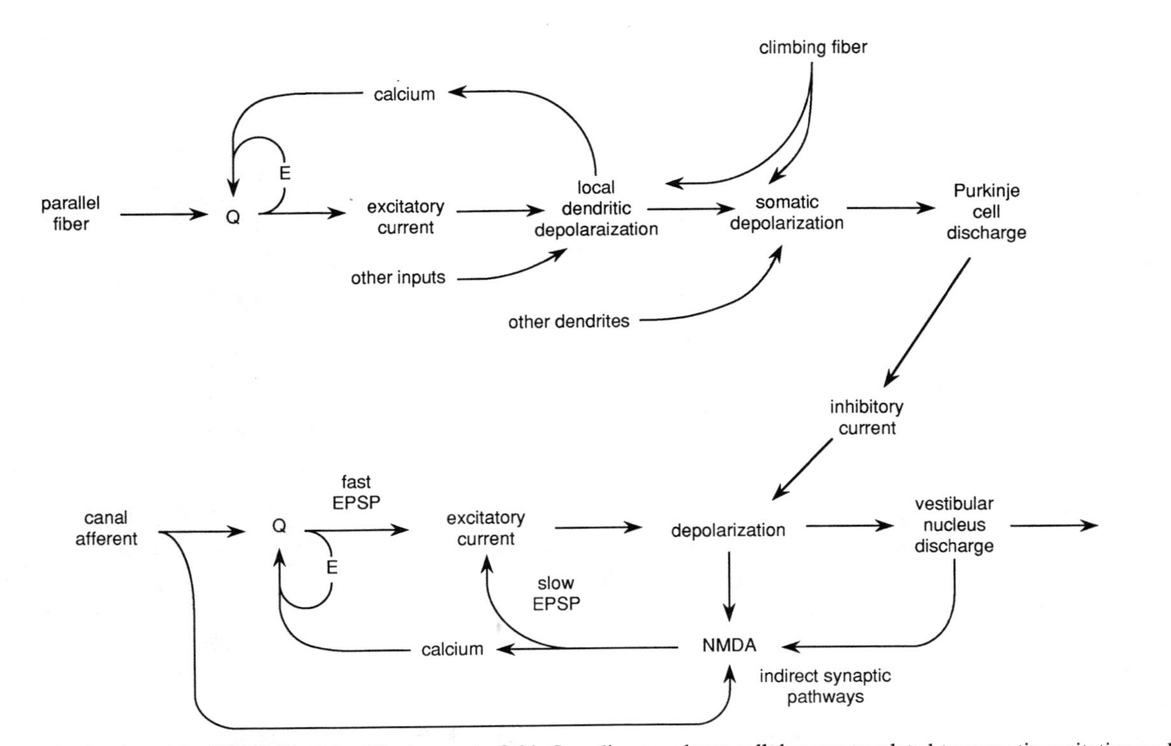

FIGURE 4. Cellular level model of VOR plasticity. The top part of this flow diagram shows cellular events related to synaptic excitation and plasticity in Purkinje cells, whereas the bottom part shows the events postulated to occur in vestibular neurons. Climbing fibers are assumed to provide specific training signals that rapidly adjust the weights of parallel fiber synapses on Purkinje cells. In contrast, we postulate a use-dependent rule with a slower time course for the adjustment of synaptic weights of canal afferents on vestibular neurons. Via the latter mechanism, the use patterns that Purkinje cells regularly impose on the vestibular neurons become established as habitual reflexes.

eligibility signal (designated E in FIG. 4) and calcium are present, parallel fiber synapses are modified by a protein kinase that desensitizes Q receptors, thus depressing synaptic transmission. Calcium levels appear to be controlled both by the local postsynaptic responses in dendrites and by the occurrences of climbing fiber action potentials. The simultaneous presence of both influences greatly enhances plateau potentials and seems to be required to produce long-term depression of the parallel fiber synapses.[24] On the other hand, long-term facilitation of parallel fiber synapses occurs when the synapse is activated in the absence of climbing fiber activity, which suggests a requirement of both eligibility and a low level of calcium. The time constant of these plastic changes is hypothesized to be 0.5 hr, corresponding to the rapid phase of VOR plasticity.[7,30,31]

The training scheme proposed for vestibular neurons and other sites along the VOR pathway is summarized in the bottom half of FIGURE 4. Activity in a semicircular canal afferent is shown to release an excitatory amino acid transmitter (probably glutamate) that activates Q receptors on vestibular neurons.[32,33] Indirect pathways to vestibular neurons are also present, and they appear to rely more on the *N*-methyl-D-aspartate (NMDA) subtype of excitatory amino acid receptor. The fast and slow excitatory postsynaptic potentials (EPSPs) produced by these reactions cause depolarization of the dendritic membrane. The dendritic trees of vestibular neurons are less branched than those of Purkinje cells, and we have lumped dendritic and somatic zones into a single site of depolarization in FIGURE 4. Membrane potential is also influenced by inhibitory input from Purkinje cells, and the net depolarization controls vestibular nucleus discharge. Because there is evidence for long-term potentiation (LTP) in the vestibular nuclei,[34] the mechanism proposed for synaptic plasticity is adapted from the hypothesis for LTP discussed in Brown *et al.*[35] According to this scheme, long-term facilitation of Q receptors depends on two factors: the synaptic activation of NMDA receptors and the level of postsynaptic depolarization in the vicinity of NMDA-activated channels. When this dual dependence is satisfied, these unique voltage- and transmitter-dependent channels permit an influx of calcium. Via a protein kinase mechanism different from the one operative at parallel fiber synapses, the sensitivity of Q receptors is enhanced.[36] This mechanism is thought to give rise to a use-dependent, Hebbian type of synaptic plasticity whereby synapses that are active when the postsynaptic neuron is active become strengthened,[35] and synapses active when the neuron is inactive become weakened.[37] We propose that this plasticity has a time course corresponding to the 8-12-hr time constant for the slow phase of VOR plasticity.[2]

Hebbian synapses are capable of organizing input systems for efficient feature detection.[38] However, we do not believe that a simple use-dependent mechanism would be capable of controlling the interactions of the motor system with the environment in a very intelligent fashion. In essence, Hebbian synapses would be effective in reinforcing habits. However, useful habits would have to be imposed by a more intelligent system that is responsive to training signals based on sensory feedback from the environment. We propose that the intelligent interactions with the environment are mediated by Purkinje cells in the cerebellar cortex, and that those patterns Purkinje cells regularly impose on the more direct VOR pathways are progressively adopted as useful habits by a Hebbian training scheme operating at the latter synapses. According to the scheme in FIGURE 4, Purkinje cells impose useful patterns by withholding inhibition, and these become habitual reflexes as a consequence of enhanced postsynaptic depolarization, which strengthens canal inputs that are simultaneously active. In contrast, existing habits that are not useful would be inhibited by Purkinje cell inhibition, and this would promote a weakening of synapses that are attempting to excite the inhibited VOR neurons. As a new habit is acquired by the VOR pathway, its dependence on cerebellar cortical control would diminish.

The adaptive mechanisms operative in the cerebellar cortex are seen to be intelligent as a consequence of two factors: 1) the enormous number and diversity of parallel fiber signals available as potential input to each Purkinje cell and 2) the specialized training inputs from climbing fibers that signal specific evaluations of motor performance. Although the present model includes five types of parallel fiber input to floccular Purkinje cells and presumes variety in the nonlinear relations of each of these types, it undoubtedly underestimates the true diversity of the input vector to Purkinje cells. The specificity of training signals is provided by the directional tuning characteristics of retinal-slip climbing fiber input to different categories of Purkinje cells. This combination of diversity of potential input with specific training signals seems ideal for the intelligent control of motor outflow.

When pitch rotation is paired with horizontal image motions, the circuits in the model undergo plastic changes that couple pitch rotations to horizontal eye movements. The sequence of events can be described as follows. 1) Downward pitch excites afferents from the RAC, which generate vertical vestibuloocular eye movements and also activate vestibular neurons that project to the flocculus as mossy fibers. The granule cells activated by these mossy fibers send their parallel fibers to both vertical and horizontal eye movement regions of the flocculus (Ito, personal communication). 2) Leftward image motion activates accessory optic pathways that excite neurons in the left inferior olive that send climbing fibers to the middle (horizontal) zone of the right flocculus. 3) Within the horizontal eye velocity module, climbing fiber spikes will coincide with parallel fiber activity produced by the RAC, causing the synaptic weight of their synaptic excitation of Purkinje cells to decrease (plastic synapse 1 in FIG. 3). This change will occur with a time constant of 0.5 hr. 4) This synaptic change will decrease the inhibition of ipsilaterally projecting horizontal VORNs during downward pitch. 5) These horizontal VORNs will therefore fire during downward pitch, causing a leftward eye movement by their inhibition of right lateral rectus motoneurons. 6) If the corrective Purkinje inhibition continues to act on the horizontal VORN for a long enough time, it will cause plastic changes in synaptic input reaching that VORN from the RAC (plastic synapse 2). These changes, which have a time constant of 8-12 hr, would cause the same cross-coupled vestibuloocular eye movements that were produced by the flocculus signal. 7) The flocculus would therefore receive less climbing fiber input correlated with downward pitch and its synapses would tend to return to their initial state so that after a day of training, the adaptive response would be mediated primarily by brainstem pathways.

If the cross-axis plastic eye movement is exactly in phase with the moving visual pattern, no change will occur in the horizontal eye position module because climbing fiber activity that is in phase with the velocity of retinal slip will be 90° out of phase with the eye position signals reaching Purkinje cells via parallel fibers. If there is an error in phase, however, climbing fiber activity will tend to coincide with either leftward or rightward position signals carried by parallel fibers, thereby reducing the synaptic weight of the signal with which they overlap (plastic synapse 3). The resulting change in activity of medial and prepositus nuclei neurons will shift the ratio of eye position and velocity signals reaching motoneurons thereby reducing the phase error. As in the eye velocity module, prolonged action of the corrective Purkinje inhibition on target neurons in the marginal zone of the medial nucleus would give rise to slower adaptive changes in synaptic weights of afferent signals acting on those neurons (plastic synapse 4). In addition to horizontal eye position signals expected in this module, the model also shows a possible convergent vertical eye velocity or position signal acting on the flocculus and medial nucleus neurons of this module. The rationale for this signal will be discussed below.

EXPERIMENTAL EVIDENCE THAT SUPPORTS THE MODEL

The model is designed to be compatible with the known physiological properties of VOR adaptation. One such property is the fact that cerebellar lesions block plasticity in mammals even when training continues for many days.[3,39] This indicates that the flocculus is either 1) a relay for adaptive changes occurring before it, 2) the site of plastic changes, or 3) the generator of error signals that elicit plasticity elsewhere. Our model hypothesizes that options 2 and 3 are true for short- and long-term plasticity, respectively. Recent observations of VOR plasticity in cerebellectomized fish[40] can be explained by assuming that after a long recovery period, signals other than flocculus output can induce plastic changes at the brainstem site proposed in the model. Flocculus lesions also impair the adjustment of the eye position component of horizontal saccadic eye movements,[41] which is consistent with the proposed role of the flocculus in the eye position module of the model.

The role of the flocculus in VOR plasticity is a matter of vigorous debate. Ito *et al.*[42] have shown that vestibular input to flocculus can be altered by long-term depression induced by pairing climbing fiber and vestibular mossy fiber input. He asserts that this is the mechanism underlying changes in VOR gain in animals that wear magnifying, reducing, or reversing lenses (*cf.* Ito[43] and Nagao[44,45]). After analyzing changes in behavior of gaze velocity Purkinje cells during VOR in the dark in monkeys with long-term changes in VOR gain, Miles *et al.*[46] concluded that such changes are secondary to alteration in eye velocity feedback and do not cause the adaptive change in VOR gain, a conclusion that is in line with the short latency of the gain change (19 msec) under these conditions.[47] We believe that a key to resolving this controversy lies in the fact that VOR plasticity occurs in two distinct phases: short-term plasticity with a time constant of ~0.5 hr and long-term plasticity with a time constant of 8-12 hr.[2,7,30,48] FIGURE 5B shows the time course of onset of short-term plastic changes produced by coupling vertical image motion to horizontal rotation. The controversy concerning the flocculus can be reconciled if we hypothesize, as shown in FIGURE 3, that *short-term plasticity occurs in the flocculus (site 1) and that the persistence of the flocculus correction signal on brainstem neurons causes long-term plastic changes in brainstem (site 2).* Once the slower learning had occurred, the synaptic weight would decrease toward normal at site 1, yielding the situation observed by Miles *et al.*[46] and Lisberger.[47] The two sites of learning are also compatible with other properties of VOR plasticity described below. Galiana *et al.*[49] proposed that learning might occur in the vestibular commissural system rather than at sites 2 and 4 in FIGURE 3. This hypothesis is not consistent with Lisberger's[47] observation that the largest changes occur in neurons that receive flocculus inhibition since Sato and Kawasaki[15] have shown that these neurons do not receive commissural inhibition.

Latency

As illustrated in FIGURE 5C, the latency of vertical eye movements induced by yaw rotation in darkness after 2-3 hr of coupling horizontal rotations with vertical pitch image motions is typically 30-40 msec, although exceptional cats reach latencies as short as 20 msec.[50] The former correspond to the 30-msec latencies Snyder and King[51] observed for short-term plasticity of cat vertical VOR. The latter resemble the 19-msec latencies Lisberger[47] obtained by adapting monkeys to 2× lenses for >7 days (long-term plasticity). Nineteen milliseconds appears to be too short to allow involvement of cerebellar circuitry, suggesting that plastic changes in brainstem elements are involved. On the other hand, latencies ≥ 30 msec that characterize short-term plasticity are consistent with cerebellar involvement as proposed by Ito.[43] Once again the data can be explained by proposing that *short-term learning occurs in the flocculus and that continued correction of brainstem circuits by flocculus causes long-term plasticity in brainstem neurons.*

Context Dependency

If a cat is trained by placing it right ear down and pairing rotation in its sagittal plane (horizontal to the earth) with left-right optokinetic image motion, a cross-coupled adaptive response develops that is maximal when the cat is right ear down and smaller when the same sagittal rotation is applied in different orientations with respect to gravity.[52] As illustrated in FIGURE 6, it is even possible to intersperse the training with alternating periods in which the cat lies on its left ear and receives the same sagittal rotation coupled with oppositely directed right-left image motion in which case the cross-axis plastic responses in left and right ear down positions are in opposite directions even though the sagittal plane rotation generates identical patterns of semicircular canal activation.[53] These data are best explained by proposing that *otolith organs provide posture- or context-dependent signals that cause a switching between two independently learned adaptive responses.* We postulate that this occurs in the cerebellar cortex where two sets of Purkinje cells are used to control the two responses. One set, which is turned on by otolith afferents active when the cat is on its left side, would learn to respond to the appropriate combination of vestibular and eye movement signals. The other set, which is turned on by otolith afferents that are active when the cat is on its left side, would learn to respond to a different combination of vestibular and eye movement signals. At intermediate positions some cells that are maximally turned on in a given position would drop out, giving intermediate responses.

We thus hypothesize that complicated properties such as context dependency of VOR plasticity take advantage of the great diversity of parallel fiber signals that are available as potential inputs to individual Purkinje cells. Through weight modification of parallel fiber synapses, different sets of Purkinje cells would become specialized for handling different situations. The adaptive process might teach Purkinje cells both to respond with appropriate sensitivity to time-varying inputs from vestibular and eye-movement mossy fibers and also to respond in a gated, or switched, fashion to tonic inputs that register different states of the body and environment. The latter process would allow Purkinje cells to recognize those specific contexts that require their particular computation. We predict that complicated responses such as these would not recover after appropriately placed cerebellar cortical lesions. The information processing required to regulate them is probably too complex for them to be transferred to brainstem sites for habitual execution.

Adaptive Control of VOR Phase

If visual pattern motion in the paradigm coupling pitch rotation to horizontal eye movement is phase shifted so that it lags motion of the rotator, the plastic vertical eye

A

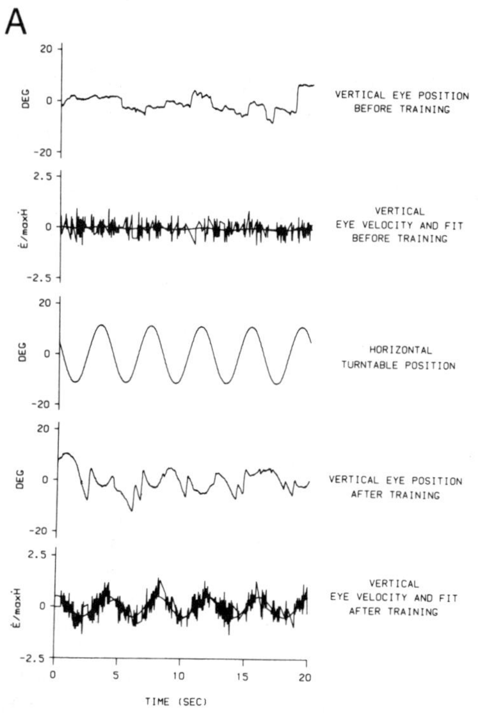

DEG
20
0
-20
VERTICAL EYE POSITION
BEFORE TRAINING
Ė/maxḢ
2.5
0
-2.5
VERTICAL
EYE VELOCITY AND FIT
BEFORE TRAINING
HORIZONTAL
TURNTABLE POSITION
VERTICAL EYE POSITION
AFTER TRAINING
VERTICAL
EYE VELOCITY AND FIT
AFTER TRAINING
0
5
10
15
20
TIME (SEC)

B

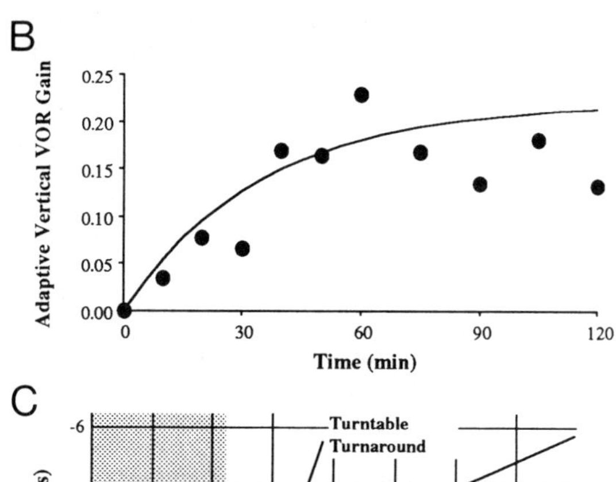

Adaptive Vertical VOR Gain
0.25
0.20
0.15
0.10
0.05
0.00
0
30
60
90
120
Time (min)

C

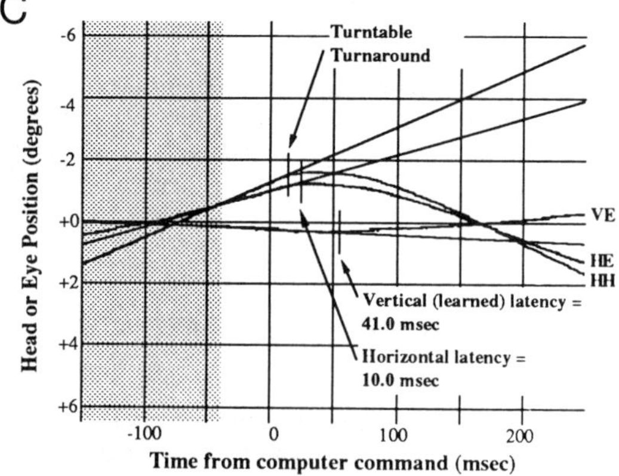

Head or Eye Position (degrees)
-6
-4
-2
+0
+2
+4
+6
Turntable
Turnaround
VE
HE
HH
Vertical (learned) latency =
41.0 msec
Horizontal latency =
10.0 msec
-100
0
100
200
Time from computer command (msec)

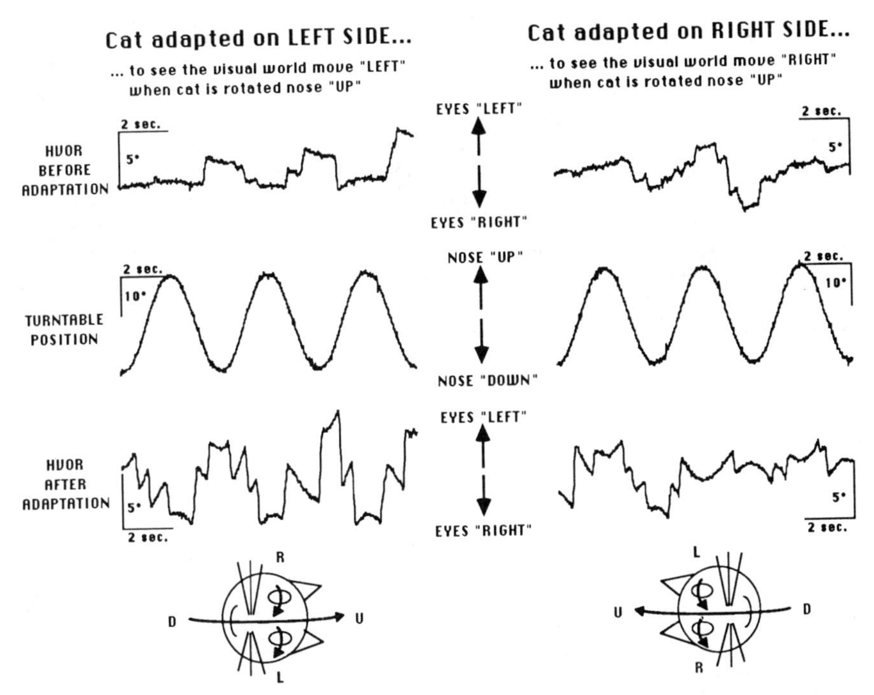

FIGURE 6. Context-dependent adaptation of VOR. Traces show horizontal eye position during on-side pitch rotation in darkness before (top traces) and after (bottom traces) adaptive training. Center trace is turntable position. Figurines show pairing of vestibular and optokinetic stimuli that was presented during alternate 10-min periods in which the cat was positioned with either its left or right ear down. When the left ear was down, upward rotation was coupled with leftward image motion. When the right ear was down, upward rotation was coupled with rightward image motion. Before training the rotation elicited little horizontal eye movement. After 4 hr of alternating training, upward rotation is coupled with leftward eye movement in the left ear down position and rightward eye movement in the right ear down position. (From Baker *et al.*[53]; used with permission.)

FIGURE 5. Time course and latency of adaptive eye movements after 2 hr of cross-axis VOR adaptation. (**A**) Vertical eye movements elicited by horizontal rotation (center trace shows platform angular position) before (upper traces) and after (lower traces) 2 hr of pairing 0.25 Hz, 18° horizontal rotation with 0.25 Hz, 27° vertical (pitch) optokinetic stimulation. Upper trace in each pair is vertical eye position, lower trace is smooth vertical eye velocity after removal of saccades. (**B**) Time course of adaptive change. Circles show gain of vertical eye movement produced by horizontal rotation in total darkness at various times during adaptive training, which consisted of combining horizontal rotation and vertical optokinetic stimulation. Solid line is an exponential curve fitted to data. Its time constant is 30 min. (From Harrison *et al.*[7]; used with permission.) (**C**) Latency of eye movements elicited by sudden change in horizontal velocity from +18°/sec to −18°/sec. Data traces indicate horizontal angular position of head (HH) and eye (HE) and vertical angular eye position (VE). A straight line is fitted to each trace in the period from the end of the hashing to time 0 at which the computer command controlling turntable velocity changed. Vertical lines indicate points at which traces deviate significantly from the fitted lines. Latencies are times from turntable turnaround to initial deviation of horizontal (10-msec) and vertical (41-msec) eye traces.

movements develop a phase that matches this phase lag.[54] This finding, which is illustrated in FIGURE 7, has led us to incorporation of a plastic eye position module in our model. In considering how a single retinal slip velocity error signal (slip velocity) could be used to adjust *both* gain and phase, we have found that the product of slip velocity and head velocity is primarily sensitive to errors in VOR *gain* while the product of slip velocity and eye or head position is primarily sensitive to *phase* errors.[55] Recent data[17,18] suggest that horizontal eye position signals are present in the medial vestibular and prepositus nuclei, which receive input from the flocculus.[56] Thus by including in the circuit an eye position module involving flocculus interactions with these nuclei, we arrive at a model that can replicate our observations of VOR plasticity. In the model coincidence of climbing fiber input with velocity or position-related signals carried by parallel fibers mediates two-stage learning. First, slip velocity signals con-

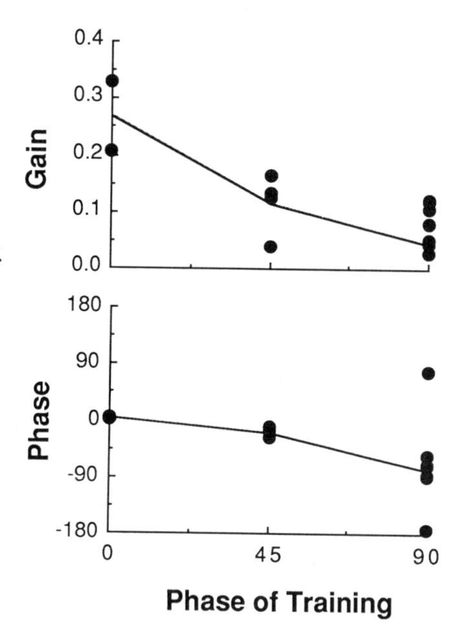

FIGURE 7. Adaptive modulation of VOR phase. Graphs indicate how gain and phase of vertical eye movements elicited by horizontal rotation varied with the phase lag between the horizontal rotation and vertical optokinetic pattern motion during 2 hr of cross-axis training.

veyed by climbing fibers cause adjustments in plastic synapse 1 to compensate for gain errors or in synapse 3 to adjust phase by changing the ratio of eye position and velocity signals reaching extraocular motor nuclei. Later, altered Purkinje cell discharge caused by these changes induces changes at sites 2 or 4, which have a similar effect on gain and phase at the brainstem level.

Two options exist for incorporating a position module in our model of cross-axis plasticity. One is to assume that when phase-shifted horizontal image motions are used to produce a horizontal VOR response to pitch rotation, alterations occur in the circuits that control integration of horizontal velocity signals to produce horizontal eye position signals. In this case, cross-axis training would alter the dynamics of the horizontal VOR. Alternatively, the horizontal eye position module might receive afferent signals related to vertical eye position or velocity, which normally cancel out and do not cause any horizontal eye motion. The adaptation process could cause these to become coupled to neurons that generate horizontal eye position signals sent to medial and lateral rectus motor nuclei. In this case the adaptation would not alter dynamics of the horizontal VOR.

Neural Recordings

Lisberger[47] showed that long-term plasticity leads to large, short latency changes in the response of the population of vestibular neurons that receive flocculus inhibition (FTNs). Although the projections of these FTNs are not identified, we have shown that some VORNs exhibit large changes in their MADs during cross-axis plastic change in VOR direction (*cf.* FIG. 2B). Others show no change. The likely conclusion is that large plastic changes occur in signals carried by FTNs that project to extraocular motor nuclei. Collectively these findings are compatible with the structure of the velocity module of the model where VOR adaptation involves short-term plastic changes in flocculus or long-term changes in synaptic input to VORNs that receive flocculus inhibition. No data are available on behavior of neurons that are likely to form part of the position module during VOR adaptation.

The MADs of VORNs shown in FIGURE 2A have another property that is required by the model. This is the significant convergent horizontal input to vertical VORNs and vertical input to horizontal VORNs (revealed by the deviation of their MAD vectors from an exactly vertical or horizontal orientation in the front view). In the normal situation, corresponding to FIGURE 2A, these cross-axis signals are approximately evenly weighted in the positive and negative directions so that they cancel out. During long-term cross-axis training, however, they would provide the substrate for changes in synaptic weights (at site 2 in FIG. 3) which would lead to a cross coupling of vertical and horizontal pathways at the brainstem level.

CONCLUSIONS

The great capacity of the cerebellar architecture for pattern recognition was recognized some years ago by Marr[57] and Albus.[58] More recently the pattern recognition capacity of Purkinje cells was combined with new ideas about adjustable pattern generation to explain how motor programs for limb movements might be represented in the CNS.[23,59] The present model attempts to extend these ideas into the domain of eye movement control.

The data summarized above require a model in which adaptive plasticity occurs in modules that control both eye velocity and eye position. It is not necessary, however, to insist that only pure position and velocity modules exist. In reality there are probably a range of flocculus-brainstem circuits that exhibit varying combinations of eye position and velocity (*cf.* Noda and Warabi[60]). Relatively little is known about the physiology of circuits that generate eye position signals. In the horizontal system, this function is likely played by medial vestibular and prepositus nucleus neurons whose connections resemble those in FIGURE 3.[18] The corresponding circuits for generating vertical eye position are less well characterized, although it appears that loops connecting the interstitial nucleus of Cajal and vestibular nuclei are somehow involved.[61] A bit of information is available concerning the spatial properties of the latter, which exhibit a wider range of MADs than VORNs.[62] Nothing is known about the spatial properties of the horizontal eye position system. We thus cannot be sure how cross coupling of vertical and horizontal signals would occur in this system.

Both the dual time constants with which the adaptive changes in the VOR develop and data indicating that their latency is shorter after long-term adaptation are consistent with the idea that short-term plasticity occurs within indirect, transcerebellar pathways while long-term plasticity involves changes in relatively direct brainstem arcs. We and Galiana[27] therefore prefer the two-site adaptation hypothesis, but other explanations exist. The two time courses could reflect changes at the same site mediated by different cellular processes. The shortening of latency could be explained by invoking a threshold process that would be reached sooner by larger adaptive components that develop with prolonged training. We nevertheless feel that our hypothesis is a good basis for developing neuronal recording studies that will resolve the issue. We have suggested possible cellular processes that could mediate changes in synaptic efficacy at the two sites.

We are also attracted to the idea that the cerebellum is the site of complex, context-dependent plasticity utilizing circuitry that can function as a neural pattern recognition and generation system. Previous work demonstrated how pattern-generating modules based on the anatomy and physiology of the intermediate cerebellum can learn to recognize visual targets and execute accurate limb movements to them.[23,63] Here we outline how similar concepts might be combined with the anatomy and physiology of the flocculus to explain context dependency and phase-specific properties of VOR cross-axis adaptation, in addition to more basic aspects of VOR plasticity. Simulation studies are now needed to explore the actual capabilities of the model, and new experiments motivated by the model are needed to test and refine the theoretical concepts.

REFERENCES

1. GONSHOR, A. & G. MELVILL-JONES. 1976. Short-term adaptive changes in the human vestibuloocular reflex arc. J. Physiol. (London) **256:** 361-379.
2. MILES, F. A. & B. B. EIGHMY. 1980. Long-term adaptive changes in primate vestibuloocular reflex. I. Behavioral observations. J. Neurophysiol. **43:** 1406-1425.
3. ROBINSON, D. A. 1976. Adaptive gain control of vestibulo-ocular reflex by the cerebellum. J. Neurophysiol. **39:** 954-969.
4. BERTHOZ, A., G. MELVILL-JONES & A. E. BEGUE. 1981. Differential visual adaptation of vertical canal-dependant vestibulo-ocular reflexes. Exp. Brain Res. **44:** 19-26.
5. SCHULTHEIS, L. W. & D. A. ROBINSON. 1981. Directional plasticity of the vestibulo-ocular reflex in the cat. Ann. N.Y. Acad. Sci. **374:** 504-512.
6. BAKER, J., R. E. W. HARRISON, N. ISU, C. WICKLAND & B. PETERSON. 1986. Dynamics of adaptive change in vestibulo-ocular reflex direction. II. Sagittal plane rotations. Brain Res. **371:** 166-170.
7. HARRISON, R. E. W., J. F. BAKER, N. ISU, C. R. WICKLAND & B. W. PETERSON. 1986. Dynamics of adaptive change in vestibulo-ocular reflex direction. I. Rotations in the horizontal plane. Brain Res. **371:** 162-165.
8. FUKUSHIMA, K., S. I. PERLMUTTER, J. F. BAKER & B. W. PETERSON. 1990. Spatial properties of second-order vestibulo-ocular relay neurons in the alert cat. Exp. Brain Res.: in press.
9. BAKER, J., C. WICKLAND, J. GOLDBERG & B. PETERSON. 1988. Motor output to lateral rectus in cats during the vestibulo-ocular reflex in three-dimensional space. Neuroscience **25:** 1-12.
10. PETERSON, B. W., J. F. BAKER, J. BANOVETZ, K. FUKUSHIMA, W. GRAF, Y. IWAMOTO, A. J. PELLIONISZ, S. I. PERLMUTTER & K. QUINN. 1990. Neuronal substrates of vestibular

reflex control of eye and head movements. *In* Vestibular and Brain Stem Control of Eye, Head and Body Movements. H. Shimazu & Y. Shinoda, Eds. Springer-Verlag. New York, NY.

11. Estes, M., R. Blanks & C. Markham. 1975. Physiologic characteristics of vestibular first-order canal neurons in the cat. I. Response plane determination and resting discharge characteristics. J. Neurophysiol. **38:** 1232-1249.
12. Graf, W., R. A. McCrea & R. Baker. 1983. Morphology of posterior canal related secondary vestibular neurons in rabbit and cat. Exp. Brain Res. **52:** 125-138.
13. Graf, W. & K. Ezure. 1986. Morphology of vertical canal related second order vestibular neurons in the cat. Exp. Brain Res. **63:** 35-48.
14. Perlmutter, S. I., K. Fukushima, B. W. Peterson & J. F. Baker. 1988. Spatial properties of second order vestibuloocular relay neurons in the alert cat. Soc. Neurosci. Abstr. **14:** 331.
15. Sato, Y. & T. Kawasaki. 1990. Operational unit in cat cerebellar flocculus responsible for plane-specific control of eye movement in three dimensional space. *In* Vestibular and Brain Stem Control of Eye, Head and Body Movements. H. Shimazu & Y. Shinoda, Eds. Springer-Verlag. New York, NY.
16. Sato, Y., K. Kanda & T. Kawasaki. 1988. Target neurons of floccular middle zone inhibition in medial vestibular nucleus. Brain Res. **446:** 225-235.
17. Cannon, S. C. & D. A. Robinson. 1987. Loss of the neural integrator of the oculomotor system from brainstem lesions in monkey. J. Neurophysiol. **57:** 1383-1409.
18. McFarland, J. L. & A. F. Fuchs. 1990. The nucleus prepositus and nearby medial vestibular nucleus and the control of simian eye movements. *In* Vestibular and Brain Stem Control of Eye, Head and Body Movements. H. Shimazu & Y. Shinoda, Eds. Springer-Verlag. New York, NY.
19. Skavenski, A. A. & D. A. Robinson. 1973. Role of abducens neurons in vestibuloocular reflex. J. Neurophysiol. **36:** 724-738.
20. Pola, J. & D. A. Robinson. 1978. Oculomotor signals in medial longitudinal fasciculus of the monkey. J. Neurophysiol. **41:** 245-258.
21. Graf, W., J. I. Simpson & C. S. Leonard. 1988. Spatial organization of visual messages of the rabbit's cerebellar flocculus. II. Complex and simple spike responses of Purkinje cells. J. Neurophysiol. **60:** 2091-2121.
22. Miles, F. A., J. H. Fuller, D. J. Braitman & B. M. Dow. 1980. Long-term adaptive changes in primate vestibuloocular reflex. III. Electrophysiological observations in flocculus of normal monkeys. J. Neurophysiol. **43:** 1437-1476.
23. Houk, J. C., S. P. Singh, C. Fisher & A. G. Barto. 1990. An adaptive sensorimotor network inspired by the anatomy and physiology of the cerebellum. *In* Neural Networks for Control. W. T. Miller, R. S. Sutton & P. J. Werbos, Eds.: 301-348. MIT Press. Cambridge, MA.
24. Ekerot, C.-F. 1984. Climbing fibre actions of Purkinje cells—plateau potentials and long-lasting depression of parallel fibre responses. *In* Cerebellar Functions. J. Bloedel *et al.*, Eds.: 268-274. Springer-Verlag. Berlin.
25. Llinás, R. & M. Sugimori. 1980. Electrophysiological properties of in vitro Purkinje cell dendrites in mammalian cerebellar slices. J. Physiol. (London) **305:** 197-213.
26. Sugimori, M. & R. Llinás. 1990. Real-time imaging of calcium influx in mammalian cerebellar Purkinje cells in vitro. Proc. Natl. Acad. Sci. USA **87:** 5084-5088.
27. Galiana, H. L. 1986. A new approach to understanding adaptive visual-vestibular interactions in the central nervous system. J. Neurophysiol. **55:** 349-374.
28. Ito, M. 1989. Long-term depression. Annu. Rev. Neurosci. **12:** 85-102.
29. Crepel, F. C. & M. Krupa. 1988. Activation of protein kinase C induces a long-term depression of glutamate sensitivity of cerebellar Purkinje cells. An in vitro study. Brain Res. **458:** 397-401.
30. Collewijn, H., A. J. Martins & R. M. Steinman. 1983. Compensatory eye movements during active and passive head movements: Fast adaptation to changes in visual magnification. J. Physiol. (London) **340:** 259-286.
31. Khater, T. T., J. F. Baker & B. W. Peterson. 1990. Dynamics of adaptive change in human vestibulo-ocular reflex direction. J. Vestib. Res. **1:** 23-29.

32. Cochran, S. L., P. Kasik & W. Precht. 1987. Pharmacological aspects of excitatory synaptic transmission to second-order vestibular neurons in the frog. Synapse **1:** 102-123.
33. Knöpfel, T. 1987. Evidence for *N*-methyl-D-aspartic acid receptor-mediated modulation of the commisural input to central vestibular neurons of the frog. Brain Res. **426:** 212-224.
34. Racine, R. J., D. A. Wilson, R. Gingell & D. Sunderland. 1986. Long-term potentiation in the interpositus and vestibular nuclei in the rat. Exp. Brain Res. **63:** 158-162.
35. Brown, T. H., A. H. Ganong, E. W. Kairiss, C. L. Keenan & S. R. Kelso. 1989. Long-term potentiation in two synaptic systems of the hippocampal brain slice. *In* Neural Models of Plasticity. J. H. Byrne & W. O. Berry, Eds.: 266-306. Academic Press. San Diego, CA.
36. Collingridge, G. L. & R. J. Lester. 1990. Excitatory amino acids in the vertebrate central nervous system. Pharmacol. Rev. **40:** 143-210.
37. Stanton, P. K. & T. J. Sejnowski. 1989. Associative long-term depression in the hippocampus induced by Hebbian covariance. Nature **339:** 215-218.
38. Linsker, R. 1988. Self-organization in a perceptual network. IEEE Computer: 105-117.
39. Lisberger, S., F. A. Miles & D. S. Zee. 1984. Signals used to compute errors in monkey vestibulo-ocular reflex: Possible role of flocculus. J. Neurophysiol. **52:** 1140-1153.
40. Weiser, M., A. M. Pastor & R. Baker. 1989. Monocular adaptive gain control of the vestibulo-ocular reflex in the goldfish. Soc. Neurosci. Abstr. **15:** 324.
41. Optican & D. A. Robinson. 1980. Cerebellar-dependent adaptive control of primate saccadic system. J. Neurophysiol. **44:** 1058-1076.
42. Ito, M., M. Sakurai & P. Tongroach. 1982. Climbing fiber induced depression of both mossy fiber responsiveness and glutamate sensitivity of cerebellar Purkinje cells. J. Physiol. (London) **324:** 113-134.
43. Ito, M. 1982. Cerebellar control of the vestibulo-ocular reflex: Around the flocculus hypothesis. Annu. Rev. Neurosci. **5:** 275-296.
44. Nagao, S. 1989. Behavior of floccular Purkinje cells correlated with adaptation of vestibulo-ocular reflex. Exp. Brain Res. **77:** 531-540.
45. Nagao, S. 1989. Role of cerebellar flocculus in the adaptive interaction between horizontal optokinetic eye movement response and vestibulo-ocular reflex. Exp. Brain Res. **77:** 541-551.
46. Miles, F. A., D. J. Braitman & B. M. Dow. 1980. Long-term adaptive changes in primate vestibulo-ocular reflex. IV. Electrophysiological observations in flocculus of adapted monkeys. J. Neurophysiol. **43:** 1477-1493.
47. Lisberger, S. 1988. The neural basis for the learning of simple motor skills. Science. **242:** 728-735.
48. Keller, E. L. & W. Precht. 1979. Adaptive modification of central vestibuloocular neurons in response to visual stimulation through reversing prisms. J. Neurophysiol. **42:** 896-911.
49. Galiana, H. L., H. Flohr & G. Melvill-Jones. 1984. A reevaluation of intervestibular nuclear coupling: Its role in vestibular compensation. J. Neurophysiol. **51:** 242-259.
50. Khater, T. T., J. F. Baker & B. W. Peterson. 1990. Horizontal and vertical VOR latency before and after cross-axis VOR plasticity. Soc. Neurosci. Abstr. **16:** 733.
51. Snyder, L. H. & W. M. King. 1988. Vertical vestibuloocular reflex in cat: Asymmetry and adaptation. J. Neurophysiol. **59:** 279-298.
52. Baker, J., C. Wickland & B. Peterson. 1987. Dependence of cat vestibulo-ocular reflex direction adaptation on animal orientation during adaptation and rotation in darkness. Brain Res. **408:** 339-343.
53. Baker, J. F., S. Perlmutter, B. W. Peterson, S. A. Rude & F. R. Robinson. 1987. Simultaneous opposing adaptive changes in cat vestibulo-ocular reflex direction for two body orientations. Exp. Brain Res. **69:** 220-224.
54. Powell, K. D., S. A. Rude, K. J. Quinn, B. W. Peterson & J. F. Baker. 1989. Vestibulo-ocular reflex (VOR) direction adaptation to a phase shifted optokinetic stimulus. Soc. Neurosci. Abstr. **15:** 211.6.
55. Quinn, K., N. Schmajuk, J. Baker & B. Peterson. 1990. Modeling the three-neuron vestibulo-ocular reflex arc: Contribution to eye movement computation. IJCNN Proceedings: II-699-II-704.

56. McCrea, R. A., R. Baker & J. Delgado-Garcia. 1979. Afferent and efferent organization of the prepositus hypoglossi nucleus. Prog. Brain Res. **50:** 653-665.
57. Marr, D. 1969. A theory of cerebellar cortex. J. Physiol. **202:** 437-470.
58. Albus, J. S. 1971. A theory of cerebellar function. Math. Biosci. **10:** 25-61.
59. Houk, J. C. 1989. Cooperative control of limb movements by the motor cortex, brainstem and cerebellum. *In* Models of Brain Function. R. M. J. Cotterill, Eds.: 309-325. Cambridge University Press. Cambridge.
60. Noda, H. & T. Warabi. 1982. Eye position signals in the flocculus of the monkey during smooth-pursuit eye movements. J. Physiol. (London) **324:** 187-202.
61. Fukushima, K., J. Fukushima, C. Harada, T. Ohashi & M. Kase. 1990. Neuronal activity related to vertical eye movement in the region of the interstitial nucleus of Cajal. Exp. Brain Res. **79:** 43-64.
62. Fukushima, K., C. Harada, J. Fukushima & Y. Suzuki. 1990. Spatial properties of vertical eye-movement-related neurons in the region of the interstitial nucleus of Cajal in awake cats. Exp. Brain Res. **79:** 25-42.
63. Barto, A., N. Berthier, S. Singh & J. C. Houk. 1990. Network model of the cerebellum and motor cortex that learns to control planar limb movement. Soc. Neurosci. Abstr. **16:** 1223.

Operantly Conditioned Plasticity in Spinal Cord[a]

JONATHAN R. WOLPAW, CHONG L. LEE, AND JONATHAN S. CARP

Wadsworth Center for Laboratories and Research
New York State Department of Health
and
State University of New York at Albany
Albany, New York 12201

LOCALIZATION OF THE SUBSTRATES OF BEHAVIORAL CHANGE

The first obstacle encountered by efforts to define the neuronal and synaptic substrates of learned changes in behavior is the need to know where these substrates, or memory traces, are located. In all animals to some degree, and in vertebrates to an extreme degree, the complexity and limited accessibility of the CNS render this primary necessity a severe impediment.

The first problem is that most behaviors are produced by intricate and incompletely defined pathways. In response, vertebrate model systems used to study memory traces focus on simple behaviors subserved by relatively well-defined and technically accessible pathways.[1,2]

The second problem is that even simple pathways are influenced at multiple points by other sites in the CNS, so that change in pathway function may actually be maintained by more fundamental modifications elsewhere. The location(s) of plasticity depend on the particular pattern of activity that constitutes the learning experience. Conditioning best suited for laboratory study of memory traces should entail a pattern of neuronal activity that focuses activity-driven change at sites where it can be studied. This requirement, as well as the need for a defined and accessible pathway, was the impetus for attempting to operantly condition the spinal stretch reflex (SSR) and its electrical analogue, the H-reflex.

THE SSR PATHWAY

The SSR, also called the tendon jerk or M1, is the initial response to sudden muscle stretch, and the simplest behavior of which the vertebrate CNS is capable.[3-6] The SSR is produced by the monosynaptic pathway made up of the Ia afferent neuron from the muscle spindle, its excitatory synapse on the alpha motoneuron, and the motoneuron

[a]This work was supported by a grant from the National Institutes of Health (NS22189), by the United Cerebral Palsy Research and Education Foundation, and by the Paralyzed Veterans of America Spinal Cord Research Foundation.

itself. The Ia afferent is excited by sudden muscle stretch, and then excites motoneurons innervating the same muscle and its synergists, causing muscle contraction that opposes the sudden stretch.[b] The H-reflex is comparable to the SSR, except that it is elicited by direct electrical stimulation of the Ia afferent fibers, so that the muscle spindle is removed from the reflex arc.[7] Thus, H-reflex amplitude depends on the Ia synapse and the alpha motoneuron, both of which are subject to influence exerted by descending pathways from supraspinal regions.[8–10]

We set out to develop a conditioning task that would require prolonged change in some part of this descending influence. The hypothesis was that this long-term change in descending activity would produce activity-driven change in the spinal cord, most probably at the Ia synapse and/or the alpha motoneuron itself. This hypothesis was supported by extensive evidence that spinal cord neurons and synapses, like those of supraspinal regions, change during development and in reaction to trauma,[11,12] and thus might also be expected to change with learning.

OPERANT CONDITIONING OF THE SSR PATHWAY

In the initial experiments, the SSR itself was conditioned in the biceps brachii of monkeys (*Macaca nemestrina*) (see references 12 and 13 for review). Subsequent experiments have used the same strategy and comparable design to condition the H-reflex in the triceps surae of the leg, thereby eliminating the muscle spindle from the reflex arc.[14,15]

The H-reflex paradigm is shown in FIGURE 1 (see references 14 and 15 for a full description). Monkeys implanted with chronic recording and stimulating electrodes learn a task that requires prolonged change in neuronal activity influencing the triceps surae SSR pathway, and therefore is likely to lead to an activity-driven change, that is, a memory trace, actually located in this pathway.[13] In this task, the animal maintains a specified level of triceps surae background EMG. At an unpredictable time, bilateral voltage pulses elicit triceps surae H-reflexes from both legs. Under the "control-mode," liquid reward always follows H-reflex elicitation. Under the "up-mode" or the "down-mode," reward follows only if the H-reflex in one leg (the conditioned leg) is above (up-mode) or below (down-mode) a criterion value. The H-reflex in the other leg is simply measured.

The crucial feature of this conditioning task is that the stimulus eliciting the H-reflex occurs at an unpredictable time. Because the reflex is the earliest possible response, the animal can alter its amplitude only by maintaining continual appropriate influence over the monosynaptic SSR pathway. By linking reward to H-reflex amplitude, the up- or down-mode operantly conditions the animal to maintain this influence. The expectation is that this continued activity will eventually change the SSR pathway, most probably the Ia synapse on the motoneuron. Because of its location, this learned change, and the activity-driven processes leading to it, should be accessible to physiologic and anatomic study.

FIGURE 2 summarizes data from two animals, one exposed to the up-mode and one exposed to the down-mode. In each instance, H-reflex amplitude changes appropriately over weeks. In the up-mode animal, it rises to about twice control size; while in the

[b] Other oligosynaptic pathways might possibly contribute to the SSR, but available data suggest that they do not play an important role in the SSR conditioning described here.[13,14,22,23]

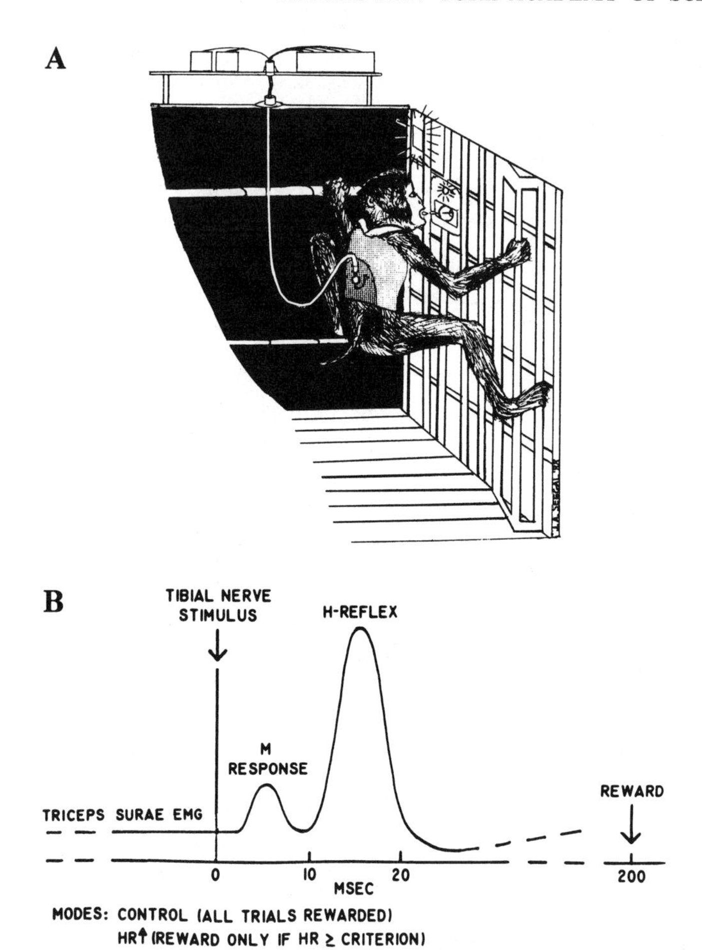

FIGURE 1. Design of the H-reflex conditioning task. (**A**) Animal wears garment with attached tether and performs computer-controlled task. Wires from chronically implanted fine-wire EMG and nerve cuff electrodes connect in posterior pocket with wires leading through flexible tether and aluminum swivel to electronic swivel, which connects to EMG amplifiers and stimulus isolation units. Large rectangular light indicates correct triceps surae background EMG in one channel of one leg. (This light is primarily an aid in initial training.) Small round light, which flashes with delivery of nerve cuff stimuli, indicates that reward squirt is (or may be) coming in 0.2 sec. (This signal is provided because cuff stimuli are probably imperceptible or barely perceptible to animal.) Solenoid-powered syringe delivers reward squirt through waterport into animal's mouth. (**B**) Task performed by animal. It keeps triceps surae background EMG in both legs within a required range for a randomly varying 1.2-1.8-sec period. Some animals, like the one shown in **A**, provide triceps surae EMG by clinging to the cage; others stand on their toes. At the end of this period, bilateral nerve cuff stimuli elicit threshold M-responses and substantial H-reflexes. Under the control mode, reward always occurs 0.2 sec after the stimuli. Under the up-mode or down-mode, reward occurs only if, in the conditioned leg, triceps surae EMG amplitude for the H-reflex interval (typically 12-20 msec after the stimuli) is above (up-mode) or below (down-mode) a specified value. The H-reflex in the other leg is simply measured. (See reference 15 for a full description.)

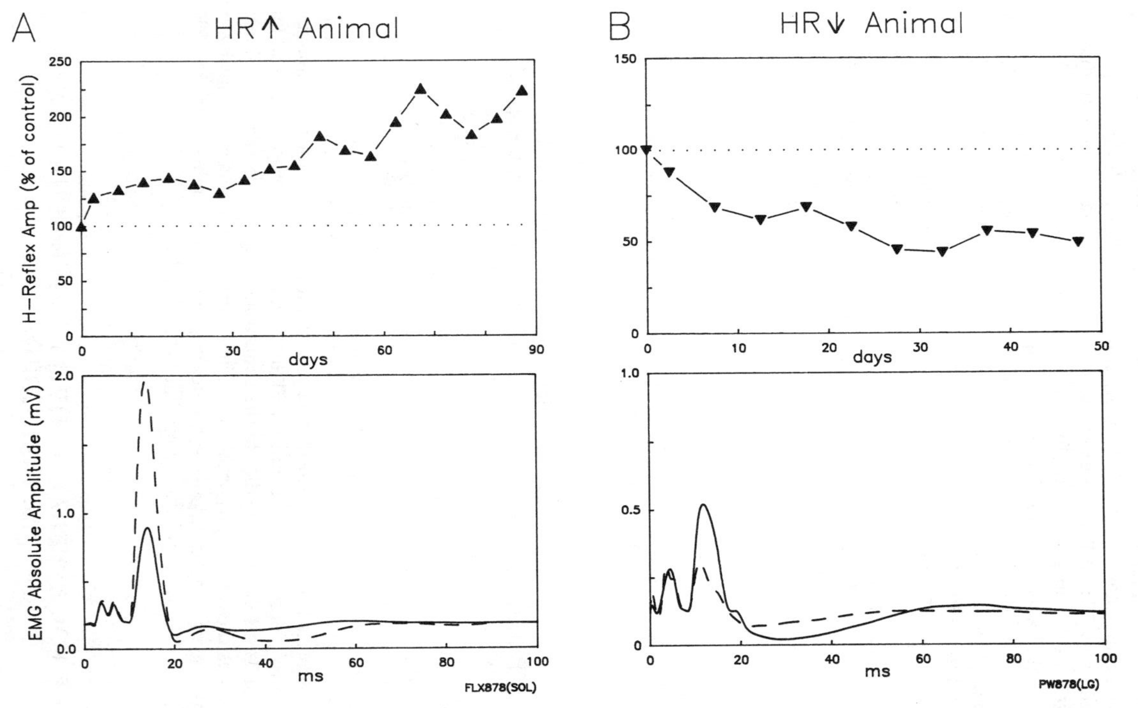

FIGURE 2. **(A)** *Top:* Average H-reflex amplitude in conditioned leg of an up-mode animal for each 5-day period for 90 days after onset of the up-mode (a percentage of control-mode H-reflex amplitude). Amplitude increases gradually with up-mode exposure. *Bottom:* Average H-reflex in conditioned leg for one control-mode day (solid) and for one day after prolonged up-mode exposure (dashed). The H-reflex is much larger after up-mode conditioning, whereas background EMG and M-response are unchanged. **(B)** *Top:* Average H-reflex amplitude in conditioned leg of a down-mode animal for each 5-day period for 50 days after onset of the down-mode (in percent of control-mode H-reflex amplitude). Amplitude decreases gradually with down-mode exposure. *Bottom:* Average H-reflex in conditioned leg for one control-mode day (solid) and for one day after prolonged down-mode exposure (dashed). The H-reflex is much smaller after down-mode conditioning. Background EMG and M-response are unchanged. (From reference 15.)

down-mode animal it falls to about half control size. The lower panels provide examples of the striking changes in the H-reflex produced by conditioning. These changes are focused on the conditioned leg; the H-reflex changes little in the opposite leg over the course of conditioning.[16]

Extensive associated studies, employing the SSR and/or the H-reflex conditioning paradigms, have defined major features of conditioning of the SSR pathway in monkeys (reviewed in references 12 and 17). Reflex increase or decrease always occurs gradually, as shown in FIGURE 2, and survives performance breaks of several weeks.[18,19] At the same time, detailed analysis reveals that SSR change begins with a rapid appropriate change of about 8% in the first 3-6 hr of exposure to the up- or down-mode.[20] This nearly immediate phase I change is followed by much slower phase II change (1-2% per day) that continues for many days.

Phase I change shows that the animal rapidly responds to the up- or down-mode with appropriate alteration in descending influence on the SSR pathway. The subsequent very slow phase II change suggests that the continuation of this influence over weeks gradually alters the Ia terminal or the motoneuron, and thereby further changes the reflex.[20,21] These findings by themselves, however, leave open the possibility that change is only supraspinal and simply exerts steadily increasing influence over the spinal pathway of the SSR.

OPERANT CONDITIONING OF THE SSR PATHWAY CHANGES THE SPINAL CORD

If the conditioning task does induce a prolonged change in the descending activity influencing the SSR pathway, and if this altered activity does change the spinal cord, evidence of the change should persist even if all descending influence is removed by transection. We evaluated this question by anesthetizing animals in which one leg's H-reflex had been increased (up-mode) or decreased (down-mode), transecting the spinal cord above the lumbosacral site of the SSR pathway, and then assessing the monosynaptic reflexes on both sides under continued anesthesia over several days.

The anesthetized transected animals showed reflex asymmetries similar to those present in the awake behaving animals.[22,23] Furthermore, these asymmetries were still evident three days after transection removed descending influence. In up-mode animals the reflex remained larger in the conditioned leg than in the control leg; and in down-mode animals it remained smaller in the conditioned leg than in the control leg. FIGURE 3 illustrates these lasting asymmetries. They indicate that, when animals learn a task that demands prolonged change in the descending activity influencing the SSR pathway, this altered activity eventually modifies the spinal cord.

EVIDENCE FOR SEVERAL ACTIVITY-DRIVEN CHANGES IN SPINAL CORD

Another effect of conditioning, besides reflex asymmetry, was found in the anesthetized transected animals. This unexpected second effect is also apparent in FIGURE 3: reflexes are larger in the control legs of down-mode animals than in the control legs of up-mode animals. This difference was not present before anesthesia and transection: control-leg reflexes did not change over the weeks of conditioning, and thus were of similar amplitude in awake behaving up-mode and down-mode animals.[16] Thus, anesthesia and transection revealed an effect of conditioning not apparent in the awake behaving animal.

This unexpected change in the spinal cord introduces an important issue. Every conditioning process has a definite goal: it is intended to produce a specific change in

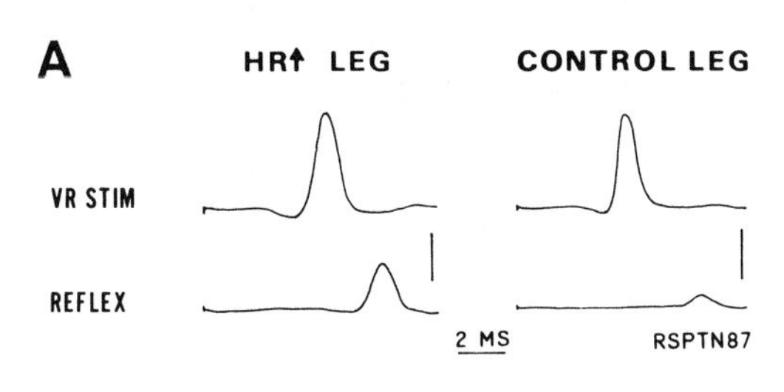

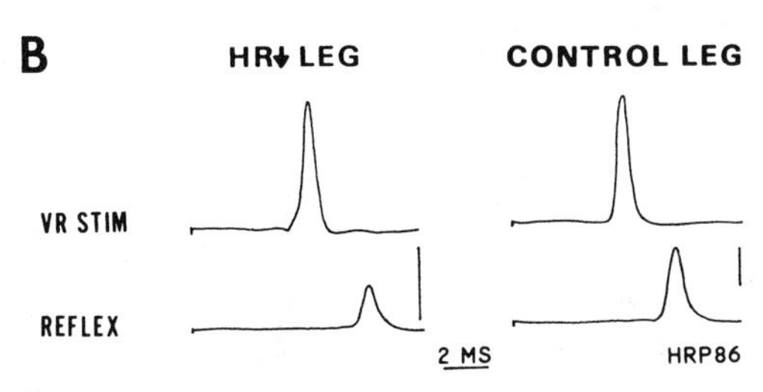

FIGURE 3. Evidence for activity-driven plasticity in spinal cord. Responses to ventral root (VR) stimulation and monosynaptic reflex responses to dorsal root stimulation from triceps surae nerve branches in conditioned and control legs of an up-mode animal and a down-mode animal, both under deep pentobarbital anesthesia, 2-3 days after thoracic spinal cord transection had eliminated supraspinal influence. To facilitate inter-leg comparison within each animal, its responses are scaled so that the responses to ventral root stimulation (that is, the maximum possible responses) are equal in height. Vertical bars: 100 μV. Mode-appropriate reflex asymmetry is present in both animals: in the up-mode animal the reflex is bigger in the conditioned leg than in the control leg, and in the down-mode animal it is smaller in the conditioned leg than in the control leg. In addition, the control-leg reflex of the down-mode animal is bigger than the control-leg reflex of the up-mode animal. Both these effects of conditioning were evident throughout the several days of posttransection study under continued anesthesia. (See reference 22 for a full presentation.)

behavior. Here the desired change in behavior is specified by the experimenter, that is, a larger or smaller H-reflex. The activity-driven plasticity responsible for this behavioral change can be called "intended plasticity." By virtue of its nature and/or location, this plasticity may compromise other behaviors that utilize the same neurons and synapses. Thus, it may lead to additional activity-driven changes, "compensatory plasticity" that permits the CNS to continue to perform these other behaviors satisfactorily. For example, if exposure to the up-mode leads to greater transmitter release at the triceps surae Ia synapse, the H-reflex would not be the only behavior affected. Other behaviors in which this synapse participates, such as walking or jumping, would also be altered. Compensatory plasticity might well be needed to ensure their proper performance (perhaps through increase of a specific inhibitory input to the motoneuron).

Compensatory plasticity may also occur with conditioning of the vestibuloocular reflex (VOR), which maintains a stable retinal image during head movement.[24–26] During VOR conditioning, the sensitivity of floccular Purkinje cells to vestibular input seems to change in a direction opposite to that expected on the basis of the change observed in VOR gain.[27] This ostensibly paradoxical change in floccular response to vestibular input may compensate for change in eye movement input to the flocculus, and thereby appropriately adjust the contribution of the cerebellum to eye movement.[26]

If compensatory plasticity occurs, it has several significant implications for attempts to delineate the activity-driven modifications responsible for specific changes in behavior. First, it implies that change in even an extremely simple behavior is associated with CNS plasticity beyond that sufficient to produce the intended behavioral change. Second, it stresses the necessity for simplicity in experimental models used to study the activity-driven modifications underlying learning. Without the comparative simplicity of the SSR and VOR pathways, possible compensatory changes would be hard to detect, and exceedingly hard to define.

RELEVANCE OF THIS MODEL SYSTEM

Initial efforts to condition the SSR pathway in the human biceps have been successful, and the major features of the phenomenon in humans appear similar to those found in monkeys.[28,29] Furthermore, the activity-driven plasticity produced by SSR conditioning is not limited to studies specially designed to produce it. Meyer-Lohmann *et al.*[30] trained monkeys to move a lever to a target, and randomly disturbed this motion with a brief torque pulse. At first, the animals's response to the disturbance was dominated by the later, M2, component of the muscle stretch response. Then, over months of training, the M1 component (that is, the SSR) slowly grew and M2 waned, so that the SSR gradually took over the role of countering the disturbance.

Similar SSR changes appear to occur outside of the laboratory. In children, gradual changes in the SSR occur with acquisition of fundamental motor skills.[31] Comparable changes seem to take place later in life during learning of skills such as ballet.[32] Slow activity-driven changes in the spinal cord and elsewhere in the CNS may account for much learning, and serve to explain why acquisition of many skills depends on prolonged practice. The need for practice is not confined to sophisticated motor behaviors: nonmotor skills also depend on it. For example, mastery of multiplication and other basic mathematical operations is a lengthy, labor-intensive undertaking.

As indicated above, the occurrence of SSR change under the up- or down-mode has a time course seen in many learning curves: rapid initial improvement followed by further improvement at a gradually decreasing rate over a long time. For SSR pathway conditioning, this form seems to result from a two-phase sequence: the rapid onset of appropriate descending activity influencing the SSR pathway causes phase I change, and the chronic persistence of this activity leads to spinal cord plasticity that is responsible for gradual phase II change (see below). Conditioning of the VOR, another simple behavior, also shows evidence of two distinct phases,[33] but acquisition of more complex behaviors usually does not. This difference is probably to be expected. Changes in more complex behaviors presumably depend on multiple adjustments in activity (that is, many phase I changes), and these adjustments probably lead to activity-driven plasticity at many sites in the CNS (that is, many phase II changes). These numerous two-phase sequences presumably overlap and obscure one another, so that the learning curve of a complex behavior usually lacks clear-cut phases.

POSSIBLE SITES OF ACTIVITY-DRIVEN PLASTICITY IN THE SPINAL CORD[c]

The activity-driven changes in the spinal cord underlying conditioning of the SSR pathway are most likely to occur at those sites in the pathway that are influenced by supraspinal structures. These sites appear in FIGURE 4. They are the Ia afferent terminal on the motoneuron and the motoneuron itself. It is unlikely that change is limited to spinal cord sites not in the SSR pathway and simply influences the pathway, because conditioned reflex changes are clear even under deep pentobarbital anesthesia,[22,23] which depresses neuronal and synaptic function and thus serves to isolate neurons from each other. If change did not occur at the Ia terminal or the motoneuron, its effect on reflex amplitude in the pentobarbital-anesthetized animal would be small.

The Ia synapse is inhibited by presynaptic inputs from several supraspinal areas.[8,9] Recent studies suggest that short-term changes in presynaptic inhibition are important in motor behavior.[10,34] Chronic alteration in this inhibition might change the Ia terminal (FIG. 4A), and thereby change the amplitude of the excitatory postsynaptic potential (EPSP) produced in the motoneuron.

Motoneuron membrane properties, like those controlling resting potential or input resistance, affect its response to any input, and could be altered (FIG. 4B) by long-term changes in descending activity. However, the constraint imposed by the need to keep constant motoneuron tone during conditioning (FIG. 1 legend), and the very significant interference that such motoneuron modifications would exert on other behaviors, make generalized membrane modifications a more complex explanation for reflex change than presynaptic (FIG. 4A) modification. A localized postsynaptic modification (FIG. 4C), such as in receptor sensitivity or dendritic architecture, might have greater specificity. However, pathways capable of producing such a selective modification are not known at present. Thus, at this point, a change in the Ia terminal driven by long-term change in presynaptic inhibition appears to be the most reasonable possibility for the spinal cord plasticity underlying conditioning of the SSR pathway.

FUTURE RESEARCH DIRECTIONS

The significance of SSR conditioning as a model for studying memory and the activity-driven processes that constitute learning is that it changes a simple, well-defined, and accessible pathway located in a readily isolated part of the vertebrate CNS. These features facilitated detection of two different conditioned changes (hypothesized

[c] This presentation discusses the likely sites of the activity-driven change responsible for the mode-appropriate reflex asymmetry. Consideration of the basis of the difference in control-leg reflexes of anesthetized transected animals is similar and appears elsewhere.[22]

to be intended and compensatory, respectively), and identification of two distinct phases in conditioning. Most important, they allow the search for the memory trace to focus on a single synapse and its postsynaptic neuron, both of which are accessible physiologically and anatomically. The ultimate value of SSR pathway conditioning as an experimental model depends on its ability to allow precise localization and description of memory traces and the activity-driven processes that create them, and on the applicability of its principles, techniques, and results to studying learned changes in other behaviors.

Intracellular recording and labeling of individual motoneurons and Ia afferents in conditioned animals should help decide among the possible trace locations shown in FIGURE 4. For example, a trace located at the Ia terminal should modify Ia EPSPs, and, given the considerable magnitude of the reflex change (for example, FIG. 2), may also have an anatomical correlate. This work could also allow investigation of the

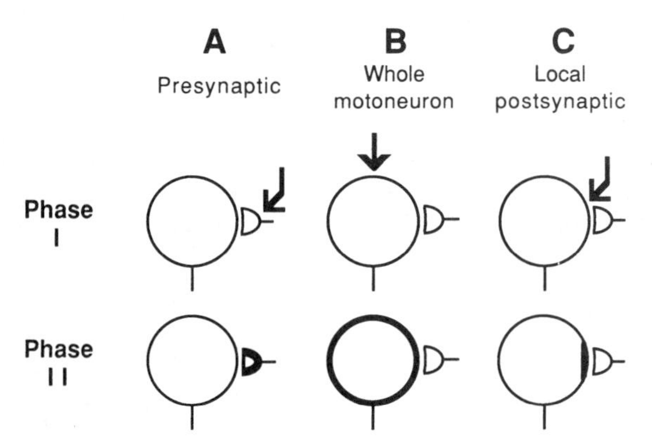

FIGURE 4. Possible locations of the spinal cord memory traces responsible for change in SSR, or H-reflex, amplitude. In each case, phase I change in descending influence on a site in the SSR pathway (indicated by the arrow) eventually produces phase II change (that is, the memory trace) at the site. (**A**) The most likely site is the Ia afferent terminal on the motoneuron. A memory trace here could be produced by prolonged change in presynaptic inhibition. (**B**) The memory trace could be a generalized change in the motoneuron that alters its response to any input. However, such a modification would have widespread effects on motoneuron function. (**C**) The trace could also be a very localized modification in the postsynaptic membrane, such as a change in receptor sensitivity or dendritic architecture. (From reference 17.)

structures and pathways responsible for creating change in the SSR pathway. In view of the apparent importance of the cerebellum in motor learning of other kinds[35–37] and its close connections to the SSR pathway, pathways mediating spinal-cerebellar interactions should be of particular interest.

Recent successes in maintaining reduced preparations of spinal cord and brainstem[38–40] suggest that it may be possible to establish the entire SSR conditioning process *in vitro* (for example, by stimulation of a specific descending pathway). This development would enhance study of the mechanisms of conditioning at both electrophysiological and biochemical levels.

Other oligosynaptic spinal cord pathways, like the three-neuron path through which Ia input inhibits antagonist muscles, should also be capable of developing memory traces in response to tasks like the one that modifies the SSR pathway. Thus, the spinal cord may provide a good environment for studying a variety of memory traces, their mechanisms, and their interactions. The knowledge gained may facilitate identification and elucidation of memory traces occurring in more complex and less accessible regions of the CNS.

SUMMARY

Recent work has shown that the monosynaptic pathway of the SSR can be operantly conditioned, and that a significant part of the plasticity responsible for the behavioral change resides in the spinal cord. The most likely sites of this activity-driven plasticity are the synapse of the Ia afferent neuron on the motoneuron and/or the motoneuron itself. Because the SSR pathway is the simplest and most accessible stimulus-response pathway in the vertebrate CNS, it may provide a valuable experimental model for elucidating activity-driven CNS changes responsible for learning.

REFERENCES

1. BYRNE, J. H. 1987. Cellular analysis of associative learning. Physiol. Rev. **67**(2): 329-439.
2. DUDAI, Y. 1989. The Neurobiology of Memory. Oxford University Press. Oxford.
3. LEE, R. G. & W. G. TATTON. 1975. Motor responses to sudden limb displacements in primates with specific CNS lesions and in human patients with motor system disorders. Can. J. Neurol. Sci. **2:** 285-293.
4. MATTHEWS, P. B. C. 1972. Mammalian Muscle Receptors and Their Central Actions. Williams & Wilkins. Baltimore, MD.
5. HENNEMAN, E. & L. M. MENDELL. 1981. Functional organization of motoneuron pool and inputs. *In* Handbook of Physiology. Sect. I. The Nervous System. Vol. II. Motor Control. Part I. V. B. Brooks, Ed.: 423-507. Williams & Wilkins. Baltimore, MD.
6. MAGLADERY, J. W., W. E. PORTER, A. M. PARK & R. D. TEASDALL. 1951. Electrophysiological studies of nerve and reflex activity in normal man. IV. The two-neuron reflex and identification of certain action potentials from spinal roots and cord. Bull. Johns Hopkins Hosp. **88:** 499-519.
7. BROWN, W. F. 1984. The Physiological and Technical Basis of Electromyography. Butterworths. Boston, MA.
8. BALDISSERA, F., H. HULTBORN & M. ILLERT. 1981. Integration in spinal neuronal systems. *In* Handbook of Physiology. Sect. I. The Nervous System. Vol. II. Motor Control. Part I. V. B. Brooks, Ed.: 509-595. Williams & Wilkins. Baltimore, MD.
9. BURKE, R. E. & P. RUDOMIN. 1978. Spinal neurons and synapses. *In* Handbook of Physiology. The Nervous System. Cellular Biology of Neurons. E. R. Kandel, Ed.: 877-944. Williams & Wilkins. Baltimore, MD.
10. CAPADAY, C. & R. B. STEIN. 1987. Difference in the amplitude of the human soleus H-reflex during walking and running. J. Physiol. **392:** 513-522.
11. MENDELL, L. M. 1984. Modifiability of spinal synapses. Physiol. Rev. **64:** 260-324.
12. WOLPAW, J. R. 1985. Adaptive plasticity in the spinal stretch reflex: An accessible substrate of memory? Cell. Mol. Neurobiol. **5:** 147-165.
13. WOLPAW, J. R., D. J. BRAITMAN & R. F. SEEGAL. 1983. Adaptive plasticity in the primate spinal stretch reflex: Initial development. J. Neurophysiol. **50:** 1296-1311.
14. WOLPAW, J. R. 1987. Operant conditioning of primate spinal reflexes: The H-reflex. J. Neurophysiol. **57:** 443-458.
15. WOLPAW, J. R. & P. A. HERCHENRODER. 1990. Operant conditioning of H-reflex in freely moving monkeys. J. Neurosci. Methods **31:** 145-152.

16. WOLPAW, J. R., C. L. LEE & J. G. CALAITGES. 1989. Operant conditioning of primate triceps surae H-reflex produces reflex asymmetry. Exp. Brain Res. **75:** 35-39.
17. WOLPAW, J. R. & J. S. CARP. 1990. Memory traces in spinal cord. Trends Neurosci. **13:** 137-142.
18. WOLPAW, J. R. 1983. Adaptive plasticity in the primate spinal stretch reflex: Reversal and redevelopment. Brain Res. **278:** 299-304.
19. WOLPAW, J. R., J. A. O'KEEFE, P. A. NOONAN & M. G. SANDERS. 1986. Adaptive plasticity in the primate spinal stretch reflex: Persistence. J. Neurophysiol. **55:** 272-279.
20. WOLPAW, J. R. & J. A. O'KEEFE. 1984. Adaptive plasticity in the primate spinal stretch reflex: Evidence for a two-phase process. J. Neurosci. **4:** 2718-2724.
21. WOLPAW, J. R., J. A. O'KEEFE, V. A. KIEFFER & M. G. SANDERS. 1985. Reduced day-to-day variation accompanies adaptive plasticity in the primate spinal stretch reflex. Neurosci. Lett. **54:** 165-171.
22. WOLPAW, J. R. & C. L. LEE. 1989. Memory traces in primate spinal cord produced by operant conditioning of H-reflex. J. Neurophysiol. **61:** 563-572.
23. WOLPAW, J. R., J. S. CARP & C. L. LEE. 1989. Memory traces in spinal cord produced by H-reflex conditioning: Effects of posttetanic potentiation. Neurosci. Lett. **103:** 113-119.
24. ITO, M. 1982. Cerebellar control of the vestibulo-ocular reflex: Around the flocculus hypothesis. Annu. Rev. Neurosci. **5:** 275-296.
25. BERTOZ, A. & G. MELVILL JONES. 1985. Adaptive Mechanisms in Gaze Control. Elsevier. New York, NY.
26. LISBERGER, S. G. 1990. The neural basis for learning of simple motor skills. Science **242:** 728-735.
27. MILES, F. A., J. H. FULLER, D. J. BRAITMAN & B. M. DOW. 1980. Long-term adaptive changes in primate vestibulo-ocular reflexes. III. Electrophysiological observations in flocculus of adapted monkeys. J. Neurophysiol. **43:** 1477-1495.
28. EVATT, M. L., S. L. WOLF & R. L. SEGAL. 1989. Modification of human spinal stretch reflexes: Preliminary studies. Neurosci. Lett. **105:** 350-355.
29. SEGAL, R. L., P. A. CATLIN, R. COOKE, L. JOHNSON, A. MCCALLUM, D. MURZIN, K. RUBEL & S. L. WOLF. 1989. Preliminary studies on modification of hyperactive spinal stretch reflexes in stroke patients. Soc. Neurosci. Abstr. **15:** 917.
30. MEYER-LOHMANN, J., C. N. CHRISTAKOS & H. WOLF. 1986. Dominance of the short-latency component in perturbation induced electromyographic responses of long-trained monkeys. Exp. Brain Res. **64:** 393-399.
31. MYKLEBUST, B. M., G. L. GOTTLIEB & G. C. AGARWAL. 1986. Stretch reflexes of the normal human infant. Dev. Med. Child Neurol. **28:** 440-449.
32. GOODE, D. J. & J. VAN HOEVEN. 1982. Loss of patellar and Achilles tendon reflex in classical ballet dancers. Arch. Neurol. **39:** 323.
33. MANDL, G., G. MELVILL JONES & M. CYNADER. 1981. Adaptability of the vestibulo-ocular reflex to vision reversal in strobe reared cats. Brain Res. **209:** 35-45.
34. CAPADAY, C. & R. B. STEIN. 1987. Amplitude modulation of the soleus H-reflex in the human during walking and standing. J. Neurosci. Methods **21:** 91-104.
35. BLOEDEL, J. R., V. BRACHA, T. M. KELLY & J. Z. WU. 1991. Substrates for motor learning: Does the cerebellum do it all? Ann. N.Y. Acad. Sci. This volume.
36. PETERSON, B. W., J. F. BAKER & J. C. HOUK. 1991. A model of adaptive control of vestibuloocular reflex based on properties of cross-axis adaptation. Ann. N.Y. Acad. Sci. This volume.
37. YEO, C. H. 1991. Cerebellum and classical conditioning of motor responses. Ann. N.Y. Acad. Sci. This volume.
38. SMITH, J. C. & J. L. FELDMAN. 1987. *In vitro* brainstem-spinal cord preparations for study of motor systems for mammalian respiration and locomotion. J. Neurosci. Methods **21:** 321-333.
39. ATSUTA, Y., E. GARCIA-RILL & R. D. SKINNER. 1988. Electrically induced locomotion in the *in vitro* brainstem-spinal cord preparation. Brain Res. **470:** 309-312.
40. KEIFER, J. & J. C. HOUK. 1989. An *in vitro* preparation for studying motor pattern generation in the cerebelloruspinal circuit of the turtle. Neurosci. Lett. **97:** 123-128.

Spontaneous Bioelectric Activity as Both Dependent and Independent Variable in Cortical Maturation

Chronic Tetrodotoxin versus Picrotoxin Effects on Spike-Train Patterns in Developing Rat Neocortex Neurons during Long-Term Culture

MICHAEL A. CORNER AND GER J. A. RAMAKERS

Netherlands Institute for Brain Research
Amsterdam, the Netherlands

Electrical discharges originating within the CNS itself constitute most of the functional activity present at early stages of development,[1] but the importance of this spontaneous bioelectric activity (SBA) for neuronal and neural network maturation is only beginning to be appreciated. *In vitro* culture of isolated fragments of tissue provides a model system that preserves endogenous activity in a wide variety of brain and spinal cord regions, often with surprisingly normal-appearing firing patterns.[2] The easy accessibility of such preparations for long-term manipulations of spontaneous firing levels that could be electrophysically monitored, as well as for high-resolution examination of the consequences, has been exploited in several lines of research. Thus, neuronal death as a result of the chronic elimination of SBA has been documented in several CNS structures, and a reduction of afferent innervation specificity has been reported in spinal cord cultures.[3]

Spontaneous neuronal discharge patterns show a characteristic developmental time course *in vitro* that could be demonstrated by a multivariate statistical analysis of single-unit spike trains.[4,5] The major trends observed (FIG. 1) were 1) a progressive decline in the incidence of strong interspike interval dependencies (expressed as the Markov value); 2) an early and persistent decrease in overall regularity of grouped spike discharges (expressed by the coefficient of variation (CV) for the burst period, that is, the time elapsing between the onset of successive spike clusters); and 3) a transient large increase in the proportion of units showing prominent, often highly regular, slow fluctuations of their mean firing level (as reflected in the CV for the number of spikes falling into successive 1-min time bins). Modal intervals calculated for individual neurons (FIG. 2) fell into distinct groups that we have defined as representing, respectively, predominantly *burst* (< 10 msec), *phasic* (10-25 msec), and *tonic* (> 25 msec) firing. The last of these three categories constituted the majority of units at 2 weeks *in vitro*, but such patterns eventually disappeared almost completely, being replaced chiefly by "bursters" (FIG. 3). As would be expected, the modal interval parameter declined with age *in vitro* as a result of the shift away from tonic firing patterns, and closely paralleled the developmental curve for the Markov value (see FIGS. 1 & 3).

Chronic exposure to tetrodotoxin (TTX), in a dose sufficient to suppress all measurable firing, led to an apparent arrest of functional ontogeny after about 2 weeks *in vitro*. Thus, interspike interval dependencies, burst regularities, and minute-order fluctuations

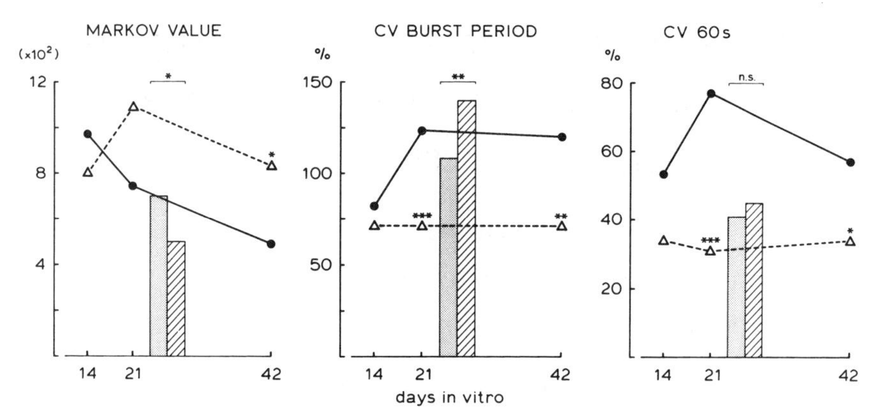

FIGURE 1. Effect of TTX (△) at different ages (days *in vitro*) on three crucial parameters describing neuronal firing with respect to the corresponding controls (●): ∗: $p < .05$, ∗∗: $p < .01$, ∗∗∗: $p < .001$ (Mann-Whitney *U* test). PTX-treated cultures (hatched bars) were compared to control cultures (toned bars) at 21 days *in vitro*: ∗: $p < .05$, ∗∗ $p < .01$.

all remained indefinitely at the level observed in the youngest cultures (FIG. 1). This retardation was visible even in raw spike-train plots, where repetitive clustered discharges are far more common in TTX-treated than in control cultures. The proportion both of spontaneously silent and of slow-firing units was very low, and the incidence of stationary spike trains relatively high, in the TTX as compared with the control group.[5] Abnormally short modal intervals and a dearth of tonic, in favor of burst, firing already at 14 days *in vitro* (FIG. 3) indicate that TTX-induced pathophysiology in fact is present earlier than indicated by the other parameters examined (see FIG. 1). On the other hand, the apparent normalcy at 42 days *in vitro* suggested by measures derived from the interspike interval histograms (FIG. 3) is belied not only by the occasional presence of visibly epileptiform activity[6] but also by the relative frequency with which bimodal distributions occurred in TTX-treated cultures (TABLE 1). In general, the interval histograms show a higher incidence of phasic, at the expense of tonic, components within the spike trains than was shown in untreated neurons.

TABLE 1. Incidence of Bimodal Interspike Interval Distributions in Experimental and Control Neurons[a]

		Interspike Interval Type			
	n/N	Burst/Phasic	Tonic/Burst	Tonic/Phasic	Tonic/Tonic
Control group	13/68	.15	.54	.25	.08
TTX-treated group	26/69	.46	.38	.15	
Control group	12/50	.25	.33	.33	.08
PTX-treated group	22/53	.09	.68		.23

[a] Note that *n/N* is the proportion of units in each of the indicated groups that showed a double peak in its interspike interval histogram. Although both of the experimental treatments favor bimodality in the histograms, the two groups differ considerably from one another ($p < .01$): TTX favors phasic firing (that is, a mode at 10-25 msec) at the expense of tonic firing (modal interval > 25 msec), whereas the reverse is true for PTX.

Chronic exposure to the GABA blocker picrotoxin (PTX) had, in most respects, the opposite effect from that produced by TTX.[7] For instance, when tested at 21 days *in vitro* both the interspike interval dependencies (Markov value) and the burst regularities (CV burst period) proved to be lower in TTX-treated than in untreated cultures (FIG. 1). In addition, an unusually high proportion of units in the PTX group

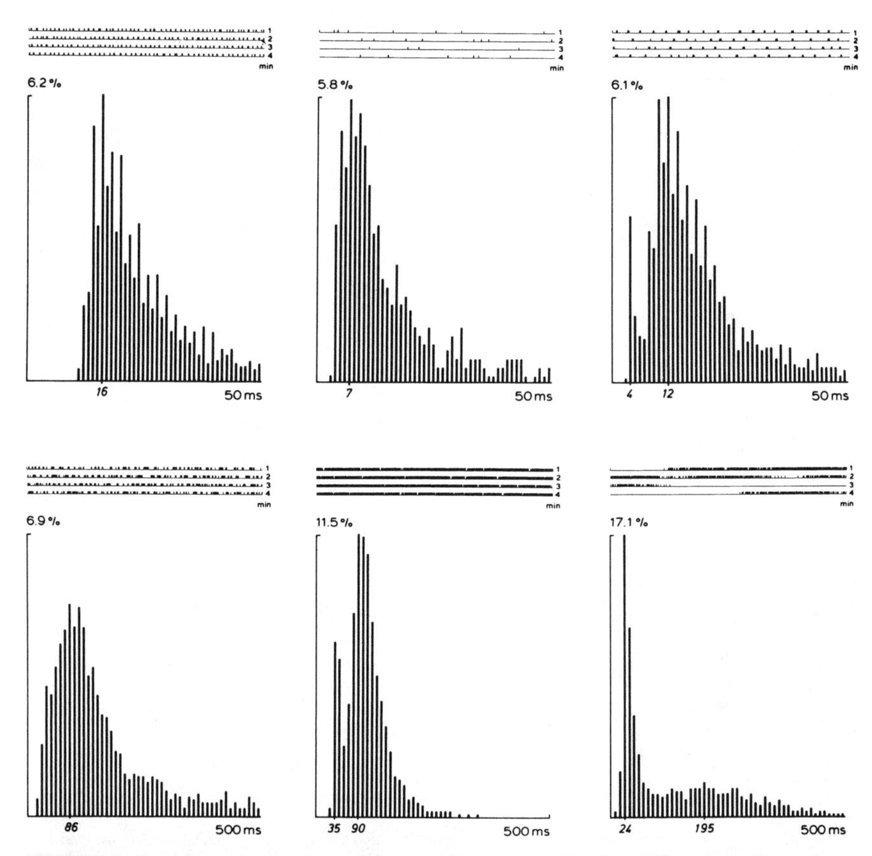

FIGURE 2. Examples of main firing patterns and corresponding interspike interval histograms encountered, in differing proportions, in all groups of cultured neocortex neurons (modal interval indicated on the abscissa). Stereotyped clustering of spikes tends to be associated with peaks in the phasic range of the histogram (upper left and right examples), whereas irregular spike clusters usually occur together with peaks in the (high-frequency) burst range (upper middle and right examples). Relatively continuous firing, in contrast, involves one or more peaks in the tonic range (bottom three examples).

were devoid of any mathematically stationary stretches of firing. Spike-train plots from PTX-treated neurons give an appearance of relatively sustained activity, an impression that was confirmed by the observation of a much lower overall incidence, in comparison with control neurons, of interspike intervals lasting more than 1 sec ($p < .01$). In addition, predominantly tonic firing (that is, modal interval > 25 msec) occurred more often in the PTX group (FIG. 3), whereas the incidence of secondary peaks in the tonic at the expense of the phasic range of the interval histograms was also clearly augmented (TABLE 1).

The complementary nature of the sequelae of TTX and PTX treatment, respectively suppressing and intensifying the bioelectric activity of developing neocortical neurons, argues in favor of SBA itself (its phasic aspects in particular) being a causal factor under these experimental conditions. Because blockage of GABAergic (inhibitory) synaptic transmission caused clear-cut pathophysiological effects, this neurotransmitter presumably serves to constrain SBA within an appropriate range for optimal network formation. Glutamic acid in turn, being the major excitatory transmitter within the cerebral cortex, is implicated as a major epigenetic factor by our SBA deprivation (TTX) results. It is of special interest that most of the electrophysiological abnormalities tend to be opposite in direction from the acute effects of the experimental intervention. Thus, cultured cortical neurons become abnormally active, sometimes even showing

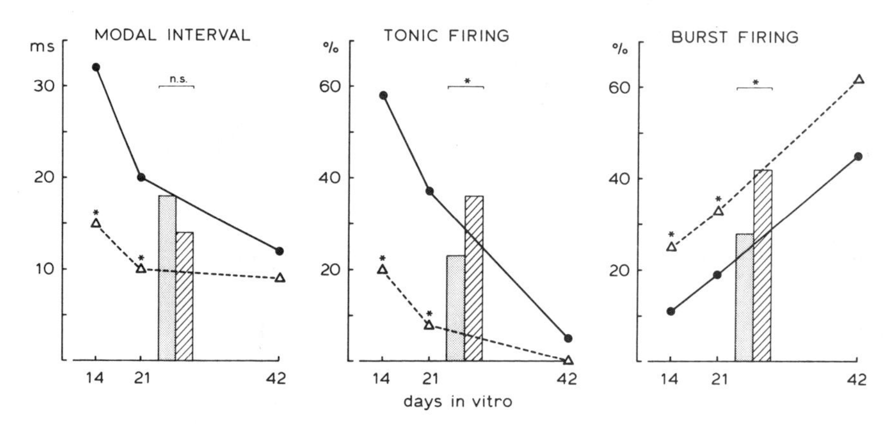

FIGURE 3. Effect of TTX (△) at different ages (days *in vitro*) on the modal intervals (ms: group median values in msec) and on the incidence (%) of tonic (that is, mode > 25 msec) and "burst" (that is, mode < 10 msec) firing patterns (*: $p < .05$ versus the corresponding control group (●); Mann-Whitney or chi-square test). PTX-treated cultures (hatched bars) versus control cultures (toned bars) at 21 days *in vitro* (*: $p < .05$).

epileptiform firing patterns following long-term deprivation of SBA (see references 5 and 6). Chronic disinhibition from GABAergic control, in contrast, results in spike trains that are almost completely devoid of the stereotyped bursting normally seen to some degree even in normally active cultures. These compensatory developmental responses suggest up- and down-regulation, respectively, of neurotransmitter receptor sensitivities and/or receptor-linked or voltage-sensitive ion channels. In addition, however, quantitative abnormalities in the relative number of excitatory versus inhibitory synapses[8,9] may well contribute to the observed functional disturbances.

REFERENCES

1. CORNER, M. A. 1990. Brainstem control of behavior: Ontogenetic aspects. *In* Brainstem Mechanisms of Behavior. W. Klemm & R. Vertes, Eds.: in press. Wiley-Liss. New York, NY.

2. CRAIN, S. M. 1976. Neurophysiologic Studies in Tissue Culture. Raven Press. New York, NY.
3. BAKER, R., M. CORNER & A. HABETS. 1984. Effects of chronic suppression of bioelectric activity on the development of sensory ganglion evoked responses in spinal cord explants. J. Neurosci. **4:** 1187-1192.
4. HABETS, A., A. VAN DONGEN, F. VAN HUIZEN & M. CORNER. 1987. Spontaneous neuronal firing patterns in fetal rat cortical networks during development *in vitro*: A quantitative analysis. Exp. Brain Res. **69:** 43-52.
5. RAMAKERS, G., M. CORNER & A. HABETS. 1989. Development in the absence of spontaneous bioelectric activity results in increased stereotyped burst firing in cultures of dissociated cerebral cortex. Exp. Brain Res. **79:** 157-166.
6. CORNER, M. & G. RAMAKERS. 1991. Epileptiform activity as dependent and independent variable in brain maturation: A comparison of effects of chronic tetrodotoxin and picrotoxin treatment on developing rat neocortex neurons cultured *in vitro*. Exp. Brain Res. (Suppl.): in press.
7. RAMAKERS, G., M. CORNER & A. HABETS. 1990. Abnormalities in the spontaneous firing patterns of cultured rat neocortical neurons after chronic exposure to picrotoxin. Brain Res. Bull. **26:** 429-432.
8. VAN HUIZEN, F., H. ROMIJN & M. CORNER. 1987. Indications for a critical period for synapse elimination in developing rat cerebral cortex cultures. Dev. Brain Res. **31:** 1-6.
9. VAN HUIZEN, F., H. ROMIJN, P. VAN DEN HOOFF & A. HABETS. 1987. Picrotoxin-induced disinhibition of spontaneous bioelectric activity accelerates synaptogenesis in rat cerebral cortex cultures. Exp. Neurol. **97:** 280-288.

Are Polyamines Modulators of the CNS?[a]

P. A. FERCHMIN

Department of Biochemistry
Universidad Central del Caribe
Bayamon, Puerto Rico 00621-6032

The polyamines (PAs) putrescine (PUT), spermidine (SPD), and spermine (SPM) are present in all cells where they interact with RNA, DNA, proteins, and phospholipids. Most of the interactions are through the cationic groups of the PAs. In some cases PAs can be replaced by Ca^{2+} or Mg^{2+} but cannot be replaced where the positive charges have to be separated by a chain of carbons. The spectrum of action of PAs is broad but not as broad as in the case of Ca^{2+} (for example, PAs are not major bone components!). PA metabolism is regulated mainly by rapid synthesis and degradation of the regulatory enzymes. These enzymes have the shortest half-lives ever described. For example, ornithine decarboxylase (ODC), the enzyme that synthesizes PUT, has a half-life of 3 to 30 min.

Recent publications show that PAs modulate regulatory processes involved in activity-driven CNS changes like long-term potentiation and seizures. Selected examples of regulatory effects of PAs follow.

ODC IS INDUCED BY INCREASED NEURONAL ACTIVITY

Electroconvulsive shock induces ODC and c-*fos* mRNA.[1,2] It was suggested that these two inductions are coupled in brain.[1] The ODC activation can be elicited by subseizure stimulation, but increased neuronal activity seems to be required because MK-801, a receptor channel blocker, decreases ODC activation.[3] A different modality of electric stimulation, amygdaloid kindling, increases PUT in the amygdala,[4] probably as a result of ODC activation. Nerve growth factor[5] and nerve crushing[6] also stimulate ODC. It seems that ODC activation exceeds the actual PUT needs because ODC inhibitors fail to stop nerve growth factor-induced neurite outgrowth[6,7] or regeneration of crushed nerves.[6] In other cases the ODC inhibitor, α-difluoromethylornithine, was effective in preventing nerve regeneration.[8]

ROLE OF PUT IN THE PATHOLOGY OF BRAIN ISCHEMIA

Recirculation after ischemia is followed by a 10-fold increase in PUT concentration.[9] The degree of PUT increase correlates well with the extent of neuronal damage,[10] indicating that it could be responsible for part of the brain damage. PUT mediates the

[a]This work was supported in part by NIH-RCMI RR03035 and NSF-EPSCOR RII-8610677.

osmotically induced reversible opening of the blood-brain barrier[11] and participates in the Ca^{2+}-dependent, K^{+}-induced release of excitatory amino acids from nerve endings.[12] Therefore, a rise in PUT could increase the release of excitatory amino acids that would cause Ca^{2+} influx into neurons followed by cell damage.

PAs AND PROTEIN KINASES REGULATE EACH OTHER

In cultures of epidermal cells, protein kinase C activation by tetradecanoylphorbol acetate or diacylglycerol may be an early event in ODC induction; protein kinase C activation increases the steady state level of ODC mRNA.[13,14] Protein kinase C inhibitors affect ODC in a complex manner: staurosporine, at protein kinase C-inhibiting concentrations, induced ODC and increased tetradecanoylphorbol acetate-dependent ODC induction. H-7 inhibited ODC induction by tetradecanoylphorbol acetate and by staurosporine.[15]

PAs (SPM > SPD > PUT) inhibit protein kinase C and Ca^{2+}/calmodulin-dependent protein kinase in brain cortex.[16] SPM, but not SPD or PUT, inhibits the proteolytically cleaved, isolated catalytic domain of brain protein kinase C.[17] Therefore, protein kinase C activates the first step of PA synthesis and is inhibited in turn by SPM, the product of this synthesis.

SPM REGULATES MITOCHONDRIAL Ca^{2+} UPTAKE

SPM regulates the cytosolic free Ca^{2+} by increasing the affinity, but not the V_{max}, of a mitochondrial Ca^{2+} uniporter.[18] Mitochondrial Ca^{2+} uptake is more responsive to the allosteric activation by SPM in cortex and hippocampus than in nontelencephalic brain areas.[19] Because intracellular Ca^{2+} increases during intense neuronal stimulation, SPM-stimulated mitochondrial uptake may serve as a Ca^{2+} sink during such episodes.

PAs REGULATE PHOSPHATIDYLINOSITOL 4-PHOSPHATE METABOLISM

Phosphatidylinositol phosphates are synthesized by phosphatidylinositol and phosphatidylinositol 4-phosphate kinases, which are activated by SPM and SPD.[20,21] PAs may act by interacting with the charges on the phospholipids and so eliminate the requirement of these kinases for supraphysiological concentrations of Mg^{2+}.[21] In addition, PAs inhibit the phosphomonoesterases that hydrolyze the polyphosphates into phosphoinositol. Therefore, it seems that PAs augment the basal concentration of phosphatidylinositol phosphates.[21] On the other hand, SPM acts as uncompetitive inhibitor of ATP for diacylglycerol kinase, so in this instance SPM would slow the resynthesis of phosphatidylinositol 4,5-bisphospate but could enhance the activation of protein kinase C by diacylglycerol.[21]

PAs MODULATE SYNAPTIC RECEPTORS

The *N*-methyl-D-aspartate receptor has a regulatory binding site for SPM and SPD; binding of PAs to this site increases the affinity for agonists, as evidenced by increased binding of MK-801 and other open-channel blockers in the presence of traces of excitatory amino acids. This modulatory SPM and SPD site is different from the glycine site, but both sites interact.[22] Ifenprodil and SL82.0715 were reported to be antagonists at the PA site,[23] but others reported a more complex interaction.[24] Both compounds were shown to be cytoprotective in animal models of ischemia,[25] probably by stabilizing the *N*-methyl-D-aspartate receptor channel in the closed form.

In our laboratory, all three PAs in nanomolar concentrations increased the carbachol-induced ion flux in electric organ vesicles rich in nicotinic acetylcholine receptor.[26] Because PAs are present in brain and in electric organ,[26] this modulation might have physiological relevance.

PUT IS A PRECURSOR AND AN INHIBITOR OF GABA SYNTHESIS

PUT is a precursor of a small pool of GABA that may be synaptic because it is released by Ca^{2+}-dependent K^+ depolarization.[27] On the other hand, intraventricularly injected PUT decreases GABA in chick diencephalon, probably by inhibiting glutamate decarboxylase, the enzyme that synthesizes the major GABA pool.[28] This inhibitory activity on glutamate decarboxylase explains why intracerebral PUT administration is convulsant in rats.[29] Direct interaction of PUT with the GABA receptor, however, cannot be ruled out.

BEHAVIORAL ACTIVITY OF PAs

PAs might have a role in experience-induced cortical plasticity and mood regulation. Rats exposed to a complex environment had significantly increased cortical, but not subcortical, SPD. The stimulatory effect of environmental complexity on cortical weight was potentiated by inhibition of ODC by α-difluoromethylornithine.[30,31]

Systemically injected PUT was sedative at high doses and apparently anxiogenic at lower doses.[32]

CONCLUSION

The information about the regulatory role of PAs in the CNS is still fragmentary and comes from such dissimilar experimental systems as purified enzymes and live animals. It is clear that PAs act on a multitude of processes, some of which seem to have opposite effects. For example, SPM increases Ca^{2+} influx by modulating the *N*-methyl-D-aspartate receptor and decreases the free cytosolic Ca^{2+} by increase the affinity of the mitochondrial uniport. Most probably these processes happen at different times or in different compartments. Much remains to be done to integrate the available information into a coherent picture.

REFERENCES

1. ZAWIA, N. H. & S. C. BONDY. 1990. Electrically stimulated rapid gene expression in the brain: Ornithine decarboxylase and c-*fos*. Mol. Brain Res. **7:** 234-247.
2. MORGAN, J. I., D. R. COHEN, J. L. HEMPSTEAD & T. CURRAN. 1987. Mapping patterns of c-*fos* expression in the central nervous system after seizure. Science **237:** 192-197.
3. ARAI, A., M. BAUDRY, U. STAUBLI, G. LYNCH & C. GALL. 1990. Induction of ornithine decarboxylase by subseizure stimulation in the hippocampus *in vivo*. Mol. Brain Res. **7:** 167-169.
4. HAYASHI, Y., Y. HATTORI, A. MORIWAKI, K. SAEKI & Y. HORI. 1989. Changes in polyamine concentrations in amigdaloid-kindled rats. J. Neurochem. **53:** 986-988.
5. GUROFF, G. & G. DICKENS. 1983. The effect of nerve growth factor on polyamine metabolism in P12 cells. J. Neurochem. **40:** 1271-1276.
6. KOHSAKA, S., A. M. HEACOCK, P. D. KLINGER, R. PORTA & B. W. AGRANOFF. 1982. Dissociation of enhanced ornithine decarboxylase activity and optic nerve regeneration in goldfish. Dev. Brain Res. **4:** 149-156.
7. GREENE, L. A. & J. C. MCGUIRE. 1978. Induction of ornithine decarboxylase by nerve growth factor dissociated from effects on survival and neurite outgrowth. Nature **276:** 191-194.
8. KANJE, M., I. FRANSSON, A. EDSTROM & B. LOWKVIST. 1986. Ornithine decarboxylase activity in dorsal root ganglia of regenerating frog sciatic nerve. Brain Res. **381:** 24-28.
9. PASCHEN, W., G. ROHN, C. O. MEESE, B. DJURICIC & R. SCHMIDT-KASTNER. 1988. Polyamine metabolism in reversible cerebral ischemia: Effect of α-difluoromethylornithine. Brain Res. **453:** 9-16.
10. ROHN, G., M. KOCHER, U. OSCHLIES, K. HOSSMANN & W. PASCHEN. 1990. Putrescine content and structural defects in isolated fractions of rat brain after reversible cerebral ischemia. Exp. Neurol. **107:** 249-255.
11. KOENIG, H., A. D. GOLDSTONE & C. Y. LU. 1989. Polyamines mediate the reversible opening of the blood-brain barrier by the intracarotid infusion of hyperosmolal mannitol. Brain Res. **483:** 110-116.
12. BONDY, S. C. & C. H. WALKER. 1986. Polyamines contribute to calcium-stimulated release of aspartate from brain particulate fractions. Brain Res. **371:** 96-100.
13. VERMA, A. K., R. C. PONG & D. ERICKSON. 1986. Involvement of protein kinase C in primary culture of newborn mouse epidermal cell and in skin tumor promotion by 12-*O*-tetradecanoylphorbol-13-acetate. Cancer Res. **46:** 6149-6155.
14. HSIEH, J. T. & A. K. VERMA. 1988. Involvement of protein kinase C in the transcriptional regulation of ornithine decarboxylase gene expression by 12-*O*-tetradecanoylphorbol-13-acetate in T24 human bladder carcinoma cells. Arch. Biochem. Biophys. **262:** 326-336.
15. KIYOTO, I., S. YAMAMOTO, E. AIZU & R. KATO. 1987. Staurosporine, a potent protein kinase C inhibitor, fails to inhibit 12-tetradecanoyl-13-acetate-caused ornithine decarboxylase induction in isolated mouse epidermal cells. Biochem. Biophys. Res. Commun. **148:** 740-746.
16. QI, D., R. C. SHCHATZMAN, G. J. MAZZEI, R. S. TURNER, R. L. RAYNOR, S. LIAO & J. F. KUO. 1983. Polyamines inhibit phospholipid-sensitive and calmodulin-sensitive Ca^{2+}-dependent protein kinases. Biochem. J. **213:** 281-288.
17. MEZZETTI, G., M. G. MONTI & M. S. MORUZZI. 1988. Polyamines and the catalytic domain of protein kinase C. Life Sci. **42:** 2293-2298.

18. NICCHITTA, C. V. & J. R. WILLIAMSON. 1984. Spermine a regulator of mitochondrial calcium cycling. J. Biol. Chem. **259:** 12978-12983.
19. JENSEN, R. J., G. LYNCH & M. BAUDRY. 1989. Allosteric activation of brain mitochondrial Ca^{2+} uptake by spermine and by Ca^{2+}: Brain regional differences. J. Neurochem. **53:** 1182-1187.
20. LUNDBERG, G. A., B. JERGIL & R. SUNDLER. 1986. Phosphatidylinositol 4-phosphate kinase from rat brain: Activation by polyamines and inhibition by phosphatidylinositol 4,5-bisphosphate. Eur. J. Biochem. **161:** 257-262.
21. SMITH, C. & R. SNYDERMAN. 1988. Modulation of inositol phospholipid metabolism by polyamines. Biochem. J. **256:** 125-130.
22. RANSOM, R. W. & N. L. STEC. 1988. Cooperative modulation of [^{3}H]MK-802 binding to the *N*-methyl-D-aspartate receptor-ion channel complex by L-glutamate, glycine, and polyamines. J. Neurochem. **51:** 830-836.
23. CARTER, C., J. P. RIVY & B. SCATTON. 1989. Ifenprodil and SL82.0715 are antagonists at the polyamine site of the *N*-methyl-D-aspartate (NMDA) receptor. Eur. J. Pharmacol. **164:** 611-612.
24. REYNOLDS, I. J. & R. J. MILLER. 1989. Ifenprodil is a novel type of *N*-methyl-D-aspartate receptor antagonist: Interaction with polyamines. Mol. Pharmacol. **36:** 758-765.
25. CARTER, C., J. BENAVIDES, P. LEGENDRE, D. J. VINCENT, F. NOEL, F. THURET, K. G. LLOYD, S. ARBILLA, B. ZIVKOVIC, E. T. MACKENZIE, B. SCATTON & S. LANGER. 1988. Ifenprodil and SL82.0715 as cerebral anti-ischemic agents. II. Evidence for *N*-methyl-D-aspartate receptor antagonist properties. J. Pharmacol. Exp. Ther. **247:** 1222-1232.
26. SZCZAWINSKA, K., R. M. HANN, P. A. FERCHMIN & V. A. ETEROVIC. 1990. Modulation of acetylcholine receptor by polyamines. Cell. Mol. Neurobiol.: submitted for publication.
27. NOTO, T., H. HASHIMOTO, J. NAKAO, H. KAMIMURA & T. NAKAJIMA. 1986. Spontaneous releases of γ-aminobutyric formed from putrescine and its enhanced Ca^{2+}-dependent release by high K^+ stimulation in the brains of freely moving rats. J. Neurochem. **46:** 1877-1880.
28. NISTICO, G., R. IENTILE, D. ROTIROTI & R. M. DI GIORGIO. 1980. GABA depletion and behavioral changes produced by intraventricular putrescine in chicks. Biochem. Pharmacol. **29:** 954-957.
29. DE SARRO, G. B., C. ASCIOTI, M. G. AUDINO, V. LIBRI, A. DE SARRO & G. NISTICO. 1989. Behavioral and electrocortical effects after injection of two ornithine decarboxylase inhibitors into several areas of the brain in the rat. Neuropharmacology **28:** 1245-1251.
30. FERCHMIN, P. A. & V. A. ETEROVIC. 1987. Role of polyamines in experience-dependent brain plasticity. Pharmacol. Biochem. Behav. **26:** 341-349.
31. FERCHMIN, P. A. & V. A. ETEROVIC. 1990. Experience affects cortical but not subcortical polyamines. Pharmacol. Biochem. Behav. **35:** 255-258.
32. FERCHMIN, P. A. & V. A. ETEROVIC. 1990. Putrescine decreases exploration of a black and white maze. Pharmacol. Biochem. Behav. **37:** 445-449.

Developmental Visual Plasticity in *Drosophila*[a]

HELMUT V. B. HIRSCH, DOREEN POTTER,
DARIUSZ ZAWIERUCHA, TANVIR CHOUDHRI,
ADRIAN GLASSER, AND DUNCAN BYERS

Neurobiology Research Center
and
Department of Biological Sciences
State University of New York at Albany
Albany, New York 12222

Neuronal activity generated by sensory stimulation can affect development of the nervous system and of behavior.[1-3] We have been studying experience-dependent development of vision in the fruitfly, *Drosophila*, by comparing visually guided responses of flies reared in darkness with those of controls reared in a normal light/dark cycle.[4] Flies were cultured from the egg stage through the adult stage either in complete darkness until 1-3 hr before testing (dark-reared), or in normal 12:12 light/dark cycling illumination (light-reared). To compare the light- and dark-reared animals, we placed flies one at a time at the center of a cylindrical arena containing four sets of black vertical stripes of different widths on a white background. Flies, which were either of the flightless strain[5] *raised,* or of the Canton-S normal laboratory strain with wings clipped under CO_2 anesthesia, were allowed to walk about on the arena floor. For each fly, we recorded whether it approached the cylinder wall (within a 2-min test period), and if so, which set of vertical bars or intervening blank wall segment was approached. This testing was done blind: the experimenter did not know the individual rearing histories, and the testing sequence of experimentals and controls followed an unpredictable randomized order.

We found that when released at the center of the test arena, both light- and dark-reared *Drosophila* adults walked toward the cylinder wall, usually reaching it within 3-10 sec. Regardless of the rearing conditions, the majority of the flies walked to the three patterns composed of the widest stripes, avoiding both the pattern composed of the narrowest stripes and the blank wall. However, light- and dark-reared flies differed in the distribution of their preferences for the four sets of vertical black stripes, implying that development of visual behavior in *Drosophila* is dependent on visual experience.

FIGURE 1 illustrates the data for flies tested at 4 days of adult age. FIGURE 1A shows the fraction of light- and dark-reared flies approaching each of the four sets of stripes within the 2-min test period. The difference in preferences between dark-reared and control flies was small but significant ($N = 1263$ flies; chi-square $= 58.44$; $df = 8$; $p < .001$). FIGURE 1B presents a summary of the data from FIGURE 1A, showing the changes due to dark-rearing. The data show that the dark-reared flies were more likely than their respective controls to approach patterns containing the two sets of wider stripes and less likely to approach patterns with the narrower ones. Thus, dark-rearing *Drosophila* influences the distribution of their responses to the visual targets presented in our arena.

[a]This work was supported by a grant from the Whitehall Foundation.

To examine the effects of varying the duration of dark-rearing and the age at testing, we compared the distributions of responses for light- and dark-reared flies at 1, 2, 4, 6, and 7-8 days of adult age (the results for 4-day-old adults are those shown in FIG. 1). FIGURE 2A shows for light- and dark-reared flies the fractions choosing the two patterns composed of wider lines at different ages. Responses of light-reared flies (striped bars) to these patterns were similar at all the ages tested. At day 1 of the adult stage, responses of dark-reared flies (solid bars) to patterns with wider lines were not different from those of light-reared controls (chi-squared = 0.004; $df = 1$; $p < .95$). By 2 days, there was a small but not significant difference between light- and dark-reared flies (chi-squared = 2.79; $df = 1$; $p < .095$). The difference was larger and significant at 4 days (chi-squared = 19.83; $df = 1$; $p < .001$), 6 days (chi-squared = 19.83; $df = 1$; $p < .001$), and 7-8 days (chi-squared = 8.98; $df = 1$; $p < .003$). The preference for wider stripes changed significantly as a function of age for the dark-reared flies ($F_{Age} = 4.00$; $df = 4$; $p < .005$) but not for the light-reared flies ($F_{Age} = 0.18$; $df = 4$; $p < .950$). The difference scores summarized in FIGURE 2B changed with age, but this change was not significant ($F_{Age} = 1.84$; $df = 4$; $p < .126$). Thus, experience-dependent shifts in preferences for wider lines resulted mainly from changes in preferences of the dark-reared flies during the first days of adult life.

To examine further how the timing of the deprivation affected visual choices of flies in our test arena, we performed a control experiment in which flies were raised in normal light/dark cycling illumination until day 4 of the adult stage and then were kept in darkness for days 4 through 10. When tested on day 10, the visual choice behavior of these flies in our test was indistinguishable from the responses of control flies raised in normal cycling illumination until adult day 10 ($N = 399$ flies; chi-squared = 3.41; $df = 8$; $p < .91$), implying that prolonged darkness by itself did not cause the behavioral changes and that the duration and/or timing of the deprivation was critical.

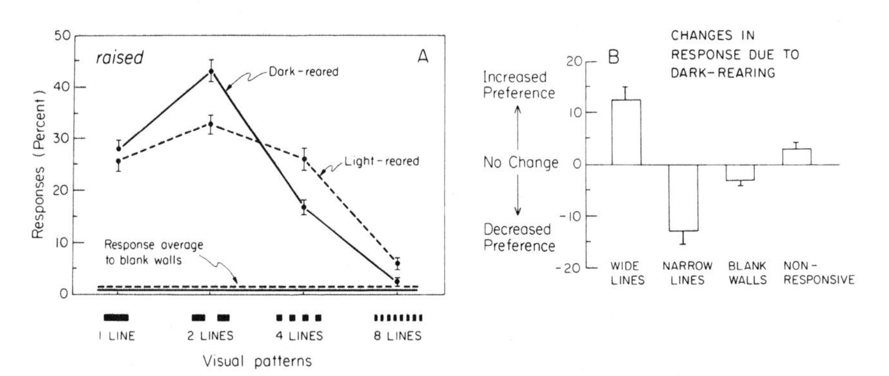

FIGURE 1. **(A)** Responses of 4-day-old light-reared and dark-reared flies of the *raised* strain ($N = 1263$ flies). Flies were tested in batches of 40. For each batch we computed the fractions approaching each of the four visual targets, the fractions approaching each section of blank wall, and the fraction not approaching the cylinder wall, and we averaged each statistic over all batches. The error bars are one SEM above and below the mean. **(B)** Summary of the data in **A**, showing dark-rearing-induced changes in the responses for the two patterns with wider lines, for the two patterns with narrower lines, for the four sections of blank wall together, and for the number of flies failing to approach the cylinder wall within the 2-min test period. The dark-reared flies responded more than their controls to patterns containing wider stripes. Reprinted with permission from Hirsch *et al.*[4]

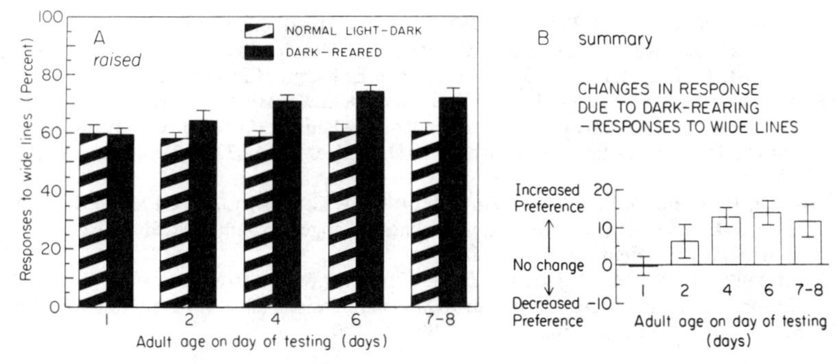

FIGURE 2. (**A**) The combined fraction of responses to the two patterns with wider stripes. The *raised* flies were tested at adult ages 1, 2, 4, 6, and 7-8 days. Responses of light-reared flies (striped bars) to patterns with wider stripes were similar at all the ages tested. Responses of dark-reared flies (solid bars) to patterns with wider stripes, however, increased with the length of the deprivation. (**B**) Summary of the differences in responses to patterns with wider stripes between light- and dark-reared flies. There was no difference after 1 day in the adult stage, a small difference at 2 days, and larger differences at 4, 6, and 7-8 days. Reprinted with permission from Hirsch *et al.*[4]

We conclude that the behavioral effects of dark-rearing were not the result of general deterioration of visual function because dark-rearing did not prevent flies from approaching the vertical stripes in the test cylinder, nor did it alter their phototaxis.[4] Furthermore, we believe that these experience-dependent behavioral changes were not caused by gross morphological changes because an examination of light- and dark-reared flies at the light-microscopic level showed that the histological appearance and morphology of the eye and visual nervous system of dark-reared flies appeared to be normal.[4]

In summary, the behavioral preferences of normal light-reared flies were stable over the first 1 to 8 days of adult life. Preferences of dark-reared flies, however, changed over the first days of adult life; flies dark-reared for longer periods responded more than did controls to the wider stripes. Because deprivation-induced change did not take place if flies had received visual exposure during the first days of adult life, visual exposure must somehow be protecting the visual system from the effects of subsequent deprivation. Furthermore, this finding that visual experience during early adult life affects the use of vision thereafter provides evidence for developmental visual plasticity in *Drosophila*.

Our paradigm makes it possible to examine the many known mutants affecting other forms of plasticity (for example, associative learning) to identify any that also alter developmental visual plasticity. Locating such mutants would help us test the possible contributions of the genes in question and their biochemical products to mechanisms of development of vision.

REFERENCES

1. HIRSCH, H. V. B. 1985. The tunable seer: Activity-dependent development of vision. *In* Handbook of Behavioral Neurobiology. E. Blass, Ed. Vol. 8: 237-295. Plenum. New York, NY.

2. HIRSCH, H. V. B. & S. B. TIEMAN. 1987. Perceptual development and experience-dependent changes in cat visual cortex. *In* Sensitive Periods in Development: Interdisciplinary Perspectives. M. H. Bornstein, Ed.: 39-79. Lawrence Erlbaum. Hillsdale, NJ.
3. SCHMIDT, J. T., L. E. EISELE, J. C. TURCOTTE & D. G. TIEMAN. 1986. Selective stabilization of retinotectal synapses by activity dependent mechanism. *In* Adaptive Processes in Visual and Oculomotor Systems. E. L. Keller & D. S. Zee, Eds.: 63-70. Pergamon. New York, NY.
4. HIRSCH, H. V. B., D. POTTER, D. ZAWIERUCHA, T. CHOUDHRI, A. GLASSER, R. K. MURPHEY & D. BYERS. 1990. Rearing in darkness changes visually-guided choice behavior in *Drosophila*. Vis. Neurosci. **5:** 281-289.
5. MAHAFFEY, J. W., M. D. COUTU, E. A. FYRBERG & W. INWOOD. 1985. The flightless *Drosophila* mutant *raised* has two distinct genetic lesions affecting accumulation of myofibrillar proteins in flight muscles. Cell **40:** 101-110.

A Cholinergic Circuit Intrinsic to Optic Tectum Modulates Retinotectal Transmission via Presynaptic Nicotinic Receptors[a]

W. MICHAEL KING AND JOHN T. SCHMIDT

Department of Biological Sciences
and
Neurobiology Research Center
State University of New York at Albany
Albany, New York 12222

The action of acetylcholine (ACh) at the nicotinic receptors of the neuromuscular junction has served as a model for study of synaptic transmission throughout the nervous system, but the function of ACh in the brain has remained obscure. Some evidence points to a presynaptic modulatory function. Nicotinic receptors are found on afferent nerve terminals in several regions of vertebrate brain,[1-4] and their stimulation causes biochemically measurable transmitter release,[5,6] but there was previously no direct evidence that activation of these receptors modulates synaptic function. We recently found that nicotinic agonists directly depolarize optic nerve terminals in goldfish tectum (assessed via sucrose gap recordings) and simultaneously enhance synaptic transmission (assessed via field potentials (FP) after optic nerve stimulation) in an *in vitro* tectal preparation.[7] With synaptic transmission blocked (high Mg^{2+}/no Ca^{2+}—to assure that the actions of nicotinic agonists were direct), either cytosine (10-500 nM), nicotine (2-10 μM), or carbachol (4 μM in presence of 10 μM atropine) depolarized retinal terminals by 6-12 mV from a resting level of about 25-30 mV (FIG. 1). In standard Ringers, 50 nM cytosine increased the amplitude of the monosynaptic FP by 43 $\pm$ 4%, eliminated the long-latency FP, and depolarized the terminals by about 5 mV. With lower calcium concentrations, augmentation of transmission was much larger.

Because nicotinic agents eliminated the long-latency FP, we tested the hypothesis that this component is generated by an intrinsic tectal circuit that uses ACh to depolarize retinal terminals past threshold and releases a second burst of transmitter. This possibility was supported by the fact that the sources and sinks of the long-latency FP matched those of the monosynaptic FP.[8] Recording from the stump of the nerve, we found that associated with the long-latency FP is a rebound compound action potential (CAP) at a latency of about 15 msec (FIG. 2). In addition, all manipulations that blocked the long-latency polysynaptic FP in tectum also blocked the retinopetal CAP in the optic nerve. When the long-latency FP was blocked by 1-5 μM carbachol, alcuronium, or curare, by high Mg^{2+}, or by increasing stimulus frequency, the retinopetal CAP in the optic nerve was also eliminated. Because the amplitude of the rebound CAP was often over 50% of the initial orthodromic CAP, it is likely that the rebound CAP reflects backfiring of many optic fibers (antidromic activity originating at retinal terminals) rather than activity in the efferents to the retina, which are very few in

[a]This work was supported by a grant from the National Institutes of Health (EY-03736).

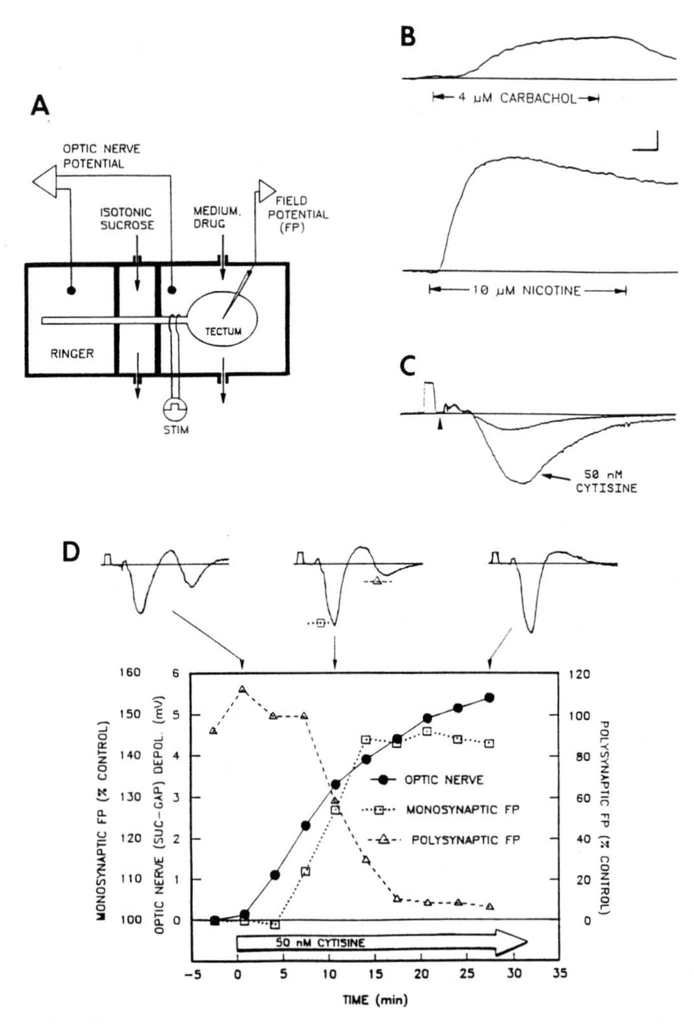

FIGURE 1. **(A)** Diagram of sucrose gap apparatus. **(B)** Sucrose gap recordings showing that nicotine and carbachol (in the presence of 10 μM atropine) directly depolarize optic nerve terminals in 4 mM Mg^{2+} medium with no added Ca^{2+}. Calibration bars: 1 mV, 1 min. **(C)** Tectal recordings showing enhancement of the monosynaptic FP in high Mg^{2+} (2.5 mM)/low Ca^{2+} (1.2 mM) medium by the nicotinic agonist cytosine. The initial square wave is a calibration pulse (1 mV, 2 msec), and the time of stimulus is indicated by an arrowhead in this and subsequent tectal recordings; stimulus artifacts were electronically suppressed. **(D)** Plot of three effects of cytosine in standard medium as a function of time after application: augmentation of the monosynaptic FP, decline of the long-latency FP, and optic terminal depolarization. Example responses are shown at the top at three time points (arrows). Reprinted with permission from reference 7.

number.[9] In addition, double stimulus CAP collision experiments also suggest "backfiring" (FIG. 3). Paired stimuli were delivered at varying interstimulus intervals. When the second stimulation occurred just before the expected long-latency FP in tectum, the rebound CAP expected from the first stimulus was blocked. In addition, the presynaptic volley and monosynaptic FP in tectum expected from the second stimulus were substantially reduced or blocked. This occlusion suggests that orthodromic action potentials of the second stimulus collided with the antidromic action potentials of the rebound CAP in the same axons. On the other hand, when the long latency FP and the rebound CAP were eliminated (increased frequency or nicotinic agents), no such occlusion of the second monosynaptic FP was apparent. This rebound CAP demonstrates that retinal terminals can be strongly depolarized by ACh released onto them from cholinergic tectal circuits and demonstrates an endogenous modulation consistent with the previous finding that exogenous carbachol and eserine enhance retinotectal synaptic transmission.[10]

Two sources of cholinergic innervation identified by choline acetyltransferase staining could supply this modulation, and both are driven by optic input. The first is a

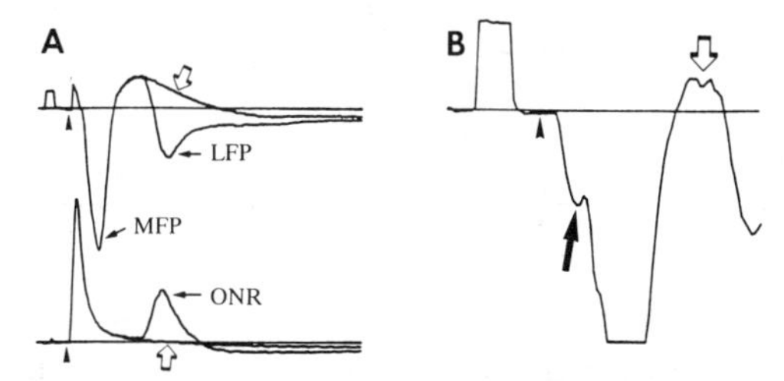

FIGURE 2. (**A**) Simultaneous recordings from tectum (field potentials, upper) and optic nerve (suction electrode, lower) before and after (open arrows) application of the nicotinic antagonist alcuronium (1 μM). Note that the long-latency FP (LFP) is temporally associated with a delayed positivity in the optic nerve (ONR) and that both are blocked by alcuronium. Initial deflection in the optic nerve recording is the compound action potential elicited by nerve shock. Calibration pulse: 1 mV, 2 msec for tectal trace and 2 mV for nerve recording. (**B**) Apparent nerve terminal spike precedes the long-latency FP. Deflection (open arrow) preceding the long-latency FP resembles the presynaptic nerve terminal spike (filled arrow) preceding the monosynaptic FP (MFP). Calibration pulse: 1 mV, 2 msec. Reprinted with permission from reference 7.

subgroup of type XIV pyriform neurons of deep tectum that have dendrites and recurrent axonal collaterals in the optic terminal layer.[11] These mediate the long-latency FP.[7] Because we have shown elsewhere that the long-latency FP spreads into tectal areas adjacent to those receiving retinal input,[8] and because the visual field maps retinotopically onto the tectum,[12] this intrinsic tectal recurrent cholinergic circuit may enhance retinotectal transmission from repeated stimuli in the same or adjacent areas of the visual field. The second cholinergic input to tectum arises from nucleus isthmi.[13] Direct stimulation of nucleus isthmi results in enhancement of the monosynaptic and especially the long-latency component of retinotectal transmission,[8] and also depolarizes optic nerve terminals (unpublished observations).

Beyond the role in visual processing, these recurrent nicotinic circuits may play an important role in the activity-driven sharpening of the retinotopic map during regeneration and development.[14] Visual synaptic inputs are apparently stabilized by patterned visual activity.[15,16] The convergence of active inputs onto a common postsyn-

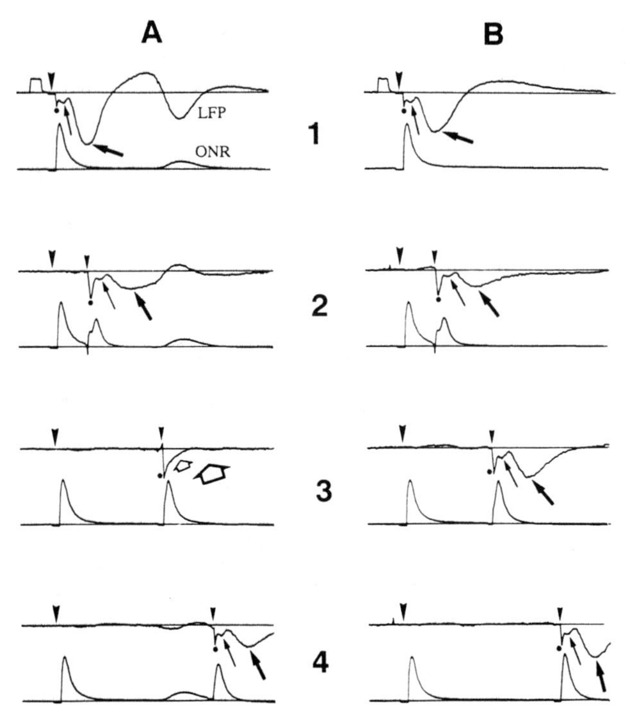

FIGURE 3. Collision experiment using paired stimuli delivered at 0.06 Hz reveals occlusion of the tectal field potential and the optic nerve rebound (ONR). Upper and lower traces in each panel are simultaneous recordings from tectum and optic nerve stump (suction electrode), respectively, while large and small arrowheads indicate time of first and second stimuli of each pair, respectively. Small arrows indicate the presynaptic component of the tectal field potential, that is, the compound action potential of the optic nerve terminals. Large arrows indicate the initial postsynaptic component of the field potential, the monosynaptic FP (MFP). Shock artifacts were only minimally suppressed electronically to insure adequate visualization of the presynaptic component of the field potentials. **(A1)** Responses to single shock. **(A2-A4)** Responses to paired stimuli with increasing interstimulus intervals. In order to better visualize the tectal response to the second shock, the response to a single shock **(A1)** was subtracted from the responses to paired stimuli. When the second shock just preceded the long-latency FP (LFP) **(A3)**, both presynaptic and postsynaptic components of the tectal response to the second stimulus were absent (open arrows), but were present when the interstimulus interval was shorter **(A2)** or longer **(A4)**. Calibration pulse shown in tectal trace of **A1** indicates 1 mV, 2 msec for both tectal and nerve recordings. **(B)** Elimination of the long-latency FP and optic nerve positivity by increasing the frequency of paired stimuli to 6 Hz eliminates occlusion. Notation as in **A**. Note the presence of both presynaptic and postsynaptic components of the tectal responses at each of the interstimulus intervals. Reprinted with permission from reference 7.

aptic neuron results in a summation of excitatory postsynaptic potentials and depolarization to a level sufficient for *N*-methyl-D-aspartate receptors to allow calcium entry as an intracellular signal. Given the dependence of this mechanism on the level of synaptic activation, it is not surprising that modulators would produce strong effects. Two reports have demonstrated that blocking nicotinic receptors locally in tectum with α-bungarotoxin results in local destabilization of retinotectal synapses.[17,18] In visual cortex, destruction of both cholinergic and adrenergic innervation interferes with binocular competition during monocular deprivation.[19] Although that study did not distinguish between nicotinic and muscarinic actions, nicotinic receptors have been found on the geniculocortical afferents.[3]

In summary, nicotinic modulation of transmitter release in the visual system may play substantial roles both in development and in ongoing visual processing.

REFERENCES

1. Henley, J. M., J. M. Lindstrom & R. E. Oswald. 1986. Science **232:** 1627-1629.
2. Prusky, G. T. & M. S. Cynader. 1988. Vis. Neurosci. **1:** 245-248.
3. Prusky, G. T., C. Shaw & M. S. Cynader. 1987. Brain Res. **412:** 131-138.
4. Swanson, L. W., D. M. Simmons, P. J. Whiting & J. Lindstrom. 1987. J. Neurosci. **7:** 3334-3342.
5. Lapchak, P. A., M. Araujo, R. Quiron & B. Collier. 1989. J. Neurochem. **53:** 1843-1851.
6. Rowel, P. P. & D. L. Winkler. 1989. J. Neurochem. **43:** 1593-1598.
7. King, W. M. 1990. Brain Res. **527:** 150-154.
8. King, W. M. & J. T. Schmidt. 1991. Neuroscience **40:** 701-712.
9. Schmidt, J. T. 1979. Proc. R. Soc. London **205:** 287-306.
10. Langdon, R. B. & J. A. Freeman. 1987. J. Neurosci. **7:** 760-773.
11. Tumosa, N., W. K. Stell, C. D. Johnson & M. L. Epstein. 1986. Brain Res. **370:** 365-369.
12. Attardi, D. G. & R. W. Sperry. 1963. Exp. Neurol. **7:** 46-64.
13. Zottoli, S. J., K. J. Rhodes, J. G. Corrodi & E. J. Mufson. 1988. J. Comp. Neurol. **273:** 385-398.
14. Schmidt, J. T. 1990. Ann. N.Y. Acad. Sci. This volume.
15. Bear, M. F. & S. M. Dudek. 1990. Ann. N.Y. Acad. Sci. This volume.
16. Udin, S. B. & W. J. Scherer. 1990. Ann. N.Y. Acad. Sci. This volume.
17. Freeman, J. A. 1977. Nature **269:** 218-222.
18. Schmidt, J. T. 1985. J. Neurophysiol. **53:** 237-257.
19. Bear, M. F. & W. Singer. 1986. Nature **320:** 172-176.

Relationship between Tubulin Delivery and Synapse Formation during Goldfish Optic Nerve Regeneration[a]

DENIS LARRIVEE

Department of Physiology
Cornell University Medical College
New York, New York 10021

In a number of neurons axotomy suppresses the content of synaptic transmitter-related proteins, including enzymes involved in the synthesis and degradation of neurotransmitters. These observations have prompted the suggestion that during regeneration the neuron modifies its delivery of proteins to the growing axons, decreasing that of synaptic proteins while increasing that of structural proteins during axonal outgrowth, and reversing this trend during synapse formation.[1] We examined this hypothesis in the cases of the growth-associated protein (GAP) and synapsin, two prominent synaptic proteins, and tubulin, a prominent structural protein, during normal and perturbed synaptogenesis.

METHODS

Protein delivery to the axon terminals was assayed in synaptosomal preparations from goldfish optic tecta 22-24 hr after injection of [^{3}H]proline.[2] During the period of labeling protein synthesis was suppressed in the optic tectum by injecting cycloheximide (5 μg) into the cranial cavity 1 hr before injecting label into the eye.

To inhibit physiological activity in the optic axons, tetrodotoxin (TTX) was injected (0.06 μg/eye) into the right eye every 3 days between 18 and 35 days of regeneration.[3]

RESULTS

All measurements were made at 7-day intervals, beginning 17 days after nerve injury and ending at 45 days, in order to include the period from the onset of synapse formation to that of synaptic refinement.[3] At 17 days of regeneration [^{3}H]proline labeling of total synaptosomal protein was 80% of maximum and rose to 92% at 24 days, thereafter slowly increasing (FIG. 1). By contrast, that of synapsin and GAP was transiently elevated near the onset of synaptogenesis. In the case of synapsin, labeling

[a] This work was supported by grants from the National Institute of Neurological and Communicative Disorders and Stroke (NS-09015 and NS-14967) and by a grant from the Spinal Cord Research Foundation (SFR-572).

rose from 51% of maximum at 17 days of regeneration to 100% at 24 days and then fell rapidly. The increase in labeling of GAP preceded that of synapsin by about 7 days, but otherwise displayed a similar time course. Tubulin labeling increased roughly in parallel with the two synaptic proteins; in contrast to these proteins, however, it decreased slowly. At 45 days, for example, it had fallen by only 39% from its maximum

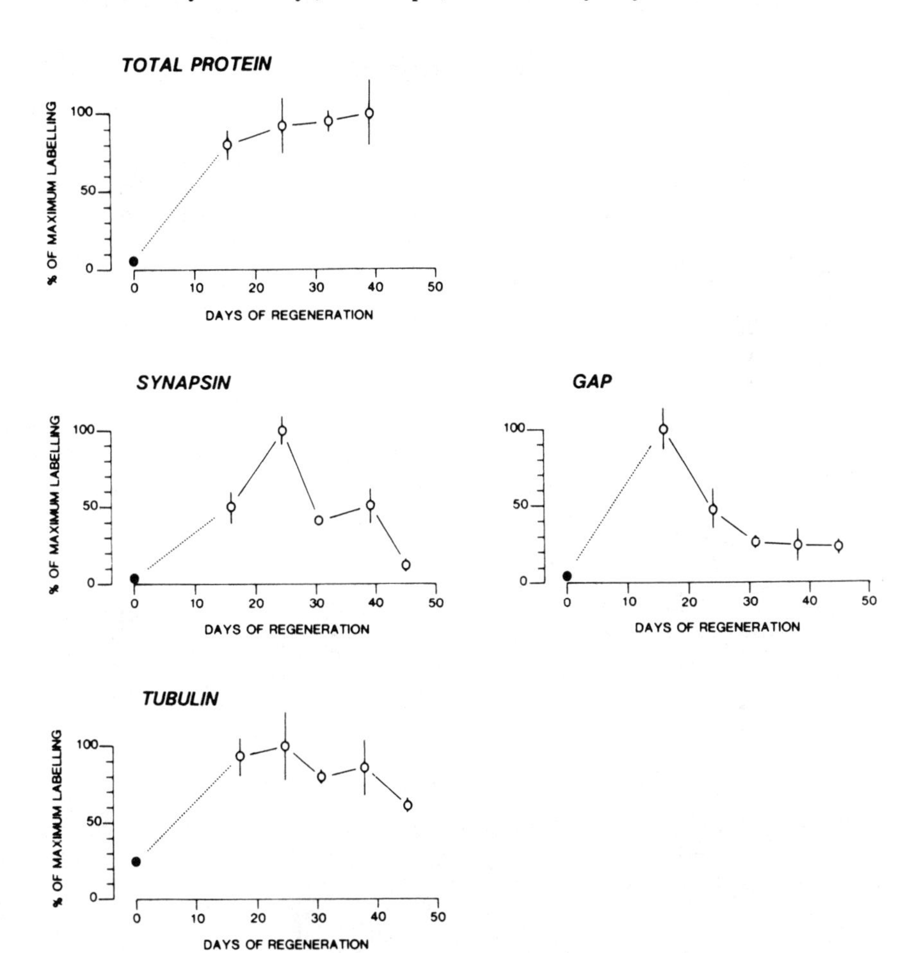

FIGURE 1. Time course of protein delivery to goldfish optic axon terminals during synaptogenesis. Protein delivery was assayed as described in Methods. Each value represents a mean ± SEM of three to four independent observations after normalization to the maximum mean value attained during synaptogenesis.

value. (We detected significant labeling of tubulin 24 hr after injecting [^{3}H]proline into the eye, consistent with previous observations that showed large increases in its synthesis and velocity of transport during regeneration.[4])

Because inhibiting physiological activity in the optic fibers is known to modify synapse formation by preventing visual field refinement in the optic tectum,[3] we monitored the amount of label in each protein following chronic, TTX-induced impulse blockade (Methods). Under these conditions labeling of synapsin and GAP underwent a small, but not statistically significant, reduction (FIG. 2). That of tubulin, however, more than doubled after TTX treatment.

DISCUSSION

The current study indicates that some prominent synaptic proteins are not selectively delivered to the optic synapses during synapse formation following axotomy. Most of the delivery of GAP and synapsin, in fact, was confined to a relatively brief interval early in synaptogenesis, whereas the delivery of tubulin was considerably more prolonged. In the case of the synaptic proteins it is possible that the number of optic synapses formed at the beginning of synaptogenesis is the primary determinant regulating the amount of protein delivered to the axon terminals, since the number of

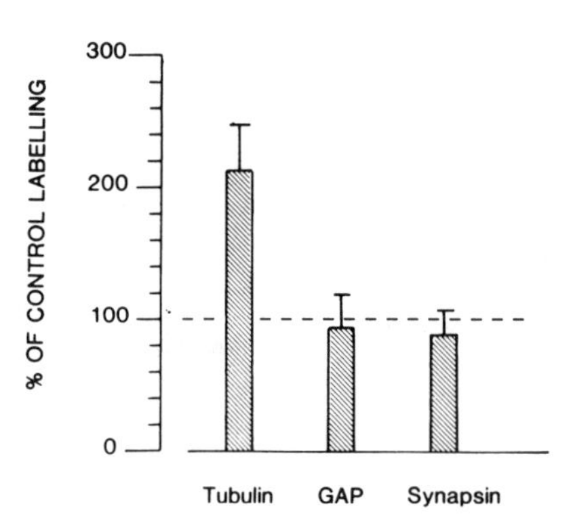

FIGURE 2. Effect of blocking physiological activity in the optic axons on protein delivery. All assays were performed at 35 days of regeneration and were preceded by chronic application of TTX as specified (Methods). Values are given as the percentage of labeling at 35 days of regeneration in control nerves ($N = 6$). Each error bar denotes an SEM.

such synapses is relatively fixed throughout this stage.[5] The prolonged delivery of tubulin, on the other hand, is likely to reflect continued axon outgrowth for which tubulin is needed.[6] Indeed, some fiber extension is almost certain to occur later in synaptogenesis in order to generate new synaptic contacts as the visual fields are gradually refined.[3,7] The extent of outgrowth, or remodeling, is probably significantly influenced by physiological activity in the optic synapses, since impulse blockade greatly enhances tubulin delivery (FIG. 2). Conversely, increased synaptic activity may be expected to suppress tubulin delivery and retard fiber outgrowth.

ACKNOWLEDGMENTS

We are grateful to Dr. Andrew Czernik and the laboratory of Dr. Paul Greengard, Rockefeller University, for providing synapsin antibody.

REFERENCES

1. Reis, D. J., R. A. Ross, G. Gilad & T. H. Joh. 1978. *In* Neuronal Plasticity. C. W. Cotman, Ed.: 197-226. Raven Press. New York, NY.
2. Goelz, S., E. Nestler & P. Greengard. 1985. J. Neurochem. **45:** 63-72.
3. Schmidt, J. & D. L. Edwards. 1983. Brain Res. **269:** 29-39.
4. Perry, G., D. Burmeister & B. Grafstein. 1987. J. Neurosci. **7:** 792-806.
5. Hayes, W. P. & R. L. Meyer. 1989. J. Neurosci. **9:** 1400-1413.
6. Bamburg, J. R., D. Bray & K. Chapman. 1986. Nature **321:** 788-790.
7. Meyer, R. L. 1983. Science **218:** 589-591.

Dual Effects of the Protein Kinase Inhibitor H-7 on CA1 Responses in the Hippocampal Slice[a]

J. CLANCY LEAHY AND MARY LOU VALLANO

Department of Pharmacology
State University of New York Health Science Center at Syracuse
Syracuse, New York 13210

Long-term potentiation (LTP) in the hippocampus is currently one of the most attractive models for mammalian learning and memory. In the pathway from CA3 to CA1, LTP can be recorded extracellularly as an enhancement in the population excitatory postsynaptic potential (EPSP) amplitude/slope as well as the population spike amplitude, following tetanization of the Schaffer-commissural fibers. The molecular mechanisms underlying LTP remain unclear, although the induced changes in EPSP and spike potentiation may be distinct.[1] Recent studies employing high concentrations (300 μM) of the protein kinase inhibitor H-7 (that is, 1-(5-isoquinolinylsulfonyl)-2-methylpiperazine) support a role for calcium-dependent kinases in both the induction and maintenance of EPSP potentiation.[2] In a separate study, however, 100 μM H-7 neither prevented the induction of synaptic LTP nor abolished previously established LTP.[3] Additionally, low concentrations (10-100 μM) of H-7 have been shown to enhance the CA1 population spike amplitude and produce multiple responses.[4] The goal of this study was to examine whether H-7 exhibits concentration-dependent effects on more than one pathway, which may explain the apparent discrepancies described above.

Experiments were performed using the *in vitro* hippocampal slice preparation from young (15-60 days postnatal) Sprague-Dawley rats. Hippocampal slices (400 μm) were placed in a submersion chamber and perfused at 4 ml/min with medium continuously bubbled with 95% O_2/5% CO_2 at 33-34 °C, pH 7.4. Standard medium contained 124 mM NaCl, 5 mM KCl, 1.25 mM KH_2PO_4, 2.0 mM $MgSO_4$, 2.0 mM $CaCl_2$, 25 mM $NaHCO_3$, and 10 mM glucose. Extracellular CA1 pyramidal cell responses were recorded with glass micropipettes containing 2 M NaCl, using conventional recording procedures. The recording electrode was positioned in stratum pyramidale, with bipolar stimulating electrodes placed on either side in stratum radiatum. The stimulus intensity at both sites was adjusted to evoke a population spike amplitude of about 1.0-1.5 mV (that is, 20-30% of maximum amplitude). Initially, test stimuli (0.1-msec pulses) were delivered to both pathways at a rate of 1/20 sec (10-sec interstimulus interval) for 10-15 min to establish a baseline response. Following the test period, a high-frequency stimulus (HFS) was delivered to the S1 pathway (100 Hz, 1.0 sec, 0.2-msec pulses). The effect of the HFS on the population spike amplitude was then monitored in both pathways for 20 min, employing the initial test stimulus pattern. At this point, the standard medium was replaced with one containing either 50 μM or 300 μM H-7 (Sigma). The medium containing H-7 was administered for 15-20 min followed by washout with standard medium for 30-60 min.

Examples of the time course of the effect of H-7 on CA1 responses in control and potentiated pathways are shown in FIGURE 1. Following the HFS in the S1 pathway,

[a]This work was supported by a grant from the U.S. Public Health Service (NS 24705).

the S1 population spike amplitude exhibited a characteristic initial increase followed by a slow decline to a level of about 200% above baseline. The population spike amplitude in the S2 pathway was either unaffected (FIG. 1) or exhibited a short-term potentiation (< 5 min, not shown). Examples of the effects of 50 μM and 300 μM H-7 on CA1 responses are shown in FIGURE 2. Addition of 50 μM H-7 20 min after the HFS typically resulted in the onset of multiple population spikes (two to four) and an increase in the population spike amplitude, particularly in the previously unpotentiated

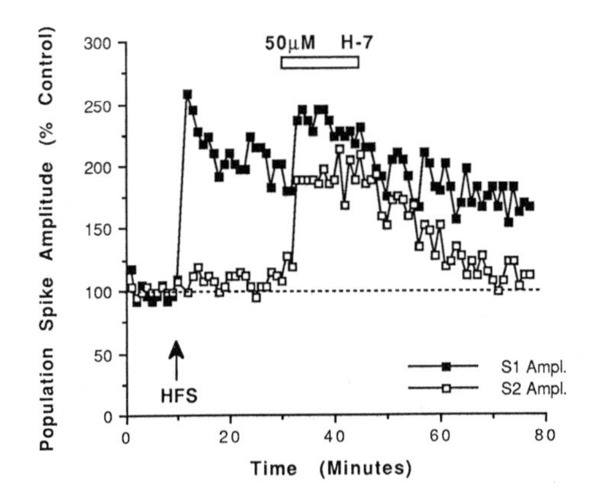

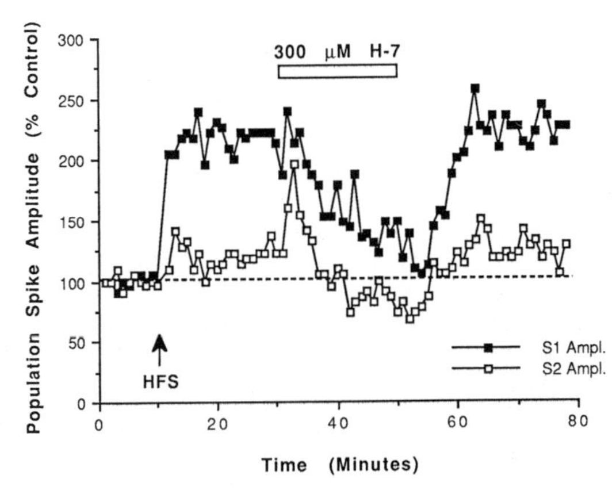

FIGURE 1. Time course of the effect of 50 μM and 300 μM H-7. Bath application of 50 μM H-7 (*top*), following the induction of LTP (HFS, arrow), resulted in an increase in the amplitude of the population spike, particularly in the previously unpotentiated (S2, control) pathway. Bath application of 300 μM H-7 (*bottom*) resulted in a marked inhibition of the population spike amplitude, particularly in the previously potentiated (S1, LTP) pathway. Both H-7 effects were reversible following washout with standard medium.

(S2, control) pathway. In the presence of 300 μM H-7, however, the population spike amplitude was reduced, particularly in the previously potentiated (S1, LTP) pathway. In two of six slices, 300 μM H-7 treatment also induced an initial, transient increase in the population spike amplitude prior to the reduction in spike amplitude (FIG. 1, S2 pathway). The effects of both concentrations of H-7 on CA1 responses were reversible following a washout period with standard medium.

These results, in conjunction with pharmacological studies using GABAergic compounds,[4] suggest that H-7 exerts concentration-dependent effects in different pathways.

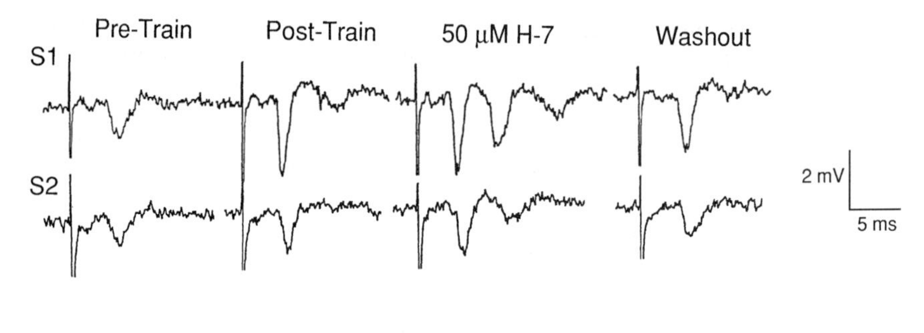

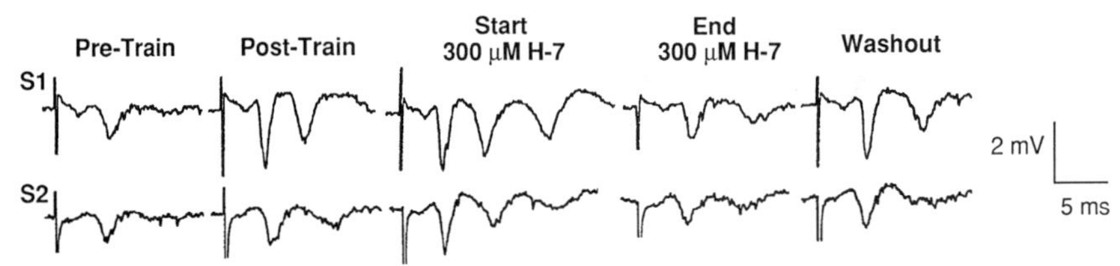

FIGURE 2. Concentration-dependent effect of H-7. *Top*: examples of responses in the LTP (S1) and control (S2) pathway from a representative experiment in which the effect of 50 μM H-7 was examined. Note the onset of multiple responses in both pathways and the enhancement of the initial spike in the S2 pathway during H-7 treatment. Negativity is downward. Population spike amplitude was measured from the peak negativity to the subsequent peak positivity. *Bottom*: same as above except that the effect of 300 μM H-7 was examined. An additional response was elicited following HFS of the S1 pathway. Note that additional responses were elicited at the start of H-7 treatment, followed by a marked inhibition particularly in the S1 pathway. Responses in both pathways returned to pretreatment levels following washout (30-60 min).

At low concentrations (≤ 50 μM), H-7 disinhibits CA1 neurons by reducing inhibition from local circuit neurons. At high concentrations (300 μM), H-7 exerts a more powerful effect in the CA3/CA1 pathway to inhibit transmitter release presynaptically and/or inhibit CA1 responsiveness postsynaptically. In contrast to previous studies demonstrating a specific inhibition of EPSP potentiation by H-7,[2,5] inhibition of the population spike amplitude was not restricted to the previously potentiated (S1) pathway in the present study.

REFERENCES

1. Taube, J. S. & P. A. Schwartzkroin. 1988. J. Neurosci. **8:** 1632-1644.
2. Malinow, R., D. V. Madison & R. W. Tsien. 1988. Nature **335:** 820-824.
3. Muller, D., J. Turnbull, M. Baudry & G. Lynch. 1988. Proc. Natl. Acad. Sci. USA **85:** 6997-7000.
4. Corradetti, R., A. M. Pugliese & N. Ropert. 1989. Br. J. Pharmacol. **98:** 1376-1382.
5. Malinow, R., H. Schulman & R. W. Tsien. 1989. Science **245:** 862-866.

Influence of Temperature on Adaptive Changes of the Vestibuloocular Reflex in the Goldfish

JAMES G. McELLIGOTT

Department of Pharmacology
Temple University School of Medicine
Philadelphia, Pennsylvania 19140

MICHAEL WEISER AND ROBERT BAKER

Department of Physiology and Biophysics
New York University Medical Center
New York, New York 10016

The vestibuloocular reflex (VOR) functions to keep the eyes fixed on a target of visual interest in the presence of head movements. Thus, the VOR minimizes blurring of a scene by reducing movement of visual stimuli across the retina. Adaptive modification of the gain of this reflex takes place to compensate for changes during development, during aging, or after injury. The gain of the VOR is defined as the slow phase movement of the eyes relative to the head during head movement. Recent experimentation[1] has shown that fishes, like other vertebrates, are capable of VOR gain modification. In addition, the neural substrates of the VOR in the fish are similar to those in the mammals. Unlike the homothermic mammals, however, fish are ectotherms and are subject to large temperature fluctuations. The aim of this study was to investigate the effect temperature has on the VOR and on adaptive VOR changes.

Goldfishes that had been acclimated to a temperature of approximately 20 °C for a period of at least 4 weeks were used for VOR modification. Each had its head restrained by placing its mouth around a respiration tube, which passed oxygenated water over its gills. Furthermore, body movements were minimized by placing each fish between two Plexiglass plates. The restraining apparatus was located in a small cylindrical test aquarium that was secured to a servo-controlled vestibular platform that oscillated the animal around the vertical axis ($\pm 20°$, $\sim \frac{1}{8}$ Hz). A set of electromagnetic field-generating coils that were used to measure eye movements were also attached to the platform. Small ocular coils (diameter = 1.5 mm) were attached to each eye so as not to obstruct vision. These coils were used to measure eye movements by the search coil technique developed by Robinson.[2]

Aligned above the fish in the same vertical axis as the platform was a second servo motor, which rotated a planetarium that displayed a random dot stimulus pattern of light on the inside painted surface of the cylindrical aquarium. Initial calibration of the VOR (gain = 1×) was achieved by oscillating the platform and the animal in the presence of earth-fixed, stationary visual stimuli. Sinusoidal oscillation of the visual stimuli 180° out of phase with the test platform has the effect of making the animal attempt to increase the gain of the VOR in order to attain a stable visual image. Decreases in VOR gain were achieved by rotating the visual stimuli and the table in phase. During the modification period, which ranged from 3.5 to 15 hr for the different experiments, the signals from the test table, the visual stimuli servomotor, and the horizontal and vertical eye position and velocity were monitored. VOR gain in the dark was measured before and after modification and at both the acclimated and reduced temperatures. Temperature of the fish and the aquarium water was kept constant (±0.1 °C) by means of a thermoregulator. Increases or decreases in water and fish temperature were achieved within 15 min by exchanging water that was respectively warmer or cooler than that in the test aquarium.

Goldfish were trained to modify VOR gain at the temperature to which they had been acclimated. Significant gain decreases and increases were generally achieved within a period of 1 to 2 hr. FIGURES 1A-1C present examples of three goldfishes that were trained to produce gain decreases (training toward VOR gain = 0×). Each histogram in the figure plots VOR gain in the dark for both eyes of the fishes. Significant adaptive

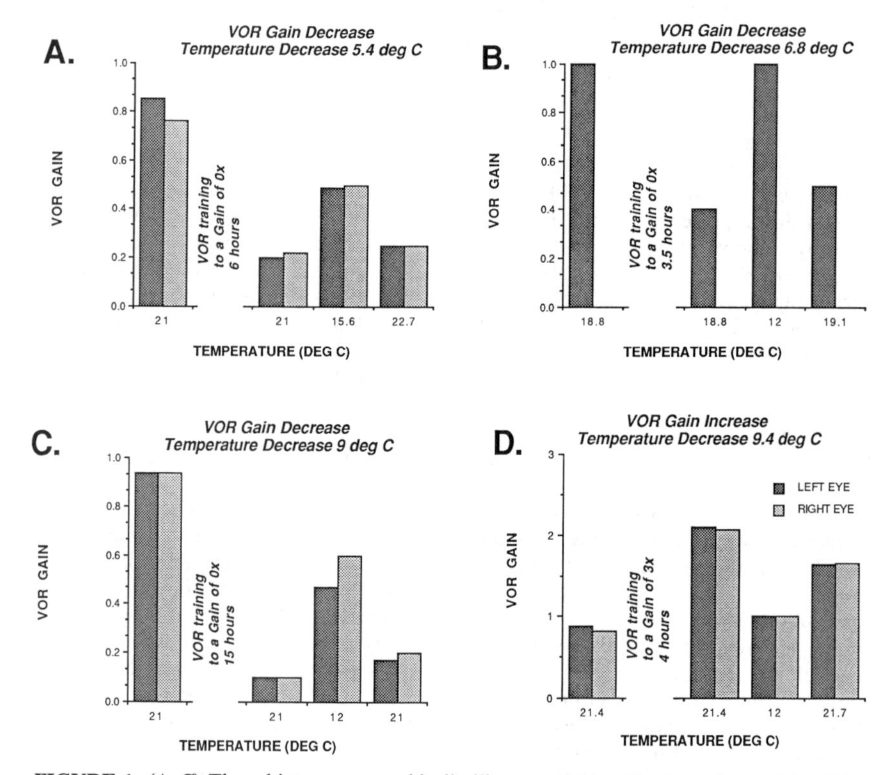

FIGURE 1. (**A-C**) These histograms graphically illustrate the inactivation of a modified VOR gain decrease in three different fishes after a temperature reduction ranging from 5.4 to 9.0 °C. VOR gain was measured for both eyes in all animals except for that presented in **B**. In each panel, the first histogram columns represent VOR gain in the dark before modification. The subsequent columns represent, respectively, the modified gain after modification at the acclimation temperature, then after a temperature reduction, and finally after restoration to the acclimation temperature. Temperature in °C is presented below each set of columns. (**D**) The same phenomenon as presented in **A-C** but for a modified VOR gain increase. Note that there is a restoration of the modified gain in all cases when the temperature is returned to its acclimation value.

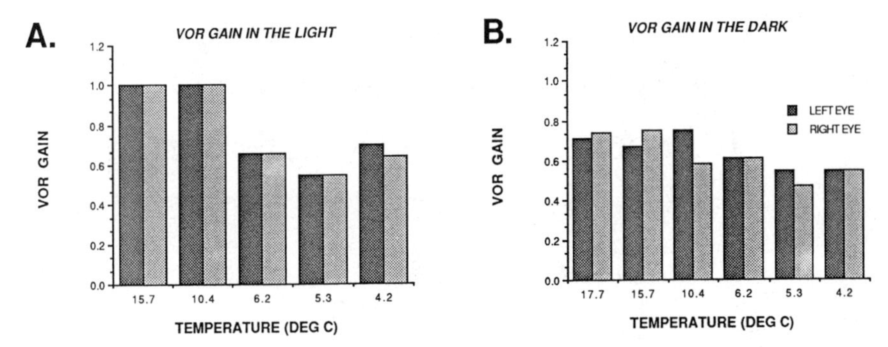

FIGURE 2. These histograms present the effect of temperature on the VOR in the light (**A**) and in the dark (**B**). Note that for a temperature decrease to 10 °C, there is no change in normal unmodified VOR gain.

VOR gain decreases (0.1 to 0.4) were produced. These newly modified VOR adaptive gain changes, which occurred in 3.5 to 15 hr, were inactivated by an acute temperature decrease of 5.4 to 9.0 °C. Thus, when the temperature was lowered, the value of the gain returned to that of an unmodified VOR. This effect was reversible, however, since return of the fishes to their acclimated temperature restored the VOR to the gain achieved after modification. This phenomenon occurs for either modified gain decreases (FIGS. 1A-1C) or increases (FIG. 1D). Control experiments show that temperature decreases to around 10 °C do not affect normal unadapted VOR gain in the light (FIG. 2A) or in the dark (FIG. 2B).

There are two significant results of these experiments. First, normal unadapted VOR is insensitive to temperature changes over a wide range. Second, adapted or modified VOR is inactivated by a decrease in temperature. This occurs for an adapted VOR whose gain has been either increased or decreased. An acute temperature drop of 5 to 10 °C will tend to change the gain of the VOR to a value approximately equal to that of the unmodified VOR. In addition, this effect is reversed when the temperature of the fish is returned to its acclimated temperature. Both adapted VOR gain decreases as well as increases are affected. Therefore temperature is influencing the adapted part of the VOR, and is not just acting in a unidirectional manner on the VOR in general. These results would suggest that the biochemical and/or neural substrate responsible for the modified part of the VOR gain is qualitatively different from that responsible for the unmodified VOR gain.

REFERENCES

1. SCHAIRER, J. O. & M. V. BENNETT. 1986. Changes in gain of the vestibuloocular reflex induced by sinusoidal visual stimulation in goldfish. Brain Res. **373:** 177-181.
2. ROBINSON, D. A. 1963. A method of measuring eye movement using a scleral search coil in a magnetic field. IEEE Trans. Biomed. Electron. **10:** 137-145.

Activity-Dependent Adaptation of Lobster Motor Neurons and Compensation of Transmitter Release by Synergistic Inputs[a]

H. BRADACS,[b] A. J. MERCIER,[c] AND
H. L. ATWOOD[d]

[b]*Institut für Zoologie*
Karl-Franzens Universität Graz
A-8010 Graz, Austria

[c]*Department of Biological Sciences*
Brock University
St. Catharines, Ontario, Canada L2S 3A1

[d]*Department of Physiology*
University of Toronto
Toronto, Ontario, Canada M5S 1A8

Earlier studies with crayfish motor neurons show that increasing neural activity over 3-14 days causes progressive changes in synaptic properties.[1] The synaptic terminals release less neurotransmitter at the beginning of a stimulus train or when stimulated at a low frequency of 0.1 impulses per second (Hz). Such activity, however, reduces the amount of synaptic depression that normally occurs at 5 Hz. Thus, although transmitter release is lower at the beginning of a stimulus train, the neurons are capable of sustaining transmitter release for a longer period of time. Neurons, therefore, are able to adjust their transmitter output to meet the demands imposed on them by daily activity. These synaptic changes are referred to as long-term adaptation (LTA) because they persist for several weeks.

The present study examined LTA in the lobster, *Homarus americanus.* The initial aim was to determine whether LTA could be elicited in motor neurons innervating lobster deep abdominal extensor muscles. Two muscles were selected for study, muscles L1 and M of the fourth abdominal segment (FIG. 1). Each muscle is innervated by a *specific* excitatory axon from the third abdominal segment (axon 1 for muscle M and axon 3 for muscle L1) and by a *common* excitatory axon (axon 2) from the fourth abdominal segment.[2] Axons in the third segment were stimulated *in situ* on the left side of the lobster using implanted electrodes; axons on the right side served as controls. The stimuli consisted of 1-sec bursts of 0.5-msec square pulses at 5 Hz, with one burst every 2 sec. We sought to determine whether the appearance of LTA is sensitive to the pattern of daily activity and whether any changes occur in unstimulated axons that share the same target as the stimulated motor neurons.

In one set of experiments, the motor neurons were stimulated for 1-4 hr per day over a 3-5 day period, for a total of 4-12 hr. At least one day after completing the stimulation, excitatory postsynaptic potentials (EPSPs) were recorded at 0.1 Hz from

[a]This work was supported by the Natural Sciences and Engineering Research Council of Canada. Additional support was provided by the Government of Austria.

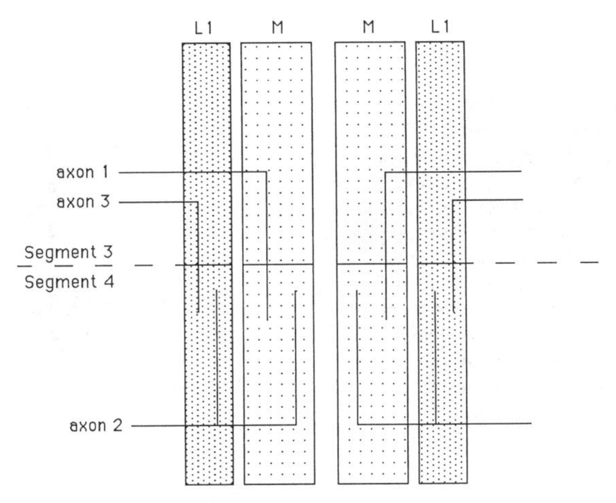

FIGURE 1. Schematic diagram depicting innervation of lobster deep abdominal extensor muscles (M and L1) by axons 1, 2, and 3. (Based on Parnas and Atwood.[2])

up to five sites per muscle. EPSPs from axons 1 and 3 were 60-65% lower on the stimulated side than on the control side (TABLE 1), which is characteristic of LTA. Interestingly, EPSPs from axon 1 of the fourth segment were greater on the stimulated side than on the control side. This neuron shares the same synaptic target as axons 1 and 3 but was not stimulated during the daily conditioning procedure. Thus, a reduction in transmitter release from one set of axons appears to be compensated for by increased synaptic efficacy by less active, synergistic inputs. Muscle cell input resistances did not change.

In a second set of experiments, stimulation *in situ* was applied for 1 hr per day for 5-6 days, but the days of stimulation were interspersed over a 9-13-day period. In this case, there was no significant change in the size of EPSPs recorded initially at 0.1 Hz. During repetitive stimulation at 5 Hz, however, the EPSPs on the stimulated side were less prone to synaptic depression, and they were significantly larger than EPSPs on the control side 5-10 min after the stimulus train was begun (FIG. 2). Thus, the reduction

TABLE 1. Effect of Stimulation on EPSPs in Lobster Muscles M and L1[a]

Axon	Muscle	EPSP (mV) Stimulated Side	EPSP (mV) Control Side	Number of Muscles Examined	Significance Level
1	M	2.6 ± 0.5	7.5 ± 1.7	15	.001
3	L1	4.4 ± 1.1	10.5 ± 1.5	16	.001
2	M	32.7 ± 3.9	18.3 ± 2.2	10	.01
2	L1	23.1 ± 2.6	15.9 ± 1.5	12	.01

[a] Stimulation was applied 1-4 hr per day over a 3-6-day period. Daily stimulation was applied to axons 1 and 3 in the third abdominal segment on one side of the animal. No stimuli were delivered to axon 2 of the fourth abdominal segment, which also innervates muscles M and L1. Standard errors of the means are indicated. The level of significance was determined using a Student's *t* test for matched pairs.

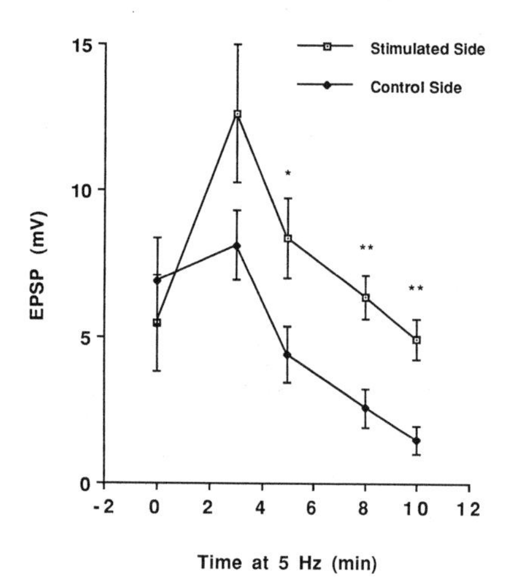

FIGURE 2. Effect of 5-6 days of stimulation for 1 hr per day over a 9-13-day period. EPSPs from axon 3 were recorded in muscle L1 at least one day after completing *in situ* stimulation. EPSPs at time 0 were measured at 0.1 Hz. Bars depict standard errors of the means. The number of preparations was 8, 8, 8, 6, and 5, respectively, for times 0, 3, 5, 8, and 10 min. Values for the stimulated and unstimulated sides were compared using a Student's *t* test: *: significance at the .05 level; **: significance at the .01 level; other differences were not significant at the .05 level.

in transmitter release and the development of fatigue resistance, both of which are characteristic of LTA, can occur separately. These two synaptic changes appear to be elicited by different stimulus paradigms, and they probably involve separate subcellular mechanisms.

REFERENCES

1. Lnenicka, G. A. & H. L. Atwood. 1985. Age-dependent long-term adaptation of crayfish phasic motor axon synapses to altered activity. J. Neurosci. **5:** 459-467.
2. Parnas, I. & H. L. Atwood. 1966. Phasic and tonic neuromuscular systems in the abdominal flexor muscles of the crayfish and rock lobster. Comp. Biochem. Physiol. **18:** 701-723.

Effect of MK801 on Long-Term Potentiation in the Hippocampal Dentate Gyrus of the Unanesthetized Rabbit[a]

G. D. REED AND G. B. ROBINSON [b]

Psychology Department
University of New Brunswick
Fredericton, New Brunswick, Canada E3B 6E4

A major goal of behavioral neuroscience research is identification of the neural mechanisms that underlie learning and memory. One of the strongest candidate neural mechanisms for mammalian learning and memory is long-term potentiation (LTP),[1] a long-lasting increase in response amplitude induced by application of brief high-frequency stimulation. It is well established that LTP and learning and memory share many characteristics, including rapid onset, associative interactions, long duration, and strengthening with repetition.[1-3]

Results from a number of studies[4,5] have suggested that both LTP and certain types of learning require activation of hippocampal *N*-methyl-D-aspartate (NMDA) receptors. These studies, however, have only examined the shorter-lasting component of LTP, whereas the longer-lasting component of LTP (LTP1 and LTP2, respectively[6]) may be most pertinent to learning and memory mechanisms. Furthermore, the reduction or elimination of LTP in animals pretreated with an NMDA antagonist may result, in part, from an increased threshold for induction.[7,8] In the following experiment, the effect of MK801, a noncompetitive NMDA antagonist, on the threshold, magnitude, and duration of LTP of the perforant path-granule cell response was examined.

Male New Zealand White rabbits (2.0-2.5 kg) were deeply anesthetized with halothane and placed on a stereotaxic device that allowed maintenance of halothane anesthesia. Both stereotaxic coordinates and electrophysiological monitoring were used to maximize electrode sites for recording of perforant path-granule cell field potentials. When the optimal sites were located, the electrodes were attached to a plastic headstage and secured to the skull with dental acrylic. The rabbits then were administered 300,000 units of penicillin and allowed a 2-week recovery period.

One hour prior to the start of testing, rabbits were administered one of four dosages of MK801 (0.0, 0.1, 0.5, or 1.0 mg/kg, sc). Test pulses (0.1 Hz, 20% of the maximum intensity) were then applied to the perforant path with high-frequency trains (400 Hz, 50 msec) given at 15-min intervals. The trains were applied at seven different intensities, from subthreshold for spike discharge to an intensity 100% greater than that needed to evoke the maximum amplitude population spike. Each train intensity was applied 10 times (0.1 Hz) to ensure saturation of any LTP to that intensity. LTP decay was monitored by comparison of pre- and posttetanization input/output measures.

[a]This work was supported by the Medical Research Council of Canada and the Natural Sciences and Engineering Research Council of Canada, both of which awarded grants to G.B.R.

[b]To whom correspondence should be addressed.

MK801 appeared to increase the threshold for LTP of the population spike but not for the excitatory postsynaptic potential (EPSP). For animals in the control ($n = 6$), 0.1 mg/kg ($n = 7$), 0.5 mg/kg ($n = 7$), and 1.0 mg/kg ($n = 5$) conditions, the average train intensity for LTP of the spike was 39.2 ± 7.5, 97.1 ± 33.7, 112.9 ± 29.8, and 137.0 ± 35.7% of the maximum, respectively. The threshold for LTP of the EPSP was approximately 30% of the maximum train intensity in all groups.

There was no effect of MK801 on the magnitude of LTP of the EPSP. LTP of the spike, however, decreased in animals pretreated with MK801 (FIG. 1). Thus, the normally observed dissociation between the magnitude of spike and EPSP LTP was reduced in the drugged animals. The decrease in spike LTP was dose-dependent beginning with the third set of trains. The 0.5 and 1.0 mg/kg groups exhibited significantly ($p < .02$) less LTP than the control group. The control and 0.1 mg/kg groups did not differ significantly in the percentage change from pretrain baseline (FIG. 1) but did differ significantly ($p < .01$) in the absolute magnitude of change (9.42 mV versus 4.73 mV, respectively) (FIG. 2).

MK801 also reduced the duration of LTP of the population spike (LTP of the EPSP had decayed within 24 hr in all four groups). In controls, the spike amplitude

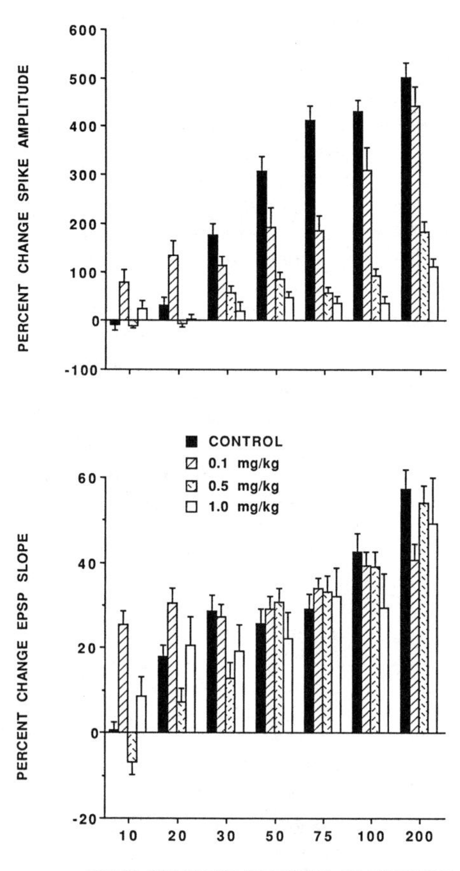

FIGURE 1. Average percentage change (±1 SEM) in perforant path-granule cell spike amplitude and EPSP slope, over the pretrain baseline, 13-15 min after each set of trains had been applied to the perforant path. MK801 reduced the magnitude of the percentage change in spike amplitude but did not affect the EPSP. Spike amplitudes were measured from the peak of the negativity to a tangent joining spike onset and offset. EPSP slopes were measured within the first millisecond of positivity.

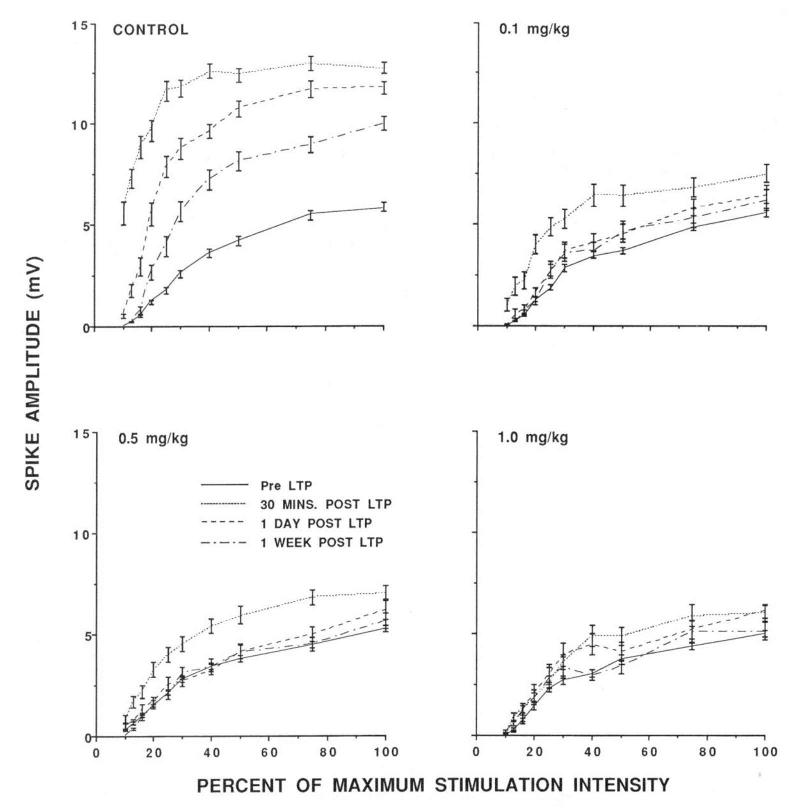

FIGURE 2. Effect of MK801 on both the magnitude and duration of LTP of the perforant path-granule cell population spike. Both the duration of LTP and the maximum spike amplitude were reduced by pretreatment with MK801. The input/output response was obtained by applying 10 pulses (0.1 Hz) to the perforant path at each of 10 intensities; illustrated are the mean (± 1 SEM) spike amplitudes evoked at each intensity.

still was significantly greater than the pretetanus amplitude 1 week following LTP induction ($p < .003$). In contrast, LTP in rabbits pretreated with MK801 had decayed within 24 hr of tetanization (FIG. 2).

In summary, MK801 increased the threshold and decreased both the magnitude and duration of LTP of the population spike but did not affect the EPSP. Thus, the magnitude of the dissociation between LTP of the spike and EPSP, normally observed in control preparations, was decreased in rabbits administered MK801. The dissociation likely reflects changes (for example, in inhibitory potentials) that alter the probability of cell discharge. Repeated tetanic stimulation, for example, may reduce the magnitude of inhibitory potentials[9] that may otherwise suppress cell discharge. MK801 pretreatment may antagonize this reduction,[10] resulting in a similar magnitude of LTP for the population spike and EPSP.

If LTP is a neural substrate of associative learning, then similar dosages of MK801 also should disrupt learning. This possibility is now being investigated using the classically conditioned rabbit nictitating membrane response.[11]

REFERENCES

1. TEYLER, T. J. & P. DISCENNA. 1984. Brain Res. **7:** 15-28.
2. MATTHIES, H. 1989. Prog. Neurobiol. **32:** 277-349.
3. MCNAUGHTON, B. L. & R. G. M. MORRIS. 1987. Trends Neurosci. **10:** 408-415.
4. ABRAHAM, W. C. & S. E. MASON. 1988. Brain Res. **462:** 40-46.
5. MORRIS, R. G. M. 1989. J. Neurosci. **9:** 3040-3057.
6. RACINE, R. J., N. W. MILGRAM & S. HAFNER. 1983. Brain Res. **260:** 217-231.
7. CAIN, D. P., F. BOON & E. L. HARGREAVES. 1989. Soc. Neurosci. Abstr. **15:** 1102.
8. ROBINSON, G. B. 1989. Soc. Neurosci. Abstr. **15:** 1102.
9. MARU, E., H. ASHIDA & J. TATSUNO. 1989. Brain Res. **478:** 112-120.
10. STELZER, A., N. T. SLATER & G. T. BRUGGENCATE. 1987. Nature **326:** 698-701.
11. STILLWELL, J. H. & G. B. ROBINSON. 1990. Soc. Neurosci. Abstr. **16:** 768.

Effects of Blocking or Synchronizing Activity on the Morphology of Individual Regenerating Arbors in the Goldfish Retinotectal Projection[a]

JOHN T. SCHMIDT

Department of Biological Sciences
and
Neurobiology Research Center
State University of New York at Albany
Albany, New York 12222

Retinotopic sharpening of a regenerating optic projection in goldfish can be prevented either by blocking activity with intraocular tetrodotoxin (TTX)[1,2] or by synchronizing activity with a xenon strobe light.[3,4] In this study, I tested, in both normal and regenerating projections, the effects of these two treatments on individual optic arbors. Arbors were stained via anterograde transport of horseradish peroxidase, drawn in camera lucida from tectal whole mounts, and analyzed by quantifying several of their features. These included the spatial extent in the plane of the retinotopic map, the order of branching, the number of branch endings, the depth of termination, and the caliber of the parent axon. In normal tectum, there were three distinct calibers of retinal axons, fine, medium, and coarse, which gave rise to small, medium, and large arbors averaging 127 μm, 211 μm, and 275 μm in horizontal extent, and which terminated at characteristic depths (FIG. 1). All three classes averaged about 21 branch endings per arbor.[5]

Optic arbors that regenerated with normal patterns of activity returned to a roughly normal appearance by 6-11 weeks postcrush (FIG. 2). The same three calibers of axons gave rise to the same three sizes of arbors, which occupied the same depths. The regenerated arbors, however, were much less stratified and were on average about 16% larger in horizontal extent.[5]

I examined the experimental arbors at this time point in regeneration. Arbors regenerated under TTX or strobe were on average 71% and 119% larger, respectively, than the control regenerated arbors, and this increase in extent was evident in all three classes (FIGS. 3 & 4). The TTX and strobe experimental arbors, being larger than the control regenerated arbors, which themselves were larger than the normal arbors, were 1.98 and 2.54 times larger than normal in linear extent, and much more in total areal extent. These great increases in extent account for the lack of retinotopic sharpening in the electrophysiologically recorded retinotopic maps of the TTX-regenerated[2] and strobe-regenerated fish.[3] In both cases, the arbors had approximately the same number of branch endings and were equally poorly stratified. Synapses formed under strobe, examined in electron micrographs, were normal in appearance, consistent with previous reports that synapses regenerated during TTX block of activity were normal.[6] Thus, the only significant effect of both strobe and TTX treatment was to enlarge the spatial extent of the arbors. A block of activity for an equivalent period of time in arbors that were not regenerating caused only a slight, but significant, enlargement (23%). Strobe

[a] This work was supported by a grant from the National Institutes of Health (EY-03736).

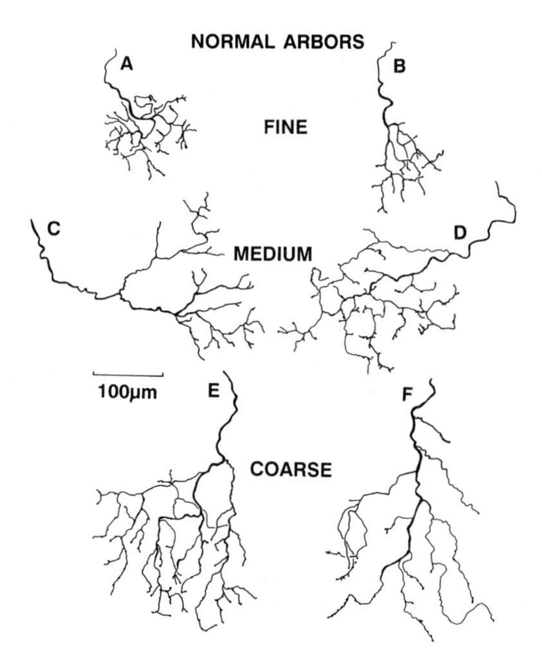

FIGURE 1. Camera lucida drawings of typical normal arbors arising from fine caliber axons (**A & B**), from medium caliber axons (**C & D**), and from coarse caliber axons (**E & F**). Rostral is toward the top in each case. Most branches are given off in the caudal direction, except for peripheral axons in caudal tectum.

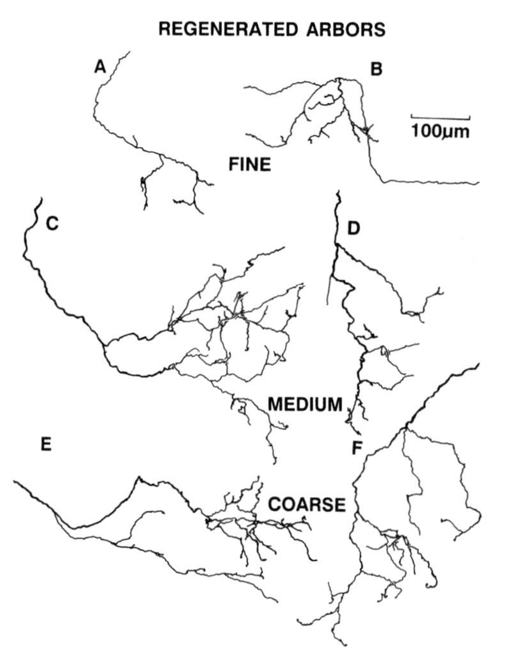

FIGURE 2. Camera lucida drawings of typical control regenerated arbors at 6 to 8 weeks postcrush. Arbors were chosen to be roughly representative for spatial extent and number of branches. Rostral is toward the top in each case. **A & B**: 6 weeks postcrush; **C**: 8 weeks; and **D, E & F**: 7 weeks.

exposures in nonregenerating fish of three times the duration also caused slight, but significant, enlargements (39%). In both cases, the arbors remained compact and normal in appearance, unlike those in the regenerating cases, and in both cases they remained significantly smaller than when the same treatment was applied during regeneration. Because previous staining showed that regenerating axons initially make wide-

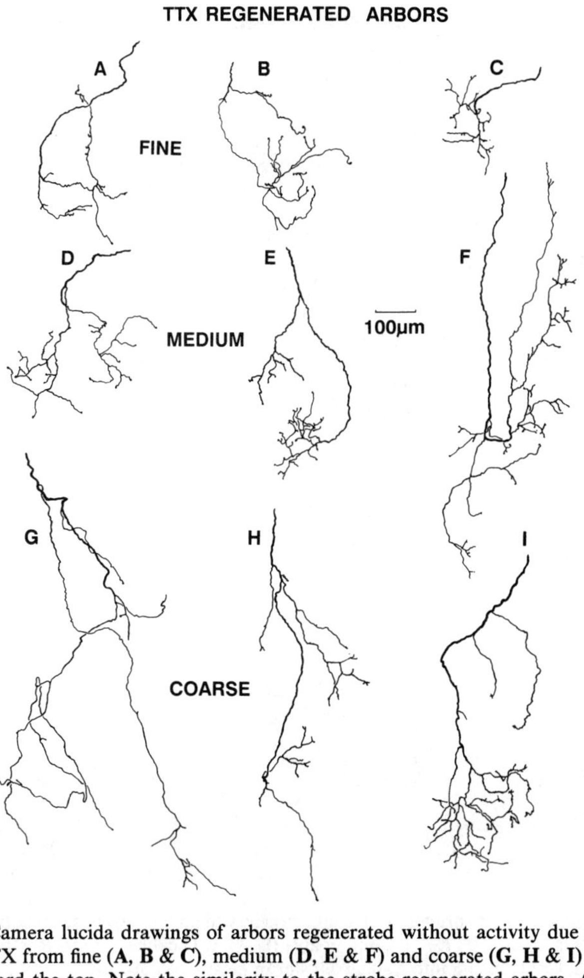

TTX REGENERATED ARBORS

FINE

MEDIUM

COARSE

100µm

FIGURE 3. Camera lucida drawings of arbors regenerated without activity due to intraocular injection of TTX from fine (**A, B & C**), medium (**D, E & F**) and coarse (**G, H & I**) caliber axons. Rostral is toward the top. Note the similarity to the strobe-regenerated arbors. Arbors from 9 weeks (**F, G & H**), 10 weeks (**C**), or 11 weeks (**A, B, D, E & I**).

spread branches and later retract many of these branches,[5,7] the present findings support the idea that blocking activity or synchronizing activity prevents retinotopic sharpening by interfering with the elimination of errant branches and redeployment of arbor branches in the retinotopically appropriate area.

The fact that the only effect in both cases was on spatial extent is interesting because the two treatments are so different. Intraocular TTX removes all activity whereas strobe illumination allows activity but synchronizes the firing of both ON and OFF ganglion cells.[3] The one thing that these treatments both disrupt is the normal pattern of activity and any information that may be carried in that pattern. Sharpening is

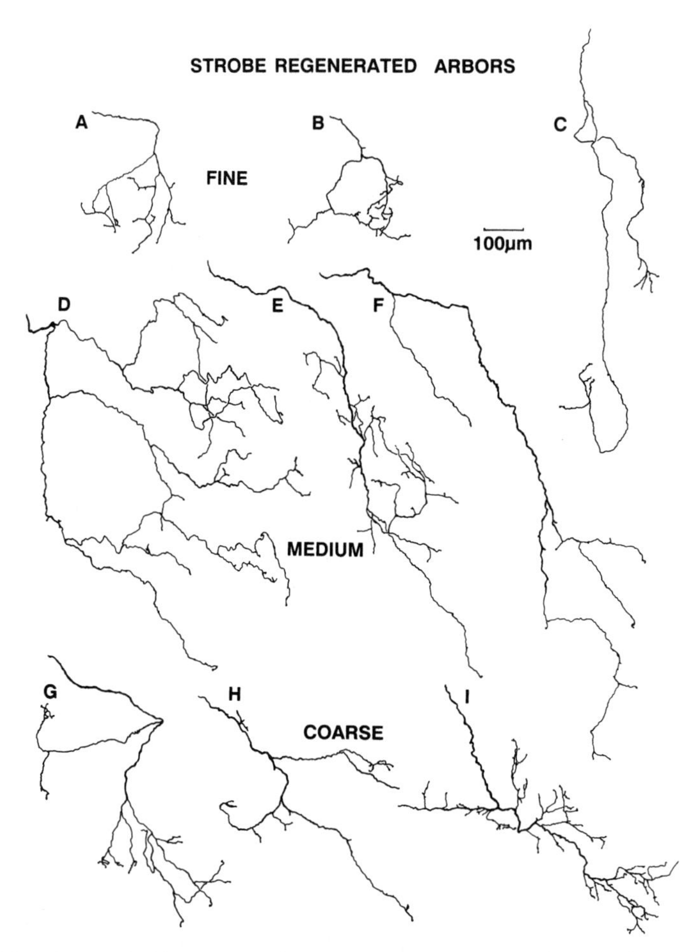

FIGURE 4. Camera lucida drawings of arbors regenerated under stroboscopic illumination from fine (**A, B** & **C**), medium (**D, E** & **F**), and coarse (**G, H** & **I**) caliber axons. Rostral is toward the top. Note the lack of focusing into a single cluster in most cases. All arbors at 9 weeks postcrush, except **G**, which is at 8 weeks.

thought to be driven by the correlated firing of neighboring ganglion cells of the same receptive field type that occurs both under normal visual conditions and even in complete darkness.[8,9] Presynaptic inputs that fire simultaneously could summate their postsynaptic responses and become more effective by a Hebbian mechanism that may be mediated via *N*-methyl-D-aspartate (NMDA) receptors.[10,11] Synaptic transmission is reestablished by the time that sharpening takes place,[12] and the retinotopically inappropriate branches of the regenerating arbors make effective synaptic connections.[13,14] Direct evidence for such a Hebbian mechanism was recently obtained with the discovery of greatly enhanced capacity for long-term potentiation (LTP) during early regeneration.[11] This LTP[11] as well as the retinotopic sharpening[10,11] can be prevented by blockers of NMDA receptors. The fact that similar changes were found in developing frog visual system[10] suggests that these NMDA receptor-driven Hebbian mechanisms may allow activity to shape the morphology of axonal arborizations and the distribution of visual connections within the visual centers. Here the result is a sharpening of the retinotopic map, but the same mechanisms could segregate receptive field types or eye-specific inputs.

REFERENCES

1. MEYER, R. L. 1983. Dev. Brain Res. **6:** 293-298.
2. SCHMIDT, J. T. & D. L. EDWARDS. 1983. Brain Res. **269:** 29-39.
3. SCHMIDT, J. T. & L. E. EISELE. 1985. Neuroscience **14:** 535-546.
4. COOK, J. E. & E. C. C. RANKIN. 1986. Exp. Brain Res. **63:** 421-430.
5. SCHMIDT, J. T., J. C. TURCOTTE, M. BUZZARD & D. G. TIEMAN. 1988. J. Comp. Neurol. **269:** 565-591.
6. HAYES, W. P. & R. L. MEYER. 1989. J. Neurosci. **9:** 1414-1423.
7. STUERMER, C. A. O. 1988. J. Comp. Neurol. **267:** 69-91.
8. ARNETT, D. W. 1978. Exp. Brain Res. **32:** 49-53.
9. MASTRONARDE, D. N. 1989. Trends Neurosci. **12:** 75-80.
10. CLINE, H. T. & M. CONSTANTINE-PATON. 1989. Neuron **3:** 413-426.
11. SCHMIDT, J. T. 1990. J. Neurosci. **10:** 233-246.
12. SCHMIDT, J. T., D. L. EDWARDS & C. A. O. STUERMER. 1983. Brain Res. **269:** 15-27.
13. ADAMSON, J. R., J. BURKE & P. GROBSTEIN. 1984. J. Neurosci. **4:** 2635-2649.
14. MATSUMOTO, N., M. KOMETANI & K. NAGANO. 1987. Neuroscience **22:** 1103-1110.

Three-Dimensional Imaging of Neurophysiologically Characterized Hippocampal Neurons by Confocal Scanning Laser Microscopy[a]

KAREN L. SMITH, JAMES N. TURNER,
DONALD H. SZAROWSKI, AND JOHN W. SWANN

Wadsworth Center for Laboratories and Research
New York State Department of Health
Albany, New York 12201

Understanding use- and age-dependent alterations in local circuit functioning in the mammalian central nervous system requires detailed anatomical knowledge of the types of cells present within networks and the physiological properties of each cell type.[1] Moreover, knowledge of the patterns of neuronal connectivity and of where chemical and electrical synaptic contacts are located can be of paramount importance in understanding network output. In recent years the injection of intracellular markers via a recording micropipette has expanded our knowledge of physiological properties of identified neurons. Detailed descriptions of neurons in three-dimensional space, however, has relied on the use of time-consuming camera lucida drawing techniques. The development of confocal laser scanning microscopy (CLSM) holds great promise in aiding in the analysis of individual neurons and their connectivity.[2] The ability to rapidly section individual fluorescently labeled cells or groups of cells in narrow optical planes, the rejection of out-of-focus light, and three-dimensional computer reconstruction of optical sections can result in detailed microanatomical information. These techniques should offer an unprecedented opportunity to explore the microstructure of neurons within neuronal networks.

In studies reported here intracellular recordings were obtained from *in vitro* slices of rat hippocampus. The slices were prepared by methods previously described.[3] Microelectrodes were pulled from fiber-filled capillary tubing and filled with 5-10% Lucifer yellow in 0.33 M LiAc. In some instances electrodes were beveled. Resistances ranged from 80 to 300 MΩ.

Upon impalement neurons were characterized physiologically by their active and passive membrane properties. Cells were categorized by characteristic features including spontaneous discharging and responses to current applied via the recording electrode. Following characterization cells were filled with Lucifer yellow by passing 0.2-2.0 nA of hyperpolarizing current (200-500 msec, 1-2 Hz for 5-30 min). Afterwards, the approximate position of the electrode tip in the cell was determined with a Wild stereomicroscope equipped with an epifluorescence illuminator. Instances of dye coupling were also noted at this time. After withdrawal of the recording electrode, the slice was removed from the chamber, fixed in 10% formalin, dehydrated in graded alcohol, and cleared with methyl salicylate. Slices were then imaged as whole mounts using a BioRad MRC-500 Confocal Scanning Laser Microscope equipped with an analog preprocessor.[4]

[a]This work was supported by grants from the National Institutes of Health (NS18309, RR02984, and RR01219).

The majority of recordings from cells impaled in the CA3 pyramidal cell body layer displayed spontaneous bursts of action potentials (FIG. 1B), and invariably had pyramidal cell type anatomical features. FIGURE 1A is a stereo image of one such cell. The three-dimensional distribution of apical and basilar dendrites is displayed. A single primary apical dendrite emanates from the cell body. Four secondary dendritic branches are present. The two most distal secondary dendrites arise from the terminal portion (see arrow) of the primary dendrite, which angles toward the slice surface. Fine

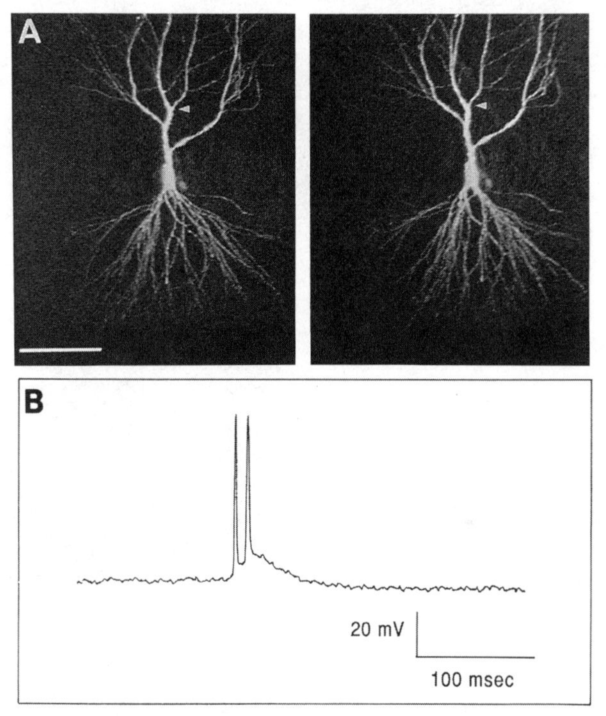

FIGURE 1. **(A)** Stereo image of a single pyramidal cell from a 24-day-old rat containing 22 optical sections collected at 4-μm intervals. A 20×, 0.8 numerical aperture oil-immersion lens was used. **(B)** Recordings from the same cell showing a spontaneous, intrinsic burst discharge. Arrow denotes the point at which the primary apical dendrite branches to two secondary dendrites. Bar: 100 μm.

third-order dendrites emanate from these two secondary dendrites and project at right angles to the principal axis of their parent dendrite.

The basilar dendrites arise from the cell body as three separate primary dendrites, which branch almost immediately into numerous fine dendritic branches. A complex latticework of dendritic processes is apparent when the basilar dendritic tree is viewed in two dimensions. With stereo imaging, however, an orderliness is apparent in dendritic morphology. Each of the three primary dendrites gives rise to a dendritic arbor that is largely localized to a restricted neighborhood in the dendritic layer. Each of these overlaps only slightly with the arbors arising from the other primary dendrites.

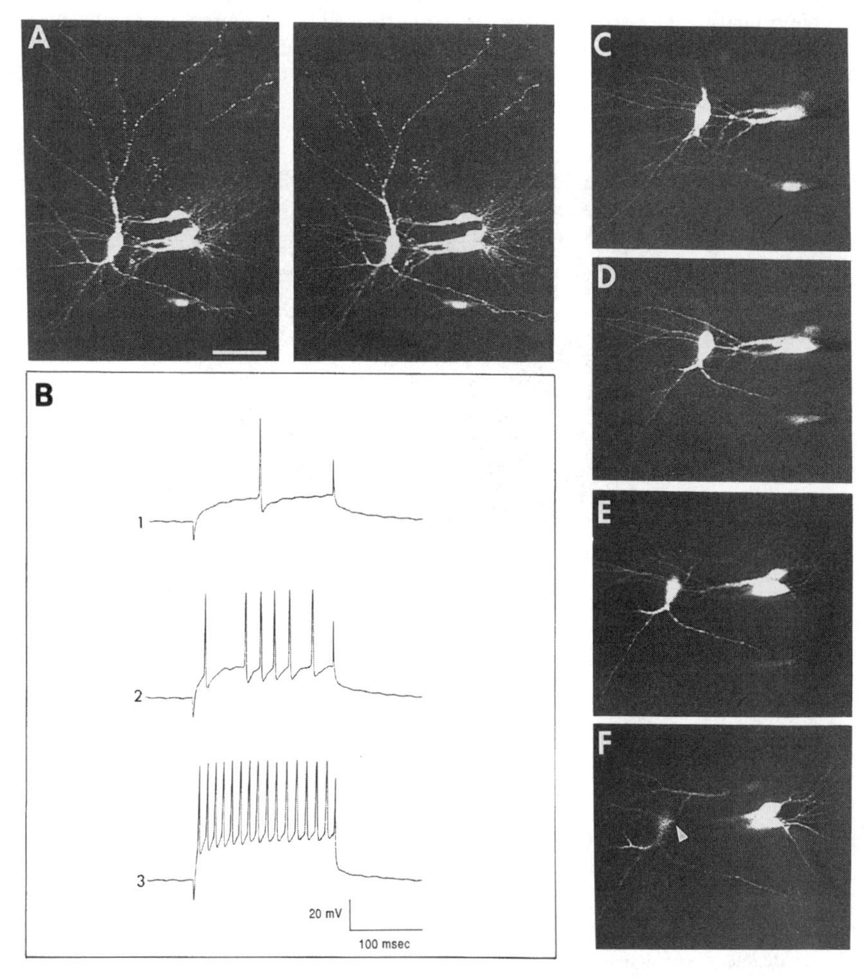

FIGURE 2. (**A**) Five dye-coupled cells (from an 11-day-old rat) in a stereo projection. The image contains 31 optical sections collected at 5-μm intervals. Data were obtained with a 20×, 0.8 numerical aperture oil-immersion objective. The impaled cell displayed spikes with rapid rates of rise and fall. (**B**) Responses to injections of depolarizing current steps. Three separate amplitudes of current were used. In the top trace the injected current produced a single fast spike with rapid afterhyperpolarization. The other two spike trains show repetitive firing and lack of spike accommodation. (**C-F**) Four of the 31 optical sections used to construct the stereo image in **A**. The arrow in **F** denotes the potential site of electrotonic coupling between the apical dendrite of a pyramidal cell and the soma of the impaled interneuron. Bar: 100 μm.

In addition to analyzing the three-dimensional morphology of single neurons, CSLM permits a detailed analysis of anatomical interrelations among neurons in local networks. FIGURE 2A shows a stereo image of five neurons in the hippocampal CA3 subfield. Visualization of the microelectrode during the recording session showed that the electrode had impaled the soma of the interneuron located in the stratum radiatum in the left-hand portion of the field of view. FIGURE 2B shows intracellular recordings. The rapid rising and falling phase of action potentials, prominent spike afterhyperpolarization, and repetitive firing with little spike accommodation are characteristic features of some hippocampal interneurons.[5] Four other neurons were filled as the interneuron was injected with Lucifer yellow. Three pyramidal cells are seen with somas to the far right of the interneuron, and basilar dendrites project to the right edge of the field of view. Many of the apical dendritic processes arising from these three cells course beneath the soma of the interneuron. This is a clear example of dye coupling, which is thought to be a reflection of electronic coupling between cells mediated by gap junctions.[6] With the three-dimensional information obtained by CSLM, a search can be undertaken for potential sites of coupling. This is facilitated by examination of individual optical sections (for example, FIGS. 2C-2F). The arrow on FIGURE 2F shows one potential site of contact where an apical dendrite of a pyramidal cell is very closely apposed to the underside of the soma of the impaled interneuron. Verification that such sites are indeed regions of electrical contact will require studies at the ultrastructural level. Indeed, methods have recently been reported that facilitate identification of the ultrastructural features of cellular elements identified in *in vitro* brain slices using CSLM methods.[7]

Although CSLM is a relatively new technology, our first glimpses at the microneuroanatomy within local networks suggest that many new details on network organization and functioning will be forthcoming. This will be especially true with the continued application of advanced image-processing techniques that allow image rotation and viewing from any desired perspective.[8] Imaging of fluorescently labeled neurons in slices during recording sessions[9] is likely to supply even further information on use-dependent alterations in local circuit functioning.

REFERENCES

1. GETTING, P. A. 1989. Emerging principles governing the operation of neural networks. Annu. Rev. Neurosci. **12:** 185-204.
2. FINE, A., W. B. AMOS, R. M. DURBIN & P. A. MCNAUGHTON. 1988. Confocal microscopy: Applications in neurobiology. Trends Neurosci. **11:** 346-351.
3. SWANN, J. W. & R. J. BRADY. 1984. Penicillin-induced epileptogenesis in immature rat CA_3 hippocampal pyramidal cells. Dev. Brain Res. **12:** 243-254.
4. SZAROWSKI, D. H., D. P. BARNARD, K. SMITH, J. W. SWANN, K. V. HOLMES & J. N. TURNER. 1990. Confocal laser scanned microscopy: Analog signal processing. Scanning **12:** 265-272.
5. SCHWARTZKROIN, P. A. & L. H. MATHERS. 1978. Physiological and morphological identification of a non-pyramidal hippocampal cell type. Brain Res. **157:** 1-10.
6. DUDEK, F. E., R. D. ANDREW, B. A. MACVICAR, R. W. SNOW & C. P. TAYLOR. 1983. Recent evidence for and possible significance of gap junctions and electronic synapses in the mammalian brain. *In* Basic Mechanisms of Neuronal Hyperexcitability. H. H. Jasper & N. M. van Gelder, Eds.: 31-73. Alan R. Liss. New York, NY.
7. DEITCH, J. S., K. L. SMITH, J. W. SWANN & J. N. TURNER. 1991. Ultrastructural investiga-

tion of neurons identified and localized using the confocal scanning laser microscope. J. Electron Microsc. Tech. **18:** 82-90.

8. TURNER, J. N., D. H. SZAROWSKI, J. S. DEITCH, K. L. SMITH & J. W. SWANN. 1991. Confocal microscopy and three-dimensional reconstruction of electrophysiologically identified neurons in thick brain slices. J. Electron Microsc. Tech. **18:** 11-23.
9. SMITH, K. L., J. N. TURNER, D. H. SZAROWSKI & J. W. SWANN. 1990. Intradendritic recordings from identified sites in CA_3 hippocampal pyramidal cells. Soc. Neurosci. Abstr. **16:** 57.

Striatal c-*fos* Induction in Neonatally Dopamine-Depleted Rats Given Transplants

ABIGAIL M. SNYDER-KELLER

Wadsworth Center for Laboratories and Research
New York State Department of Health
Albany, New York 12201

Induction of the protooncogene c-*fos* may be the first in a series of transcriptional events that underlie a cell's adaptive response to impinging neural activity. Immunocytochemistry using antibodies to the *fos* protein can be used to map cells in the CNS that respond to various stimuli by c-*fos* induction.[1] In the striatum, the amphetamine-stimulated release of dopamine (DA), or direct activation of DA receptors of the D1 subtype, leads to c-*fos* induction.[2] This response was used to study the interaction of DA-rich transplants with the denervated host striatum in animals given bilateral DA-depleting lesions and unilateral transplants as infants.

Three-day-old Sprague-Dawley rat pups were pretreated with desmethylimipramine (25 mg/kg, sc) and then given bilateral injections of 6-hydroxydopamine (6-HDA) (120 μg in a total dose of 10 μl) into the lateral ventricles, in order to deplete striatal DA. Transplants of DA-rich tissue were made into the striatum of 6-HDA-treated rats in one of two ways. Some animals were given unilateral injections of a suspension prepared from E14 ventral mesencephalon (containing dopaminergic nigral cells) on postnatal day 6. Other animals were given unilateral injections of a suspension prepared from a combination of nigral and striatal tissue taken from E14 embryos. Cells were injected into the rostral striatum of 12-16-day-old rats, and remained as a bolus within the denervated host tissue.

At 2-3 months of age, animals were injected with *d*-amphetamine sulfate (AMPH) (5 mg/kg, ip), or given a "stress" consisting of tail pinch or an intraperitoneal saline injection. Some animals were given the *N*-methyl-D-aspartate (NMDA) antagonist MK-801 (Merck Sharp & Dohme; 3 mg/kg, ip) or the D1 antagonist SCH-23390 (Schering; 0.4 mg/kg, ip) 30 min prior to AMPH injection. Two hours following the test stimulus animals were deeply anesthetized and perfused through the ascending aorta with 4% paraformaldehyde in phosphate buffer. Thirty-micron coronal sections through the striatum were incubated for 40 hr in sheep polyclonal anti-*fos* antisera from Cambridge Research Biochemicals (OA-11-823; 3000:1). Alternate sections pretreated with normal goat serum were incubated in rabbit antibodies to tyrosine hydroxylase (TH) (Eugene Tech; 6000:1) or calbindin (gift of Dr. Sylvia Christakos; 1500:1). Antibody binding was revealed using the ABC-peroxidase technique (Vector Labs), using nickel-intensified diaminobenzidine as the substrate.

In comparison to the lack of *fos* immunostaining in the striatum of untreated unstressed rats, following amphetamine injection numerous cells with *fos*-immunoreactive nuclei were found homogeneously distributed throughout the medial and central sectors of the striatum (FIG. 1A). Laterally, however, patches of *fos*-immunoreactive cells were seen on a background of negative staining. In order to determine whether the patches of *fos*-immunoreactive cells corresponded to the intrinsic patches of the striatum, adjacent sections were reacted for calbindin immunocytochem-

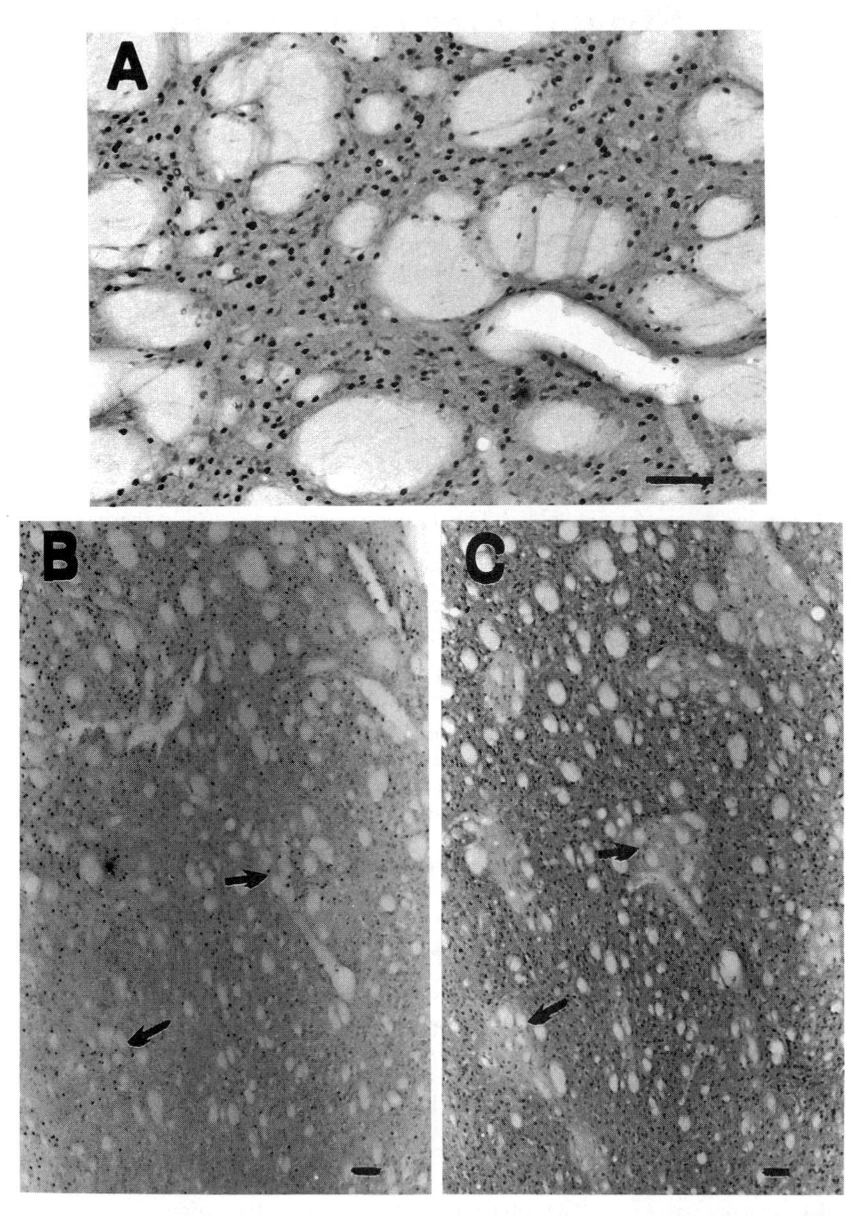

FIGURE 1. (**A**) High-power photomicrograph of *fos*-immunoreactive cells in the medial striatum of a control rat injected with 5 mg/kg AMPH. Large unstained zones are bundles of corticofugal fibers coursing through the striatum. (**B & C**) Adjacent sections taken from the lateral striatum of a control rat given AMPH, stained with antisera against *fos* (**B**) and calbindin (**C**). Arrows indicate corresponding patches of *fos*-immunoreactive cells and patches devoid of calbindin immunoreactivity. Lateral border is to the right. Scale bars: 100 μm.

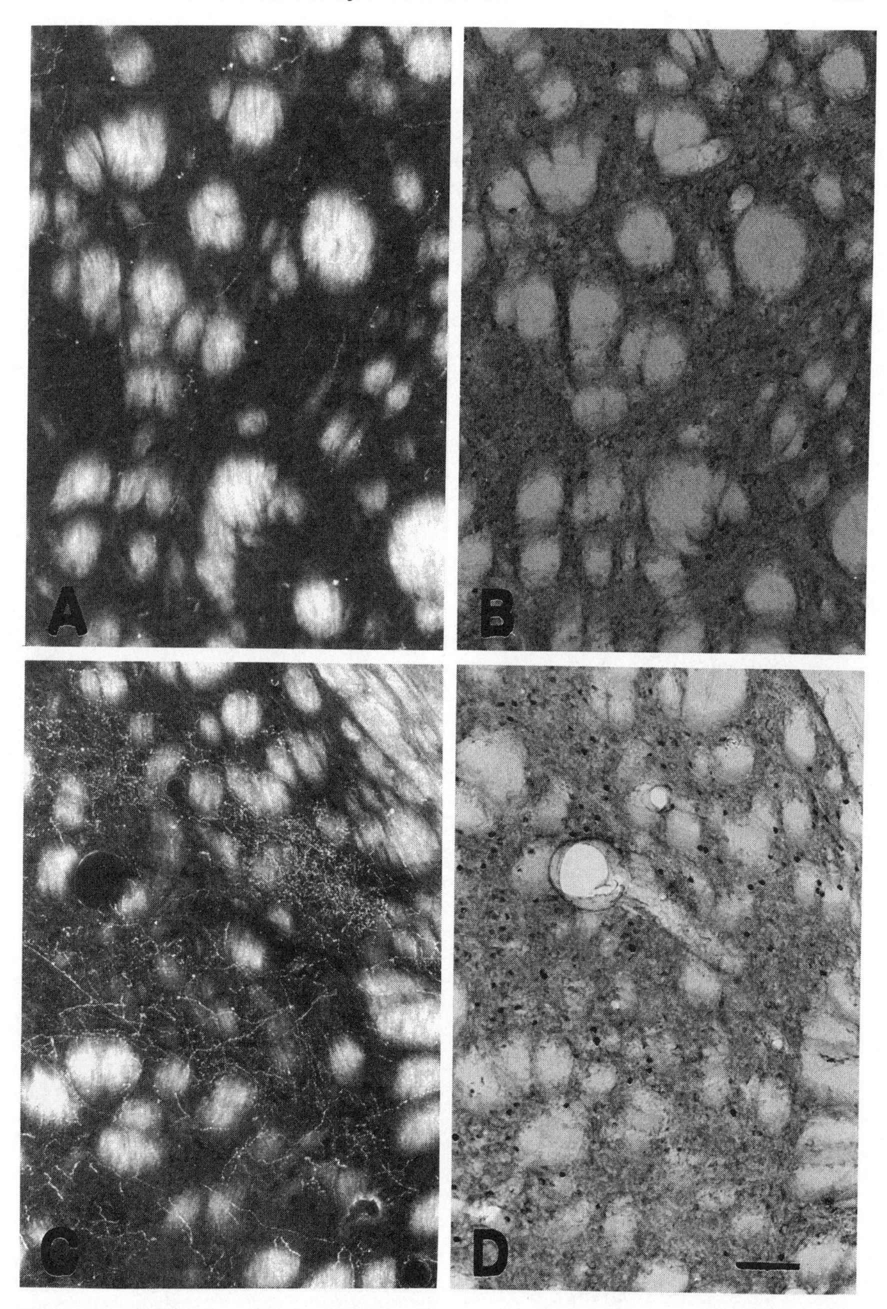

FIGURE 2. High-power photomicrographs of TH-immunoreactive fibers photographed in dark-field (**A** & **C**), and *fos*-immunoreactive neurons (**B** & **D**) in adjacent sections taken from the striatum of a lesioned rat (**A** & **B**) and a lesioned rat with a DA transplant (**C** & **D**). Both animals were given AMPH (5 mg/kg, ip) 2 hr prior to analysis. Few DA fibers and *fos*-immunoreactive neurons were found in the lesioned rat, but in areas reinnervated by the transplant, c-*fos* induction appeared intermediate between lesioned and control rats (see FIG. 1A). Scale bar: 100 μm for all.

istry. Calbindin-immunoreactive neurons were found exclusively in the striatal matrix exclusive of the patches, and patches devoid of calbindin immunoreactivity in the lateral striatum were found to correspond to patches of *fos*-immunoreactive neurons (FIGS. 1B & 1C).

Two observations were made that supported the hypothesis that AMPH induction of c-*fos* was mediated by DA stimulation of D1 receptors. First, the distribution of *fos*-immunoreactive neurons after AMPH injection was roughly correlated with the extent of DA innervation in neonatally 6-HDA-treated rats. In zones where a complete DA denervation was evident by TH immunocytochemistry, very few *fos*-immunoreactive neurons were present (FIGS. 2A & 2B). In animals given DA transplants as infants, both the density of TH-immunoreactive fibers within the striatum and the number of *fos*-immunoreactive striatal neurons after AMPH treatment were intermediate between controls and lesioned animals (FIGS. 2C & 2D). Finally, in all groups of animals, AMPH induction of c-*fos* was blocked by pretreatment with the D1 antagonist SCH-23390.

In order to determine the contribution of excitatory amino acid receptor activation to the AMPH induction of c-*fos*, animals were injected with the noncompetitive NMDA antagonist MK-801 30 min prior to AMPH. As previously reported,[3] this dose of MK-801 blocked striatal c-*fos* induction within the medial two-thirds of the striatum, but enhanced c-*fos* induction within the lateral region of normal animals, where a homogeneously dense band of *fos*-immunoreactive cells resulted. Whereas the blockade of c-*fos* induction also occurred in the medial striatum of lesioned animals with transplants, no increase in *fos* immunoreactivity was observed in the lateral portion when this was devoid of DA fibers.

Due to the unilateral reinnervation of the striatum by the transplant, prolonged contraversive turning is seen in these animals in response to AMPH, and transient turning (2-3 min) in response to a mild stress such as tail pinch or intraperitoneal saline injection.[4] These stressors produced sporadic c-*fos* induction in the medial striatum that was no greater than that which occurred in control animals or 6-HDA-treated animals not given transplants. Stress failed to induce c-*fos* within the transplanted cells, whereas this was occasionally seen in animals with combination transplants that were given AMPH. This intratransplant c-*fos* induction, however, occurred in regions of the transplant that received DA innervation from cotransplanted DA cells.

These findings demonstrate that DA released from DA-rich transplants has the ability to induce c-*fos* in striatal neurons in much the same way as occurs in normal animals. The ability of AMPH, but not stress, to produce widespread *fos* immunostaining of striatal neurons may be related to the intensity of the stimulus. The AMPH-stimulated induction of c-*fos* was mediated by a combination of DA D1 and NMDA receptor activation in the medial striatum, but a different NMDA/DA interaction appears to exist in the lateral striatum. Whereas the long-term consequences of striatal c-*fos* induction remain to be elucidated, they may include the sensitization of stress-induced turning that occurs in these transplant animals following AMPH exposure.[5]

REFERENCES

1. SAGAR, S. M., F. R. SHARP & T. CURRAN. 1988. Science **240:** 1328-1331.
2. ROBERTSON, H. A., M. R. PETERSON, K. MURPHY & G. S. ROBERTSON. 1989. Brain Res. **503:** 346-349.
3. JOHNSON, K. & H. A. ROBERTSON. 1989. Soc. Neurosci. Abstr. **15:** 156.
4. SNYDER-KELLER, A. M., R. K. CARDER & R. D. LUND. 1989. Neuroscience **30:** 779-794.
5. SNYDER-KELLER, A. M. & R. D. LUND. 1990. Brain Res. **514:** 143-146.

Index of Contributors